■ **TABLE 1.6**
Approximate Physical Properties of Some Common Gases at Standard Atmospheric Pressure (BG Units)

Gas	Temperature (°F)	Density ρ (slugs/ft³)	Specific Weight, γ (lb/ft³)	Dynamic Viscosity, μ (lb·s/ft)	Kinematic Viscosity, ν (ft²/s)	Gas Constant,[a] R (ft·lb/slug·°R)	Specific Heat Ratio,[b] k
Air (standard)	59	2.38 E − 3	7.65 E − 2	3.74 E − 7	1.57 E − 4	1.716 E + 3	1.40
Carbon dioxide	68	3.55 E − 3	1.14 E − 1	3.07 E − 7	8.65 E − 5	1.130 E + 3	1.30
Helium	68	3.23 E − 4	1.04 E − 2	4.09 E − 7	1.27 E − 3	1.242 E + 4	1.66
Hydrogen	68	1.63 E − 4	5.25 E − 3	1.85 E − 7	1.13 E − 3	2.466 E + 4	1.41
Methane (natural gas)	68	1.29 E − 3	4.15 E − 2	2.29 E − 7	1.78 E − 4	3.099 E + 3	1.31
Nitrogen	68	2.26 E − 3	7.28 E − 2	3.68 E − 7	1.63 E − 4	1.775 E + 3	1.40
Oxygen	68	2.58 E − 3	8.31 E − 2	4.25 E − 7	1.65 E − 4	1.554 E + 3	1.40

[a]Values of the gas constant are independent of temperature.
[b]Values of the specific heat ratio depend only slightly on temperature.

■ **TABLE 1.7**
Approximate Physical Properties of Some Common Gases at Standard Atmospheric Pressure (SI Units)

Gas	Temperature (°C)	Density ρ (kg/m³)	Specific Weight, γ (N/m³)	Dynamic Viscosity, μ (N·s/m)	Kinematic Viscosity, ν (m²/s)	Gas Constant,[a] R (J/kg·K)	Specific Heat Ratio,[b] k
Air (standard)	15	1.23 E + 0	1.20 E + 1	1.79 E − 5	1.46 E − 5	2.869 E + 2	1.40
Carbon dioxide	20	1.83 E + 0	1.80 E + 1	1.47 E − 5	8.03 E − 6	1.889 E + 2	1.30
Helium	20	1.66 E − 1	1.63 E + 0	1.94 E − 5	1.15 E − 4	2.077 E + 3	1.66
Hydrogen	20	8.38 E − 2	8.22 E − 1	8.84 E − 6	1.05 E − 4	4.124 E + 3	1.41
Methane (natural gas)	20	6.67 E − 1	6.54 E + 0	1.10 E − 5	1.65 E − 5	5.183 E + 2	1.31
Nitrogen	20	1.16 E + 0	1.14 E + 1	1.76 E − 5	1.52 E − 5	2.968 E + 2	1.40
Oxygen	20	1.33 E + 0	1.30 E + 1	2.04 E − 5	1.53 E − 5	2.598 E + 2	1.40

[a]Values of the gas constant are independent of temperature.
[b]Values of the specific heat ratio depend only slightly on temperature.

A Brief Introduction to Fluid Mechanics

Fourth Edition

DONALD F. YOUNG
BRUCE R. MUNSON

Department of Aerospace Engineering
and Engineering Mechanics

THEODORE H. OKIISHI

Department of Mechanical Engineering
Iowa State University
Ames, Iowa, USA

WADE W. HUEBSCH

Department of Mechanical and Aerospace Engineering
West Virginia University
Morgantown, West Virginia, USA

BICENTENNIAL
1807
WILEY
2007
BICENTENNIAL

John Wiley & Sons, Inc.

ACQUISITIONS EDITOR   Jennifer Welter

MARKETING MANAGER   Phyllis Cerys

PRODUCTION EDITOR   Lea Radick

DESIGNER   Hope Miller

COVER PHOTO   A pair of Black-browed Albatrosses photographed over the Southern ocean by Michael Gore.
See the "Fluids in the News" item in Section 9.2.1 for more information on the albatross.

This book was typeset in 10/12 Times by The GTS Companies/York, PA and printed and bound by
R. R. Donnelley Willard. The cover was printed by Phoenix Color.

100508855451

ISBN 13: 978-0-470-03962-5

Printed in the United States of America.

10 9 8 7 6 5 4 3 2 1

*A*bout the Authors

Donald F. Young, Anson Marston Distinguished Professor Emeritus in Engineering, is a faculty member in the Department of Aerospace Engineering and Engineering Mechanics at Iowa State University. Dr. Young received his B.S. degree in mechanical engineering, his M.S. and Ph.D. degrees in theoretical and applied mechanics from Iowa State, and has taught both undergraduate and graduate courses in fluid mechanics for many years. In addition to being named a Distinguished Professor in the College of Engineering, Dr. Young has also received the Standard Oil Foundation Outstanding Teacher Award and the Iowa State University Alumni Association Faculty Citation. He has been engaged in fluid mechanics research for more than 45 years, with special interests in similitude and modeling and the interdisciplinary field of biomedical fluid mechanics. Dr. Young has contributed to many technical publications and is the author or coauthor of two textbooks on applied mechanics. He is a Fellow of The American Society of Mechanical Engineers.

Bruce R. Munson, Professor Emeritus of Engineering Mechanics, has been a faculty member at Iowa State University since 1974. He received his B.S. and M.S. degrees from Purdue University and his Ph.D. degree from the Aerospace Engineering and Mechanics Department of the University of Minnesota in 1970.

From 1970 to 1974, Dr. Munson was on the mechanical engineering faculty of Duke University. From 1964 to 1966, he worked as an engineer in the jet engine fuel control department of Bendix Aerospace Corporation, South Bend, Indiana.

Dr. Munson's main professional activity has been in the area of fluid mechanics education and research. He has been responsible for the development of many fluid mechanics courses for studies in civil engineering, mechanical engineering, engineering science, and agricultural engineering and is the recipient of an Iowa State University Superior Engineering Teacher Award and the Iowa State University Alumni Association Faculty Citation.

He has authored and coauthored many theoretical and experimental technical papers on hydrodynamic stability, low Reynolds number flow, secondary flow, and the applications of viscous incompressible flow. He is a member of The American Society of Mechanical

Engineers, The American Physical Society, and The American Society for Engineering Education.

Theodore H. Okiishi, Associate Dean of Engineering and past Chair of Mechanical Engineering at Iowa State University, has taught fluid mechanics courses there since 1967. He received his undergraduate and graduate degrees at Iowa State.

From 1965 to 1967, Dr. Okiishi served as a U.S. Army officer with duty assignments at the National Aeronautics and Space Administration Lewis Research Center, Cleveland, Ohio, where he participated in rocket nozzle heat transfer research, and at the Combined Intelligence Center, Saigon, Republic of South Vietnam, where he studied seasonal river flooding problems.

Professor Okiishi is active in research on turbomachinery fluid dynamics. He and his graduate students and other colleagues have written a number of journal articles based on their studies. Some of these projects have involved significant collaboration with government and industrial laboratory researchers with one technical paper winning the ASME Melville Medal.

Dr. Okiishi has received several awards for teaching. He has developed undergraduate and graduate courses in classical fluid dynamics as well as the fluid dynamics of turbomachines.

He is a licensed professional engineer. His technical society activities include having been chair of the board of directors of The American Society of Mechanical Engineers (ASME) International Gas Turbine Institute. He is a fellow member of the ASME and the technical editor of the *Journal of Turbomachinery*.

Wade W. Huebsch has been a faculty member in the Department of Mechanical and Aerospace Engineering at West Virginia University since 2001. He received his B.S. degree in aerospace engineering from San Jose State University where he played college baseball. He received his M.S. degree in mechanical engineering and his Ph.D. in aerospace engineering from Iowa State University in 2000.

Dr. Huebsch specializes in computational fluid dynamics research and has authored multiple journal articles in the areas of aircraft icing, roughness-induced flow phenomena, and boundary layer flow control. He has taught both undergraduate and graduate courses in fluid mechanics and has developed a new undergraduate course in computational fluid dynamics. He has received multiple teaching awards such as Outstanding Teacher and Teacher of the Year from the College of Engineering and Mineral Resources at WVU as well as the Ralph R. Teetor Educational Award from SAE. He was also named as the Young Researcher of the Year from WVU. He is a member of the American Institute of Aeronautics and Astronautics, the Sigma Xi research society, the Society of Automotive Engineers, and the American Society of Engineering Education.

Also by these authors

Fundamentals of Fluid Mechanics, 5e
0-471-67582-2

Complete in-depth coverage of basic fluid mechanics principles, including compressible flow, for use in either a one- or two-semester course.

Preface

A Brief Introduction to Fluid Mechanics, fourth edition, is an abridged version of a more comprehensive treatment found in *Fundamentals of Fluid Mechanics* by Munson, Young, and Okiishi. Although this latter work continues to be successfully received by students and colleagues, it is a large volume containing much more material than can be covered in a typical one-semester undergraduate fluid mechanics course. A consideration of the numerous fluid mechanics texts that have been written during the past several decades reveals that there is a definite trend toward larger and larger books. This trend is understandable because the knowledge base in fluid mechanics has increased, along with the desire to include a broader scope of topics in an undergraduate course. Unfortunately, one of the dangers in this trend is that these large books can become intimidating to students who may have difficulty, in a beginning course, focusing on basic principles without getting lost in peripheral material. It is with this background in mind that the authors felt that a shorter but comprehensive text, covering the basic concepts and principles of fluid mechanics in a modern style, was needed. In this abridged version there is still more than ample material for a one-semester undergraduate fluid mechanics course. We have made every effort to retain the principal features of the original book while presenting the essential material in a more concise and focused manner that will be helpful to the beginning student.

This fourth edition has been prepared by the authors after several years of using the previous editions for an introductory course in fluid mechanics. Based on this experience, along with suggestions from reviewers, colleagues, and students, we have made a number of changes and additions in this new edition.

New to This Edition

Website—The electronic assets for this book (including videos, lab problems, and much more!) can be accessed on the website for this book. Access is free-of-charge with the registration code included in the front of every new book.

Fluids in the News—63 short news stories illustrate some of the current, important, and novel ways that fluid mechanics affects our lives.

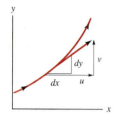

Extended Examples—Many of the example problems have been extended to illustrate what happens if one or more of the parameters is changed. This gives the user a better feel for some of the basic principles involved.

Simple Figures—To help students better understand and visualize some of the basic concepts, additional margin figures of the type shown in the left margin have been added at appropriate places throughout the text.

Learning Objectives—Each chapter begins with a set of learning objectives to provide the student with a brief preview of the topics covered in the chapter.

Problems—Approximately 30% new homework problems have been added for this edition, including new problems based on the "Fluids In the News" topics and simple CFD problems that can be solved using FlowLab.

Computational Fluid Dynamics (CFD)—Owing to the growing importance of CFD in engineering design and analysis, material on this subject is included in Appendix A. Material related to using FlowLab, a CFD software package, is included in Appendices I, J, and K.

Fundamentals of Engineering Exam—A set of FE exam questions is available on the book web site.

Key Features

Examples

One of our aims is to represent fluid mechanics as it really is—an exciting and useful discipline. To this end, we include analyses of numerous everyday examples of fluid-flow phenomena to which students and faculty can easily relate. In the fourth edition 99 examples are presented that provide detailed solutions to a variety of problems. A *new* feature of this edition is the fact that many of the examples have been extended to illustrate what happens if one or more of the parameters is changed. This gives the user a better feel for some of the basic principles involved. Also, all of the examples are *newly* outlined and carried out with the problem solving methodology of "Given, Find, Solution, and Comment." (See Note to User before Example 1.1.) *New* to this edition is a set of 63 short "Fluids in the News" stories that illustrates some of the current, important, and novel ways that fluid mechanics affects our lives. Many of these stories have homework problems associated with them. Also, to help the student better visualize and understand some of the basic concepts, the set of simple figures of the type shown in the left margin has been expanded.

Videos

V2.3 Hoover dam

There are 75 video segments illustrating many interesting and practical applications of real-world fluid phenomena. Each video segment is identified at the appropriate location in the text material by an icon of the type shown in the left margin. There are many homework problems that are directly related to the topics in the videos.

Problems

A generous set of more than 800 homework problems stresses the practical application of principles. The end-of-the-chapter homework problems (of which approximately 30% are *new* to this edition) are *newly* grouped and identified according to topic. The following types of problems are included: (1) "standard" problems, (2) computer problems, (3) discussion problems, (4) supply-your-own-data problems, (5) review problems with solutions, (6) *new* problems based on the "Fluids in the News" topics, (7) problems based on the fluid videos, (8) Excel-based lab problems, and (9) *new* simple CFD problems to be solved using FlowLab.

Lab Problems—There are 30 extended, laboratory-type problems that involve actual experimental data for simple experiments of the type that are often found in the laboratory portion of many introductory fluid mechanics courses. The data for these problems are provided in Excel format.

Review Problems—There is a set of 163 review problems covering most of the main topics in the book. Complete, detailed solutions to these problems can be found in the *Student Solution Manual for A Brief Introduction to Fluid Mechanics,* by Young, et al. (© 2007 John Wiley and Sons, Inc.).

Well-Paced Concept and Problem-Solving Development

Since this is an introductory text, we have designed the presentation of material to allow for the gradual development of student confidence in fluid mechanics problem solving. Each important concept or notion is considered in terms of simple and easy-to-understand circumstances before more complicated features are introduced. A *new* brief Learning Objectives section is provided at the beginning of each chapter. It should be helpful to read through this list prior to reading the chapter to gain a "big picture" idea of what key knowledge is to be gained from the chapter. Upon completion of the chapter, it would be beneficial to look back at the original learning objectives to ensure that a satisfactory level of understanding has been acquired for each item. Additional reinforcement of these learning objectives is provided in the form of the Chapter Summary and Study Guide at the end of each chapter.

Systems of Units

Two systems of units continue to be used throughout most of the text: the British Gravitational System (pounds, slugs, feet, and seconds), and the International System of Units (newtons, kilograms, meters, and seconds). About one-half of the examples and homework problems are in each set of units.

Topical Organization

In the first four chapters the student is made aware of some fundamental aspects of fluid motion, including important fluid properties, regimes of flow, pressure variations in fluids at rest and in motion, fluid kinematics, and methods of flow description and analysis. The Bernoulli equation is introduced in Chapter 3 to draw attention, early on, to some of the interesting effects of fluid motion on the distribution of pressure in a flow field. We believe that this timely consideration of elementary fluid dynamics increases student enthusiasm for the more complicated material that follows. In Chapter 4 we convey the essential elements of kinematics, including Eulerian and Lagrangian mathematical descriptions of flow phenomena, and indicate the vital relationship between the two views. For teachers who wish to consider kinematics in detail before the material on elementary fluid dynamics, Chapters 3 and 4 can be interchanged without loss of continuity.

Chapters 5, 6, and 7 expand on the basic analysis methods generally used to solve or to begin solving fluid mechanics problems. Emphasis is placed on understanding how flow phenomena are described mathematically and on when and how to use infinitesimal and finite control volumes. The effects of fluid friction on pressure and velocity distributions are also considered in some detail. A formal course in thermodynamics is not required to understand the various portions of the text that consider some elementary aspects of the thermodynamics of fluid flow. Chapter 7 features the advantages of using dimensional analysis and similitude for organizing test data and for planning experiments and the basic techniques involved.

Chapters 8 through 11 offer students opportunities for the further application of the principles learned early in the text. Also, where appropriate, additional important notions such as boundary layers, transition from laminar to turbulent flow, turbulence modeling, and flow separation are introduced. Practical concerns such as pipe flow, open-channel flow, flow measurement, drag and lift, and the fluid mechanics fundamentals associated with turbomachines are included.

Owing to the growing importance of computational fluid dynamics (CFD) in engineering design and analysis, *new* material on this subject is included in Appendix A. This material may be omitted without any loss of continuity to the rest of the text. This introductory CFD overview includes examples and problems of various interesting flow situations that are to be solved using FlowLab software.

Students who study this text and who solve a representative set of the exercises provided should acquire a useful knowledge of the fundamentals of fluid mechanics. Faculty who use this text are provided with numerous topics to select from in order to meet the objectives of their own courses. All are reminded of the fine collection of supplementary material. We have cited throughout the text the articles, books, and DVDs that are available for enrichment.

Student and Instructor Resources

Student Solution Manual, by Young, et al. (© 2007 John Wiley and Sons, Inc.)—This short paperback book is available as a supplement for the text. It provides detailed solutions to the Review Problems. This supplement is available through your local bookstore, or you may purchase it on the Wiley website at www.wiley.com/college/young.

Student Companion Site—The student section of the book website at www.wiley.com/college/young contains the assets listed below. Access is free-of-charge with the registration code included in the front of every new book.

Video Library	CFD Drive Cavity Example
Review Problems with Answers	FlowLab Tutorial and User's Guide
Lab Problems	FlowLab Problems
Comprehensive Table of Conversion Factors	FE Exam Questions

Instructor Companion Site—The instructor section of the book website at www.wiley.com/college/young contains the assets in the Student Companion Site, as well as the following, which are available only to professors who adopt this book for classroom use:

- Instructor Solutions Manual, containing complete, detailed solutions to all of the problems in the text.
- Figures from the text and homework problems appropriate for use in lecture slides.

These instructor materials are password-protected. Visit the Instructor Companion Site to register for a password.

FlowLab®—In cooperation with Wiley, Fluent, Inc. is offering to instructors who adopt this text the option to have FlowLab software installed in their department lab free of charge. (This offer is available in the Americas only. Fees vary by geographic region outside the Americas). FlowLab is a CFD package that allows students to solve fluid dynamics problems without requiring a long training period. This software introduces CFD technology to undergraduates and uses CFD to excite students about fluid dynamics and learning more about transport phenomena of all kinds. To learn more about FlowLab, and how to have it installed in your department, visit the Instructor Companion Site at www.wiley.com/college/young.

Acknowledgments

We express our thanks to the many colleagues who have helped in the development of this text, including our *new* co-author, Prof. Wade Huebsch of West Virginia University. We wish to express our gratitude to the many persons who provided suggestions for this and previous editions through reviews and surveys. Finally, we thank our families for their continued encouragement during the writing of this fourth edition.

Working with students over the years has taught us much about fluid mechanics education. We have tried in earnest to draw from this experience for the benefit of users of this book. Obviously we are still learning, and we welcome any suggestions and comments from you.

BRUCE R. MUNSON
DONALD F. YOUNG
THEODORE H. OKIISHI
WADE W. HUEBSCH

Featured in this Book

FLUIDS IN THE NEWS

Throughout the book are many brief news stories involving current, sometimes novel, applications of fluid phenomena. Many of these stories have homework problems associated with them.

4.5   Chapter Summary and Study Guide

field representation
velocity field
Eulerian method
Lagrangian method
one-, two-, and three-dimensional flow
steady and unsteady flow
streamline
streakline
pathline
acceleration field
material derivative
local acceleration
convective acceleration
system
control volume
Reynolds transport theorem

This chapter considered several fundamental concepts of fluid kinematics. That is, various aspects of fluid motion are discussed without regard to the forces needed to produce this motion. The concepts of a field representation of a flow and the Eulerian and Lagrangian approaches to describing a flow are introduced, as are the concepts of velocity and acceleration fields.

The properties of one-, two-, or three-dimensional flows and steady or unsteady flows are introduced along with the concepts of streamlines, streaklines, and pathlines. Streamlines, which are lines tangent to the velocity field, are identical to streaklines and pathlines if the flow is steady. For unsteady flows, they need not be identical.

As a fluid particle moves about, its properties (i.e., velocity, density, temperature) may change. The rate of change of these properties can be obtained by using the material derivative, which involves both unsteady effects (time rate of change at a fixed location) and convective effects (time rate of change due to the motion of the particle from one location to another).

The concepts of a control volume and a system are introduced, and the Reynolds transport theorem is developed. By using these ideas, the analysis of flows can be carried out using a control volume (a fixed volume through which the fluid flows), whereas the governing principles are stated in terms of a system (a flowing portion of fluid).

The following checklist provides a study guide for this chapter. When your study of the entire chapter and end-of-chapter exercises has been completed you should be able to

- write out the meanings of the terms listed here in the margin and understand each of the related concepts. These terms are particularly important and are set in color and bold type in the text.
- understand the concept of the field representation of a flow and the difference between Eulerian and Lagrangian methods of describing a flow.
- explain the differences among streamlines, streaklines, and pathlines.
- calculate and plot streamlines for flows with given velocity fields.
- use the concept of the material derivative, with its unsteady and convective effects, to determine time rate of change of a fluid property.
- determine the acceleration field for a flow with a given velocity field.
- understand the properties of and differences between a system and a control volume.
- interpret, physically and mathematically, the concepts involved in the Reynolds transport theorem.

MARGIN FIGURES

A set of simple figures in the margins is provided to help the students visualize concepts being described.

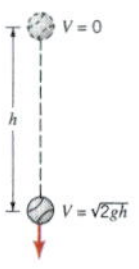

FLUID VIDEOS

A set of videos illustrating interesting and practical applications of fluid phenomena is provided on the book website. An icon in the margin identifies each video. Approximately 85 homework problems are tied to the videos.

xiv

CHAPTER SUMMARY AND STUDY GUIDE

At the end of each chapter is a brief summary of key concepts and principles introduced in the chapter along with key terms involved.

BOXED EQUATIONS

Important equations are boxed to help the user identify them.

3.6   Examples of Use of the Bernoulli Equation

Between any two points, (1) and (2), on a streamline in steady, inviscid, incompressible flow the Bernoulli equation (Eq. 3.6) can be applied in the form

$$\boxed{p_1 + \tfrac{1}{2}\rho V_1^2 + \gamma z_1 = p_2 + \tfrac{1}{2}\rho V_2^2 + \gamma z_2} \qquad (3.14)$$

The use of this equation is discussed in this section.

3.6.1   Free Jets

Consider flow of a liquid from a large reservoir as is shown in Fig. 3.7. A jet of liquid of diameter d flows from the nozzle with velocity V. Application of Eq. 3.14 between points (1) and (2) on the streamline shown gives

$$\gamma h = \tfrac{1}{2}\rho V^2$$

We have used the facts that $z_1 = h$, $z_2 = 0$, the reservoir is large ($V_1 = 0$), open to the atmosphere ($p_1 = 0$ gage), and the fluid leaves as a *"free jet"* ($p_2 = 0$). Thus, we obtain

$$V = \sqrt{2\,\frac{\gamma h}{\rho}} = \sqrt{2gh} \qquad (3.15)$$

Be careful when applying Eq. 3.15. It is valid only if the reservoir is large enough so that the initial fluid speed is negligible ($V_1 = 0$), the beginning and ending pressures are the same ($p_1 = p_2$), and viscous effects are absent. Note that as shown in the figure in the margin, a ball starting from rest will also obtain a speed of $V = \sqrt{2gh}$ after it has dropped through a distance h, provided friction is negligible.

Once outside the nozzle, the stream continues to fall as a free jet with zero pressure throughout ($p_5 = 0$) and as seen by applying Eq. 3.14 between points (1) and (5), the speed increases according to

$$V = \sqrt{2g(h + H)}$$

where H is the distance the fluid has fallen outside the nozzle.

Equation 3.15 could also be obtained by writing the Bernoulli equation between points (3) and (4) using the fact that $z_4 = 0$, $z_3 = \ell$. Also, $V_3 = 0$ since it is far from the nozzle, and from hydrostatics, $p_3 = \gamma(h - \ell)$.

If the exit of the tank shown in Fig. 3.7 is not a smooth, well-contoured nozzle, the diameter of the jet, d_j, will be less than the diameter of the hole, d_h. This phenomenon, called a *vena contracta* effect, is a result of the inability of the fluid to turn a sharp 90° corner.

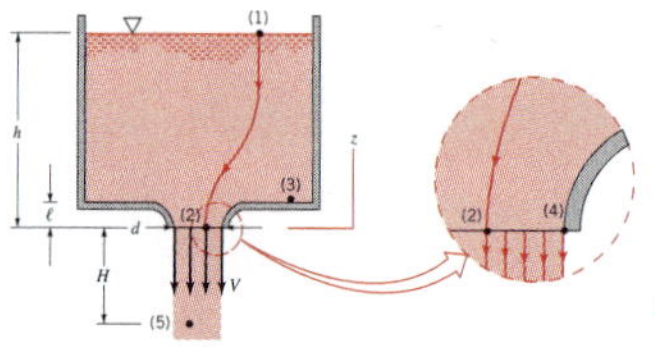

$V = 0$

$V = \sqrt{2gh}$

V3.5 Flow from a tank

■ **FIGURE 3.7**
Vertical flow from a tank.

EXAMPLE 3.6 Pitot-Static Tube

GIVEN An airplane flies 100 mi/hr at an elevation of 10,000 ft in a standard atmosphere as shown in Fig. E3.6a.

FIND Determine the pressure at point (1) far ahead of the airplane, the pressure at the stagnation point on the nose of the airplane, point (2), and the pressure difference indicated by a Pitot-static probe attached to the fuselage.

$V_1 = 100$ mi/hr (2)

(1)

Pitot-static tube

■ **FIGURE E3.6a**

SOLUTION

From Table C.1 we find that the static pressure at the altitude given is

$$p_1 = 1456 \text{ lb/ft}^2 \text{ (abs)} = 10.11 \text{ psia} \quad \text{(Ans)}$$

Also, the density is $\rho = 0.001756$ slug/ft³.

If the flow is steady, inviscid, and incompressible and elevation changes are neglected, Eq. 3.6 becomes

$$p_2 = p_1 + \frac{\rho V_1^2}{2}$$

With $V_1 = 100$ mi/hr $= 146.7$ ft/s and $V_2 = 0$ (since the coordinate system is fixed to the airplane) we obtain

$$p_2 = 1456 \text{ lb/ft}^2 + (0.001756 \text{ slugs/ft}^3)(146.7 \text{ ft/s})^2/2$$
$$= (1456 + 18.9) \text{ lb/ft}^2 \text{ (abs)}$$

Hence, in terms of gage pressure

$$p_2 = 18.9 \text{ lb/ft}^2 = 0.1313 \text{ psi} \quad \text{(Ans)}$$

Thus, the pressure difference indicated by the Pitot-static tube is

$$p_2 - p_1 = \frac{\rho V_1^2}{2} = 0.1313 \text{ psi} \quad \text{(Ans)}$$

COMMENTS Note that it is very easy to obtain incorrect results by using improper units. Do not add lb/in.² and lb/ft². Note that (slug/ft³)(ft²/s²) = (slug·ft/s²)/(ft²) = lb/ft².

It was assumed that the flow is incompressible—the density remains constant from (1) to (2). However, because $\rho = p/RT$, a change in pressure (or temperature) will cause a change in density. For this relatively low speed, the ratio of the absolute pressures is nearly unity [i.e., $p_1/p_2 = (10.11$ psia)/(10.11 + 0.1313 psia) = 0.987] so that the density change is negligible. However, by repeating the calculations for various values of the speed, V_1, the results shown in Fig. E3.6b are obtained. Clearly at the 500 to 600 mph speeds normally flown by commercial airliners, the pressure ratio is such that density changes are important. In such situations it is necessary to use compressible flow concepts to obtain accurate results.

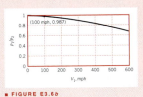

(100 mph, 0.987)

p_1/p_2

V_1, mph

■ **FIGURE E3.6b**

Lab Problems for Chapter 5 **13**

Lab Problems for Chapter 5

5.126 Force from a Jet of Air Deflected by a Flat Plate

Objective: A jet of a fluid striking a flat plate as shown in Fig. P5.126 exerts a force on the plate. It is the equal and opposite force of the plate on the fluid that causes the fluid momentum change that accompanies such a flow. The purpose of this experiment is to compare the theoretical force on the plate with the experimentally measured force.

Equipment: Air source with an adjustable flowrate and a flow meter; nozzle to produce a uniform air jet; balance beam with an attached flat plate; weights; barometer; thermometer.

Experimental Procedure: Adjust the counter weight so that the beam is level when there is no mass, m, on the beam and no flow through the nozzle. Measure the diameter, d, of the nozzle outlet. Record the barometer reading, H_{atm}, in inches of mercury and the air temperature, T, so that the air density can be calculated by use of the perfect gas law. Place a known mass, m, on the flat plate and adjust the fan speed control to produce the necessary flowrate, Q, to make the balance beam level again. The flowrate is related to the flow meter manometer reading, h, by the equation $Q = 0.358 \, h^{1/2}$, where Q is in ft³/s and h is in inches of water. Repeat the measurements for various masses on the plate.

Calculations: For each flowrate, Q, calculate the weight, $W = mg$, needed to balance the beam and use the continuity equation, $Q = VA$, to determine the velocity, V, at the nozzle exit. Use the momentum equation for this problem, $W = \rho V^2 A$, to determine the theoretical relationship between velocity and weight.

Graph: Plot the experimentally measured force on the plate, W, as ordinates and air speed, V, as abscissas.

Results: On the same graph, plot the theoretical force as a function of air speed.

Data: To proceed, print this page for reference when you work the problem and *click here* to bring up an EXCEL page with the data for this problem.

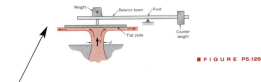

■ **FIGURE P5.126**

LAB PROBLEMS

On the book website is a set of lab problems in Excel format involving actual data for experiments of the type found in many introductory fluid mechanics labs.

EXAMPLE PROBLEMS

A set of example problems provides the student detailed solutions and comments for interesting, real-world situations.

REVIEW PROBLEMS

On the book web site are nearly 200 Review Problems covering most of the main topics in the book. Complete, detailed solutions to these problems are found in the supplement *Student Solution Manual for A Brief Introduction to Fundamentals of Fluid Mechanics*, by Young, et al. (© 2007 John Wiley and Sons, Inc.).

Review Problems for Chapter 4

Click on the answers of the review problems to go to the detailed solutions.

4.1R (Streamlines) The velocity field in a flow is given by $\mathbf{V} = x^2 y \hat{\mathbf{i}} + x^2 t \hat{\mathbf{j}}$. (a) Plot the streamline through the origin at times $t = 0$, $t = 1$, and $t = 2$. (b) Do the streamlines plotted in part (a) coincide with the path of particles through the origin? Explain.
(ANS: $y^2/2 = tx + C$; no)

4.2R (Streamlines) A velocity field is given by $u = y - 1$ and $v = y - 2$, where u and v are in m/s and x and y are in meters. Plot the streamline that passes through the point $(x, y) = (4, 3)$. Compare this streamline with the streamline through the point $(x, y) = (4, 3)$.
(ANS: $x = y + \ln(y - 2) + 1$)

4.3R (Material derivative) The pressure in the pipe near the discharge of a reciprocating pump fluctuates according to $p = [200 + 40 \sin(8t)]$ kPa, where t is in seconds. If the fluid speed in the pipe is 5 m/s, determine the maximum rate of change of pressure experienced by a fluid particle.
(ANS: 320 kPa/s)

4.4R (Acceleration) A shock wave is a very thin layer (thickness $= \ell$) in a high-speed (supersonic) gas flow across which the flow properties (velocity, density, pressure, etc.) change from state (1) to state (2) as shown in Fig. P4.4R. If $V_1 = 1800$ fps, $V_2 = 700$ fps, and $\ell = 10^{-4}$ in., estimate the average deceleration of the gas as it flows across the shock wave. How many g's deceleration does this represent?
(ANS: -1.65×10^{11} ft/s²; -5.12×10^9)

V_1 V_2

Shock wave ℓ

V_1 V

V_2

ℓ x

■ **FIGURE P4.4R**

4.5R (Acceleration) Air flows through a pipe with a uniform velocity of $\mathbf{V} = 5 \, t^2 \hat{\mathbf{i}}$ ft/s, where t is in seconds. Determine the acceleration at time $t = -1$, 0, and 1 s.
(ANS: $-10 \, \hat{\mathbf{i}}$ ft/s²; 0; 10 $\hat{\mathbf{i}}$ ft/s²)

4.6R (Acceleration) A fluid flows steadily along the streamline as shown in Fig. P4.6R. Determine the acceleration at point A. At point A what is the angle between the acceleration and the x axis? At point A what is the angle between the acceleration and the streamline?
(ANS: $10 \, \hat{\mathbf{n}} + 30 \, \hat{\mathbf{s}}$ ft/s²; 48.5 deg; 18.5 deg)

30°

$\mathcal{R} = 10$ ft

$V = 10$ ft/s

$\frac{\partial V}{\partial s} = 3$ s⁻¹

A s

x

■ **FIGURE P4.6R**

4.7R (Acceleration) In the conical nozzle shown in Fig. P4.7R the streamlines are essentially radial lines emanating from point A and the fluid velocity is given approximately by $V = C/r^2$, where C is a constant. The fluid velocity is 2 m/s along the centerline at the beginning of the nozzle ($x = 0$). Determine the acceleration along the nozzle centerline as a function of x. What is the value of the acceleration at $x = 0$ and $x = 0.3$ m?
(ANS: $1.037/(0.6 - x)^5 \, \hat{\mathbf{i}}$ m/s²; 13.3 $\hat{\mathbf{i}}$ m/s²; 427 $\hat{\mathbf{i}}$ m/s²)

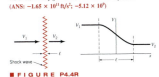

0.6 m

0.3 m

Q V A

x

■ **FIGURE P4.7R**

4.8R (Reynolds transport theorem) A sanding operation injects 10^5 particles/s into the air in a room as shown in

STUDENT SOLUTION MANUAL

A brief paperback book titled *Student Solution Manual for A Brief Introduction to Fluid Mechanics*, by Young, et al. (© 2007 John Wiley and Sons, Inc.), is available. It contains detailed solutions to the Review Problems.

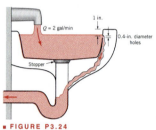

■ FIGURE P3.24

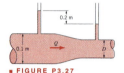

■ FIGURE P3.27

3.28 Water flows steadily with negligible viscous effects through the pipe shown in Fig. P3.28. Determine the diameter, D, of the pipe at the outlet (a free jet) if the velocity there is 20 ft/s.

PROBLEMS

A generous set of homework problems at the end of each chapter stresses the practical applications of fluid mechanics principles. Over 800 homework problems are included.

3.25 Air is drawn into a wind tunnel used for testing automobiles as shown in Fig. P3.25. **(a)** Determine the manometer reading, h, when the velocity in the test section is 60 mph. Note that there is a 1-in. column of oil on the water in the manometer. **(b)** Determine the difference between the stagnation pressure on the front of the automobile and the pressure in the test section.

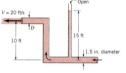

■ FIGURE P3.28

3.29 The circular stream of water from a faucet is observed to taper from a diameter of 20 to 10 mm in a distance of 50 cm. Determine the flowrate.

3.30 Water is siphoned from the tank shown in Fig. P3.30. The water barometer indicates a reading of 30.2 ft. Determine the maximum value of h allowed without cavitation occurring. Note that the pressure of the vapor in the closed end of the barometer equals the vapor pressure.

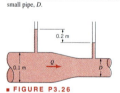

■ FIGURE P3.25

3.26 Water flows through the pipe contraction shown in Fig. P3.26. For the given 0.2-m difference in the manometer level, determine the flowrate as a function of the diameter of the small pipe, D.

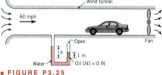

■ FIGURE P3.26

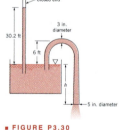

3.27 Water flows through the pipe contraction shown in Fig. P3.27. For the given 0.2-m difference in the manometer level, determine the flowrate as a function of the diameter of the small pipe, D.

■ FIGURE P3.30

3.31 An inviscid fluid flows steadily through the contraction shown in Fig. P3.31. Derive an expression for the fluid

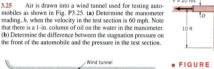

Axial Velocity

Legend
- ■ inlet
- ■ x = 0.5d
- ■ x = 1d
- ■ x = 5d
- ■ x = 10d
- ? x = 25d
- ? outlet

CFD and FlowLab

For those who wish to become familiar with the basic concepts of computational fluid dynamics, a new overview to CFD is provided in Appendices A and I. In addition, the use of FlowLab software to solve interesting flow problems is described in Appendices J and K.

Contents

10
OPEN-CHANNEL FLOW   376

11
TURBOMACHINES   410

A
COMPUTATIONAL FLUID DYNAMICS AND FLOWLAB   454

B
PHYSICAL PROPERTIES OF FLUIDS   469

C
PROPERTIES OF THE U.S. STANDARD ATMOSPHERE   475

D
REYNOLDS TRANSPORT THEOREM   477

E
COMPREHENSIVE TABLE OF CONVERSION FACTORS   483

ONLINE APPENDIX LIST   487

F
VIDEO LIBRARY

See book web site, www.wiley.com/ college/young, for this material.

G
REVIEW PROBLEMS

See book web site, www.wiley.com/ college/young, for this material.

H
LABORATORY PROBLEMS

See book web site, www.wiley.com/ college/young, for this material.

I
CFD DRIVEN CAVITY EXAMPLE

See book web site, www.wiley.com/ college/young, for this material.

J
FLOWLAB TUTORIAL AND USER'S GUIDE

See book web site, www.wiley.com/ college/young, for this material.

K
FLOWLAB PROBLEMS

See book web site, www.wiley.com/ college/young, for this material.

Introduction

Viscous shear stresses between a moving layer of fluid and fluid within a cavity produce a swirl motion within the cavity. (Particles in oil; time exposure.) (Photograph by B. R. Munson.)

LEARNING OBJECTIVES

After completing this chapter, you should be able to:
- determine the dimensions and units of physical quantities.
- identify the key fluid properties used in the analysis of fluid behavior.
- calculate common fluid properties given appropriate information.
- explain effects of fluid compressibility.
- use the concepts of viscosity, vapor pressure and surface tension.

Fluid mechanics is that discipline within the broad field of applied mechanics concerned with the behavior of liquids and gases at rest or in motion. This field of mechanics obviously encompasses a vast array of problems that may vary from the study of blood flow in the capillaries (which are only a few microns in diameter) to the flow of crude oil across Alaska through an 800-mile-long, 4-ft-diameter pipe. Fluid mechanics principles are needed to explain why airplanes are made streamlined with smooth surfaces for the most efficient flight, whereas golf balls are made with rough surfaces (dimpled) to increase their efficiency. It is very likely that during your career as an engineer you will be involved in the analysis and design of systems that require a good understanding of fluid mechanics. It is hoped that this introductory text will provide a sound foundation of the fundamental aspects of fluid mechanics.

1.1 Some Characteristics of Fluids

One of the first questions we need to explore is—what is a fluid? Or we might ask—what is the difference between a solid and a fluid? We have a general, vague idea of the difference. A solid is "hard" and not easily deformed, whereas a fluid is "soft" and is easily deformed (we can readily move through air). Although quite descriptive, these casual observations of the differences between solids and fluids are not very satisfactory from a scientific or engineering point of view. A more specific distinction is based on how materials deform under the action of an external load. *A **fluid** is defined as a substance that deforms continuously when acted on by a shearing stress of any magnitude.* A shearing stress (force per unit area) is created whenever a tangential force acts on a surface. When common solids such as steel or other metals are acted on by a shearing stress, they will initially deform (usually a very small deformation), but they will not continuously deform (flow). However, common fluids such as water, oil, and air satisfy the definition of a fluid—that is, they will flow when acted on by a shearing stress. Some materials, such as slurries, tar, putty, toothpaste, and so on, are not easily classified since they will behave as a solid if the applied shearing stress is small, but if the stress exceeds some critical value, the substance will flow. The study of such materials is called *rheology* and does not fall within the province of classical fluid mechanics.

Although the molecular structure of fluids is important in distinguishing one fluid from another, because of the large number of molecules involved, it is not possible to study the behavior of individual molecules when trying to describe the behavior of fluids at rest or in motion. Rather, we characterize the behavior by considering the average, or macroscopic, value of the quantity of interest, where the average is evaluated over a small volume containing a large number of molecules.

We thus assume that all the fluid characteristics we are interested in (pressure, velocity, etc.) vary continuously throughout the fluid—that is, we treat the fluid as a *continuum.* This concept will certainly be valid for all the circumstances considered in this text.

1.2 Dimensions, Dimensional Homogeneity, and Units

Since we will be dealing with a variety of fluid characteristics in our study of fluid mechanics, it is necessary to develop a system for describing these characteristics both *qualitatively* and *quantitatively.* The qualitative aspect serves to identify the nature, or type, of the characteristics (such as length, time, stress, and velocity), whereas the quantitative aspect provides a numerical measure of the characteristics. The quantitative description requires both a number and a standard by which various quantities can be compared. A standard for length might be a meter or foot, for time an hour or second, and for mass a slug or kilogram. Such standards are called **units,** and several systems of units are in common use as described in the following section. The qualitative description is conveniently given in terms of certain *primary quantities,* such as length, L, time, T, mass, M, and temperature, Θ. These primary quantities can then be used to provide a qualitative description of any other *secondary quantity,* for example, area $\doteq L^2$, velocity $\doteq LT^{-1}$, density $\doteq ML^{-3}$, and so on, where the symbol $\doteq$ is used to indicate the *dimensions* of the secondary quantity in terms of the primary quantities. Thus, to describe qualitatively a velocity, V, we would write

$$V \doteq LT^{-1}$$

and say that "the dimensions of a velocity equal length divided by time." The primary quantities are also referred to as *basic dimensions.*

For a wide variety of problems involving fluid mechanics, only the three basic dimensions, L, T, and M, are required. Alternatively, L, T, and F could be used, where F is the basic dimension of force. Since Newton's law states that force is equal to mass times acceleration, it follows that $F \doteq MLT^{-2}$ or $M \doteq FL^{-1}T^2$. Thus, secondary quantities expressed in terms of M can be expressed in terms of F through the relationship just given. For example, stress, σ, is a force per unit area, so that $\sigma \doteq FL^{-2}$, but an equivalent dimensional equation is $\sigma \doteq ML^{-1}T^{-2}$. Table 1.1 provides a list of dimensions for a number of common physical quantities.

All theoretically derived equations are *dimensionally homogeneous*—that is, the dimensions of the left side of the equation must be the same as those on the right side, and all additive separate terms must have the same dimensions. We accept as a fundamental premise

■ **TABLE 1.1**
Dimensions Associated with Common Physical Quantities

	FLT System	MLT System
Acceleration	LT^{-2}	LT^{-2}
Angle	$F^0L^0T^0$	$M^0L^0T^0$
Angular acceleration	T^{-2}	T^{-2}
Angular velocity	T^{-1}	T^{-1}
Area	L^2	L^2
Density	$FL^{-4}T^2$	ML^{-3}
Energy	FL	ML^2T^{-2}
Force	F	MLT^{-2}
Frequency	T^{-1}	T^{-1}
Heat	FL	ML^2T^{-2}
Length	L	L
Mass	$FL^{-1}T^2$	M
Modulus of elasticity	FL^{-2}	$ML^{-1}T^{-2}$
Moment of a force	FL	ML^2T^{-2}
Moment of inertia (area)	L^4	L^4
Moment of inertia (mass)	FLT^2	ML^2
Momentum	FT	MLT^{-1}
Power	FLT^{-1}	ML^2T^{-3}
Pressure	FL^{-2}	$ML^{-1}T^{-2}$
Specific heat	$L^2T^{-2}\Theta^{-1}$	$L^2T^{-2}\Theta^{-1}$
Specific weight	FL^{-3}	$ML^{-2}T^{-2}$
Strain	$F^0L^0T^0$	$M^0L^0T^0$
Stress	FL^{-2}	$ML^{-1}T^{-2}$
Surface tension	FL^{-1}	ML^{-2}
Temperature	Θ	Θ
Time	T	T
Torque	FL	$ML^{-2}T^{-2}$
Velocity	LT^{-1}	LT^{-1}
Viscosity (dynamic)	$FL^{-2}T$	$ML^{-1}T^{-1}$
Viscosity (kinematic)	L^2T^{-1}	L^2T^{-1}
Volume	L^3	L^3
Work	FL	ML^2T^{-2}

that all equations describing physical phenomena must be dimensionally homogeneous. For example, the equation for the velocity, V, of a uniformly accelerated body is

$$V = V_0 + at \tag{1.1}$$

where V_0 is the initial velocity, a the acceleration, and t the time interval. In terms of dimensions the equation is

$$LT^{-1} \doteq LT^{-1} + LT^{-1}$$

and thus Eq. 1.1 is dimensionally homogeneous.

Some equations that are known to be valid contain constants having dimensions. The equation for the distance, d, traveled by a freely falling body can be written as

$$d = 16.1t^2 \tag{1.2}$$

and a check of the dimensions reveals that the constant must have the dimensions of LT^{-2} if the equation is to be dimensionally homogeneous. Actually, Eq. 1.2 is a special form of the well-known equation from physics for freely falling bodies,

$$d = \frac{gt^2}{2} \tag{1.3}$$

in which g is the acceleration of gravity. Equation 1.3 is dimensionally homogeneous and valid in any system of units. For $g = 32.2$ ft/s² the equation reduces to Eq. 1.2, and thus Eq. 1.2 is valid only for the system of units using feet and seconds. Equations that are restricted to a particular system of units can be denoted as *restricted homogeneous equations,* as opposed to equations valid in any system of units, which are *general homogeneous equations.* The concept of dimensions also forms the basis for the powerful tool of *dimensional analysis,* which is considered in detail in Chapter 7.

Note to the users of this text. All of the examples in the text use a consistent problem-solving methodology which is similar to that in other engineering courses such as statics. Each example highlights the key elements of analysis: *Given, Find, Solution,* and *Comment.*

The *Given* and *Find* are steps that ensure the user understands what is being asked in the problem and explicitly list the items provided to help solve the problem.

The *Solution* step is where the equations needed to solve the problem are formulated and the problem is actually solved. In this step, there are typically several other tasks that help to set up the solution and are required to solve the problem. The first is a drawing of the problem; where appropriate, it is always helpful to draw a sketch of the problem. Here the relevant geometry and coordinate system to be used as well as features such as control volumes, forces and pressures, velocities and mass flow rates are included. This helps in gaining a visual understanding of the problem. Making appropriate assumptions to solve the problem is the second task. In a realistic engineering problem-solving environment, the necessary assumptions are developed as an integral part of the solution process. Assumptions can provide appropriate simplifications or offer useful constraints, both of which can help in solving the problem. Throughout the examples in this text, the necessary assumptions are embedded within the *Solution* step, as they are in solving a real-world problem. This provides a realistic problem-solving experience.

The final element in the methodology is the *Comment.* For the examples in the text, this section is used to provide further insight into the problem or the solution. It can also be a point in the analysis at which certain questions are posed. For example: Is the answer reasonable, and does it make physical sense? Are the final units correct? If a certain parameter were changed, how would the answer change? Adopting the above type of methodology will aid in the development of problem-solving skills for fluid mechanics, as well as other engineering disciplines.

EXAMPLE 1.1   Restricted and General Homogeneous Equations

GIVEN A commonly used equation for determining the volume rate of flow, Q, of a liquid through an orifice located in the side of a tank as shown in Fig. E1.1 is

$$Q = 0.61\, A\sqrt{2gh}$$

where A is the area of the orifice, g is the acceleration of gravity, and h is the height of the liquid above the orifice.

FIND Investigate the dimensional homogeneity of this formula.

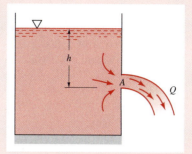

■ **FIGURE E1.1**

SOLUTION

The dimensions of the various terms in the equation are

$$Q = \text{volume/time} \doteq L^3 T^{-1}$$
$$A = \text{area} \doteq L^2$$
$$g = \text{acceleration of gravity} \doteq LT^{-2}$$
$$h = \text{height} \doteq L$$

These terms, when substituted into the equation, yield the dimensional form

$$(L^3 T^{-1}) \doteq (0.61)(L^2)(\sqrt{2})(LT^{-2})^{1/2}(L)^{1/2}$$

or

$$(L^3 T^{-1}) \doteq [(0.61)\sqrt{2}](L^3 T^{-1})$$

It is clear from this result that the equation is dimensionally homogeneous (both sides of the formula have the same dimensions of $L^3 T^{-1}$), and the numbers (0.61 and $\sqrt{2}$) are dimensionless.

COMMENT If we were going to use this relationship repeatedly, we might be tempted to simplify it by replacing g with its standard value of 32.2 ft/s^2 and rewriting the formula as

$$Q = 4.90\, A\sqrt{h} \qquad \qquad \textbf{(1)}$$

A quick check of the dimensions reveals that

$$L^3 T^{-1} \doteq (4.90)(L^{5/2})$$

and, therefore, the equation expressed as Eq. 1 can only be dimensionally correct if the number, 4.90, has the dimensions of $L^{1/2}T^{-1}$. Whenever a number appearing in an equation or formula has dimensions, it means that the specific value of the number will depend on the system of units used. Thus, for the case being considered with feet and seconds used as units, the number 4.90 has units of ft$^{1/2}$/s. Equation 1 will only give the correct value for Q (in ft^3/s) when A is expressed in square feet and h in feet. Thus, Eq. 1 is a *restricted homogeneous equation*, whereas the original equation is a *general* homogeneous equation that would be valid for any consistent system of units. A quick check of the dimensions of the various terms in an equation is a useful practice and will often be helpful in eliminating errors—that is, as noted previously, all physically meaningful equations must be dimensionally homogeneous. We have briefly alluded to units in this example, and this important topic will be considered in more detail in the next section.

1.2.1   Systems of Units

In addition to the qualitative description of the various quantities of interest, it is generally necessary to have a quantitative measure of any given quantity. For example, if we measure the width of this page in the book and say that it is 10 units wide, the statement has no meaning until the unit of length is defined. If we indicate that the unit of length is a meter, and define the meter as some standard length, a unit system for length has been established (and a numerical value can be given to the page width). In addition to length, a unit must be established for each of the remaining basic quantities (force, mass, time, and temperature). There are several systems of units in use and we shall consider two systems that are commonly used in engineering.

British Gravitational (BG) System. In the BG system the unit of length is the foot (ft), the time unit is the second (s), the force unit is the pound (lb), and the temperature unit is the degree Fahrenheit (°F), or the absolute temperature unit is the degree Rankine (°R), where

$$°R = °F + 459.67$$

The mass unit, called the *slug,* is defined from Newton's second law (force = mass × acceleration) as

$$1 \text{ lb} = (1 \text{ slug})(1 \text{ ft/s}^2)$$

This relationship indicates that a 1-lb force acting on a mass of 1 slug will give the mass an acceleration of 1 ft/s².

The weight, $\mathcal{W}$ (which is the force due to gravity, g) of a mass, $m,$ is given by the equation

$$\mathcal{W} = mg$$

and in BG units

$$\mathcal{W} \text{ (lb)} = m \text{ (slugs) } g \text{ (ft/s}^2)$$

Since the earth's standard gravity is taken as $g = 32.174$ ft/s² (commonly approximated as 32.2 ft/s²), it follows that a mass of 1 slug weighs 32.2 lb under standard gravity.

F l u i d s   i n   t h e   N e w s

How long is a foot? Today, in the United States, the common length *unit* is the *foot,* but throughout antiquity the unit used to measure length has quite a history. The first length units were based on the lengths of various body parts. One of the earliest units was the Egyptian cubit, first used around 3000 B.C. and defined as the length of the arm from elbow to extended fingertips. Other measures followed with the foot simply taken as the length of a man's foot. Since this length obviously varies from person to person it was often "standardized" by using the length of the current reigning royalty's foot. In 1791 a special French commission proposed that a new universal length unit called a meter (metre) be defined as the distance of one-quarter of the earth's meridian (north pole to the equator) divided by 10 million. Although controversial, the meter was accepted in 1799 as the standard. With the development of advanced technology, the length of a meter was redefined in 1983 as the distance traveled by light in a vacuum during the time interval of 1/299,792,458 s. The foot is now defined as 0.3048 meters. Our simple rulers and yardsticks indeed have an intriguing history. ■

International System (SI).   In 1960, the Eleventh General Conference on Weights and Measures, the international organization responsible for maintaining precise uniform standards of measurements, formally adopted the *International System of Units* as the international standard. This system, commonly termed SI, has been adopted worldwide and is widely used (although certainly not exclusively) in the United States. It is expected that the long-term trend will be for all countries to accept SI as the accepted standard, and it is imperative that engineering students become familiar with this system. In SI the unit of length is the meter (m), the time unit is the second (s), the mass unit is the kilogram (kg), and the temperature unit is the kelvin (K). Note that there is no degree symbol used when expressing a temperature in kelvin units. The Kelvin temperature scale is an absolute scale and is related to the Celsius (centigrade) scale (°C) through the relationship

$$K = {}^{\circ}C + 273.15$$

Although the Celsius scale is not in itself part of SI, it is common practice to specify temperatures in degrees Celsius when using SI units.

The force unit, called the newton (N), is defined from Newton's second law as

$$1 \text{ N} = (1 \text{ kg})(1 \text{ m/s}^2)$$

Thus, a 1-N force acting on a 1-kg mass will give the mass an acceleration of 1 m/s². Standard gravity in SI is 9.807 m/s² (commonly approximated as 9.81 m/s²) so that a 1-kg mass weighs 9.81 N under standard gravity. Note that weight and mass are different, both qualitatively and

■ **TABLE 1.2**
Conversion Factors from BG Units to SI Units

(See inside of back cover.)

■ **TABLE 1.3**
Conversion Factors from SI Units to BG Units

(See inside of back cover.)

quantitatively! The unit of *work* in SI is the joule (J), which is the work done when the point of application of a 1-N force is displaced through a 1-m distance in the direction of the force. Thus,

$$1 \text{ J} = 1 \text{ N} \cdot \text{m}$$

The unit of *power* is the watt (W) defined as a joule per second. Thus,

$$1 \text{ W} = 1 \text{ J/s} = 1 \text{ N} \cdot \text{m/s}$$

Prefixes for forming multiples and fractions of SI units are commonly used. For example, the notation kN would be read as "kilonewtons" and stands for 10^3N. Similarly, mm would be read as "millimeters" and stands for 10^{-3}m. The centimeter is not an accepted unit of length in the SI system, and for most problems in fluid mechanics in which SI units are used, lengths will be expressed in millimeters or meters.

In this text we will use the BG system and SI for units. Approximately one-half the problems and examples are given in BG units and one-half in SI units. Tables 1.2 and 1.3 provide conversion factors for some quantities that are commonly encountered in fluid mechanics, and these tables are located on the inside of the back cover. Note that in these tables (and others) the numbers are expressed by using computer exponential notation. For example, the number 5.154 E + 2 is equivalent to 5.154×10^2 in scientific notation, and the number 2.832 E − 2 is equivalent to 2.832×10^{-2}. More extensive tables of conversion factors for a large variety of unit systems can be found in Appendix E.

F l u i d s   i n   t h e   N e w s

Units and space travel A NASA spacecraft, the Mars Climate Orbiter, was launched in December 1998 to study the Martian geography and weather patterns. The spacecraft was slated to begin orbiting Mars on September 23, 1999. However, NASA officials lost communication with the spacecraft early that day, and it is believed that the spacecraft broke apart or overheated because it came too close to the surface of Mars.

Errors in the maneuvering commands sent from earth caused the Orbiter to sweep within 37 miles of the surface rather than the intended 93 miles. The subsequent investigation revealed that the errors were due to a simple mix-up in *units*. One team controlling the Orbiter used SI units whereas another team used BG units. This costly experience illustrates the importance of using a consistent system of units. ■

1.3   Analysis of Fluid Behavior

The study of fluid mechanics involves the same fundamental laws you have encountered in physics and other mechanics courses. These laws include Newton's laws of motion, conservation of mass, and the first and second laws of thermodynamics. Thus, there are strong similarities between the general approach to fluid mechanics and to rigid-body and deformable-body solid mechanics.

The broad subject of fluid mechanics can be generally subdivided into *fluid statics,* in which the fluid is at rest, and *fluid dynamics,* in which the fluid is moving. In subsequent chapters we will consider both of these areas in detail. Before we can proceed, however, it will be necessary to define and discuss certain fluid *properties* that are intimately related to fluid behavior. In the following several sections, the properties that play an important role in the analysis of fluid behavior are considered.

1.4   Measures of Fluid Mass and Weight

1.4.1   Density

The *density* of a fluid, designated by the Greek symbol ρ (rho), is defined as its mass per unit volume. Density is typically used to characterize the mass of a fluid system. In the BG system, ρ has units of slugs/ft^3 and in SI the units are kg/m^3.

The value of density can vary widely between different fluids, but for liquids, variations in pressure and temperature generally have only a small effect on the value of ρ. The small change in the density of water with large variations in temperature is illustrated in Fig. 1.1. Tables 1.4 and 1.5 list values of density for several common liquids. The density of water at 60° F is 1.94 slugs/ft^3 or 999 kg/m^3. The large difference between those two values illustrates the importance of paying attention to units! Unlike liquids, the density of a gas is strongly influenced by both pressure and temperature, and this difference is discussed in the next section.

The *specific volume, v,* is the *volume* per unit mass and is therefore the reciprocal of the density—that is,

$$v = \frac{1}{\rho} \tag{1.4}$$

This property is not commonly used in fluid mechanics but is used in thermodynamics.

1.4.2   Specific Weight

The *specific weight* of a fluid, designated by the Greek symbol γ (gamma), is defined as its *weight* per unit volume. Thus, specific weight is related to density through the equation

$$\gamma = \rho g \tag{1.5}$$

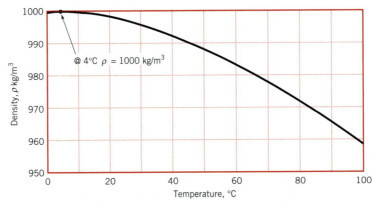

■ **FIGURE 1.1**   **Density of water as a function of temperature.**

■ **TABLE 1.4**
Approximate Physical Properties of Some Common Liquids (BG Units)

(See inside of front cover.)

■ **TABLE 1.5**
Approximate Physical Properties of Some Common Liquids (SI Units)

(See inside of front cover.)

where g is the local acceleration of gravity. Just as density is used to characterize the mass of a fluid system, the specific weight is used to characterize the weight of the system. In the BG system, γ has units of lb/ft^3 and in SI the units are N/m^3. Under conditions of standard gravity ($g = 32.174$ ft/s^2 = 9.807 m/s^2), water at 60° F has a specific weight of 62.4 lb/ft^3 and 9.80 kN/m^3. Tables 1.4 and 1.5 list values of specific weight for several common liquids (based on standard gravity). More complete tables for water can be found in Appendix B (Tables B.1 and B.2).

1.4.3   Specific Gravity

The *specific gravity* of a fluid, designated as *SG*, is defined as the ratio of the density of the fluid to the density of water at some specified temperature. Usually the specified temperature is taken as 4° C (39.2° F), and at this temperature the density of water is 1.94 slugs/ft^3 or 1000 kg/m ^{3}. In equation form specific gravity is expressed as

$$SG = \frac{\rho}{\rho_{H_2O@4°C}} \tag{1.6}$$

and since it is the *ratio* of densities, the value of *SG* does not depend on the system of units used. For example, the specific gravity of mercury at 20° C is 13.55, and the density of mercury can thus be readily calculated in either BG or SI units through the use of Eq. 1.6 as

$$\rho_{Hg} = (13.55)(1.94 \text{ slugs/ft}^3) = 26.3 \text{ slugs/ft}^3$$

or

$$\rho_{Hg} = (13.55)(1000 \text{ kg/m}^3) = 13.6 \times 10^3 \text{ kg/m}^3$$

It is clear that density, specific weight, and specific gravity are all interrelated, and from a knowledge of any one of the three the others can be calculated.

1.5   Ideal Gas Law

Gases are highly compressible in comparison to liquids, with changes in gas density directly related to changes in pressure and temperature through the equation

$$p = \rho RT \tag{1.7}$$

where p is the absolute pressure, ρ the density, T the absolute temperature,[1] and R is a gas constant. Equation 1.7 is commonly termed the *perfect* or *ideal gas law,* or the *equation of*

---

[1]We will use T to represent temperature in thermodynamic relationships, although T is also used to denote the basic dimension of time.

■ **TABLE 1.6**

Approximate Physical Properties of Some Common Gases at Standard Atmospheric Pressure (BG Units)

(See inside of front cover.)

■ **TABLE 1.7**

Approximate Physical Properties of Some Common Gases at Standard Atmospheric Pressure (SI Units)

(See inside of front cover.)

state for an ideal gas. It is known to closely approximate the behavior of real gases under normal conditions when the gases are not approaching liquefaction.

Pressure in a fluid at rest is defined as the normal force per unit area exerted on a plane surface (real or imaginary) immersed in a fluid, and is created by the bombardment of the surface with the fluid molecules. From the definition, pressure has the dimension of FL^{-2}, and in BG units is expressed as lb/ft² (psf) or lb/in.² (psi) and in SI units as N/m². In SI, 1 N/m² is defined as a *pascal,* abbreviated as Pa, and pressures are commonly specified in pascals. The pressure in the ideal gas law must be expressed as an ***absolute pressure,*** which means that it is measured relative to absolute zero pressure (a pressure that would only occur in a perfect vacuum). Standard sea-level atmospheric pressure (by international agreement) is 14.696 psi (abs) or 101.33 kPa (abs). For most calculations, these pressures can be rounded to 14.7 psi and 101 kPa, respectively. In engineering, it is common practice to measure pressure relative to the local atmospheric pressure; when measured in this fashion it is called ***gage pressure.*** Thus, the absolute pressure can be obtained from the gage pressure by adding the value of the atmospheric pressure. For example, as shown by the figure in the margin, a pressure of 30 psi (gage) in a tire is equal to 44.7 psi (abs) at standard atmospheric pressure. Pressure is a particularly important fluid characteristic, and it will be discussed more fully in the next chapter.

The gas constant, R, which appears in Eq. 1.7, depends on the particular gas and is related to the molecular weight of the gas. Values of the gas constant for several common gases are listed in Tables 1.6 and 1.7. Also in these tables the gas density and specific weight are given for standard atmospheric pressure and gravity and for the temperature listed. More complete tables for air at standard atmospheric pressure can be found in Appendix B (Tables B.3 and B.4).

EXAMPLE 1.2   Ideal Gas Law

GIVEN   A compressed air tank has a volume of 0.84 ft³. The tank is filled with air at a gage pressure of 50 psi and a temperature of 70° F. The atmospheric pressure is 14.7 psi (abs).

FIND   Determine the density of the air and the weight of air in the tank.

SOLUTION

The air density can be obtained from the ideal gas law (Eq. 1.7) expressed as

$$\rho = \frac{p}{RT}$$

so that

$$\rho = \frac{(50 \text{ lb/in.}^2 + 14.7 \text{ lb/in.}^2)(144 \text{ in.}^2/\text{ft}^2)}{(1716 \text{ ft·lb/slug·°R})[(70 + 460)°R]}$$

$$= 0.0102 \text{ slugs/ft}^3 \qquad \textbf{(Ans)}$$

COMMENT   Note that both the pressure and the temperature were changed to absolute values.
The weight, $\mathcal{W}$, of the air is equal to

$$\mathcal{W} = \rho g \times (\text{volume})$$
$$= (0.0102 \text{ slugs/ft}^3)(32.2 \text{ ft/s}^2)(0.84 \text{ ft}^3)$$
$$= 0.276 \text{ slug} \cdot \text{ft/s}^2$$

so that

$$\mathcal{W} = 0.276 \text{ lb} \qquad \textbf{(Ans)}$$

since $1 \text{ lb} = 1 \text{ slug} \cdot \text{ft/s}^2$.

COMMENT   By repeating the calculations for various values of the pressure, p, the results shown in Fig. E1.2 are obtained. Note that doubling the gage pressure does not double the amount

of air in the tank, but doubling the absolute pressure does. Thus, a scuba diving tank at a gage pressure of 100 psi does not contain twice the amount of air as when the gage reads 50 psi.

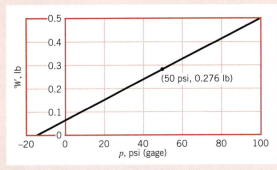

(50 psi, 0.276 lb)

■ **FIGURE E1.2**

1.6   Viscosity

V1.1 Viscous fluids

V1.2 No-slip condition

The properties of density and specific weight are measures of the "heaviness" of a fluid. It is clear, however, that these properties are not sufficient to uniquely characterize how fluids behave, as two fluids (such as water and oil) can have approximately the same value of density but behave quite differently when flowing. There is apparently some additional property that is needed to describe the "fluidity" of the fluid (i.e., how easily it flows).

To determine this additional property, consider a hypothetical experiment in which a material is placed between two very wide parallel plates as shown in Fig. 1.2. The bottom plate is rigidly fixed, but the upper plate is free to move.

When the force P is applied to the upper plate, it will move continuously with a velocity U (after the initial transient motion has died out) as illustrated in Fig. 1.2. This behavior is consistent with the definition of a fluid—that is, if a shearing stress is applied to a fluid it will deform continuously. A closer inspection of the fluid motion between the two plates would reveal that the fluid in contact with the upper plate moves with the plate velocity, U, and the fluid in contact with the bottom fixed plate has a zero velocity. The fluid between the two plates moves with velocity $u = u(y)$ that would be found to vary linearly, $u = Uy/b$, as illustrated in Fig. 1.2. Thus, a *velocity gradient, du/dy,* is developed in the fluid between the plates. In this particular case the velocity gradient is a constant, as $du/dy = U/b$, but in more complex flow situations this would not be true. The experimental observation that the fluid "sticks" to the solid boundaries is a very important one in fluid mechanics and is usually referred to as the ***no-slip condition.*** All fluids, both liquids and gases, satisfy this condition.

In a small time increment δt, an imaginary vertical line AB in the fluid (see Fig. 1.2) would rotate through an angle, $\delta \beta$, so that

$$\tan \delta \beta \approx \delta \beta = \frac{\delta a}{b}$$

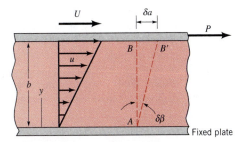

■ **FIGURE 1.2**   **Behavior of a fluid placed between two parallel plates.**

Since $\delta a = U \, \delta t$ it follows that

$$\delta \beta = \frac{U \, \delta t}{b}$$

Note that in this case, $\delta \beta$ is a function not only of the force P (which governs U) but also of time. We consider the *rate* at which $\delta \beta$ is changing, and define the *rate of shearing strain,* $\dot{\gamma}$, as

$$\dot{\gamma} = \lim_{\delta t \to 0} \frac{\delta \beta}{\delta t}$$

which in this instance is equal to

$$\dot{\gamma} = \frac{U}{b} = \frac{du}{dy}$$

A continuation of this experiment would reveal that as the shearing stress, τ, is increased by increasing P (recall that $\tau = P/A$), the rate of shearing strain is increased in direct proportion—that is

$$\tau \propto \dot{\gamma}$$

or

$$\tau \propto \frac{du}{dy}$$

This result indicates that for common fluids, such as water, oil, gasoline, and air, the shearing stress and rate of shearing strain (velocity gradient) can be related with a relationship of the form

$$\tau = \mu \frac{du}{dy} \qquad (1.8)$$

V1.3 Capillary tube viscometer

V1.4 Non-Newtonian behavior

where the constant of proportionality is designated by the Greek symbol μ (mu) and is called the ***absolute viscosity,*** *dynamic viscosity,* or simply the *viscosity* of the fluid. In accordance with Eq. 1.8, plots of τ versus du/dy should be linear with the slope equal to the viscosity as illustrated in Fig. 1.3. The actual value of the viscosity depends on the particular fluid, and for a particular fluid the viscosity is also highly dependent on temperature as illustrated in Fig. 1.3 with the two curves for water. Fluids for which the shearing stress is *linearly* related to the rate of shearing strain (also referred to as rate of angular deformation) are designated as ***Newtonian fluids.*** Fortunately, most common fluids, both liquids and gases, are Newtonian. A more general formulation of Eq. 1.8, which applies to more complex flows of Newtonian fluids, is given in Section 6.8.1.

Fluids for which the shearing stress is not linearly related to the rate of shearing strain are designated as *non-Newtonian fluids.* It is beyond the scope of this book to consider the behavior of such fluids, and we will only be concerned with Newtonian fluids.

F l u i d s   i n   t h e   N e w s

A vital fluid In addition to air and water, another fluid that is essential for human life is blood. Blood is an unusual fluid consisting of red blood cells that are disk-shaped, about 8 microns in diameter, suspended in plasma. As you would suspect, since blood is a suspension, its mechanical behavior is that of a *non-Newtonian* fluid. Its density is only slightly higher than that of water, but its typical apparent *viscosity* is significantly higher than that of water at the same temperature. It is difficult to measure the viscosity of blood since it is a non-Newtonian fluid and the viscosity is a function of the shear rate. As the shear rate is increased from a low value, the apparent viscosity decreases and approaches asymptotically a constant value at high shear rates. The "asymptotic" value of the viscosity of normal blood is 3 to 4 times the viscosity of water. The viscosity of blood is not routinely measured like some biochemical properties such as cholesterol and triglycerides, but there is some evidence indicating that the viscosity of blood may play a role in the development of cardiovascular disease. If this proves to be true, viscosity could become a standard variable to be routinely measured. (See Problem 1.24.) ■

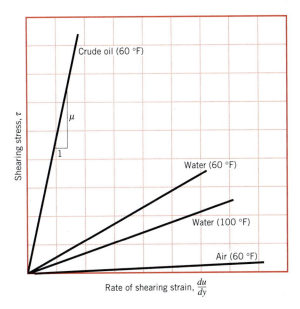

■ **FIGURE 1.3** Linear variation of shearing stress with rate of
shearing strain for common fluids.

From Eq. 1.8 it can be readily deduced that the dimensions of viscosity are FTL^{-2}. Thus, in BG units viscosity is given as lb·s/ft² and in SI units as N·s/m². Values of viscosity for several common liquids and gases are listed in Tables 1.4 through 1.7. A quick glance at these tables reveals the wide variation in viscosity among fluids. Viscosity is only mildly dependent on pressure, and the effect of pressure is usually neglected. However, as mentioned previously, and as illustrated in Appendix B (Figs. B.1 and B.2), viscosity is very sensitive to temperature.

Quite often viscosity appears in fluid flow problems combined with the density in the form

$$\nu = \frac{\mu}{\rho}$$

This ratio is called the ***kinematic viscosity*** and is denoted with the Greek symbol ν (nu). The dimensions of kinematic viscosity are L^2/T, and the BG units are ft²/s and SI units are m²/s. Values of kinematic viscosity for some common liquids and gases are given in Tables 1.4 through 1.7. More extensive tables giving both the dynamic and kinematic viscosities for water and air can be found in Appendix B (Tables B.1 through B.4), and graphs showing the variation in both dynamic and kinematic viscosity with temperature for a variety of fluids are also provided in Appendix B (Figs. B.1 and B.2).

Although in this text we are primarily using BG and SI units, dynamic viscosity is often expressed in the metric CGS (centimeter-gram-second) system with units of dyne·s/cm². This combination is called a *poise,* abbreviated P. In the CGS system, kinematic viscosity has units of cm²/s, and this combination is called a *stoke,* abbreviated St.

EXAMPLE 1.3   Viscosity and Dimensionless Quantities

GIVEN   A dimensionless combination of variables that is important in the study of viscous flow through pipes is called the *Reynolds number,* Re, defined as $\rho VD/\mu$ where ρ is the fluid density, V the mean fluid velocity, D the pipe diameter, and μ the fluid viscosity. A Newtonian fluid having a viscosity of 0.38 N·s/m² and a specific gravity of 0.91

flows through a 25-mm-diameter pipe with a velocity of 2.6 m/s.

FIND   Determine the value of the Reynolds number using

(a)   SI units and

(b)   BG units.

SOLUTION

(a) The fluid density is calculated from the specific gravity as

$$\rho = SG\, \rho_{H_2O@4°C} = 0.91\,(1000 \text{ kg/m}^3) = 910 \text{ kg/m}^3$$

and from the definition of the Reynolds number

$$\text{Re} = \frac{\rho VD}{\mu} = \frac{(910 \text{ kg/m}^3)(2.6 \text{ m/s})(25 \text{ mm})(10^{-3} \text{ m/mm})}{0.38 \text{ N·s/m}^2}$$

$$= 156\,(\text{kg·m/s}^2)/\text{N}$$

However, since 1 N = 1 kg·m/s² it follows that the Reynolds number is unitless (dimensionless)—that is,

$$\text{Re} = 156 \qquad \text{(Ans)}$$

COMMENT The value of any dimensionless quantity does not depend on the system of units used if all variables that make up the quantity are expressed in a consistent set of units. To check this we will calculate the Reynolds number using BG units.

(b) We first convert all the SI values of the variables appearing in the Reynolds number to BG values by using the conversion factors from Table 1.3. Thus,

$$\rho = (910 \text{ kg/m}^3)(1.940 \times 10^{-3}) = 1.77 \text{ slugs/ft}^3$$
$$V = (2.6 \text{ m/s})(3.281) = 8.53 \text{ ft/s}$$
$$D = (0.025 \text{ m})(3.281) = 8.20 \times 10^{-2} \text{ ft}$$
$$\mu = (0.38 \text{ N·s/m}^2)(2.089 \times 10^{-2}) = 7.94 \times 10^{-3} \text{ lb·s/ft}^2$$

and the value of the Reynolds number is

$$\text{Re} = \frac{(1.77 \text{ slugs/ft}^3)(8.53 \text{ ft/s})(8.20 \times 10^{-2} \text{ ft})}{7.94 \times 10^{-3} \text{ lb·s/ft}^2}$$

$$= 156\,(\text{slug·ft/s}^2)/\text{lb} = 156 \qquad \text{(Ans)}$$

since 1 lb = 1 slug·ft/s².

COMMENT The values from part (a) and part (b) are the same, as expected. Dimensionless quantities play an important role in fluid mechanics, and the significance of the Reynolds number, as well as other important dimensionless combinations, will be discussed in detail in Chapter 7. It should be noted that in the Reynolds number it is actually the ratio μ/ρ that is important, and this is the property that we have defined as the kinematic viscosity.

EXAMPLE 1.4   Newtonian Fluid Shear Stress

GIVEN The velocity distribution for the flow of a Newtonian fluid between two wide, parallel plates (see Fig. E1.4) is given by the equation

$$u = \frac{3V}{2}\left[1 - \left(\frac{y}{h}\right)^2\right]$$

where V is the mean velocity. The fluid has a viscosity of 0.04 lb·s/ft². Also, $V = 2$ ft/s and $h = 0.2$ in.

FIND Determine

(a) the shearing stress acting on the bottom wall and

(b) the shearing stress acting on a plane parallel to the walls and passing through the centerline (midplane).

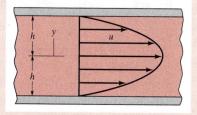

■ **FIGURE E1.4**

SOLUTION

For this type of parallel flow the shearing stress is obtained from Eq. 1.8.

$$\tau = \mu\frac{du}{dy} \qquad (1)$$

Thus, if the velocity distribution, $u = u(y)$, is known, the shearing stress can be determined at all points by evaluating the velocity gradient, du/dy. For the distribution given

$$\frac{du}{dy} = -\frac{3Vy}{h^2} \qquad (2)$$

(a) Along the bottom wall $y = -h$ so that (from Eq. 2)

$$\frac{du}{dy} = \frac{3V}{h}$$

and therefore the shearing stress is

$$\tau_{\text{bottom wall}} = \mu\left(\frac{3V}{h}\right) = \frac{(0.04 \text{ lb·s/ft}^2)(3)(2 \text{ ft/s})}{(0.2 \text{ in.})(1 \text{ ft/12 in.})}$$

$$= 14.4 \text{ lb/ft}^2 \text{ (in direction of flow)} \qquad \text{(Ans)}$$

COMMENT This stress creates a drag on the wall. Since the velocity distribution is symmetrical, the shearing stress along the upper wall would have the same magnitude and direction.

(b)   Along the midplane where $y = 0$, it follows from Eq. 2 that

$$\frac{du}{dy} = 0$$

and thus the shearing stress is

$$\tau_{\text{midplane}} = 0 \qquad \textbf{(Ans)}$$

COMMENT   From Eq. 2 we see that the velocity gradient (and therefore the shearing stress) varies linearly with y and in this particular example varies from 0 at the center of the channel to 14.4 lb/ft^2 at the walls. For the more general case the actual variation will, of course, depend on the nature of the velocity distribution.

1.7   Compressibility of Fluids

1.7.1   Bulk Modulus

An important question to answer when considering the behavior of a particular fluid is how easily can the volume (and thus the density) of a given mass of the fluid be changed when there is a change in pressure? That is, how compressible is the fluid? A property that is commonly used to characterize compressibility is the *bulk modulus, E_v*, defined as

$$E_v = -\frac{dp}{d\mathcal{V}/\mathcal{V}} \qquad \textbf{(1.9)}$$

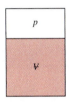

where dp is the differential change in pressure needed to create a differential change in volume, $d\mathcal{V}$, of a volume $\mathcal{V}$, as shown by the figure in the margin. The negative sign is included since an increase in pressure will cause a decrease in volume. Because a decrease in volume of a given mass, $m = \rho\mathcal{V}$, will result in an increase in density, Eq. 1.9 can also be expressed as

$$E_v = \frac{dp}{d\rho/\rho} \qquad \textbf{(1.10)}$$

The bulk modulus (also referred to as the *bulk modulus of elasticity*) has dimensions of pressure, FL^{-2}. In BG units values for E_v are usually given as lb/in.^{2} (psi) and in SI units as N/m^2 (Pa). Large values for the bulk modulus indicate that the fluid is relatively incompressible—that is, it takes a large pressure change to create a small change in volume. As expected, values of E_v for common liquids are large (see Tables 1.4 and 1.5).

Since such large pressures are required to effect a change in volume, we conclude that liquids can be considered as *incompressible* for most practical engineering applications. As liquids are compressed the bulk modulus increases, but the bulk modulus near atmospheric pressure is usually the one of interest. The use of bulk modulus as a property describing compressibility is most prevalent when dealing with liquids, although the bulk modulus can also be determined for gases.

F l u i d s   i n   t h e   N e w s

This water jet is a blast   Usually liquids can be treated as incompressible fluids. However, in some applications the *compressibility* of a liquid can play a key role in the operation of a device. For example, a water pulse generator using compressed water has been developed for use in mining operations. It can fracture rock by producing an effect comparable to a conventional explosive such as gunpowder. The device uses the energy stored in a water-filled accumulator to generate an ultrahigh-pressure water pulse ejected through a 10- to 25-mm-diameter discharge valve. At the ultrahigh pressures used (300 to 400 MPa, or 3000 to 4000 atmospheres), the water is compressed (i.e., the volume reduced) by about 10 to 15%. When a fast-opening valve within the pressure vessel is opened, the water expands and produces a jet of water that upon impact with the target material produces an effect similar to the explosive force from conventional explosives. Mining with the water jet can eliminate various hazards that arise with the use of conventional chemical explosives such as those associated with the storage and use of explosives and the generation of toxic gas by-products that require extensive ventilation. (See Problem 1.43.) ∎

1.7.2 Compression and Expansion of Gases

When gases are compressed (or expanded), the relationship between pressure and density depends on the nature of the process. If the compression or expansion takes place under constant temperature conditions (*isothermal process*), then from Eq. 1.7

$$\frac{p}{\rho} = \text{constant} \tag{1.11}$$

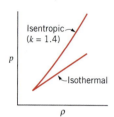

If the compression or expansion is frictionless and no heat is exchanged with the surroundings (*isentropic process*), then

$$\frac{p}{\rho^k} = \text{constant} \tag{1.12}$$

where k is the ratio of the specific heat at constant pressure, c_p, to the specific heat at constant volume, c_v (i.e., $k = c_p/c_v$). The two specific heats are related to the gas constant, R, through the equation $R = c_p - c_v$. The pressure-density variations for isothermal and isentropic conditions are illustrated in the margin figure. As was the case for the ideal gas law, the pressure in both Eqs. 1.11 and 1.12 must be expressed as an absolute pressure. Values of k for some common gases are given in Tables 1.6 and 1.7, and for air over a range of temperatures, in Appendix B (Tables B.3 and B.4). It is clear that in dealing with gases greater attention will need to be given to the effect of compressibility on fluid behavior. However, as discussed in Section 3.8, gases can often be treated as incompressible fluids if the changes in pressure are small.

EXAMPLE 1.5 | Isentropic Compression of a Gas

GIVEN A cubic foot of air at an absolute pressure of 14.7 psi is compressed isentropically to $\frac{1}{2}$ ft^3.

FIND What is the final pressure?

SOLUTION

For an isentropic compression

$$\frac{p_i}{\rho_i^k} = \frac{p_f}{\rho_f^k}$$

where the subscripts i and f refer to initial and final states, respectively. Since we are interested in the final pressure, p_f, it follows that

$$p_f = \left(\frac{\rho_f}{\rho_i}\right)^k p_i$$

As the volume, V, is reduced by one-half, the density must double, since the mass, $m = \rho V$, of the gas remains constant. Thus, with $k = 1.40$ for air

$$p_f = (2)^{1.40}(14.7 \text{ psi}) = 38.8 \text{ psi (abs)} \qquad \textbf{(Ans)}$$

COMMENT By repeating the calculations for various values of the ratio of the final volume to the initial volume, V_f/V_i, the results shown in Fig. E1.5 are obtained. Note that even though air is often considered to be easily compressed (at least compared to liquids), it takes considerable pressure to significantly reduce a given volume of air as is done in an automobile engine where the compression ratio is on the order of $V_f/V_i = 1/8 = 0.125$.

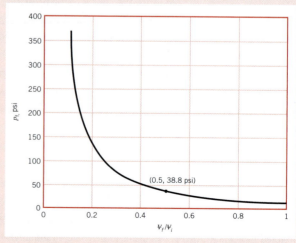

■ **FIGURE E1.5**

1.7.3   Speed of Sound

Another important consequence of the compressibility of fluids is that disturbances introduced at some point in the fluid propagate at a finite velocity. For example, if a fluid is flowing in a pipe and a valve at the outlet is suddenly closed (thereby creating a localized disturbance), the effect of the valve closure is not felt instantaneously upstream. It takes a finite time for the increased pressure created by the valve closure to propagate to an upstream location. Similarly, a loud speaker diaphragm causes a localized disturbance as it vibrates, and the small change in pressure created by the motion of the diaphragm is propagated through the air with a finite velocity. The velocity at which these small disturbances propagate is called the *acoustic velocity* or the ***speed of sound,*** *c*. It can be shown that the speed of sound is related to changes in pressure and density of the fluid medium through the equation

$$c = \sqrt{\frac{dp}{d\rho}} \tag{1.13}$$

or in terms of the bulk modulus defined by Eq. 1.10

$$c = \sqrt{\frac{E_v}{\rho}} \tag{1.14}$$

Because the disturbance is small, there is negligible heat transfer and the process is assumed to be isentropic. Thus, the pressure–density relationship used in Eq. 1.13 is that for an isentropic process.

For gases undergoing an isentropic process, $E_v = kp$, so that

$$c = \sqrt{\frac{kp}{\rho}}$$

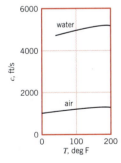

and making use of the ideal gas law, it follows that

$$c = \sqrt{kRT} \tag{1.15}$$

Thus, for ideal gases the speed of sound is proportional to the square root of the absolute temperature. The speed of sound in air at various temperatures can be found in Appendix B (Tables B.3 and B.4). Equation 1.14 is also valid for liquids, and values of E_v can be used to determine the speed of sound in liquids. As shown by the figure in the margin, the speed of sound in water is much higher than in air. If a fluid were truly incompressible ($E_v = \infty$) the speed of sound would be infinite. The speed of sound in water for various temperatures can be found in Appendix B (Tables B.1 and B.2).

EXAMPLE 1.6   Speed of Sound and Mach Number

GIVEN   A jet aircraft flies at a speed of 550 mph at an altitude of 35,000 ft, where the temperature is −66° F and the specific heat ratio is $k = 1.4$.

FIND   Determine the ratio of the speed of the aircraft, V, to that of the speed of sound, c, at the specified altitude.

SOLUTION

From Eq. 1.15 the speed of sound can be calculated as

$$c = \sqrt{kRT}$$
$$= \sqrt{(1.40)(1716 \text{ ft·lb/slug·°R})(-66 + 460)\text{°R}}$$
$$= 973 \text{ ft/s}$$

Since the air speed is

$$V = \frac{(550 \text{ mi/hr})(5280 \text{ ft/mi})}{(3600 \text{ s/hr})} = 807 \text{ ft/s}$$

the ratio is

$$\frac{V}{c} = \frac{807 \text{ ft/s}}{973 \text{ ft/s}} = 0.829 \qquad \textbf{(Ans)}$$

COMMENT This ratio is called the *Mach number*, Ma. If Ma < 1.0 the aircraft is flying at *subsonic* speeds, whereas for Ma > 1.0 it is flying at *supersonic* speeds. The Mach number is an important dimensionless parameter used in the study of the flow of gases at high speeds and will be further discussed in Chapters 7 and 9.

By repeating the calculations for different temperatures, the results shown in Fig. E1.6 are obtained. Because the speed of sound increases with increasing temperature, for a constant airplane speed, the Mach number decreases as the temperature increases.

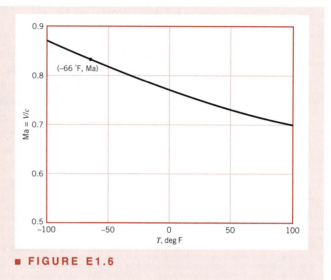

■ **FIGURE E1.6**

1.8 Vapor Pressure

It is a common observation that liquids such as water and gasoline will evaporate if they are simply placed in a container open to the atmosphere. Evaporation takes place because some liquid molecules at the surface have sufficient momentum to overcome the intermolecular cohesive forces and escape into the atmosphere. As shown in the figure in the margin, if the lid on a completely liquid-filled, closed container is raised (without letting any air in), a pressure will develop in the space as a result of the vapor that is formed by the escaping molecules. When an equilibrium condition is reached so that the number of molecules leaving the surface is equal to the number entering, the vapor is said to be saturated and the pressure the vapor exerts on the liquid surface is termed the ***vapor pressure,*** p_v.

Since the development of a vapor pressure is closely associated with molecular activity, the value of vapor pressure for a particular liquid depends on temperature. Values of vapor pressure for water at various temperatures can be found in Appendix B (Tables B.1 and B.2), and the values of vapor pressure for several common liquids at room temperatures are given in Tables 1.4 and 1.5. *Boiling*, which is the formation of vapor bubbles within a fluid mass, is initiated when the absolute pressure in the fluid reaches the vapor pressure.

An important reason for our interest in vapor pressure and boiling lies in the common observation that in flowing fluids it is possible to develop very low pressure due to the fluid motion, and if the pressure is lowered to the vapor pressure, boiling will occur. For example, this phenomenon may occur in flow through the irregular, narrowed passages of a valve or pump. When vapor bubbles are formed in a flowing liquid, they are swept along into regions of higher pressure where they suddenly collapse with sufficient intensity to actually cause structural damage. The formation and subsequent collapse of vapor bubbles in a flowing liquid, called *cavitation,* is an important fluid flow phenomenon to be given further attention in Chapters 3 and 7.

1.9 Surface Tension

At the interface between a liquid and a gas, or between two immiscible liquids, forces develop in the liquid surface that cause the surface to behave as if it were a "skin" or "membrane" stretched over the fluid mass. Although such a skin is not actually present, this conceptual

V1.5 Floating razor blades

analogy allows us to explain several commonly observed phenomena. For example, a steel needle or razor blade will float on water if placed gently on the surface because the tension developed in the hypothetical skin supports these objects. Small droplets of mercury will form into spheres when placed on a smooth surface because the cohesive forces in the surface tend to hold all the molecules together in a compact shape. Similarly, discrete water droplets will form when placed on a newly waxed surface.

These various types of surface phenomena are due to the unbalanced cohesive forces acting on the liquid molecules at the fluid surface. Molecules in the interior of the fluid mass are surrounded by molecules that are attracted to each other equally. However, molecules along the surface are subjected to a net force toward the interior. The apparent physical consequence of this unbalanced force along the surface is to create the hypothetical skin or membrane. A tensile force may be considered to be acting in the plane of the surface along any line in the surface. The intensity of the molecular attraction per unit length along any line in the surface is called the ***surface tension*** and is designated by the Greek symbol σ (sigma). Surface tension is a property of the liquid and depends on temperature as well as the other fluid it is in contact with at the interface. The dimensions of surface tension are FL^{-1} with BG units of lb/ft and SI units of N/m. Values of surface tension for some common liquids (in contact with air) are given in Tables 1.4 and 1.5 and in Appendix B (Tables B.1 and B.2) for water at various temperatures. The value of the surface tension decreases as the temperature increases.

F l u i d s   i n   t h e   N e w s

Walking on water Water striders are insects commonly found on ponds, rivers, and lakes that appear to "walk" on water. A typical length of a water strider is about 0.4 in., and they can cover 100 body lengths in one second. It has long been recognized that it is *surface tension* that keeps the water strider from sinking below the surface. What has been puzzling is how they propel themselves at such a high speed. They can't pierce the water surface or they would sink. A team of mathematicians and engineers from the Massachusetts Institute of Technology (MIT) applied conventional flow visualization techniques and high-speed

video to examine in detail the movement of the water striders. They found that each stroke of the insect's legs creates dimples on the surface with underwater swirling vortices sufficient to propel it forward. It is the rearward motion of the vortices that propels the water strider forward. To further substantiate their explanation the MIT team built a working model of a water strider, called Robostrider, which creates surface ripples and underwater vortices as it moves across a water surface. Waterborne creatures, such as the water strider, provide an interesting world dominated by surface tension. (See Problem 1.49.) ∎

Among common phenomena associated with surface tension is the rise (or fall) of a liquid in a capillary tube. If a small open tube is inserted into water, the water level in the tube will rise above the water level outside the tube as is illustrated in Fig. 1.4a. In this situation we have a liquid–gas–solid interface. For the case illustrated there is an attraction (adhesion) between the wall of the tube and liquid molecules, which is strong enough to overcome the mutual attraction (cohesion) of the molecules and pull them up to the wall. Hence, the liquid is said to *wet* the solid surface.

The height, h, is governed by the value of the surface tension, σ, the tube radius, R, the specific weight of the liquid, γ, and the *angle of contact, θ*, between the fluid and tube. From the free-body diagram of Fig. 1.4b we see that the vertical force due to the surface tension is equal to $2\pi R\sigma \cos \theta$ and the weight is $\gamma\pi R^2 h$ and these two forces must balance for equilibrium. Thus,

$$\gamma\pi R^2 h = 2\pi R\sigma \cos \theta$$

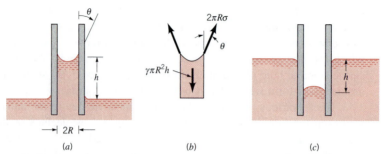

■ **FIGURE 1.4** Effect of capillary action in small tubes. (*a*) Rise of column for a liquid that wets the tube. (*b*) Free-body diagram for calculating column height. (*c*) Depression of column for a nonwetting liquid.

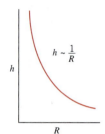

so that the height is given by the relationship

$$h = \frac{2\sigma \cos \theta}{\gamma R} \tag{1.16}$$

The angle of contact is a function of both the liquid and the surface. For water in contact with clean glass $\theta \approx 0°$. It is clear from Eq. 1.16 (and shown by the figure in the margin) that the height is inversely proportional to the tube radius. Therefore, the rise of a liquid in a tube as a result of capillary action becomes increasingly pronounced as the tube radius is decreased.

EXAMPLE 1.7   Capillary Rise in a Tube

GIVEN Pressures are sometimes determined by measuring the height of a column of liquid in a vertical tube.

FIND What diameter of clean glass tubing is required so that the rise of water at 20° C in a tube due to a capillary action (as opposed to pressure in the tube) is less than 1.0 mm?

SOLUTION

From Eq. 1.16

$$h = \frac{2\sigma \cos \theta}{\gamma R}$$

so that

$$R = \frac{2\sigma \cos \theta}{\gamma h}$$

For water at 20° C (from Table B.2), $\sigma = 0.0728$ N/m and $\gamma = 9.789$ kN/m³. Since $\theta \approx 0°$ it follows that for $h = 1.0$ mm,

$$R = \frac{2(0.0728 \text{ N/m})(1)}{(9.789 \times 10^3 \text{ N/m}^3)(1.0 \text{ mm})(10^{-3} \text{ m/mm})}$$

$$= 0.0149 \text{ m}$$

and the minimum required tube diameter, D, is

$$D = 2R = 0.0298 \text{ m} = 29.8 \text{ mm} \qquad \textbf{(Ans)}$$

COMMENT By repeating the calculations for various values of the capillary rise, h, the results shown in Fig. E1.7 are obtained. Note that as the allowable capillary rise is decreased, the diameter of the tube must be significantly increased. There is always some capillarity effect, but it can be minimized by using a large enough diameter tube.

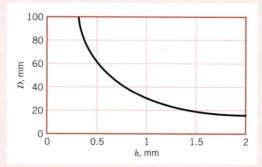

■ **FIGURE E1.7**

If adhesion of molecules to the solid surface is weak compared to the cohesion between molecules, the liquid will not wet the surface and the level in a tube placed in a nonwetting liquid will actually be depressed, as shown in Fig. 1.4c. Mercury is a good example of a nonwetting liquid when it is in contact with a glass tube. For nonwetting liquids the angle of contact is greater than 90°, and for mercury in contact with clean glass $\theta \approx 130°$.

Surface tension effects play a role in many mechanics problems, including the movement of liquids through soil and other porous media, flow of thin films, formation of drops and bubbles, and the breakup of liquid jets. Surface phenomena associated with liquid–gas, liquid–liquid, liquid–gas–solid interfaces are exceedingly complex, and a more detailed and rigorous discussion of them is beyond the scope of this text. Fortunately, in many fluid mechanics problems, surface phenomena, as characterized by surface tension, are not important, as inertial, gravitational, and viscous forces are much more dominant.

F l u i d s   i n   t h e   N e w s

Spreading of oil spills. With the large traffic in oil tankers there is great interest in the prevention of and response to oil spills. As evidenced by the famous *Exxon Valdez* oil spill in Prince William Sound in 1989, oil spills can create disastrous environmental problems. It is not surprising that much attention is given to the rate at which an oil spill spreads. When spilled, most oils tend to spread horizontally into a smooth and slippery surface, called a slick. There are many factors which influence the ability of an oil slick to spread, including the size of the spill, wind speed and direction, and the physical properties of the oil. These properties include *surface tension, specific gravity,* and *viscosity.* The higher the surface tension the more likely a spill will remain in place. Since the specific gravity of oil is less than one it floats on top of the water, but the specific gravity of an oil can increase if the lighter substances within the oil evaporate. The higher the viscosity of the oil the greater the tendency to stay in one place. ■

1.10   Chapter Summary and Study Guide

fluid
units
basic dimensions
dimensionally
  homogeneous
density
specific weight
specific gravity
ideal gas law
absolute pressure
gage pressure
no-slip condition
absolute viscosity
Newtonian fluid
kinematic viscosity
bulk modulus
speed of sound
vapor pressure
surface tension

This introductory chapter discussed several fundamental aspects of fluid mechanics. Methods for describing fluid characteristics both quantitatively and qualitatively are considered. For a quantitative description, units are required, and in this text, two system of units are used: the British Gravitational (BG) System (pounds, slugs, feet, and seconds) and the International (SI) System (newtons, kilograms, meters, and seconds). For the qualitative description the concept of dimensions is introduced in which basic dimensions such as length, L, time, T, and mass, M, are used to provide a description of various quantities of interest. The use of dimensions is helpful in checking the generality of equations, as well as serving as the basis for the powerful tool of dimensional analysis discussed in detail in Chapter 7.

Various important fluid properties are defined, including fluid density, specific weight, specific gravity, viscosity, bulk modulus, speed of sound, vapor pressure, and surface tension. The ideal gas law is introduced to relate pressure, temperature, and density in common gases, along with a brief discussion of the compression and expansion of gases. The distinction between absolute and gage pressure is introduced, and this important idea is explored more fully in Chapter 2.

The following checklist provides a study guide for this chapter. When your study of the entire chapter and end-of-chapter exercises has been completed, you should be able to

- write out meanings of the terms listed here in the margin and understand each of the related concepts. These terms are particularly important and are set in color and bold type in the text.

- determine the dimensions of common physical quantities.

- determine whether an equation is a general or restricted homogeneous equation.
- use both BG and SI systems of units.
- calculate the density, specific weight, or specific gravity of a fluid from a knowledge of any two of the three.
- calculate the density, pressure, or temperature of an ideal gas (with a given gas constant) from a knowledge of any two of the three.
- relate the pressure and density of a gas as it is compressed or expanded using Eqs. 1.11 and 1.12.
- use the concept of viscosity to calculate the shearing stress in simple fluid flows.
- calculate the speed of sound in fluids using Eq. 1.14 for liquids and Eq. 1.15 for gases.
- determine whether boiling or cavitation will occur in a liquid using the concept of vapor pressure.
- use the concept of surface tension to solve simple problems involving liquid–gas or liquid–solid–gas interfaces.

Review Problems

Go to Appendix F for a set of review problems with answers. Detailed solutions can be found in *Student Solution Manual for a Brief Introduction to Fluid Mechanics,* by Young et al. (© 2006 John Wiley and Sons, Inc.).

Problems

Note: Unless specific values of required fluid properties are given in the statement of the problem, use the values found in the tables on the inside of the front cover. Problems designated with an (*) are intended to be solved with the aid of a programmable calculator or a computer. Problems designated with a (†) are "open-ended" problems and require critical thinking in that to work them one must make various assumptions and provide the necessary data. There is not a unique answer to these problems.

Answers to the even-numbered problems are listed at the end of the book. Access to the videos that accompany problems can be obtained through the book's web site, www.wiley.com/college/young. The lab-type problems and FE problems can also be accessed on this web site.

Section 1.2 Dimensions, Dimensional Homogeneity, and Units

1.1   Verify the dimensions, in both the *FLT* and *MLT* systems, of the following quantities, which appear in Table 1.1: **(a)** angular velocity, **(b)** energy, **(c)** moment of inertia (area), **(d)** power, and **(e)** pressure.

1.2   Determine the dimensions, in both the *FLT* system and *MLT* system, for **(a)** the product of force times volume, **(b)** the product of pressure times mass divided by area, and **(c)** moment of a force divided by velocity.

1.3   If V is a velocity, ℓ a length, and ν a fluid property having dimensions of L^2T^{-1}, which of the following combinations are dimensionless: **(a)** $V\ell\nu$, **(b)** $V\ell/\nu$, **(c)** $V^2\nu$, **(d)** $V/\ell\nu$?

1.4   Dimensionless combinations of quantities (commonly called dimensionless parameters) play an important role in fluid

mechanics. Make up five possible dimensionless parameters by using combinations of some of the quantities listed in Table 1.1.

1.5   The volume rate of flow, Q, through a pipe containing a slowly moving liquid is given by the equation

$$Q = \frac{\pi R^4 \Delta p}{8 \mu \ell}$$

where R is the pipe radius, Δp the pressure drop along the pipe, μ a fluid property called viscosity $(FL^{-2}T)$, and ℓ the length of pipe. What are the dimensions of the constant $\pi/8$? Would you classify this equation as a general homogeneous equation? Explain.

1.6   The pressure difference, Δp, across a partial blockage in an artery (called a *stenosis*) is approximated by the equation

$$\Delta p = K_v \frac{\mu V}{D} + K_u \left(\frac{A_0}{A_1} - 1\right)^2 \rho V^2$$

where V is the blood velocity, μ the blood viscosity $(FL^{-2}T)$, ρ the blood density (ML^{-3}), D the artery diameter, A_0 the area of the unobstructed artery, and A_1 the area of the stenosis. Determine the dimensions of the constants K_v and K_u. Would this equation be valid in any system of units?

1.7   According to information found in an old hydraulics book, the energy loss per unit weight of fluid flowing through a nozzle connected to a hose can be estimated by the formula

$$h = (0.04 \text{ to } 0.09)(D/d)^4 V^2/2g$$

where h is the energy loss per unit weight, D the hose diameter, d the nozzle tip diameter, V the fluid velocity in the hose, and g

the acceleration of gravity. Do you think this equation is valid in any system of units? Explain.

†1.8 Cite an example of a restricted homogeneous equation contained in a technical article found in an engineering journal in your field of interest. Define all terms in the equation, explain why it is a restricted equation, and provide a complete journal citation (title, date, etc.).

1.9 Make use of Table 1.2 to express the following quantities in SI units: **(a)** 10.2 in./min, **(b)** 4.81 slugs, **(c)** 3.02 lb, **(d)** 73.1 ft/s^2, **(e)** 0.0234 lb·s/ft^2.

1.10 Make use of Table 1.3 to express the following quantities in BG units: **(a)** 14.2 km, **(b)** 8.14 N/m^3, **(c)** 1.61 kg/m^3, **(d)** 0.0320 N·m/s, **(e)** 5.67 mm/hr.

1.11 Clouds can weigh thousands of pounds due to their liquid water content. Often this content is measured in grams per cubic meter (g/m^3). Assume that a cumulus cloud occupies a volume of one cubic kilometer, and its liquid water content is 0.2 g/m^3. **(a)** What is the volume of this cloud in cubic miles? **(b)** How much does the water in the cloud weigh in pounds?

1.12 An important dimensionless parameter in certain types of fluid flow problems is the *Froude number* defined as $V/\sqrt{g\ell}$, where V is a velocity, g the acceleration of gravity, and ℓ a length. Determine the value of the Froude number for $V = 10$ ft/s, $g = 32.2$ ft/s^2, and $\ell = 2$ ft. Recalculate the Froude number using SI units for V, g, and ℓ. Explain the significance of the results of these calculations.

Section 1.4 Measures of Fluid Mass and Weight

1.13 The density of a certain liquid is 2.15 slugs/ft^3. Determine its specific weight and specific gravity.

1.14 A *hydrometer* is used to measure the specific gravity of liquids. (See **Video V2.6.**) For a certain liquid a hydrometer reading indicates a specific gravity of 1.15. What is the liquid's density and specific weight? Express your answer in SI units.

†1.15 Estimate the number of pounds of mercury it would take to fill your bath tub. List all assumptions and show all calculations.

1.16 When poured into a graduated cylinder, a liquid is found to weigh 6 N when occupying a volume of 500 ml (milliliters). Determine its specific weight, density, and specific gravity.

*1.17 The variation in the density of water, ρ, with temperature, T, in the range of $20° C \le T \le 50° C$, is given in the following table.

Density (kg/m^3)	998.2	997.1	995.7	994.1	992.2	990.2	988.1
Temperature (°C)	20	25	30	35	40	45	50

Use these data to determine an empirical equation of the form $\rho = c_1 + c_2 T + c_3 T^2$, which can be used to predict the density over the range indicated. Compare the predicted values with the data given. What is the density of water at 42.1° C?

†1.18 Estimate the number of kilograms of water consumed per day for household purposes in your city. List all assumptions and show all calculations.

Section 1.5 Ideal Gas Law

1.19 Some experiments are being conducted in a laboratory in which the air temperature is 27° C and the atmospheric pressure is 14.3 psia. Determine the density of the air. Express your answers in slugs/ft^3 and in kg/m^3.

1.20 A closed tank having a volume of 2 ft^3 is filled with 0.30 lb of a gas. A pressure gage attached to the tank reads 12 psi when the gas temperature is 80° F. There is some question as to whether the gas in the tank is oxygen or helium. Which do you think it is? Explain how you arrived at your answer.

†1.21 Estimate the volume of car exhaust produced per day by automobiles in the United States. List all assumptions and show calculations.

1.22 A tire having a volume of 3 ft^3 contains air at a gage pressure of 26 psi and a temprature of 70 °F. Determine the density of the air and the weight of the air contained in the tire.

Section 1.6 Viscosity (also see Lab Problems 1.52 and 1.53)

1.23 A liquid has a specific weight of 59 lb/ft^3 and a dynamic viscosity of 2.75 lb·s/ft^2. Determine its kinematic viscosity.

1.24 (See Fluids in the News article titled "A vital fluid," Section 1.6.) Some measurements on a blood sample at 37 °C (98.6 °F) indicate a shearing stress of 0.52 N/m^2 for a corresponding rate of shearing strain of 200 s^{-1}. Determine the apparent viscosity of the blood and compare it with the viscosity of water at the same temperature.

1.25 The time, t, it takes to pour a liquid from a container depends on several factors, including the kinematic viscosity, ν, of the liquid. (See **Video V1.1.**) In some laboratory tests, various oils having the same density but different viscosities were poured at a fixed tipping rate from small 150-ml beakers. The time required to pour 100 ml of the oil was measured, and it was found that an approximate equation for the pouring time in seconds was $t = 1 + 9 \times 10^2 \nu + 8 \times 10^3 \nu^2$ with ν in m^2/s. **(a)** Is this a general homogeneous equation? Explain. **(b)** Compare the time it would take to pour 100 ml of SAE 30 oil from a 150-ml beaker at 0° C to the corresponding time at a temperature of 60° C. Make use of Fig. B.2 in Appendix B for viscosity data.

1.26 SAE 30 oil at 60° F flows through a 2-in.-diameter pipe with a mean velocity of 5 ft/s. Determine the value of the Reynolds number (see Example 1.3).

1.27 Calculate the Reynolds number for the flow of water and for air through a 3-mm-diameter tube if the mean velocity is 2 m/s and the temperature is 30° C in both cases (see Example 1.3). Assume the air is at standard atmospheric pressure.

1.28 A Newtonian fluid having a specific gravity of 0.92 and a kinematic viscosity of 4×10^{-4} m^2/s flows past a fixed surface. Due to the no-slip condition, the velocity at the fixed surface is zero (as shown in **Video V1.2**), and the velocity profile near the surface is shown in Fig. P1.28. Determine the magnitude and direction of the shearing stress developed on the plate. Express your answer in terms of U and δ, with U and δ expressed in units of meters per second and meters, respectively.

$$\frac{u}{U} = \frac{3}{2}\frac{y}{\delta} - \frac{1}{2}\left(\frac{y}{\delta}\right)^3$$

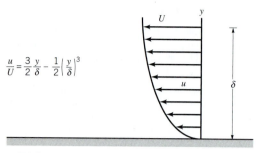

■ **FIGURE P1.28**

1.29 As shown in **Video V1.2**, the "no-slip" condition means that a fluid "sticks" to a solid surface. This is true for both fixed and moving surfaces. Let two layers of fluid be dragged along by the motion of an upper plate as shown in Fig. P1.29. The bottom plate is stationary. The top fluid puts a shear stress on the upper plate, and the lower fluid puts a shear stress on the botton plate. Determine the ratio of these two shear stresses.

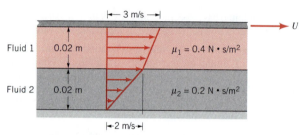

■ **FIGURE P1.29**

1.30 When a viscous fluid flows past a thin sharp-edged plate, a thin layer adjacent to the plate surface develops in which the velocity, u, changes rapidly from zero to the approach velocity, U, in a small distance, δ. This layer is called a *boundary layer*. The thickness of this layer increases with the distance x along the plate as shown in Fig. P1.30. Assume that $u = Uy/\delta$ and $\delta = 3.5\sqrt{vx/U}$ where v is the kinematic viscosity of the fluid. Determine an expression for the force (drag) that would be developed on one side of the plate of length ℓ and width b. Express your answer in terms of ℓ, b, v and ρ, where ρ is the fluid density.

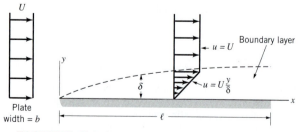

■ **FIGURE P1.30**

1.31 The sled shown in Fig. P1.31 slides along on a thin horizontal layer of water between the ice and the runners. The horizontal force that the water puts on the runners is equal to 1.2 lb when the sled's speed is 50 ft/s. The total area of both runners in

contact with the water is 0.08 ft², and the viscosity of the water is 3.5×10^{-5} lb s/ft². Determine the thickness of the water layer under the runners. Assume a linear velocity distribution in the water layer.

■ **FIGURE P1.31**

1.32 A 40-lb, 0.8-ft-diameter, 1-ft-tall cylindrical tank slides slowly down a ramp with a constant speed of 0.1 ft/s as shown in Fig. P1.32. The uniform-thickness oil layer on the ramp has a viscosity of 0.2 lb · s/ft². Determine the angle, θ, of the ramp.

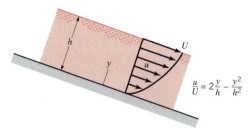

■ **FIGURE P1.32**

1.33 A layer of water flows down an inclined fixed surface with the velocity profile shown in Fig. P1.33. Determine the magnitude and direction of the shearing stress that the water exerts on the fixed surface for $U = 3$ m/s and $h = 0.1$ m.

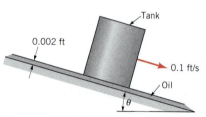

$$\frac{u}{U} = 2\frac{y}{h} - \frac{y^2}{h^2}$$

■ **FIGURE P1.33**

1.34 Water flows near a flat surface, and some measurements of the water velocity, u, parallel to the surface, at different heights, y, above the surface are obtained. At the surface $y = 0$. After an analysis of data, the lab technician reports that the velocity distribution in the range $0 < y < 0.1$ ft is given by the equation

$$u = 0.81 + 9.2y + 4.1 \times 10^3 y^3$$

with u in ft/s when y is in ft. **(a)** Do you think that this equation would be valid in any system of units? Explain. **(b)** Do you think this equation is correct? Explain. You may want to look at **Video 1.2** to help you arrive at your answer.

1.35 The viscosity of liquids can be measured through the use of a *rotating cylinder viscometer* of the type illustrated in Fig. P1.35. In this device the outer cylinder is fixed and the inner cylinder is rotated with an angular velocity, ω. The torque $\mathcal{T}$ required to develop ω is measured, and the viscosity is calculated from these two measurements. Develop an equation relating μ, ω, $\mathcal{T}$, ℓ, R_o and R_i. Neglect end effects and assume the velocity distribution in the gap is linear.

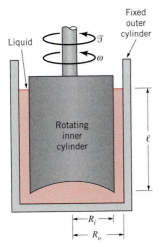

■ **FIGURE P1.35**

1.36 The viscosity of a soft drink was determined by using a capillary tube viscometer similar to that shown in Fig. P1.36 and **Video V1.3.** For this device the kinematic viscosity, v, is directly proportional to the time, t, that it takes for a given amount of liquid to flow through a small capillary tube. That is, $v = Kt$. The following data were obtained from regular pop and diet pop. The corresponding measured specific gravities are also given. Based on these data, by what percent is the absolute viscosity, μ, of regular pop greater than that of diet pop?

	Regular pop	Diet pop
$t(s)$	377.8	300.3
SG	1.044	1.003

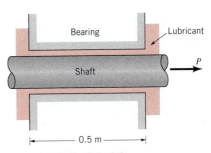

■ **FIGURE P1.36**

1.37 One type of rotating cylinder viscometer, called a *Stormer viscometer*, uses a falling weight, $\mathcal{W}$, to cause the cylinder to rotate with an angular velocity, ω, as illustrated in Fig. P1.37. For this device the viscosity, μ, of the liquid is related to $\mathcal{W}$ and ω through the equation $\mathcal{W} = K\mu\omega$, where K is a constant that depends only on the geometry (including the liquid depth) of the viscometer. The value of K is usually determined by using a calibration liquid (a liquid of known viscosity).

(a) Some data for a particular Stormer viscometer, obtained using glycerin at 20° C as a calibration liquid, are given below. Plot values of the weight as ordinates and values of the angular velocity as abscissae. Draw the best curve through the plotted points and determine K for the viscometer.

$\mathcal{W}$ (lb)	0.22	0.66	1.10	1.54	2.20
ω (rev/s)	0.53	1.59	2.79	3.83	5.49

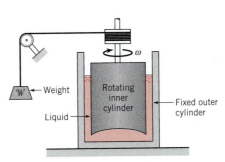

■ **FIGURE P1.37**

(b) A liquid of unknown viscosity is placed in the same viscometer used in part (a), and the following data are obtained. Determine the viscosity of this liquid.

$\mathcal{W}$ (lb)	0.04	0.11	0.22	0.33	0.44
ω (rev/s)	0.72	1.89	3.73	5.44	7.42

1.38 A 25-mm-diameter shaft is pulled through a cylindrical bearing as shown in Fig. P1.38. The lubricant that fills the 0.3-mm gap between the shaft and bearing is an oil having a kinematic viscosity of 8.0×10^{-4} m²/s and a specific gravity of 0.91. Determine the force P required to pull the shaft at a velocity of 3 m/s. Assume the velocity distribution in the gap is linear.

■ **FIGURE P1.38**

1.39 There are many fluids that exhibit non-Newtonian behavior (see, for example, **Video V1.4**). For a given fluid the distinction between Newtonian and non-Newtonian behavior is usually based on measurements of shear stress and rate of shearing strain. Assume that the viscosity of blood is to be determined by measurements of shear stress, τ, and rate of shearing strain, du/dy, obtained from a small blood sample tested in a suitable viscometer. Based on the data given, determine if the blood is a Newtonian or non-Newtonian fluid. Explain how you arrived at your answer.

τ (N/m^2)	0.04	0.06	0.12	0.18	0.30	0.52	1.12	2.10
du/dy (s^{-1})	2.25	4.50	11.25	22.5	45.0	90.0	225	450

Section 1.7 Compressibility of Fluids

1.40 An important dimensionless parameter concerned with very high speed flow is the *Mach number,* defined as V/c, where V is the speed of the object, such as an airplane or projectile, and c is the speed of sound in the fluid surrounding the object. For a projectile traveling at 800 mph through air at 50° F and standard atmospheric pressure, what is the value of the Mach number?

1.41 Often the assumption is made that the flow of a certain fluid can be considered as incompressible flow if the density of the fluid changes by less than 2%. If air is flowing through a tube such that the air gage pressure at one section is 9.0 psi and at a downstream section it is 8.6 psi at the same temperature, do you think that this flow could be considered an incompressible flow? Support your answer with the necessary calculations. Assume standard atmospheric pressure.

1.42 Oxygen at 30 °C and 300 kPa absolute pressure expands isothermally to an absolute pressure of 140 kPa. Determine the final density of the gas.

1.43 (See Fluids in the News article titled "This water jet is a blast," Section 1.7.1.) By what percent is the volume of water decreased if its pressure is increased to an equivalent to 3000 atmospheres (44,100 psi)?

1.44 Determine the speed of sound at 20 °C in **(a)** air, **(b)** helium, and **(c)** natural gas. Express your answer in m/s.

Section 1.8 Vapor Pressure

1.45 When a fluid flows through a sharp bend, low pressures may develop in localized regions of the bend. Estimate the minimum absolute pressure (in psi) that can develop without causing cavitation if the fluid is water at 160 °F.

1.46 When water at 70 °C flows through a converging section of pipe, the pressure is reduced in the direction of flow. Estimate the minimum absolute pressure that can develop without causing cavitation. Express your answer in both BG and SI units.

1.47 At what atmospheric pressure will water boil at 35 °C? Express your answer in both SI and BG units.

Section 1.9 Surface Tension

1.48 An open, clean glass tube ($\theta = 0°$) is inserted vertically into a pan of water. What tube diameter is needed if the water level in the tube is to rise one tube diameter (due to surface tension)?

1.49 (See Fluids in the News article titled "Walking on water," Section 1.9.) **(a)** The water strider bug shown in Fig. P1.49 is supported on the surface of a pond by surface tension acting along the interface between the water and the bug's legs. Determine the minimum length of this interface needed to support the bug. Assume the bug weighs 10^{-4} N and the surface tension force acts vertically upwards. **(b)** Repeat part (a) if surface tension were to support a person weighing 750 N.

■ **FIGURE P1.49**

1.50 As shown in **Video V1.5,** surface tension forces can be strong enough to allow a double-edge steel razor blade to "float" on water, but a single-edge blade will sink. Assume that the surface tension forces act at an angle θ relative to the water surface as shown in Fig. P1.50. **(a)** The mass of the double-edge blade is 0.64×10^{-3} kg, and the total length of its sides is 206 mm. Determine the value of θ required to maintain equilibrium between the blade weight and the resultant surface tension force. **(b)** The mass of the single-edge blade is 2.61×10^{-3} kg, and the total length of its sides is 154 mm. Explain why this blade sinks. Support your answer with the necessary calculations.

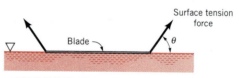

Surface tension force

Blade

θ

■ **FIGURE P1.50**

1.51 Under the right conditions, it is possible, due to surface tension, to have metal objects float on water. (See **Video V1.5.**) Consider placing a short length of a small diameter steel (sp. wt. = 490 lb/ft^3) rod on a surface of water. What is the maximum diameter that the rod can have before it will sink? Assume that the surface tension forces act vertically upward. *Note:* A standard paper clip has a diameter of 0.036 in. Partially unfold a paper clip and see if you can get it to float on water. Do the results of this experiment support your analysis?

Lab Problems

1.52   This problem involves the use of a Stormer viscometer to determine whether a fluid is a Newtonian or a non-Newtonian fluid. To proceed with this problem, go to the book's web site, www.wiley.com/college/young.

1.53   This problem involves the use of a capillary tube viscometer to determine the kinematic viscosity of water as a function of temperature. To proceed with this problem, go to the book's web site, www.wiley.com/college/young.

FE Exam Problems

Sample FE (Fundamentals of Engineering) exam questions for fluid mechanics are provided on the book's web site, www.wiley.com/college/young.

CHAPTER 2

Fluid Statics

Floating ice: According to fluid statics principles, even though the glass is brim full and the floating ice protrudes above the free surface, when the ice melts the water level in the glass will remain constant and no water will spill over the lip of the glass (Photograph by B. R. Munson).

LEARNING OBJECTIVES

After completing this chapter, you should be able to:
- determine the pressure at various locations in a fluid at rest.
- explain the concept of manometers and apply appropriate equations to determine pressures.
- calculate the hydrostatic pressure force on a plane or curved submerged surface.
- calculate the buoyant force and discuss the stability of floating or submerged objects.

In this chapter we will consider an important class of problems in which the fluid is either at rest or moving in such a manner that there is no relative motion between adjacent particles. In both instances there will be no shearing stresses in the fluid, and the only forces that develop on the surfaces of the particles will be due to the pressure. Thus, our principal concern is to investigate pressure and its variation throughout a fluid, and the effect of pressure on submerged surfaces.

2.1   Pressure at a Point

As discussed briefly in Chapter 1, the term *pressure* is used to indicate the normal force per unit area at a given point acting on a given plane within the fluid mass of interest. A question that immediately arises is how the pressure at a point varies with the orientation of the plane passing through the point. To answer this question, consider the free-body diagram, illustrated in Fig. 2.1, that was obtained by removing a small triangular wedge of fluid from some arbitrary location within a fluid mass. Since we are considering the situation in which there are no shearing stresses, the only external forces acting on the wedge are due to the pressure and the weight. For simplicity the forces in the x direction are not shown, and the z axis is taken as the vertical axis so the weight acts in the negative z direction. Although we are primarily interested in fluids at rest, to make the analysis as general as possible, we will allow the fluid element to have accelerated motion. The assumption of zero shearing stresses will still be valid so long as the fluid element moves as a rigid body; that is, there is no relative motion between adjacent elements.

The equations of motion (Newton's second law, $\mathbf{F} = m\mathbf{a}$) in the y and z directions are, respectively,

$$\Sigma F_y = p_y\, \delta x\, \delta z - p_s\, \delta x\, \delta s \sin \theta = \rho \frac{\delta x\, \delta y\, \delta z}{2} a_y$$

$$\Sigma F_z = p_z\, \delta x\, \delta y - p_s\, \delta x\, \delta s \cos \theta - \gamma \frac{\delta x\, \delta y\, \delta z}{2} = \rho \frac{\delta x\, \delta y\, \delta z}{2} a_z$$

where p_x, p_y, and p_z are the average pressures on the faces, γ and ρ are the fluid specific weight and density, respectively, and a_y, a_z the accelerations. It follows from the geometry that

$$\delta y = \delta s \cos \theta \qquad \delta z = \delta s \sin \theta$$

so that the equations of motion can be rewritten as

$$p_y - p_s = \rho a_y \frac{\delta y}{2}$$

$$p_z - p_s = (\rho a_z + \gamma) \frac{\delta z}{2}$$

Since we are really interested in what is happening at a point, we take the limit as δx, δy, and δz approach zero (while maintaining the angle θ), and it follows that

$$p_y = p_s \qquad p_z = p_s$$

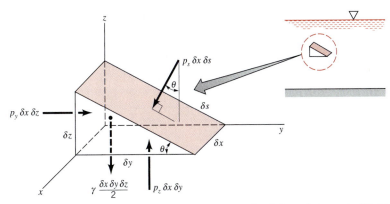

■ **FIGURE 2.1**   **Forces on an arbitrary wedged-shaped element of fluid.**

or $p_s = p_y = p_z$. The angle θ was arbitrarily chosen so we can conclude that *the pressure at a point in a fluid at rest, or in motion, is independent of direction as long as there are no shearing stresses present.* This important result is known as ***Pascal's law.***

In Chapter 6 it will be shown that for moving fluids in which there is relative motion between particles (so that shearing stresses develop) the normal stress at a point, which corresponds to pressure in fluids at rest, is not necessarily the same in all directions. In such cases the pressure is defined as the *average* of any three mutually perpendicular normal stresses at the point.

2.2 Basic Equation for Pressure Field

Although we have answered the question of how the pressure at a point varies with direction, we are now faced with an equally important question—how does the pressure in a fluid in which there are no shearing stresses vary from point to point? To answer this question, consider a small rectangular element of fluid removed from some arbitrary position within the mass of fluid of interest as illustrated in Fig. 2.2. There are two types of forces acting on this element: ***surface forces*** due to the pressure, and a ***body force*** equal to the weight of the element.

If we let the pressure at the center of the element be designated as p, then the average pressure on the various faces can be expressed in terms of p and its derivatives as shown in Fig. 2.2. For simplicity the surface forces in the x direction are not shown. The resultant surface force in the y direction is

$$\delta F_y = \left(p - \frac{\partial p}{\partial y}\frac{\delta y}{2} \right) \delta x\,\delta z - \left(p + \frac{\partial p}{\partial y}\frac{\delta y}{2} \right) \delta x\,\delta z$$

or

$$\delta F_y = -\frac{\partial p}{\partial y}\,\delta x\,\delta y\,\delta z$$

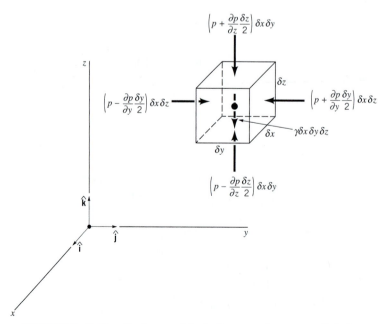

■ **FIGURE 2.2**   Surface and body forces acting on small fluid element.

Similarly, for the x and z directions the resultant surface forces are

$$\delta F_x = -\frac{\partial p}{\partial x}\,\delta x\,\delta y\,\delta z \qquad \delta F_z = -\frac{\partial p}{\partial z}\,\delta x\,\delta y\,\delta z$$

The resultant surface force acting on the element can be expressed in vector form as

$$\delta\mathbf{F}_s = \delta F_x\hat{\mathbf{i}} + \delta F_y\hat{\mathbf{j}} + \delta F_z\hat{\mathbf{k}}$$

or

$$\delta\mathbf{F}_s = -\left(\frac{\partial p}{\partial x}\,\hat{\mathbf{i}} + \frac{\partial p}{\partial y}\hat{\mathbf{j}} + \frac{\partial p}{\partial z}\hat{\mathbf{k}}\right)\delta x\,\delta y\,\delta z \tag{2.1}$$

where $\hat{\mathbf{i}}, \hat{\mathbf{j}}$, and $\hat{\mathbf{k}}$ are the unit vectors along the coordinate axes shown in Fig. 2.2. The group of terms in parentheses in Eq. 2.1 represents in vector form the *pressure gradient* and can be written as

$$\frac{\partial p}{\partial x}\hat{\mathbf{i}} + \frac{\partial p}{\partial y}\hat{\mathbf{j}} + \frac{\partial p}{\partial z}\hat{\mathbf{k}} = \nabla p$$

where

$$\nabla(\;) = \frac{\partial(\;)}{\partial x}\hat{\mathbf{i}} + \frac{\partial(\;)}{\partial y}\hat{\mathbf{j}} + \frac{\partial(\;)}{\partial z}\hat{\mathbf{k}}$$

and the symbol ∇ is the *gradient* or "del" vector operator. Thus, the resultant surface force per unit volume can be expressed as

$$\frac{\delta\mathbf{F}_s}{\delta x\,\delta y\,\delta z} = -\nabla p$$

Because the z axis is vertical, the weight of the element is

$$-\delta\mathcal{W}\hat{\mathbf{k}} = -\gamma\,\delta x\,\delta y\,\delta z\,\hat{\mathbf{k}}$$

where the negative sign indicates that the force due to the weight is downward (in the negative z direction). Newton's second law, applied to the fluid element, can be expressed as

$$\Sigma\,\delta\mathbf{F} = \delta m\,\mathbf{a}$$

where $\Sigma\,\delta\mathbf{F}$ represents the resultant force acting on the element, $\mathbf{a}$ is the acceleration of the element, and δm is the element mass, which can be written as $\rho\,\delta x\,\delta y\,\delta z$. It follows that

$$\Sigma\,\delta\mathbf{F} = \delta\mathbf{F}_s - \delta\mathcal{W}\hat{\mathbf{k}} = \delta m\,\mathbf{a}$$

or

$$-\nabla p\,\delta x\,\delta y\,\delta z - \gamma\,\delta x\,\delta y\,\delta z\,\hat{\mathbf{k}} = \rho\,\delta x\,\delta y\,\delta z\,\mathbf{a}$$

and, therefore,

$$-\nabla p - \gamma\hat{\mathbf{k}} = \rho\mathbf{a} \tag{2.2}$$

Equation 2.2 is the general equation of motion for a fluid in which there are no shearing stresses. Although Eq. 2.2 applies to both fluids at rest and moving fluids, we will primarily restrict our attention to fluids at rest.

2.3 Pressure Variation in a Fluid at Rest

For a fluid at rest $\mathbf{a} = 0$ and Eq. 2.2 reduces to

$$\nabla p + \gamma\hat{\mathbf{k}} = 0$$

or in component form

$$\frac{\partial p}{\partial x} = 0 \qquad \frac{\partial p}{\partial y} = 0 \qquad \frac{\partial p}{\partial z} = -\gamma \qquad \text{(2.3)}$$

These equations show that the pressure does not depend on x or y. Thus, as we move from point to point in a horizontal plane (any plane parallel to the $x - y$ plane), the pressure does not change. Since p depends only on z, the last of Eqs. 2.3 can be written as the ordinary differential equation

$$\frac{dp}{dz} = -\gamma \qquad \text{(2.4)}$$

Equation 2.4 is the fundamental equation for fluids at rest and can be used to determine how pressure changes with elevation. This equation indicates that the pressure gradient in the vertical direction is negative; that is, the pressure decreases as we move upward in a fluid at rest. There is no requirement that γ be a constant. Thus, it is valid for fluids with constant specific weight, such as liquids, as well as fluids whose specific weight may vary with elevation, such as air or other gases. However, to proceed with the integration of Eq. 2.4 it is necessary to stipulate how the specific weight varies with z.

2.3.1 Incompressible Fluid

Since the specific weight is equal to the product of fluid density and acceleration of gravity ($\gamma = \rho g$), changes in γ are caused by a change in either ρ or g. For most engineering applications the variation in g is negligible, so our main concern is with the possible variation in the fluid density. In general, a fluid with constant density is called an *incompressible fluid.* For liquids the variation in density is usually negligible, even over large vertical distances, so that the assumption of constant specific weight when dealing with liquids is a good one. For this instance, Eq. 2.4 can be directly integrated

$$\int_{p_1}^{p_2} dp = -\gamma \int_{z_1}^{z_2} dz$$

to yield

$$p_1 - p_2 = \gamma(z_2 - z_1) \qquad \text{(2.5)}$$

where p_1 and p_2 are pressures at the vertical elevations z_1 and z_2, as is illustrated in Fig. 2.3.
    Equation 2.5 can be written in the compact form

$$p_1 - p_2 = \gamma h \qquad \text{(2.6)}$$

or

$$p_1 = \gamma h + p_2 \qquad \text{(2.7)}$$

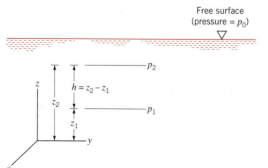

■ **FIGURE 2.3** Notation for pressure variation in a fluid at rest with a free surface.

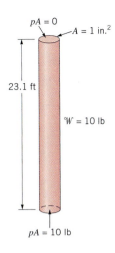

where h is the distance, $z_2 - z_1$, which is the depth of fluid measured downward from the location of p_2. This type of pressure distribution is commonly called a ***hydrostatic pressure distribution,*** and Eq. 2.7 shows that in an incompressible fluid at rest the pressure varies linearly with depth. The pressure must increase with depth to "hold up" the fluid above it.

It can also be observed from Eq. 2.6 that the pressure difference between two points can be specified by the distance h since

$$h = \frac{p_1 - p_2}{\gamma}$$

In this case h is called the ***pressure head*** and is interpreted as the height of a column of fluid of specific weight γ required to give a pressure difference $p_1 - p_2$. For example, a pressure difference of 10 psi can be specified in terms of pressure head as 23.1 ft of water ($\gamma = 62.4 \text{ lb/ft}^3$), or 518 mm of Hg ($\gamma = 133 \text{ kN/m}^3$). As illustrated by the figure in the margin, a 23.1-ft-tall column of water with a coss-sectional area of 1 in.^{2} weighs 10 lb.

When one works with liquids there is often a free surface, as is illustrated in Fig. 2.3, and it is convenient to use this surface as a reference plane. The reference pressure p_0 would correspond to the pressure acting on the free surface (which would frequently be atmospheric pressure), and thus, if we let $p_2 = p_0$ in Eq. 2.7, it follows that the pressure p at any depth h below the free surface is given by the equation:

$$p = \gamma h + p_0 \qquad\qquad \textbf{(2.8)}$$

As is demonstrated by Eq. 2.7 or 2.8, the pressure in a homogeneous, incompressible fluid at rest depends on the depth of the fluid relative to some reference plane, and it is *not* influenced by the *size* or *shape* of the tank or container in which the fluid is held.

F l u i d s   i n   t h e   N e w s

Giraffe's blood pressure A giraffe's long neck allows it to graze up to 6 m above the ground. It can also lower its head to drink at ground level. Thus, in the circulatory system there is a significant *hydrostatic pressure* effect due to this elevation change. To maintain blood to its head throughout this change in elevation, the giraffe must maintain a relatively high blood pressure at heart level—approximately two and a half times that in humans. To prevent rupture of blood vessels in the high-pressure lower leg regions, giraffes have a tight sheath of thick skin over their lower limbs which acts like an elastic bandage in exactly the same way as do the g-suits of fighter pilots. In addition, valves in the upper neck prevent backflow into the head when the giraffe lowers its head to ground level. It is also thought that blood vessels in the giraffe's kidney have a special mechanism to prevent large changes in filtration rate when blood pressure increases or decreases with its head movement. (See Problem 2.8.) ■

EXAMPLE 2.1   Pressure–Depth Relationship

GIVEN   Because of a leak in a buried gasoline storage tank, water has seeped in to the depth shown in Fig. E2.1. The specific gravity of the gasoline is $SG = 0.68$.

FIND   Determine the pressure at the gasoline–water interface and at the bottom of the tank. Express the pressure in units of lb/ft^2, lb/in.^{2}, and as a pressure head in feet of water.

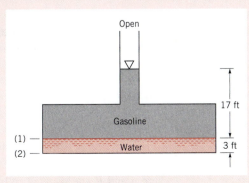

■ **FIGURE E2.1**

SOLUTION

Since we are dealing with liquids at rest, the pressure distribution will be hydrostatic, and therefore the pressure variation can be found from the equation

$$p = \gamma h + p_0$$

With p_0 corresponding to the pressure at the free surface of the gasoline, then the pressure at the interface is

$$p_1 = SG\gamma_{H_2O}h + p_0$$
$$= (0.68)(62.4 \text{ lb/ft}^3)(17 \text{ ft}) + p_0$$
$$= 721 + p_0 \text{ (lb/ft}^2)$$

If we measure the pressure relative to atmospheric pressure (gage pressure), it follows that $p_0 = 0$, and therefore

$$p_1 = 721 \text{ lb/ft}^2 \qquad \textbf{(Ans)}$$
$$p_1 = \frac{721 \text{ lb/ft}^2}{144 \text{ in.}^2/\text{ft}^2} = 5.01 \text{ lb/in.}^2 \qquad \textbf{(Ans)}$$
$$\frac{p_1}{\gamma_{H_2O}} = \frac{721 \text{ lb/ft}^2}{62.4 \text{ lb/ft}^3} = 11.6 \text{ ft} \qquad \textbf{(Ans)}$$

It is noted that a rectangular column of water 11.6 ft tall and 1 ft^2 in cross section weighs 721 lb. A similar column with a 1-in.^{2} cross section weighs 5.01 lb.

We can now apply the same relationship to determine the pressure at the tank bottom; that is,

$$p_2 = \gamma_{H_2O}h_{H_2O} + p_1$$
$$= (62.4 \text{ lb/ft}^3)(3 \text{ ft}) + 721 \text{ lb/ft}^2$$
$$= 908 \text{ lb/ft}^2 \qquad \textbf{(Ans)}$$
$$p_2 = \frac{908 \text{ lb/ft}^2}{144 \text{ in.}^2/\text{ft}^2} = 6.31 \text{ lb/in.}^2 \qquad \textbf{(Ans)}$$
$$\frac{p_2}{\gamma_{H_2O}} = \frac{908 \text{ lb/ft}^2}{62.4 \text{ lb/ft}^3} = 14.6 \text{ ft} \qquad \textbf{(Ans)}$$

COMMENT   Observe that if we wish to express these pressures in terms of *absolute* pressure, we would have to add the local atmospheric pressure (in appropriate units) to the previous results. A further discussion of gage and absolute pressure is given in Section 2.5.

2.3.2   Compressible Fluid

We normally think of gases such as air, oxygen, and nitrogen as being *compressible fluids* since the density of the gas can change significantly with changes in pressure and temperature. Thus, although Eq. 2.4 applies at a point in a gas, it is necessary to consider the possible variation in γ before the equation can be integrated. However, as discussed in Chapter 1, the specific weights of common gases are small when compared with those of liquids. For example, the specific weight of air at sea level and 60° F is 0.0763 lb/ft^3, whereas the specific weight of water under the same conditions is 62.4 lb/ft^3. Since the specific weights of gases are comparatively small, it follows from Eq. 2.4 that the pressure gradient in the vertical direction is correspondingly small, and even over distances of several hundred feet the pressure will remain essentially constant for a gas. This means we can neglect the effect of elevation changes on the pressure in gases in tanks, pipes, and so forth in which the distances involved are small.

For those situations in which the variations in heights are large, on the order of thousands of feet, attention must be given to the variation in the specific weight. As is described in Chapter 1, the equation of state for an ideal (or perfect) gas is

$$p = \rho RT$$

where p is the absolute pressure, R is the gas constant, and T is the absolute temperature. This relationship can be combined with Eq. 2.4 to give

$$\frac{dp}{dz} = -\frac{gp}{RT}$$

and by separating variables

$$\int_{p_1}^{p_2} \frac{dp}{p} = \ln\frac{p_2}{p_1} = -\frac{g}{R}\int_{z_1}^{z_2}\frac{dz}{T} \qquad \textbf{(2.9)}$$

where g and R are assumed to be constant over the elevation change from z_1 to z_2.

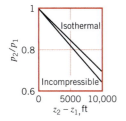

Before completing the integration, one must specify the nature of the variation of temperature with elevation. For example, if we assume that the temperature has a constant value T_0 over the range z_1 to z_2 (*isothermal* conditions), it then follows from Eq. 2.9 that

$$p_2 = p_1 \exp\left[-\frac{g(z_2 - z_1)}{RT_0}\right] \tag{2.10}$$

This equation provides the desired pressure–elevation relationship for an isothermal layer. As shown in the margin figure, even for 10,000 ft altitude change the difference between the constant temperature (isothermal) and the constant density (incompressible) results are relatively minor. For nonisothermal conditions a similar procedure can be followed if the temperature–elevation relationship is known.

2.4 Standard Atmosphere

An important application of Eq. 2.9 relates to the variation in pressure in the earth's atmosphere. Ideally, we would like to have measurements of pressure versus altitude over the specific range for the specific conditions (temperature, reference pressure) for which the pressure is to be determined. However, this type of information is usually not available. Thus, a "standard atmosphere" has been determined that can be used in the design of aircraft, missiles, and spacecraft, and in comparing their performance under standard conditions.

The currently accepted standard atmosphere is based on a report published in 1962 and updated in 1976 (see Refs. 1 and 2), defining the so-called *U.S. standard atmosphere,* which is an idealized representation of middle-latitude, year-round mean conditions of the earth's atmosphere. Several important properties for standard atmospheric conditions at *sea level* are listed in Table 2.1.

Tabulated values for temperature, acceleration of gravity, pressure, density, and viscosity for the U.S. standard atmosphere are given in Tables C.1 and C.2 in Appendix C.

2.5 Measurement of Pressure

Since pressure is a very important characteristic of a fluid field, it is not surprising that numerous devices and techniques are used in its measurement. As noted briefly in Chapter 1, the pressure at a point within a fluid mass will be designated as either an *absolute pressure* or a *gage pressure.* Absolute pressure is measured relative to a perfect vacuum (absolute zero pressure), whereas gage pressure is measured relative to the local atmospheric pressure. Thus, a gage pressure of zero corresponds to a pressure that is equal to the local atmospheric

■ **TABLE 2.1**
Properties of U.S. Standard Atmosphere at Sea Level[a]

Property	SI Units	BG Units
Temperature, T	288.15 K (15 °C)	518.67 °R (59.00 °F)
Pressure, p	101.33 kPa (abs)	2116.2 lb/ft^2 (abs)
		[14.696 lb/in.^{2} (abs)]
Density, ρ	1.225 kg/m^3	0.002377 slugs/ft^3
Specific weight, γ	12.014 N/m^3	0.07647 lb/ft^3
Viscosity, μ	1.789×10^{-5} N·s/m^2	3.737×10^{-7} lb·s/ft^2

[a]Acceleration of gravity at sea level = 9.807 m/s^2 = 32.174 ft/s^2.

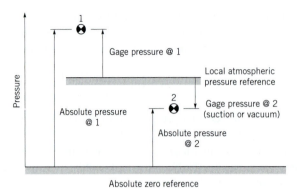

■ **FIGURE 2.4** **Graphical representation of gage and absolute pressure.**

pressure. Absolute pressures are always positive, but gage pressure can be either positive or negative depending on whether the pressure is above atmospheric pressure (a positive value) or below atmospheric pressure (a negative value). A negative gage pressure is also referred to as a *suction* or *vacuum pressure.* For example, 10 psi (abs) could be expressed as −4.7 psi (gage), if the local atmospheric pressure is 14.7 psi, or alternatively 4.7 psi suction or 4.7 psi vacuum. The concept of gage and absolute pressure is illustrated graphically in Fig. 2.4 for two typical pressures located at points 1 and 2.

In addition to the reference used for the pressure measurement, the *units* used to express the value are obviously of importance. As described in Section 1.5, pressure is a force per unit area, and the units in the BG system are lb/ft^2 or $lb/in.^2$, commonly abbreviated psf or psi, respectively. In the SI system the units are N/m^2; this combination is called the pascal and is written as Pa ($1 N/m^2 = 1$ Pa). As noted earlier, pressure can also be expressed as the height of a column of liquid. Then, the units will refer to the height of the column (in., ft, mm, m, etc.), and in addition, the liquid in the column must be specified (H_2O, Hg, etc.). For example, standard atmospheric pressure can be expressed as 760 mm Hg (abs). *In this text, pressures will be assumed to be gage pressures unless specifically designated absolute.* For example, 10 psi or 100 kPa would be gage pressures, whereas 10 psia or 100 kPa (abs) would refer to absolute pressures. It is to be noted that because *pressure differences* are independent of the reference, no special notation is required in this case.

The measurement of atmospheric pressure is usually accomplished with a mercury *barometer,* which in its simplest form consists of a glass tube closed at one end with the open end immersed in a container of mercury as shown in Fig. 2.5. The tube is initially filled with mercury (inverted with its open end up) and then turned upside down (open end down) with the open end in the container of mercury. The column of mercury will come to an equilibrium

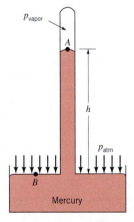

■ **FIGURE 2.5** **Mercury barometer.**

position where its weight plus the force due to the vapor pressure (which develops in the space above the column) balances the force due to the atmospheric pressure. Thus,

$$p_{atm} = \gamma h + p_{vapor} \tag{2.11}$$

where γ is the specific weight of mercury. For most practical purposes the contribution of the vapor pressure can be neglected since it is very small [for mercury, $p_{vapor} = 0.000023$ lb/in.^{2} (abs) at a temperature of 68° F] so that $p_{atm} \approx \gamma h$. It is convenient to specify atmospheric pressure in terms of the height, h, in millimeters or inches of mercury.

F l u i d s   i n   t h e   N e w s

Weather, barometers, and bars One of the most important indicators of weather conditions is *atmospheric pressure.* In general, a falling or low pressure indicates bad weather; rising or high pressure, good weather. During the evening TV weather report in the United States, atmospheric pressure is given as so many inches (commonly around 30 in.). This value is actually the height of the mercury column in a mercury *barometer* adjusted to sea level. To determine the true atmospheric pressure at a particular location, the elevation relative to sea level must be known. Another unit used by meteorologists to indicate atmospheric pressure is the *bar,* first used in weather reporting in 1914, and defined as 10^5 N/m^2. The definition of a bar is probably related to the fact that standard sea-level pressure is 1.0133×10^5 N/m^2, that is, only slightly larger than one bar. For typical weather patterns "sea-level equivalent" atmospheric pressure remains close to one bar. However, for extreme weather conditions associated with tornadoes, hurricanes, or typhoons, dramatic changes can occur. The lowest atmospheric pressure ever recorded was associated with a typhoon, Typhoon Tip, in the Pacific Ocean on October 12, 1979. The value was 0.870 bars (25.8 in. Hg). (See Problem 2.10.) ∎

2.6   Manometry

A standard technique for measuring pressure involves the use of liquid columns in vertical or inclined tubes. Pressure measuring devices based on this technique are called ***manometers.*** The mercury barometer is an example of one type of manometer, but there are many other configurations possible, depending on the particular application. Three common types of manometers include the piezometer tube, the U-tube manometer, and the inclined-tube manometer.

2.6.1   Piezometer Tube

The simplest type of manometer, called a piezometer tube, consists of a vertical tube, open at the top, and attached to the container in which the pressure is desired, as illustrated in Fig. 2.6. Because manometers involve columns of fluids at rest, the fundamental equation describing their use is Eq. 2.8

$$p = \gamma h + p_0$$

which gives the pressure at any elevation within a homogeneous fluid in terms of a reference pressure p_0 and the vertical distance h between p and p_0. Remember that in a fluid at rest pressure will *increase* as we move *downward,* and will decrease as we move *upward.* Application of this equation to the piezometer tube of Fig. 2.6 indicates that the pressure p_A can be determined by a measurement of h_1 through the relationship

$$p_A = \gamma_1 h_1$$

where γ_1 is the specific weight of the liquid in the container. Note that since the tube is open at the top, the pressure p_0 can be set equal to zero (we are now using gage pressure),

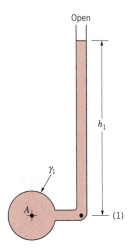

■ **FIGURE 2.6** Piezometer tube.

with the height h_1 measured from the meniscus at the upper surface to point (1). Because point (1) and point A within the container are at the same elevation, $p_A = p_1$.

Although the piezometer tube is a very simple and accurate pressure measuring device, it has several disadvantages. It is only suitable if the pressure in the container is greater than atmospheric pressure (otherwise air would be sucked into the system), and the pressure to be measured must be relatively small so the required height of the column is reasonable. Also, the fluid in the container in which the pressure is to be measured must be a liquid rather than a gas.

2.6.2   U-Tube Manometer

To overcome the difficulties noted previously, another type of manometer that is widely used consists of a tube formed into the shape of a U as is shown in Fig. 2.7. The fluid in the manometer is called the *gage fluid*. To find the pressure p_A in terms of the various column heights, we start at one end of the system and work our way around to the other end, simply utilizing Eq. 2.8. Thus, for the U-tube manometer shown in Fig. 2.7, we will start at point A and work around to the open end. The pressure at points A and (1) are the same, and as we move from point (1) to (2) the pressure will increase by $\gamma_1 h_1$. The pressure at point (2) is equal to the pressure at point (3), since the pressures at equal elevations in a continuous mass of fluid at rest must be the same. Note that we could not simply "jump across" from point (1) to a point at the same elevation in the right-hand tube since these would not be points within the same continuous mass of fluid. With the pressure at point

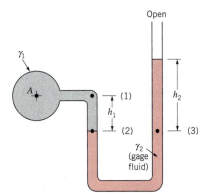

■ **FIGURE 2.7**   Simple U-tube manometer.

V2.1 Blood pressure measurement

(3) specified we now move to the open end where the pressure is zero. As we move vertically upward the pressure decreases by an amount $\gamma_2 h_2$. In equation form these various steps can be expressed as

$$p_A + \gamma_1 h_1 - \gamma_2 h_2 = 0$$

and, therefore, the pressure p_A can be written in terms of the column heights as

$$p_A = \gamma_2 h_2 - \gamma_1 h_1 \qquad (2.12)$$

A major advantage of the U-tube manometer lies in the fact that the gage fluid can be different from the fluid in the container in which the pressure is to be determined. For example, the fluid in A in Fig. 2.7 can be either a liquid or a gas. If A does contain a gas, the contribution of the gas column, $\gamma_1 h_1$, is almost always negligible so that $p_A \approx p_2$ and, in this instance, Eq. 2.12 becomes

$$p_A = \gamma_2 h_2$$

EXAMPLE 2.2   Simple U-Tube Manometer

GIVEN A closed tank contains compressed air and oil ($SG_{oil} = 0.90$) as is shown in Fig. E2.2. A U-tube manometer using mercury ($SG_{Hg} = 13.6$) is connected to the tank as shown. The column heights are $h_1 = 36$ in., $h_2 = 6$ in., and $h_3 = 9$ in.

FIND Determine the pressure reading (in psi) of the gage.

SOLUTION

Following the general procedure of starting at one end of the manometer system and working around to the other, we will start at the air–oil interface in the tank and proceed to the open end where the pressure is zero. The pressure at level (1) is

$$p_1 = p_{air} + \gamma_{oil}(h_1 + h_2)$$

This pressure is equal to the pressure at level (2), as these two points are at the same elevation in a homogeneous fluid at rest. As we move from level (2) to the open end, the pressure must decrease by $\gamma_{Hg}h_3$, and at the open end the pressure is zero. Thus, the manometer equation can be expressed as

$$p_{air} + \gamma_{oil}(h_1 + h_2) - \gamma_{Hg}h_3 = 0$$

or

$$p_{air} + (SG_{oil})(\gamma_{H_2O})(h_1 + h_2) - (SG_{Hg})(\gamma_{H_2O})h_3 = 0$$

For the values given

$$p_{air} = -(0.9)(62.4 \text{ lb/ft}^3)\left(\frac{36 + 6}{12}\text{ ft}\right)$$
$$+ (13.6)(62.4 \text{ lb/ft}^3)\left(\frac{9}{12}\text{ ft}\right)$$

so that

$$p_{air} = 440 \text{ lb/ft}^2$$

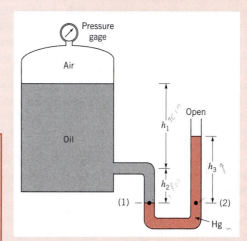

■ **FIGURE E2.2**

Because the specific weight of the air above the oil is much smaller than the specific weight of the oil, the gage should read the pressure we have calculated; that is,

$$p_{gage} = \frac{440 \text{ lb/ft}^2}{144 \text{ in.}^2/\text{ft}^2} = 3.06 \text{ psi} \qquad \textbf{(Ans)}$$

COMMENT Assume that the gage pressure remains at 3.06 psi, but the manometer is altered so that it contains only oil. That is, the mercury is replaced by oil. A simple calculation shows that in this case the vertical oil-filled tube would need to be $h_3 = 11.3$ ft tall, rather than the original $h_3 = 9$ in. There is an obvious advantage of using a heavy fluid such as mercury in manometers.

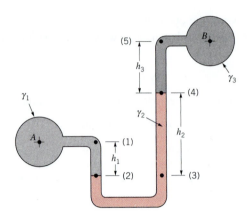

■ **FIGURE 2.8**   **Differential U-tube manometer.**

The U-tube manometer is also widely used to measure the *difference* in pressure between two containers or two points in a given system. Consider a manometer connected between containers A and B as is shown in Fig. 2.8. The difference in pressure between A and B can be found by again starting at one end of the system and working around to the other end. For example, at A the pressure is p_A, which is equal to p_1, and as we move to point (2) the pressure increases by $\gamma_1 h_1$. The pressure at p_2 is equal to p_3, and as we move upward to point (4) the pressure decreases by $\gamma_2 h_2$. Similarly, as we continue to move upward from point (4) to (5) the pressure decreases by $\gamma_3 h_3$. Finally, $p_5 = p_B$, as they are at equal elevations. Thus,

$$p_A + \gamma_1 h_1 - \gamma_2 h_2 - \gamma_3 h_3 = p_B$$

and the pressure difference is

$$p_A - p_B = \gamma_2 h_2 + \gamma_3 h_3 - \gamma_1 h_1$$

EXAMPLE 2.3   U-Tube Manometer

GIVEN   As is discussed in Chapter 3, the volume rate of flow, Q, through a pipe can be determined by a means of a flow nozzle located in the pipe as illustrated in Fig. E2.3a. The nozzle creates a pressure drop, $p_A - p_B$, along the pipe, which is related to the flow through the equation, $Q = K\sqrt{p_A - p_B}$, where K is a constant depending on the pipe and nozzle size. The pressure drop is frequently measured with a differential U-tube manometer of the type illustrated.

(b)   For $\gamma_1 = 9.80$ kN/m³, $\gamma_2 = 15.6$ kN/m³, $h_1 = 1.0$ m, and $h_2 = 0.5$ m, what is the value of the pressure drop, $p_A - p_B$?

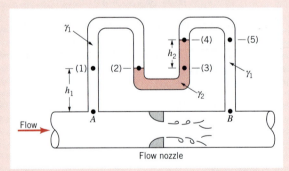

■ **FIGURE E2.3a**

FIND

(a)   Determine an equation for $p_A - p_B$ in terms of the specific weight of the flowing fluid, γ_1, the specific weight of the gage fluid, γ_2, and the various heights indicated.

SOLUTION

(a)   Although the fluid in the pipe is moving, fluids in the columns of the manometer are at rest so that the pressure variation in the manometer tubes is hydrostatic. If we start at point A and move vertically upward to level (1), the pressure will decrease by $\gamma_1 h_1$ and will be equal to the pressure at (2) and at (3). We can now move from (3) to (4) where the pressure has

been further reduced by $\gamma_2 h_2$. The pressures at levels (4) and (5) are equal, and as we move from (5) to B the pressure will increase by $\gamma_1(h_1 + h_2)$. Thus, in equation form

$$p_A - \gamma_1 h_1 - \gamma_2 h_2 + \gamma_1(h_1 + h_2) = p_B$$

or

$$p_A - p_B = h_2(\gamma_2 - \gamma_1) \qquad \textbf{(Ans)}$$

COMMENT   It is to be noted that the only column height of importance is the differential reading, h_2. The differential manometer could be placed 0.5 or 5.0 m above the pipe ($h_1 = 0.5$ m or $h_1 = 5.0$ m), and the value of h_2 would remain the same.

(b)   The specific value of the pressure drop for the data given is

$$p_A - p_B = (0.5\ \text{m})(15.6\ \text{kN/m}^3 - 9.80\ \text{kN/m}^3)$$
$$= 2.90\ \text{kPa} \qquad \textbf{(Ans)}$$

COMMENT   By repeating the calculations for manometer fluids with different specific weights, γ_2, the results shown in Fig. E2.3b are obtained. Note that relatively small pressure differences can be measured if the manometer fluid has nearly the same specific weight as the flowing fluid. It is the difference in the specific weights, $\gamma_2 - \gamma_1$, that is important.

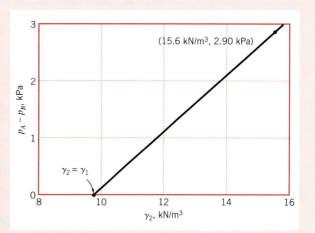

■ **FIGURE E2.3b**

2.6.3   Inclined-Tube Manometer

To measure small pressure changes, a manometer of the type shown in Fig. 2.9 is frequently used. One leg of the manometer is inclined at an angle θ, and the differential reading ℓ_2 is measured along the inclined tube. The difference in pressure $p_A - p_B$ can be expressed as

$$p_A + \gamma_1 h_1 - \gamma_2 \ell_2 \sin\theta - \gamma_3 h_3 = p_B$$

or

$$p_A - p_B = \gamma_2 \ell_2 \sin\theta + \gamma_3 h_3 - \gamma_1 h_1 \qquad \textbf{(2.13)}$$

where it is to be noted the pressure difference between points (1) and (2) is due to the *vertical* distance between the points, which can be expressed as $\ell_2 \sin\theta$. Thus, for relatively small angles the differential reading along the inclined tube can be made large even for small pressure differences. The inclined-tube manometer is often used to measure small differences in gas pressures so that if pipes A and B contain a gas then

$$p_A - p_B = \gamma_2 \ell_2 \sin\theta$$

or

$$\ell_2 = \frac{p_A - p_B}{\gamma_2 \sin\theta} \qquad \textbf{(2.14)}$$

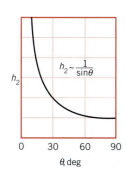

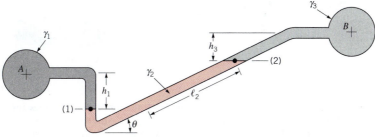

■ **FIGURE 2.9**   Inclined-tube manometer.

where the contributions of the gas columns h_1 and h_3 have been neglected. As shown by Eq. 2.14 and the figure in the margin of the previous page, the differential reading ℓ_2 (for a given pressure difference) of the inclined-tube manometer can be increased over that obtained with a conventional U-tube manometer by the factor $1/\sin\theta$. Recall that $\sin\theta \to 0$ as $\theta \to 0$.

2.7 Mechanical and Electronic Pressure Measuring Devices

V2.2 Bourdon gage

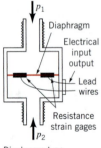

p_1

Diaphragm

Electrical input output

Lead wires

Resistance strain gages

p_2

Diaphragm-type electrical pressure tranducer

Although manometers are widely used, they are not well suited for measuring very high pressures, or pressures that are changing rapidly with time. In addition, they require the measurement of one or more column heights, which, although not particularly difficult, can be time consuming. To overcome some of these problems, numerous other types of pressure measuring instruments have been developed. Most of these make use of the idea that when a pressure acts on an elastic structure, the structure will deform, and this deformation can be related to the magnitude of the pressure. Probably the most familiar device of this kind is the ***Bourdon pressure gage,*** which is shown in Fig. 2.10a. The essential mechanical element in this gage is the hollow, elastic curved tube (Bourdon tube), which is connected to the pressure source as shown in Fig. 2.10b. As the pressure within the tube increases, the tube tends to straighten, and although the deformation is small, it can be translated into the motion of a pointer on a dial as illustrated. Since it is the difference in pressure between the outside of the tube (atmospheric pressure) and the inside of the tube that causes the movement of the tube, the indicated pressure is gage pressure. The Bourdon gage must be calibrated so that the dial reading can directly indicate the pressure in suitable units such as psi, psf, or pascals. A zero reading on the gage indicates that the measured pressure is equal to the local atmospheric pressure. This type of gage can be used to measure a negative gage pressure (vacuum) as well as positive pressures.

For many applications in which pressure measurements are required, the pressure must be measured with a device that converts the pressure into an electrical output. For example, it may be desirable to continuously monitor a pressure that is changing with time. This type of pressure measuring device is called a *pressure transducer,* and many different designs are used. A diaphragm-type electrical pressure transducer is shown in the figure in the margin.

(a) (b)

■ **FIGURE 2.10** *(a)* **Liquid-filled Bourdon pressure gages for various pressure ranges.** *(b)* **Internal elements of Bourdon gages. The "C-shaped" Bourdon tube is shown on the left, and the "coiled spring" Bourdon tube for high pressures of 1000 psi and above is shown on the right. (Photographs courtesy of Weiss Instruments, Inc.)**

2.8   Hydrostatic Force on a Plane Surface

V2.3 Hoover dam

When a surface is submerged in a fluid, forces develop on the surface due to the fluid. The determination of these forces is important in the design of storage tanks, ships, dams, and other hydraulic structures. For fluids at rest we know that the force must be *perpendicular* to the surface since there are no shearing stresses present. We also know that the pressure will vary linearly with depth if the fluid is incompressible. For a horizontal surface, such as the bottom of a liquid-filled tank (Fig. 2.11), the magnitude of the resultant force is simply $F_R = pA$, where p is the uniform pressure on the bottom and A is the area of the bottom. For the open tank shown, $p = \gamma h$. Note that if atmospheric pressure acts on both sides of the bottom, as is illustrated, the *resultant* force on the bottom is simply due to the liquid in the tank. Because the pressure is constant and uniformly distributed over the bottom, the resultant force acts through the centroid of the area as shown in Fig. 2.11.

For the more general case in which a submerged plane surface is inclined, as is illustrated in Fig. 2.12, the determination of the resultant force acting on the surface is more involved. For the present we will assume that the fluid surface is open to the atmosphere. Let the plane in which the surface lies intersect the free surface at 0 and make an angle θ with this surface as in Fig. 2.12. The x–y coordinate system is defined so that 0 is the origin and $y = 0$ (i.e., the x-axis) is directed along the surface as shown. The area can have an arbitrary shape as shown. We wish to determine the direction, location, and magnitude of the resultant force acting on one side of this area due to the liquid in contact with the area. At any given depth, h, the force acting on dA (the differential area of Fig. 2.12) is $dF = \gamma h\, dA$ and is perpendicular to the surface. Thus, the magnitude of the resultant force can be found by summing these differential forces over the entire surface. In equation form

$$F_R = \int_A \gamma h\, dA = \int_A \gamma y \sin \theta\, dA$$

where $h = y \sin \theta$. For constant γ and θ

$$F_R = \gamma \sin \theta \int_A y\, dA \qquad (2.15)$$

Because the integral appearing in Eq. 2.15 is the *first moment of the area* with respect to the x axis, we can write

$$\int_A y\, dA = y_c A$$

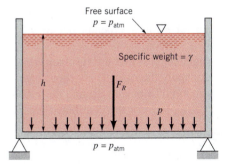

Free surface
$p = p_{atm}$

Specific weight $= \gamma$

h

F_R

p

$p = p_{atm}$

■ **FIGURE 2.11**   **Pressure and resultant hydrostatic force developed on the bottom of an open tank.**

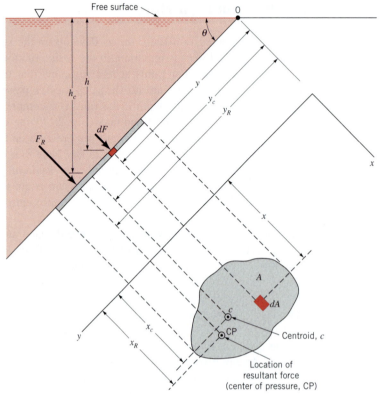

■ **FIGURE 2.12** Notation for hydrostatic force on an inclined plane surface of arbitrary shape.

where y_c is the y coordinate of the centroid of the area A measured from the x axis, which passes through 0. Equation 2.15 can thus be written as

$$F_R = \gamma A y_c \sin \theta$$

or more simply as

$$\boxed{F_R = \gamma h_c A}$$

(2.16)

where h_c is the vertical distance from the fluid surface to the centroid of the area. Note that the magnitude of the force is independent of the angle θ and depends only on the specific weight of the fluid, the total area, and the depth of the centroid of the area below the surface. In effect, Eq. 2.16 indicates that the magnitude of the resultant force is equal to the pressure at the centroid of the area multiplied by the total area. Since all the differential forces that were summed to obtain F_R are perpendicular to the surface, the resultant F_R must also be perpendicular to the surface.

Although our intuition might suggest that the resultant force should pass through the centroid of the area, this is not actually the case. The y coordinate, y_R, of the resultant force can be determined by the summation of moments around the x axis. That is, the moment of the resultant force must equal the moment of the distributed pressure force, or

$$F_R y_R = \int_A y \, dF = \int_A \gamma \sin \theta \, y^2 \, dA$$

and, therefore, since $F_R = \gamma A y_c \sin\theta$

$$y_R = \frac{\int_A y^2 \, dA}{y_c A}$$

The integral in the numerator is the *second moment of the area (moment of inertia), I_x,* with respect to an axis formed by the intersection of the plane containing the surface and the free surface (*x* axis). Thus, we can write

$$y_R = \frac{I_x}{y_c A}$$

Use can now be made of the parallel axis theorem to express I_x as

$$I_x = I_{xc} + A y_c^2$$

where I_{xc} is the second moment of the area with respect to an axis passing through its *centroid* and parallel to the *x* axis. Thus,

$$y_R = \frac{I_{xc}}{y_c A} + y_c \qquad (2.17)$$

Equation 2.17 clearly shows that the resultant force does not pass through the centroid but for non-horizontal surfaces is always *below* it, since $I_{xc}/y_c A > 0$.

The *x* coordinate, x_R, for the resultant force can be determined in a similar manner by summing moments about the *y* axis. It follows that

$$x_R = \frac{I_{xyc}}{y_c A} + x_c \qquad (2.18)$$

where I_{xyc} is the product of inertia with respect to an orthogonal coordinate system passing through the *centroid* of the area and formed by a translation of the *x–y* coordinate system. If the submerged area is symmetrical with respect to an axis passing through the centroid and parallel to either the *x* or *y* axes, the resultant force must lie along the line $x = x_c$, since I_{xyc} is identically zero in this case. The point through which the resultant force acts is called the *center of pressure.* It is to be noted from Eqs. 2.17 and 2.18 that as y_c increases the center of pressure moves closer to the centroid of the area. Because $y_c = h_c/\sin\theta$, the distance y_c will increase if the depth of submergence, h_c, increases, or, for a given depth, the area is rotated so that the angle, θ, decreases. Centroidal coordinates and moments of inertia for some common areas are given in Fig. 2.13.

F l u i d s   i n   t h e   N e w s

The Three Gorges Dam The Three Gorges Dam being constructed on China's Yangtze River will contain the world's largest hydroelectric power plant when in full operation. The dam is of the concrete gravity type having a length of 2309 meters with a height of 185 meters. The main elements of the project include the dam, two power plants, and navigation facilities consisting of a ship lock and lift. The power plants will contain 26 Francis type turbines, each with a capacity of 700 megawatts. The spillway section, which is the center section of the dam, is 483 meters long with 23 bottom outlets and 22 surface sluice gates. The maximum discharge capacity is 102,500 cubic meters per second. After more than 10 years of construction the dam gates were finally closed, and on June 10, 2003, the reservoir had been filled to its interim level of 135 meters. Due to the large depth of water at the dam and the huge extent of the storage pool, *hydrostatic pressure forces* have been a major factor considered by engineers. When filled to its normal pool level of 175 meters, the total reservoir storage capacity is 39.3 billion cubic meters. The project is scheduled for completion in 2009. (See Problem 2.28.) ∎

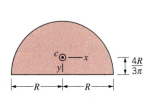

$$A = ba$$
$$I_{xc} = \frac{1}{12} ba^3$$
$$I_{yc} = \frac{1}{12} ab^3$$
$$I_{xyc} = 0$$

(a) Rectangle

$$A = \pi R^2$$
$$I_{xc} = I_{yc} = \frac{\pi R^4}{4}$$
$$I_{xyc} = 0$$

(b) Circle

$$A = \frac{\pi R^2}{2}$$
$$I_{xc} = 0.1098R^4$$
$$I_{yc} = 0.3927R^4$$
$$I_{xyc} = 0$$

(c) Semicircle

$$A = \frac{ab}{2} \qquad I_{xc} = \frac{ba^3}{36}$$
$$I_{xyc} = \frac{ba^2}{72}(b - 2d)$$

(d) Triangle

$$A = \frac{\pi R^2}{4}$$
$$I_{xc} = I_{yc} = 0.05488R^4$$
$$I_{xyc} = -0.01647R^4$$

(e) Quarter-circle

■ **FIGURE 2.13**   **Geometric properties of some common shapes.**

EXAMPLE 2.4 | Hydrostatic Force on a Plane Circular Surface

GIVEN The 4-m-diameter circular gate of Fig. E2.4a is located in the inclined wall of a large reservoir containing water ($\gamma = 9.80$ kN/m^3). The gate is mounted on a shaft along its horizontal diameter, and the water depth is 10 m above the shaft.

FIND Determine

(a) the magnitude and location of the resultant force exerted on the gate by the water and

(b) the moment that would have to be applied to the shaft to open the gate.

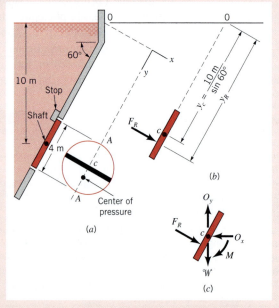

■ **FIGURE E2.4**

SOLUTION

(a) To find the magnitude of the force of the water we can apply Eq. 2.16,

$$F_R = \gamma h_c A$$

and because the vertical distance from the fluid surface to the centroid of the area is 10 m it follows that

$$F_R = (9.80 \times 10^3 \text{ N/m}^3)(10 \text{ m})(4\pi \text{ m}^2)$$
$$= 1230 \times 10^3 \text{ N} = 1.23 \text{ MN} \quad \textbf{(Ans)}$$

To locate the point (center of pressure) through which F_R acts, we use Eqs. 2.17 and 2.18,

$$x_R = \frac{I_{xyc}}{y_c A} + x_c \qquad y_R = \frac{I_{xc}}{y_c A} + y_c$$

For the coordinate system shown, $x_R = 0$ since the area is symmetrical, and the center of pressure must lie along the diameter A-A. To obtain y_R, we have from Fig. 2.13

$$I_{xc} = \frac{\pi R^4}{4}$$

and y_c is shown in Fig. E2.4b. Thus,

$$y_R = \frac{(\pi/4)(2 \text{ m})^4}{(10 \text{ m/sin } 60°)(4\pi \text{ m}^2)} + \frac{10 \text{ m}}{\sin 60°}$$
$$= 0.0866 \text{ m} + 11.55 \text{ m} = 11.6 \text{ m}$$

and the distance (along the gate) below the shaft to the center of pressure is

$$y_R - y_c = 0.0866 \text{ m} \quad \textbf{(Ans)}$$

COMMENT We can conclude from this analysis that the force on the gate due to the water has a magnitude of 1.23 MN and acts through a point along its diameter A–A at a distance of 0.0866 m (along the gate) below the shaft. The force is perpendicular to the gate surface as shown.

By repeating the calculations for various values of the depth to the centroid, h_c, the results shown in Fig. E2.4d are obtained. Note that as the depth increases the distance between the center of pressure and the centroid decreases.

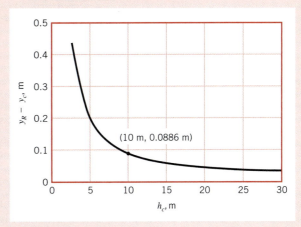

■ **FIGURE E2.4d**

(b) The moment required to open the gate can be obtained with the aid of the free-body diagram of Fig. E2.4c. In this diagram, $\mathcal{W}$ is the weight of the gate and O_x and O_y are the horizontal and vertical reactions of the shaft on the gate. We can now sum moments about the shaft

$$\Sigma M_c = 0$$

and, therefore,

$$M = F_R(y_R - y_c)$$
$$= (1230 \times 10^3 \text{ N})(0.0866 \text{ m})$$
$$= 1.07 \times 10^5 \text{ N·m} \quad \textbf{(Ans)}$$

2.9 Pressure Prism

An informative and useful graphical interpretation can be made for the force developed by a fluid acting on a plane rectangular area. Consider the pressure distribution along a vertical wall of a tank of constant width b, which contains a liquid having a specific weight γ. Since the pressure must vary linearly with depth, we can represent the variation as is shown in Fig. 2.14a, where the pressure is equal to zero at the upper surface and equal to γh at the bottom. It is apparent from this diagram that the average pressure occurs at the depth $h/2$, and therefore the resultant force acting on the rectangular area $A = bh$ is

$$F_R = p_{av}A = \gamma \left(\frac{h}{2}\right) A$$

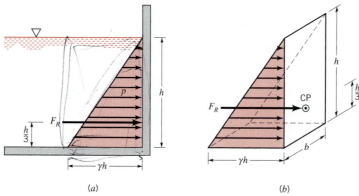

■ **FIGURE 2.14** Pressure prism for vertical rectangular area.

which is the same result as obtained from Eq. 2.16. The pressure distribution shown in Fig. 2.14a applies across the vertical surface so we can draw the three-dimensional representation of the pressure distribution as shown in Fig. 2.14b. The base of this "volume" in pressure–area space is the plane surface of interest, and its altitude at each point is the pressure. This volume is called the *pressure prism,* and it is clear that the magnitude of the resultant force acting on the rectangular surface is equal to the volume of the pressure prism. Thus, for the prism of Fig. 2.14b the fluid force is

$$F_R = \text{volume} = \frac{1}{2}(\gamma h)(bh) = \gamma\left(\frac{h}{2}\right)A$$

where bh is the area of the rectangular surface, A.

The resultant force must pass through the *centroid* of the pressure prism. For the volume under consideration the centroid is located along the vertical axis of symmetry of the surface, and at a distance of $h/3$ above the base (since the centroid of a triangle is located at $h/3$ above its base). This result can readily be shown to be consistent with that obtained from Eqs. 2.17 and 2.18.

If the surface pressure of the liquid is different from atmospheric pressure (such as might occur in a closed tank), the resultant force acting on a submerged area, A, will be changed in magnitude from that caused simply by hydrostatic pressure by an amount $p_s A$, where p_s is the gage pressure at the liquid surface (the outside surface is assumed to be exposed to atmospheric pressure).

EXAMPLE 2.5   Use of the Pressure Prism Concept

GIVEN A pressurized tank contains oil ($SG = 0.90$) and has a square, 0.6-m by 0.6-m plate bolted to its side, as is illustrated in Fig. E2.5a. The pressure gage on the top of the tank reads 50 kPa, and the outside of the tank is at atmospheric pressure.

FIND What is the magnitude and location of the resultant force on the attached plate?

SOLUTION

The pressure distribution acting on the inside surface of the plate is shown in Fig. E2.5b. The pressure at a given point on the plate is due to the air pressure, p_s, at the oil surface, and the pressure due to the oil, which varies linearly with depth as is shown in the figure. The resultant force on the plate (having an area A) is due to the components, F_1 and F_2, where F_1 and F_2 are due to the rectangular and triangular portions of the pressure distribution, respectively. Thus

$$F_1 = (p_s + \gamma h_1)A$$
$$= [50 \times 10^3 \ \text{N/m}^2$$
$$+ (0.90)(9.81 \times 10^3 \ \text{N/m}^3)(2 \ \text{m})](0.36 \ \text{m}^2)$$
$$= 24.4 \times 10^3 \ \text{N}$$

and

$$F_2 = \gamma \left(\frac{h_2 - h_1}{2} \right) A$$

$$= (0.90)(9.81 \times 10^3 \ \text{N/m}^3) \left(\frac{0.6 \ \text{m}}{2} \right) (0.36 \ \text{m}^2)$$

$$= 0.954 \times 10^3 \ \text{N}$$

The magnitude of the resultant force, F_R, is therefore

$$F_R = F_1 + F_2 = 25.4 \times 10^3 \ \text{N} = 25.4 \ \text{kN} \quad \textbf{(Ans)}$$

The vertical location of F_R can be obtained by summing moments around an axis through point O so that

$$F_R y_O = F_1(0.3 \ \text{m}) + F_2(0.2 \ \text{m})$$

or

$$y_O = \frac{(24.4 \times 10^3 \text{N})(0.3 \ \text{m}) + (0.954 \times 10^3 \text{N})(0.2 \ \text{m})}{25.4 \times 10^3 \text{N}}$$

$$= 0.296 \ \text{m} \quad \textbf{(Ans)}$$

Thus, the force acts at a distance of 0.296 m above the bottom of the plate along the vertical axis of symmetry.

COMMENT Note that the air pressure used in the calculation of the force was gage pressure. Atmospheric pressure does not affect the resultant force (magnitude or location), as it acts on both sides of the plate, thereby canceling its effect.

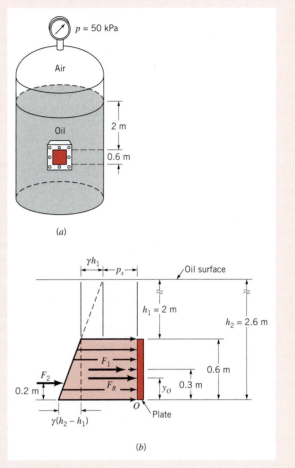

(a)

(b)

■ **FIGURE E2.5**

2.10 Hydrostatic Force on a Curved Surface

The equations developed in Section 2.8 for the magnitude and location of the resultant force acting on a submerged surface only apply to plane surfaces. However, many surfaces of interest (such as those associated with dams, pipes, and tanks) are nonplanar. Although the resultant fluid force can be determined by integration, as was done for the plane surfaces, this is generally a rather tedious process and no simple, general formulas can be developed. As an alternative approach we will consider the equilibrium of the fluid volume enclosed by the curved surface of interest and the horizontal and vertical projections of this surface.

For example, consider the curved section BC of the open tank of Fig. 2.15a. We wish to find the resultant fluid force acting on this section, which has a unit length perpendicular to the plane of the paper. We first isolate a volume of fluid that is bounded by the surface of interest, in this instance section BC, and the horizontal plane surface AB and the vertical plane surface AC. The free-body diagram for this volume is shown in Fig. 2.15b.

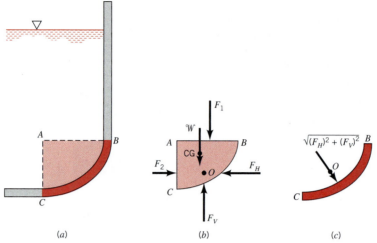

■ **FIGURE 2.15** Hydrostatic force on a curved surface.

The magnitude and location of forces F_1 and F_2 can be determined from the relationships for planar surfaces. The weight, $\mathcal{W}$, is simply the specific weight of the fluid times the enclosed volume and acts through the center of gravity (CG) of the mass of fluid contained within the volume. Forces F_H and F_V represent the components of the force that the tank *exerts on the fluid.*

V2.4 Pop bottle

In order for this force system to be in equilibrium, the horizontal component F_H must be equal in magnitude and collinear with F_2, and the vertical component F_V equal in magnitude and collinear with the resultant of the vertical forces F_1 and $\mathcal{W}$. This follows since the three forces acting on the fluid mass (F_2, the resultant of F_1 and $\mathcal{W}$, and the resultant force that the tank exerts on the mass) must form a *concurrent* force system. That is, from the principles of statics, it is known that when a body is held in equilibrium by three non-parallel forces they must be concurrent (their lines of action intersect at a common point), and coplanar. Thus,

$$F_H = F_2$$

$$F_V = F_1 + \mathcal{W}$$

and the magnitude of the resultant is obtained from the equation

$$F_R = \sqrt{(F_H)^2 + (F_V)^2}$$

The resultant F_R passes through the point O, which can be located by summing moments about an appropriate axis. The resultant force of the fluid acting *on the curved surface BC* is equal and opposite in direction to that obtained from the free-body diagram of Fig. 2.15b. The desired fluid force is shown in Fig. 2.15c.

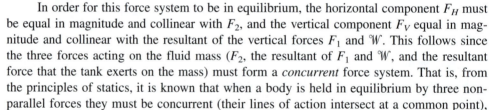

EXAMPLE 2.6 Hydrostatic Pressure Force on a Curved Surface

GIVEN The 6-ft-diameter drainage conduit of Fig. E2.6a is half full of water at rest.

FIND Determine the magnitude and line of action of the resultant force that the water exerts on the curved portion *BC* for a section of the conduit that is 1-ft long.

SOLUTION

We first isolate a volume of fluid bounded by the curved section *BC*, the horizontal surface *AB*, and the vertical surface *AC*, as shown in Fig. E2.6b. The volume has a length of 1 ft. Forces acting on the volume are the horizontal force, F_1, which acts on the vertical surface *AC*, the weight, $\mathcal{W}$, of the fluid contained within the volume, and the horizontal and vertical components of the force of the conduit wall on the fluid, F_H and F_V, respectively.

The magnitude of F_1 is found from the equation

$$F_1 = \gamma h_c A = (62.4 \text{ lb/ft}^3)(\tfrac{3}{2} \text{ ft})(3 \text{ ft}^2) = 281 \text{ lb}$$

and this force acts 1 ft above *C* as shown. The weight, $\mathcal{W}$, is

$$\mathcal{W} = \gamma \text{ vol} = (62.4 \text{ lb/ft}^3)(9\pi/4 \text{ ft}^2)(1 \text{ ft}) = 441 \text{ lb}$$

and acts through the center of gravity of the mass of fluid, which according to Fig. 2.13 is located 1.27 ft to the right of *AC* as shown. Therefore, to satisfy equilibrium

$$F_H = F_1 = 281 \text{ lb}$$
$$F_V = \mathcal{W} = 441 \text{ lb}$$

and the magnitude of the resultant force is

$$F_R = \sqrt{(F_H)^2 + (F_V)^2}$$
$$= \sqrt{(281 \text{ lb})^2 + (441 \text{ lb})^2} = 523 \text{ lb} \quad \textbf{(Ans)}$$

The force the water exerts *on* the conduit wall is equal, but *opposite in direction,* to the forces F_H and F_V shown in Fig. E2.6b. Thus, the resultant force *on the conduit wall* is shown in Fig. E2.6c. This force acts through the point *O* at the angle shown.

COMMENT   An inspection of this result will show that the line of action of the resultant force passes through the center of the conduit. In retrospect, this is not a surprising result, as at

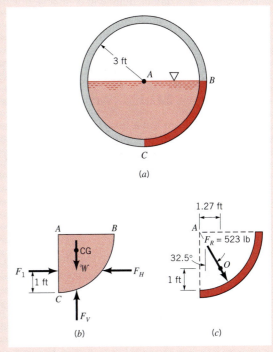

■ FIGURE E2.6

each point on the curved surface of the conduit, the elemental force due to the pressure is normal to the surface, and each line of action must pass through the center of the conduit. It therefore follows that the resultant of this concurrent force system must also pass through the center of concurrence of the elemental forces that make up the system.

The same general approach can also be used for determining the force on curved surfaces of pressurized, closed tanks. If these tanks contain a gas, the weight of the gas is usually negligible in comparison with the forces developed by the pressure. Thus, the forces (such as F_1 and F_2 in Fig. 2.15b) on horizontal and vertical projections of the curved surface of interest can simply be expressed as the internal pressure times the appropriate projected area.

2.11   Buoyancy, Flotation, and Stability

2.11.1   Archimedes' Principle

V2.5 Cartesian
Diver

When a body is completely submerged in a fluid, or floating so that it is only partially submerged, the resultant fluid force acting on the body is called the *buoyant force.* A net upward vertical force results because pressure increases with depth and the pressure forces acting from below are larger than the pressure forces acting from above.

It is well known from elementary physics that the buoyant force, F_B, is given by the equation

$$F_B = \gamma \forall$$

(2.19)

where γ is the specific weight of the fluid and $\forall$ is the volume of the body. Thus, the buoyant force has a magnitude equal to the weight of the fluid displaced by the body and is directed vertically upward. This result is commonly referred to as *Archimedes' principle.* It is easily derived by using the principles discussed in Section 2.10. The *buoyant force passes through the centroid of the displaced volume,* and the point through which the buoyant force acts is called the *center of buoyancy.*

V2.6 Hydrometer

These same results apply to floating bodies that are only partially submerged, if the specific weight of the fluid above the liquid surface is very small compared with the liquid in which the body floats. Because the fluid above the surface is usually air, for practical purposes this condition is satisfied.

F l u i d s   i n   t h e   N e w s

Concrete canoes A solid block of concrete thrown into a pond or lake will obviously sink. But, if the concrete is formed into the shape of a canoe it can be made to float. Of course the reason the canoe floats is the development of the *buoyant force* due to the displaced volume of water. With the proper design, this vertical force can be made to balance the weight of the canoe plus passengers—the canoe floats. Each year since 1988 there is a National Concrete Canoe Competition for university teams. It's jointly sponsored by the American Society of Civil Engineers and Master Builders Inc. The canoes must be 90% concrete and are typically designed with the aid of a computer by civil engineering students. Final scoring depends on four components: racing, a design paper, a business presentation, and a canoe that passes the floatation test. For the 2005 competition, the University of Wisconsin's team won its third consecutive national championship with a 175-lb, 21.5-ft canoe. (See Problem 2.52.) ■

EXAMPLE 2.7   Buoyant Force on a Submerged Object

GIVEN   A spherical buoy has a diameter of 1.5 m, weighs 8.50 kN, and is anchored to the sea floor with a cable as is shown in Fig. E2.7a. Although the buoy normally floats on the surface, at certain times the water depth increases so that the buoy is completely immersed as illustrated.

FIND   For this condition what is the tension of the cable?

SOLUTION

We first draw a free-body diagram of the buoy as shown in Fig. E2.7b, where F_B is the buoyant force acting on the buoy, $\mathcal{W}$ is the weight of the buoy, and T is the tension in the cable. For equilibrium it follows that

$$T = F_B - \mathcal{W}$$

From Eq. 2.19

$$F_B = \gamma \mathcal{V}$$

and for seawater with $\gamma = 10.1 \text{ kN/m}^3$ and $\mathcal{V} = \pi d^3/6$, then

$$F_B = (10.1 \times 10^3 \text{ N/m}^3)[(\pi/6)(1.5 \text{ m})^3]$$
$$= 1.785 \times 10^4 \text{ N}$$

The tension in the cable can now be calculated as

$$T = 1.785 \times 10^4 \text{ N} - 0.850 \times 10^4 \text{ N} = 9.35 \text{ kN} \quad \textbf{(Ans)}$$

COMMENT Note that we replaced the effect of the hydrostatic pressure force on the body by the buoyant force, F_B. Another correct free-body diagram of the buoy is shown in Fig. E2.7c. The net effect of the pressure forces on the surface

of the buoy is equal to the upward force of magnitude F_B (the buoyant force). Do not include both the buoyant force and the hydrostatic pressure effects in your calculations—use one or the other.

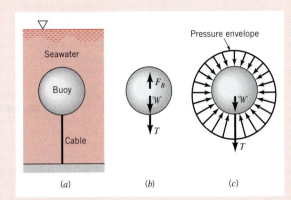

■ **FIGURE E2.7**

F l u i d s   i n   t h e   N e w s

Protecting Venice with flood gates The numerous canals of Venice are a tourist attraction and a residents' concern. Flooding due to high water is increasing in frequency. Early in the twentieth century St. Mark's Square, a low point in Venice, flooded less than 10 times a year. By the 1980s flooding occurred 40 times a year, and currently 60 times a year is the norm. The relative shift of ground to sea level and lagoon structural changes are considered contributors to this growing prob-

lem. One proposed solution involves 80 large hollow gates submerged at the three inlets to the Venice lagoon. When high tides threaten the city, compressed air will displace water out of the gates, producing a *buoyant force* that will cause them to float up into place and prevent water from entering the lagoon. The large cost of installation and annual maintenance of this specific solution is subject to considerable debate. Some think that other, less costly solutions should be made available. ■

2.11.2   Stability

Another interesting and important problem associated with submerged or floating bodies is concerned with the stability of the bodies. As indicated by the figure in the margin, a body is said to be in a *stable equilibrium* position if, when displaced, it returns to its equilibrium position. Conversely, it is in an *unstable equilibrium* position if, when displaced (even slightly), it moves to a new equilibrium position. Stability considerations are particularly important for submerged or floating bodies since the centers of buoyancy and gravity do not necessarily coincide. A small rotation can result in either a restoring or overturning couple.

For example, for a *completely* submerged body with a center of gravity below the center of buoyancy, a rotation from its equilibrium position will create a restoring couple formed by the weight, $\mathcal{W}$, and the buoyant force, F_B, which causes the body to rotate

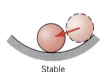

Stable

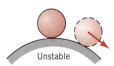

Unstable

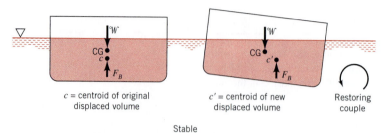

c = centroid of original
displaced volume

c′ = centroid of new
displaced volume

Restoring
couple

Stable

■ **FIGURE 2.16** **Stability of a floating body—stable configuration.**

V2.7 Stability of a
model barge

back to its original position. Thus, for this configuration the body is stable. It is to be noted that as long as the center of gravity falls *below* the center of buoyancy, this will always be true; that is, the body is in a *stable equilibrium* position with respect to small rotations. However, if the center of gravity of a completely submerged object is above the center of buoyancy, the resulting couple formed by the weight and the buoyant force will cause the body to overturn and move to a new equilibrium position. Thus, a completely submerged body with its center of gravity *above* its center of buoyancy is in an *unstable equilibrium* position.

For *floating* bodies the stability problem is more complicated, because as the body rotates the location of the center of buoyancy (which passes through the centroid of the displaced volume) may change. As is shown in Fig. 2.16, a floating body such as a barge that rides low in the water can be stable even though the center of gravity lies above the center of buoyancy. This is true since as the body rotates the buoyant force, F_B, shifts to pass through the centroid of the newly formed displaced volume and, as illustrated, combines with the weight, W, to form a couple, which will cause the body to return to its original equilibrium position. However, for the relatively tall, slender body shown in Fig. 2.17, a small rotational displacement can cause the buoyant force and the weight to form an overturning couple as illustrated.

It is clear from these simple examples that determination of the stability of submerged or floating bodies can be difficult since the analysis depends in a complicated fashion on the particular geometry and weight distribution of the body. The problem can be further complicated by the necessary inclusion of other types of external forces, such as those induced by wind gusts or currents. Stability considerations are obviously of great importance in the design of ships, submarines, bathyscaphes, and so forth, and such considerations play a significant role in the work of naval architects.

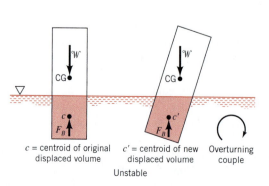

c = centroid of original
displaced volume

c′ = centroid of new
displaced volume

Overturning
couple

Unstable

■ **FIGURE 2.17** **Stability of a floating body—unstable configuration.**

2.12 Pressure Variation in a Fluid with Rigid-Body Motion

Although in this chapter we have been primarily concerned with fluids at rest, the general equation of motion (Eq. 2.2)

$$-\nabla p - \gamma\hat{\mathbf{k}} = \rho\mathbf{a}$$

was developed for both fluids at rest and fluids in motion, with the only stipulation being that there were no shearing stresses present.

A general class of problems involving fluid motion in which there are no shearing stresses occurs when a mass of fluid undergoes rigid-body motion. For example, if a container of fluid accelerates along a straight path, the fluid will move as a rigid mass (after the initial sloshing motion has died out) with each particle having the same acceleration. Since there is no deformation, there will be no shearing stresses and, therefore, Eq. 2.2 applies. Similarly, if a fluid is contained in a tank that rotates about a fixed axis, the fluid will simply rotate with the tank as a rigid body, and again Eq. 2.2 can be applied to obtain the pressure distribution throughout the moving fluid.

2.13 Chapter Summary and Study Guide

Pascal's law
surface force
body force
incompressible fluid
hydrostatic pressure distribution
pressure head
compressible fluid
U.S. standard atmosphere
absolute pressure
gage pressure
vacuum pressure
barometer
manometer
Bourdon pressure gage
center of pressure
buoyant force
Archimedes' principle
center of buoyancy

In this chapter the pressure variation in a fluid at rest is considered, along with some important consequences of this type of pressure variation. It is shown that for incompressible fluids at rest the pressure varies linearly with depth. This type of variation is commonly referred to as hydrostatic pressure distribution. For compressible fluids at rest the pressure distribution will not generally be hydrostatic, but Eq. 2.4 remains valid and can be used to determine the pressure distribution if additional information about the variation of the specific weight is specified. The distinction between absolute and gage pressure is discussed along with a consideration of barometers for the measurement of atmospheric pressure.

Pressure measuring devices called manometers, which utilize static liquid columns, are analyzed in detail. A brief discussion of mechanical and electronic pressure gages is also included. Equations for determining the magnitude and location of the resultant fluid force acting on a plane surface in contact with a static fluid are developed. A general approach for determining the magnitude and location of the resultant fluid force acting on a curved surface in contact with a static fluid is described. For submerged or floating bodies the concept of the buoyant force and the use of Archimedes' principle are reviewed.

The following checklist provides a study guide for this chapter. When your study of the entire chapter and end-of-chapter exercises has been completed, you should be able to

- write out meanings of the terms listed here in the margin and understand each of the related concepts. These terms are particularly important and are set in color and bold type in the text.
- calculate the pressure at various locations within an incompressible fluid at rest.
- calculate the pressure at various locations within a compressible fluid at rest using Eq. 2.4 if the variation in the specific weight is specified.
- use the concept of a hydrostatic pressure distribution to determine pressures from measurements using various types of manometers.

■ determine the magnitude, direction, and location of the resultant hydrostatic force acting on a plane surface.

■ determine the magnitude, direction, and location of the resultant hydrostatic force acting on a curved surface.

■ use Archimedes' principle to calculate the resultant hydrostatic force acting on floating or submerged bodies.

References

1. *The U.S. Standard Atmosphere, 1962.* U.S. Government Printing Office, Washington, D.C., 1962.

2. *The U.S. Standard Atmosphere, 1976.* U.S. Government Printing Office, Washington, D.C., 1976.

3. Hasler, A. F., Pierce, H., Morris, K. R., and Dodge, J., "Meteorological Data Fields 'In Perspective'." *Bulletin of the American Meteorological Society,* Vol. 66, No. 7, July 1985.

Review Problems

Go to Appendix F for a set of review problems with answers. Detailed solutions can be found in *Student Solution Manual for a Brief Introduction to Fluid Mechanics*, by Young et al. (© 2006 John Wiley and Sons, Inc.).

Problems

Note: Unless otherwise indicated use the values of fluid properties found in the tables on the inside of the front cover. Problems designated with an (*) are intended to be solved with the aid of a programmable calculator or a computer. Problems designated with a (†) are "open-ended" problems and require critical thinking in that to work them one must make various assumptions and provide the necessary data. There is not a unique answer to these problems.

Answers to the even-numbered problems are listed at the end of the book. Access to the videos that accompany problems can be obtained through the book's web site, www.wiley.com/college/young. The lab-type problems and FE problems can also be accessed on this web site.

Section 2.3 Pressure Variation in a Fluid at Rest

2.1 What pressure, expressed in pascals, will a skin diver be subjected to at a depth of 50 m in seawater?

2.2 In a certain chemical plant, a closed tank contains ethyl alcohol to a depth of 50 ft. Air at a pressure of 25 psi fills the gap at the top of the tank. Determine the pressure at a closed valve attached to the tank 10 ft above its bottom.

2.3 An unknown immiscible liquid seeps into the bottom of an open oil tank. Some measurements indicate that the depth of the unknown liquid is 1.5 m and the depth of the oil (specific weight = 8.5 kN/m³) floating on top is 5.0 m. A pressure gage connected to the bottom of the tank reads 65 kPa. What is the specific gravity of the unknown liquid?

2.4 The closed tank of Fig. P2.4 is filled with water and is 5 ft long. The pressure gage on the tank reads 7 psi. Determine **(a)** the height, h, in the open water column, **(b)** the gage pressure acting on the bottom tank surface AB, and **(c)** the absolute pressure of the air in the top of the tank if the local atmospheric pressure is 14.7 psia.

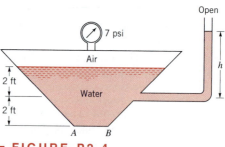

■ **FIGURE P2.4**

2.5 Blood pressure is commonly measured with a cuff placed around the arm, with the cuff pressure (which is a measure of the arterial blood pressure) indicated with a mercury manometer (see **Video 2.1**). A typical value for the maximum value of blood pressure (systolic pressure) is 120 mm Hg. Why wouldn't it be simpler and less expensive to use water in the manometer rather than mercury? Explain and support your answer with the necessary calculations.

*2.6 In a certain liquid at rest, measurements of the specific weight at various depths show the following variation:

h (ft)	γ (lb/ft^3)
0	70
10	76
20	84
30	91
40	97
50	102
60	107
70	110
80	112
90	114
100	115

The depth $h = 0$ corresponds to a free surface at atmospheric pressure. Determine, through numerical integration of Eq. 2.4, the corresponding variation in pressure and show the results on a plot of pressure (in psf) versus depth (in feet).

*2.7 Under normal conditions the temperature of the atmosphere decreases with increasing elevation. In some situations, however, a temperature inversion may exist so that the air temperature increases with elevation. A series of temperature probes on a mountain give the elevation–temperature data shown in Table P2.7. If the barometric pressure at the base of the mountain is 12.1 psia, determine (by means of numerical integration of Eq 2.4) the pressure at the top of the mountain.

Elevation (ft)	Temperature (°F)
5000	50.1 (base)
5500	55.2
6000	60.3
6400	62.6
7100	67.0
7400	68.4
8200	70.0
8600	69.5
9200	68.0
9900	67.1 (top)

TABLE P2.7

2.8 (See Fluids in the News article titled "Giraffe's blood pressure," Section 2.3.1.) **(a)** Determine the change in hydrostatic pressure in a giraffe's head as it lowers its head from eating leaves 6 m above the ground to getting a drink of water at ground level as shown in Fig. P2.8. Assume the specific gravity of blood is $SG = 1$. **(b)** Compare the pressure change calculated in part **(a)** to the normal 120 mm of mercury pressure in a human's heart.

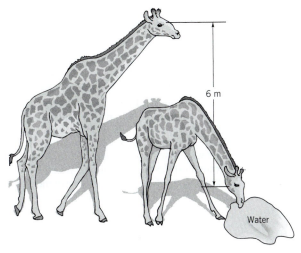

6 m

Water

■ **FIGURE P2.8**

Section 2.5 Measurement of Pressure

2.9 For an atmospheric pressure of 101 kPa (abs) determine the heights of the fluid columns in barometers containing one of the following liquids: **(a)** mercury, **(b)** water, and **(c)** ethyl alcohol. Calculate the heights, including the effect of vapor pressure, and compare the results with those obtained neglecting vapor pressure. Do these results support the widespread use of mercury for barometers? Why?

2.10 (See Fluids in the News article titled "Weather, barometers, and bars," Section 2.5.) The record low sea-level barometric pressure ever recorded is 25.8 in. of mercury. At what altitude in the standard atmosphere is the pressure equal to this value?

2.11 An absolute pressure of 7 psia corresponds to what gage pressure for standard atmospheric pressure of 14.7 psia?

2.12 Bourdon gages (see **Video V2.2** and Fig. P2.12) are commonly used to measure pressure. When such a gage is attached

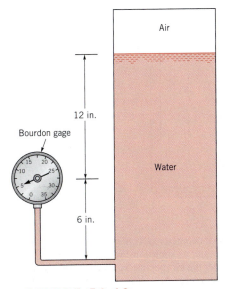

Air

12 in.

Bourdon gage

Water

6 in.

■ **FIGURE P2.12**

to the closed water tank of Fig. P2.12 the gage reads 5 psi. What is the absolute air pressure in the tank? Assume standard atmospheric pressure of 14.7 psi.

2.13   On the suction side of a pump a Bourdon pressure gage reads 40-kPa vacuum. What is the corresponding absolute pressure if the local atmospheric pressure is 100 kPa (abs)?

2.14   A closed cylindrical tank filled with water has a hemispherical dome and is connected to an inverted piping system as shown in Fig. P2.14. The liquid in the top part of the piping system has a specific gravity of 0.8, and the remaining parts of the system are filled with water. If the pressure gage reading at *A* is 60 kPa, determine **(a)** the pressure in pipe *B* and **(b)** the pressure head, in millimeters of mercury, at the top of the dome (point *C*).

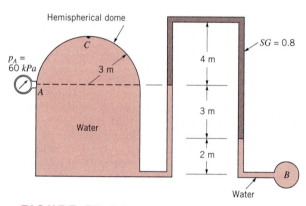

■ **FIGURE P2.14**

2.15   A U-tube manometer contains oil, mercury, and water as shown in Fig. P2.15. For the column heights indicated, what is the pressure differential between pipes *A* and *B*?

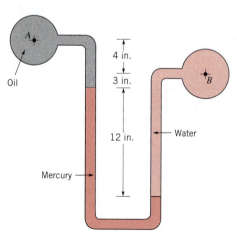

■ **FIGURE P2.15**

2.16   A U-tube manometer is connected to a closed tank containing air and water as shown in Fig. P2.16. At the closed end of the manometer the air pressure is 16 psia. Determine the reading on the pressure gage for a differential reading of 4 ft on the manometer. Express your answer in psi (gage). Assume

standard atmospheric pressure and neglect the weight of the air columns in the manometer.

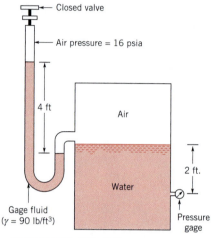

■ **FIGURE P2.16**

2.17   For the configuration shown in Fig. P2.17 what must be the value of the specific weight of the unknown fluid? Express your answer in lb/ft^3.

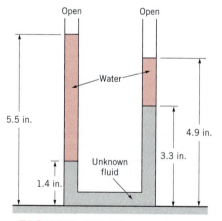

■ **FIGURE P2.17**

2.18   For the inclined-tube manometer of Fig. P2.18 the pressure in pipe *A* is 0.8 psi. The fluid in both pipes *A* and *B* is water, and the gage fluid in the manometer has a specific gravity of 2.6. What is the pressure in pipe *B* corresponding to the differential reading shown?

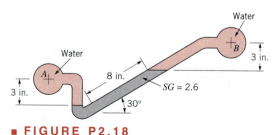

■ **FIGURE P2.18**

2.19 The differential mercury manometer of Fig. P2.19 is connected to pipe A containing gasoline (SG = 0.65) and to pipe B containing water. Determine the differential reading, h, corresponding to a pressure in A of 20 kPa and a vacuum of 150 mm Hg in B.

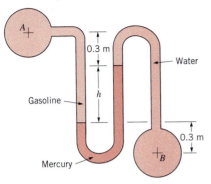

■ **FIGURE P2.19**

2.20 An inverted U-tube manometer containing oil (SG = 0.8) is located between two reservoirs as shown in Fig. P2.20. The reservoir on the left, which contains carbon tetrachloride, is closed and pressurized to 9 psi. the reservoir on the right contains water and is open to the atmosphere. With the given data, determine the depth of water, h, in the right reservoir.

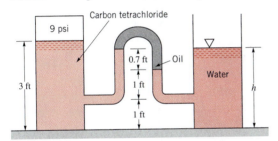

■ **FIGURE P2.20**

2.21 An inverted open tank is held in place by a force R as shown in Fig. P2.21. If the specific gravity of the manometer fluid is 2.5, determine the value of h.

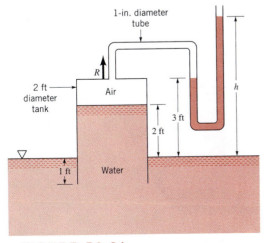

■ **FIGURE P2.21**

2.22 A suction cup is used to support a plate of weight $\mathcal{W}$ as shown in Fig. P2.22. For the conditions shown, determine $\mathcal{W}$.

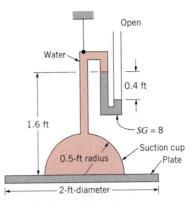

■ **FIGURE P2.22**

2.23 The cyclindrical tank with hemispherical ends shown in Fig. P2.23 contains a volatile liquid and its vapor. The liquid density is 800 kg/m³, and its vapor density is negligible. The pressure in the vapor is 120 kPa (abs), and the atmospheric pressure is 101 kPa (abs). Determine (a) the gage pressure reading on the pressure gage and (b) the height, h, of the mercury manometer.

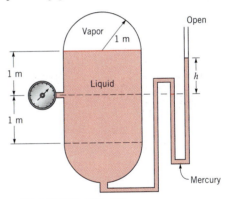

■ **FIGURE P2.23**

2.24 In Fig. P2.24 pipe A contains gasoline (SG = 0.7), pipe B contains oil (SG = 0.9), and the manometer fluid is mercury. Determine the new differential reading if the pressure in pipe A is decreased 25 kPa, and the pressure in pipe B remains constant. The initial differential reading is 0.30 m as shown.

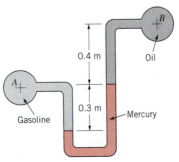

■ **FIGURE P2.24**

2.25 A piston having a cross-sectional area of 0.09 m² is located in a cylinder containing water as shown in Fig. P2.25. An open U-tube manometer is connected to the cylinder as shown. For $h_1 = 60$ mm and $h = 100$ mm, what is the value of the applied force, P, acting on the piston? The weight of the piston is negligible.

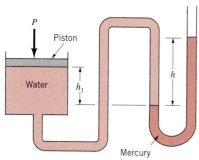

■ FIGURE P2.25

2.26 A 0.02-m-diameter manometer tube is connected to a 6-m-diameter full tank as shown in Fig. P2.26. Determine the density of the unknown liquid in the tank.

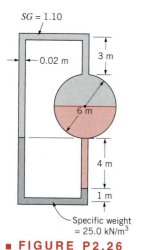

■ FIGURE P2.26

Section 2.8 Hydrostatic Force on a Plane Surface (also see Lab Problems 2.61, 2.62, 2.63, and 2.64)

2.27 A rectangular gate having a width of 4 ft is located in the sloping side of a tank as shown in Fig. P2.27. The gate is hinged along its top edge and is held in position by the force P. Friction at the hinge and the weight of the gate can be neglected. Determine the required value of P.

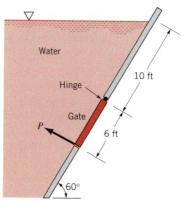

■ FIGURE P2.27

2.28 (See Fluids in the News article titled "The Three Gorges Dam," Section 2.8.) **(a)** Determine the horizontal hydrostatic force on the 2309-m-long Three Gorges Dam when the average depth of the water against it is 175 m. **(b)** If all of the 6.4 billion people on Earth were to push horizontally against the Three Gorges Dam, could they generate enough force to hold it in place? Support your answer with appropriate calculations.

2.29 A large, open tank contains water and is connected to a 6-ft-diameter conduit as shown in Fig. P2.29. A circular plug is used to seal the conduit. Determine the magnitude, direction, and location of the force of the water on the plug.

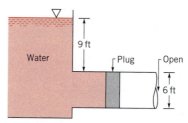

■ FIGURE P2.29

2.30 A homogeneous, 4-ft-wide, 8-ft-long rectangular gate weighing 800 lb is held in place by a horizontal flexible cable as shown in Fig. P2.30. Water acts against the gate, which is

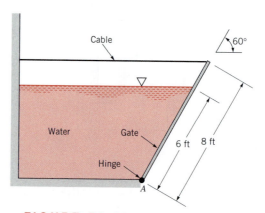

■ FIGURE P2.30

hinged at point *A*. Friction in the hinge is negligible. Determine the tension in the cable.

†**2.31**   A rubber stopper covers the drain in your bathtub. Estimate the force that the water exerts on the stopper. List all assumptions and show all calculations. Is this the force that is actually needed to lift the stopper? Explain.

2.32   An area in the form of an isosceles triangle with a base width of 6 ft and an altitude of 8 ft lies in the plane forming one side of a tank that contains a liquid having a specific weight of 79.8 lb/ft³. The side slopes upward, making an angle of 60° with the horizontal. The base of the triangle is horizontal and the vertex is above the base. Determine the resultant force the fluid exerts on the area when the fluid depth is 20 ft above the base of the triangular area. Show, with the aid of a sketch, where the center of pressure is located.

2.33   Solve Problem 2.32 if the isosceles triangle is replaced with a right triangle having the same base width and altitude as the isosceles triangle.

2.34   Two square gates close two openings in a conduit connected to an open tank of water as shown in Fig. P2.34. When the water depth, *h*, reaches 5 m it is desired that both gates open at the same time. Determine the weight of the homogeneous horizontal gate and the horizontal force, *R*, acting on the vertical gate that is required to keep the gates closed until this depth is reached. The weight of the vertical gate is negligible, and both gates are hinged at one end as shown. Friction in the hinges is negligible.

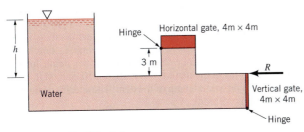

■ **FIGURE P2.34**

2.35   A 4-ft by 3-ft massless rectangular gate is used to close the end of the water tank shown in Fig. P2.35. A 200-lb weight attached to the arm of the gate at a distance ℓ from the frictionless hinge is just sufficient to keep the gate closed when the water depth is 2 ft, that is, when the water fills the semicircular lower portion of the tank. If the water were deeper the gate would open. Determine the distance ℓ.

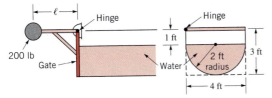

■ **FIGURE P2.35**

2.36   A gate having the cross section shown in Fig. P2.36 closes an opening 5 ft wide and 4 ft high in a water reservoir. The gate weighs 400 lb, and its center of gravity is 1 ft to the left of *AC*

and 2 ft above *BC*. Determine the horizontal reaction that is developed on the gate at *C*.

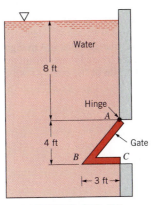

■ **FIGURE P2.36**

*2.37   A 200-lb homogeneous gate of 10 ft width and 5 ft long is hinged at point *A* and held in place by a 12-ft-long brace as shown in Fig. P2.37. As the bottom of the brace is moved to the right, the water level remains at the top of the gate. The line of action of the force that the brace exerts on the gate is along the brace. **(a)** Plot the magnitude of the force exerted on the gate by the brace as a function of the angle of the gate, *θ*, for 0 ≤ *θ* ≤ 90°. **(b)** Repeat the calculations for the case in which the weight of the gate is negligible. Comment on the results as *θ* → 0.

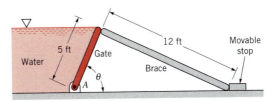

■ **FIGURE P2.37**

2.38   A rectangular gate that is 2 m wide is located in the vertical wall of a tank containing water as shown in Fig. P2.38. It is desired to have the gate open automatically when the depth

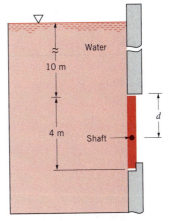

■ **FIGURE P2.38**

of water above the top of the gate reaches 10 m. **(a)** At what distance, d, should the frictionless horizontal shaft be located? **(b)** What is the magnitude of the force on the gate when it opens?

2.39   An open tank has a vertical partition and on one side contains gasoline with a density $\rho = 700$ kg/m^3 at a depth of 4 m, as shown in Fig. P2.39. A rectangular gate that is 4 m high and 2 m wide and hinged at one end is located in the partition. Water is slowly added to the empty side of the tank. At what depth, h, will the gate start to open?

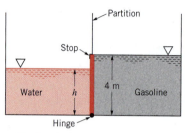

■ **FIGURE P2.39**

2.40   A thin 4-ft-wide, right-angle gate with negligible mass is free to pivot about a frictionless hinge at point O, as shown in Fig. P2.40. The horizontal portion of the gate covers a 1-ft-diameter drain pipe, which contains air at atmospheric pressure. Determine the minimum water depth, h, at which the gate will pivot to allow water to flow into the pipe.

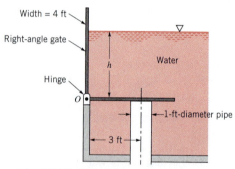

■ **FIGURE P2.40**

*2.41   An open rectangular settling tank contains a liquid suspension that at a given time has a specific weight that varies approximately with depth according to the following data:

h (m)	γ (kN/m^3)
0	10.0
0.4	10.1
0.8	10.2
1.2	10.6
1.6	11.3
2.0	12.3
2.4	12.7
2.8	12.9
3.2	13.0
3.6	13.1

The depth $h = 0$ corresponds to the free surface. By means of numerical integration, determine the magnitude and location of the resultant force that the liquid suspension exerts on a vertical wall of the tank that is 6 m wide. The depth of fluid in the tank is 3.6 m.

2.42   The concrete dam of Fig. P2.42 weighs 23.6 kN/m^3 and rests on a solid foundation. Determine the minimum coefficient of friction between the dam and the foundation required to keep the dam from sliding at the water depth shown. Assume no fluid uplift pressure along the base. Base your analysis on a unit length of the dam.

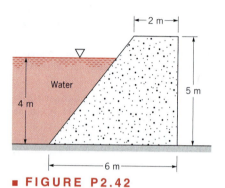

■ **FIGURE P2.42**

*2.43   Water backs up behind a concrete dam as shown in Fig. P2.43. Leakage under the foundation gives a pressure distribution under the dam as indicated. If the water depth, h, is too great, the dam will topple over about its toe (point A). For the dimensions given, determine the maximum water depth for the following widths of the dam: $\ell = 20, 30, 40, 50,$ and 60 ft. Base your analysis on a unit length of the dam. The specific weight of the concrete is 150 lb/ft^3.

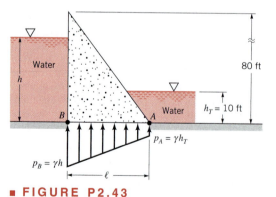

■ **FIGURE P2.43**

Section 2.10 Hydrostatic Force on a Curved Surface

2.44   A 3-m-long curved gate is located in the side of a reservoir containing water as shown in Fig. P2.44. Determine the magnitude of the horizontal and vertical components of the force of the water on the gate. Will this force pass through point A? Explain.

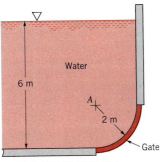

■ FIGURE P2.44

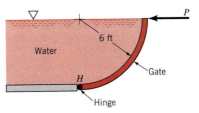

■ FIGURE P2.46

2.45 A 3-m-diameter open cylindrical tank contains water and has a hemispherical bottom as shown in Fig. P2.45. Determine the magnitude, line of action, and direction of the force of the water on the curved bottom.

2.47 An open tank containing water has a bulge in its vertical side that is semicircular in shape as shown in Fig. P2.47. Determine the horizontal and vertical components of the force that the water exerts on the bulge. Base your analysis on a 1-ft length of the bulge.

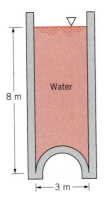

■ FIGURE P2.45

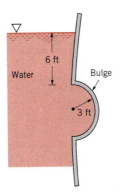

■ FIGURE P2.47

2.46 The 20-ft-long gate of Fig. P2.46 is a quarter circle and is hinged at *H*. Determine the horizontal force, *P*.

2.48 Hoover Dam (see **Video 2.3**) is the highest arch-gravity type of dam in the United States. A plan view and cross section of the dam are shown in Fig. P2.48a. The walls of the canyon in which the dam is located are sloped, and just upstream of the dam the vertical plane shown in Fig. P2.48b approximately represents the cross section of the water acting on the dam. Use this vertical cross section to estimate the resultant horizontal force of the water on the dam, and show where this force acts.

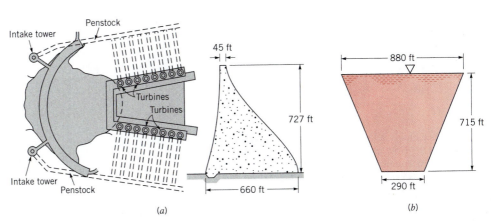

■ FIGURE P2.48

2.49 A tank wall has the shape shown in Fig. P2.49. Determine the horizontal and vertical components of the force of the water on a 1-ft width (normal to the figure) of the curved section *AB*.

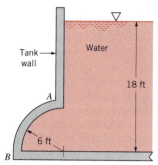

■ **FIGURE P2.49**

2.50 If the bottom of a pop bottle similar to that shown in **Video V2.4** were changed so that it was hemispherical, as in Fig. P2.50, what would be the magnitude, line of action, and direction of the resultant force acting on the hemispherical bottom? The air pressure in the top of the bottle is 40 psi, and the pop has approximately the same specific gravity as that of water. Assume that the volume of pop remains at 2 liters.

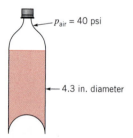

■ **FIGURE P2.50**

Section 2.11 Buoyancy, Flotation, and Stability

2.51 A solid cube floats in water with a 0.5-ft-thick oil layer on top as shown in Fig. P2.51. Determine the weight of the cube.

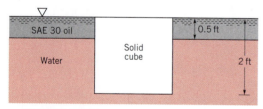

■ **FIGURE P2.51**

2.52 (See Fluids in the News article titled "Concrete canoe," Section 2.11.1.) How much extra water does a 175-lb concrete canoe displace compared to an ultra-lightweight 38-lb Kevlar canoe of the same size carrying the same load?

2.53 When the Tucurui Dam was constructed in northern Brazil, the lake that was created covered a large forest of valuable hardwood trees. It was found that even after 15 years underwater the trees were perfectly preserved and underwater logging was started. During the logging process a tree is selected, trimmed, and anchored with ropes to prevent it from shooting to the surface like a missile when cut. Assume that a typical large tree can be approximated as a truncated cone with a base diameter of 8 ft, a top diameter of 2 ft, and a height of 100 ft. Determine the resultant vertical force that the ropes must resist when the completely submerged tree is cut. The specific gravity of the wood is approximately 0.6.

†2.54 Estimate the minimum water depth needed to float a canoe carrying two people and their camping gear. List all assumptions and show all calculations.

2.55 An inverted test tube partially filled with air floats in a plastic water-filled soft drink bottle as shown in **Video V2.5** and Fig. P2.55. The amount of air in the tube has been adjusted so that it just floats. The bottle cap is securely fastened. A slight squeezing of the plastic bottle will cause the test tube to sink to the bottom of the bottle. Explain this phenomenon.

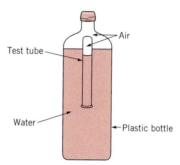

■ **FIGURE P2.55**

2.56 A 1-ft-diameter, 2-ft-long cylinder floats in an open tank containing a liquid having a specific weight γ. A U-tube manometer is connected to the tank as shown in Fig. P2.56. When the pressure in pipe *A* is 0.1 psi below atmospheric pressure, the various fluid levels are as shown. Determine the weight of the cylinder. Note that the top of the cylinder is flush with the fluid surface.

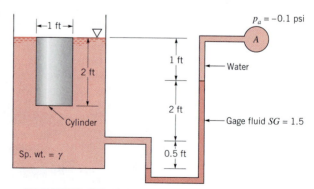

■ **FIGURE P2.56**

2.57 A 1-m-diameter cylindrical mass, *M,* is connected to a 2-m-wide rectangular gate as shown in Fig. P2.57. The gate is to open when the water level, *h,* drops below 2.5 m. Determine the required value for *M*. Neglect friction at the gate hinge and the pulley.

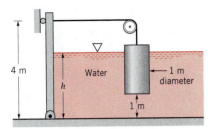

■ **FIGURE P2.57**

2.58   The hydrometer shown in **Video V2.6** and Fig. P2.58 has a mass of 0.045 kg, and the cross-sectional area of its stem is 290 mm². Determine the distance between graduations (on the stem) for specific gravities of 1.00 and 0.90.

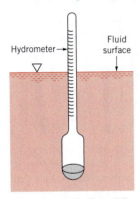

■ **FIGURE P2.58**

Section 2.12 Pressure Variation in a Fluid with Rigid-Body Motion

2.59   A 5-gal, cylindrical open container with a bottom area of 120 in.² is filled with glycerin and rests on the floor of an elevator. **(a)** Determine the fluid pressure at the bottom of the container when the elevator has an upward acceleration of 3 ft/s². **(b)** What resultant force does the container exert on the floor of the elevator during this acceleration? The weight of the container is negligible. (Note: 1 gal = 231 in.³)

2.60   A closed cylindrical tank that is 8 ft in diameter and 24 ft long is completely filled with gasoline. The tank, with its long axis horizontal, is pulled by a truck along a horizontal surface. Determine the pressure difference between the ends (along the long axis of the tank) when the truck undergoes an acceleration of 5 ft/s².

Lab Problems

2.61   This problem involves the force needed to open a gate that covers an opening in the side of a water-filled tank. To proceed with this problem go to the book's web site, www.wiley.com/college/young.

2.62   This problem involves the use of a cleverly designed apparatus to investigate the hydrostatic pressure force on a submerged rectangle. To proceed with this problem, go to the book's web site, www.wiley.com/college/young.

2.63   This problem involves determining the weight needed to hold down an open-bottom box that has slanted sides when the box is filled with water. To proceed with this problem, go to the book's web site, www.wiley.com/college/young.

2.64   This problem involves the use of a pressurized air pad to provide the vertical force to support a given load. To proceed with this problem, go to the book's web site, www.wiley.com/college/young.

FE Exam Problems

Sample FE (Fundamentals of Engineering) exam questions for fluid mechanics are provided on the book's web site, www.wiley.com/college/young.

CHAPTER 3

Elementary Fluid Dynamics— The Bernoulli Equation

Flow past a blunt body: On any object placed in a moving fluid there is a stagnation point on the front of the object where the velocity is zero. This location has a relatively large pressure and divides the flow field into two portions—one flowing to the left of the body and one flowing to the right of the body. (Dye in water) (Photograph by B. R. Munson).

LEARNING OBJECTIVES

After completing this chapter, you should be able to:

- discuss the application of Newton's second law to fluid flows.
- explain the development, uses, and limitations of the Bernoulli equation.
- use the Bernoulli equation (stand-alone or in combination with the continuity equation) to solve simple flow problems.
- apply the concepts of static, stagnation, dynamic and total pressures.
- calculate various flow properties using the energy and hydraulic grade lines.

In this chapter we investigate some typical fluid motions (fluid dynamics) in an elementary way. We will discuss in some detail the use of Newton's second law ($\mathbf{F} = m\mathbf{a}$) as it is applied to fluid particle motion that is "ideal" in some sense. We will obtain the celebrated Bernoulli equation and apply it to various flows. Although this equation is one of the oldest in fluid mechanics and the assumptions involved in its derivation are numerous, it can be used effectively to predict and analyze a variety of flow situations.

3.1 Newton's Second Law

According to Newton's second law of motion, the net force acting on the fluid particle under consideration must equal its mass times its acceleration,

$$\mathbf{F} = m\mathbf{a}$$

In this chapter we consider the motion of inviscid fluids. That is, the fluid is assumed to have zero viscosity.

We assume that the fluid motion is governed by pressure and gravity forces only and examine Newton's second law as it applies to a fluid particle in the form:

(Net pressure force on a particle) + (net gravity force on particle) =

(particle mass) × (particle acceleration)

The results of the interaction among the pressure, gravity, and acceleration provide numerous applications in fluid mechanics.

We consider two-dimensional motion like that confined to the x–z plane as is shown in Fig. 3.1a. The motion of each fluid particle is described in terms of its velocity vector, $\mathbf{V}$, which is defined as the time rate of change of the position of the particle. The particle's velocity is a vector quantity with a magnitude (the speed, $V = |\mathbf{V}|$) and direction. As the particle moves about, it follows a particular path, the shape of which is governed by the velocity of the particle.

If the flow is *steady* (i.e., nothing changes with time at a given location in the flow field), each particle slides along its path, and its velocity vector is everywhere tangent to the path. The lines that are tangent to the velocity vectors throughout the flow field are called *streamlines.* The particle motion is described in terms of its distance, $s = s(t)$, along the streamline from some convenient origin and the local radius of curvature of the streamline, $\mathcal{R} = \mathcal{R}(s)$. The distance along the streamline is related to the particle's speed by $V = ds/dt$, and the radius of curvature is related to the shape of the streamline. In addition to the coordinate along the streamline, s, the coordinate normal to the streamline, n, as is shown in Fig. 3.1b, will be of use.

By definition, acceleration is the time rate of change of the velocity of the particle, $\mathbf{a} = d\mathbf{V}/dt$. The acceleration has two components—one along the streamline, a_s, streamwise acceleration, and one normal to the streamline, a_n, normal acceleration.

By use of the chain rule of differentiation, the s component of the acceleration is given by $a_s = dV/dt = (\partial V/\partial s)(ds/dt) = (\partial V/\partial s)V$, as shown in the figure in the margin. We have used the fact that the speed is the time rate of change of distance along the streamline, $V = ds/dt$. The normal component of acceleration, centrifugal acceleration, is given in terms of the particle speed and the radius of curvature of its path as $a_n = V^2/\mathcal{R}$. Thus, the components of acceleration in the s and n directions, a_s and a_n, for steady flow are given by

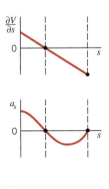

$$a_s = V\frac{\partial V}{\partial s}, \qquad a_n = \frac{V^2}{\mathcal{R}} \tag{3.1}$$

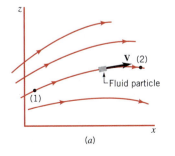

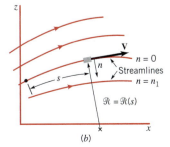

■ **FIGURE 3.1**
(*a*) **Flow in the $x - z$ plane.** (*b*) **Flow in terms of streamline and normal coordinates.**

3.2 $\quad$ F = *m*a Along a Streamline

We consider the free-body diagram of a small fluid particle as is shown in Fig. 3.2. The small fluid particle is of size δs by δn in the plane of the figure and δy normal to the figure as shown in the free-body diagram of Fig. 3.3. Unit vectors along and normal to the streamline are denoted by $\hat{s}$ and $\hat{n}$, respectively. For steady flow, the component of Newton's second law along the streamline direction, s, can be written as

$$\sum \delta F_s = \delta m\, a_s = \delta m\, V \frac{\partial V}{\partial s} = \rho\, \delta \forall\, V \frac{\partial V}{\partial s} \tag{3.2}$$

where $\sum \delta F_s$ represents the sum of the s components of all the forces acting on the particle, which has mass $\delta m = \rho\, \delta \forall$, and $V \partial V/\partial s$ is the acceleration in the s direction. Here, $\delta \forall = \delta s \delta n \delta y$ is the particle volume.

The gravity force (weight) on the particle can be written as $\delta W = \gamma\, \delta \forall$, where $\gamma = \rho g$ is the specific weight of the fluid (lb/ft^3 or N/m^3). Hence, the component of the weight force in the direction of the streamline is

$$\delta W_s = -\delta W \sin \theta = -\gamma\, \delta \forall \sin \theta$$

If the streamline is horizontal at the point of interest, then $\theta = 0$, and there is no component of the particle weight along the streamline to contribute to its acceleration in that direction.

If the pressure at the center of the particle shown in Fig. 3.3 is denoted as p, then its average value on the two end faces that are perpendicular to the streamline are $p + \delta p_s$ and $p - \delta p_s$. Because the particle is "small," we can use a one-term Taylor series expansion for the pressure field to obtain

$$\delta p_s \approx \frac{\partial p}{\partial s} \frac{\delta s}{2}$$

Thus, if δF_{ps} is the net pressure force on the particle in the streamline direction, it follows that

$$\delta F_{ps} = (p - \delta p_s)\delta n \delta y - (p + \delta p_s)\delta n \delta y = -2\delta p_s \delta n \delta y$$

$$= -\frac{\partial p}{\partial s}\delta s \delta n \delta y = -\frac{\partial p}{\partial s}\delta \forall$$

Thus, the net force acting in the streamline direction on the particle shown in Fig. 3.3 is given by

$$\sum \delta F_s = \delta W_s + \delta F_{ps} = \left(-\gamma \sin \theta - \frac{\partial p}{\partial s}\right)\delta \forall \tag{3.3}$$

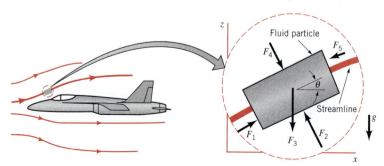

■ **FIGURE 3.2** $\quad$ **Isolation of a small fluid particle in a flow field.**

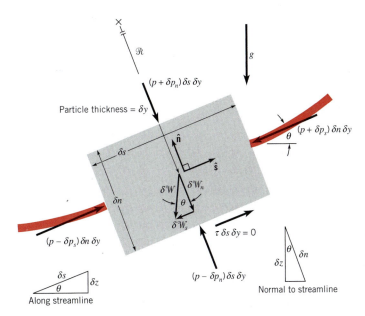

Free-body diagram of a fluid particle for which the important forces are those due to pressure and gravity.

By combining Eqs. 3.2 and 3.3 we obtain the following equation of motion along the streamline direction:

$$-\gamma \sin \theta - \frac{\partial p}{\partial s} = \rho V \frac{\partial V}{\partial s} \tag{3.4}$$

The physical interpretation of Eq. 3.4 is that a change in fluid particle speed is accomplished by the appropriate combination of pressure and particle weight along the streamline.

EXAMPLE 3.1   **Pressure Variation Along a Streamline**

GIVEN   Consider the inviscid, incompressible, steady flow along the horizontal streamline A–B in front of the sphere of radius a as shown in Fig. E3.1a. From a more advanced theory of flow past a sphere, the fluid velocity along this streamline is

$$V = V_0\left(1 + \frac{a^3}{x^3}\right)$$

FIND   Determine the pressure variation along the streamline from point A far in front of the sphere ($x_A = -\infty$ and $V_A = V_0$) to point B on the sphere ($x_B = -a$ and $V_B = 0$).

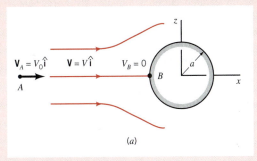

(a)

■ **FIGURE E3.1a**

SOLUTION

Since the flow is steady and inviscid, Eq. 3.4 is valid. In addition, because the streamline is horizontal, $\sin \theta = \sin 0° = 0$ and the equation of motion along the streamline reduces to

$$\frac{\partial p}{\partial s} = -\rho V \frac{\partial V}{\partial s} \tag{1}$$

With the given velocity variation along the streamline, the acceleration term is

$$V\frac{\partial V}{\partial s} = V\frac{\partial V}{\partial x} = V_0\left(1 + \frac{a^3}{x^3}\right)\left(-\frac{3V_0a^3}{x^4}\right)$$

$$= -3V_0^2\left(1 + \frac{a^3}{x^3}\right)\frac{a^3}{x^4}$$

where we have replaced s by x since the two coordinates are identical (within an additive constant) along streamline A–B. It follows that $V\partial V/\partial s < 0$ along the streamline. The fluid slows down from V_0 far ahead of the sphere to zero velocity on the "nose" of the sphere ($x = -a$).

Thus, according to Eq. 1, to produce the given motion the pressure gradient along the streamline is

$$\frac{\partial p}{\partial x} = \frac{3\rho a^3 V_0^2(1 + a^3/x^3)}{x^4} \qquad (2)$$

This variation is indicated in Fig. E3.1b. It is seen that the pressure increases in the direction of flow ($\partial p/\partial x > 0$) from point A to point B. The maximum pressure gradient ($0.610\rho V_0^2/a$) occurs just slightly ahead of the sphere ($x = -1.205a$). It is the pressure gradient that slows the fluid down from $V_A = V_0$ to $V_B = 0$.

The pressure distribution along the streamline can be obtained by integrating Eq. 2 from $p = 0$ (gage) at $x = -\infty$ to pressure p at location x. The result, plotted in Fig. E3.1c, is

$$p = -\rho V_0^2\left[\left(\frac{a}{x}\right)^3 + \frac{(a/x)^6}{2}\right] \qquad \textbf{(Ans)}$$

COMMENT The pressure at B, a stagnation point since $V_B = 0$, is the highest pressure along the streamline ($p_B = \rho V_0^2/2$). As shown in Chapter 9, this excess pressure on

the front of the sphere (i.e., $p_B > 0$) contributes to the net drag force on the sphere. Note that the pressure gradient and pressure are directly proportional to the density of the fluid, a representation of the fact that the fluid inertia is proportional to its mass.

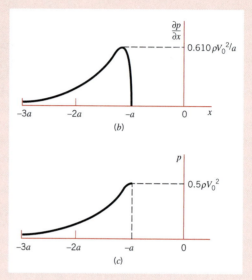

■ **FIGURES E3.1 b and c**

Equation 3.4 can be rearranged and integrated as follows. First, we note from Fig. 3.3 that along the streamline $\sin\theta = dz/ds$. Also, we can write $V\,dV/ds = \frac{1}{2}d(V^2)/ds$. Finally, along the streamline $\partial p/\partial s = dp/ds$. These ideas combined with Eq. 3.4 give the following result valid along a streamline

$$-\gamma\frac{dz}{ds} - \frac{dp}{ds} = \frac{1}{2}\rho\frac{d(V^2)}{ds}$$

This simplifies to

$$dp + \frac{1}{2}\rho d(V^2) + \gamma\,dz = 0 \qquad \text{(along a streamline)} \qquad \textbf{(3.5)}$$

which, for constant density and specific weight, can be integrated to give

$$p + \tfrac{1}{2}\rho V^2 + \gamma z = \text{constant along streamline}$$

(3.6)

V3.1 Balancing ball   This is the celebrated ***Bernoulli equation***—a very powerful tool in fluid mechanics.

EXAMPLE 3.2   The Bernoulli Equation

GIVEN Consider the flow of air around a bicyclist moving through still air with velocity V_0 as is shown in Fig. E3.2.

FIND Determine the difference in the pressure between points (1) and (2).

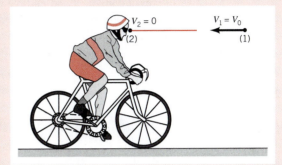

SOLUTION

In a coordinate system fixed to the bike, it appears as though the air is flowing steadily toward the bicyclist with speed V_0. If the assumptions of Bernoulli's equation are valid (steady, incompressible, inviscid flow), Eq. 3.6 can be applied as follows along the streamline that passes through (1) and (2).

$$p_1 + \tfrac{1}{2}\rho V_1^2 + \gamma z_1 = p_2 + \tfrac{1}{2}\rho V_2^2 + \gamma z_2$$

We consider (1) to be in the free stream so that $V_1 = V_0$ and (2) to be at the tip of the bicyclist's nose and assume that $z_1 = z_2$ and $V_2 = 0$ (both of which, as is discussed in Section 3.5, are reasonable assumptions). It follows that the pressure of (2) is greater than that at (1) by an amount

$$p_2 - p_1 = \tfrac{1}{2}\rho V_1^2 = \tfrac{1}{2}\rho V_0^2 \qquad \text{(Ans)}$$

COMMENTS A similar result was obtained in Example 3.1 by integrating the pressure gradient, which was known because the velocity distribution along the streamline, $V(s)$, was known.

■ **FIGURE E3.2**

The Bernoulli equation is a general integration of **F** = *m***a**. To determine $p_2 - p_1$, knowledge of the detailed velocity distribution is not needed—only the "boundary conditions" at (1) and (2) are required. Of course, knowledge of the value of V along the streamline is needed to determine the pressure at points between (1) and (2). Note that if we measure $p_2 - p_1$ we can determine the speed, V_0. As discussed in Section 3.5, this is the principle upon which many velocity measuring devices are based.

If the bicyclist were accelerating or decelerating, the flow would be unsteady (i.e., $V_0 \ne$ constant) and the aforementioned analysis would be incorrect, as Eq. 3.6 is restricted to steady flow.

3.3   F = *m*a Normal to a Streamline

We again consider the force balance on the fluid particle shown in Fig. 3.3. This time, however, we consider components in the normal direction, **n̂,** and write Newton's second law in this direction as

$$\sum \delta F_n = \frac{\delta m\, V^2}{\mathcal{R}} = \frac{\rho\, \delta \forall\, V^2}{\mathcal{R}}$$

(3.7)

where $\sum \delta F_n$ represents the sum of the normal components of all the forces acting on the particle. We assume the flow is steady with a normal acceleration $a_n = V^2/\mathcal{R}$, where $\mathcal{R}$ is the local radius of curvature of the streamlines.

We again assume that the only forces of importance are pressure and gravity. Using the method of Section 3.2 for determining forces along the streamline, the net force acting in the normal direction on the particle shown in Fig. 3.3 is determined to be

$$\sum \delta F_n = \delta \mathcal{W}_n + \delta F_{pn} = \left(-\gamma \cos \theta - \frac{\partial p}{\partial n}\right)\delta \forall$$

(3.8)

where $\partial p/\partial n$ is the pressure gradient normal to the streamline. By combining Eqs. 3.7 and 3.8 and using the fact that along a line normal to the streamline $\cos \theta = dz/dn$ (see Fig. 3.3), we obtain the following equation of motion along the normal direction

$$-\gamma \frac{dz}{dn} - \frac{\partial p}{\partial n} = \frac{\rho V^2}{\mathcal{R}} \tag{3.9}$$

The physical interpretation of Eq. 3.9 is that a change in the direction of flow of a fluid particle (i.e., a curved path, $\mathcal{R} < \infty$) is accomplished by the appropriate combination of pressure gradient and particle weight normal to the streamline. By integration of Eq. 3.9, the final form of Newton's second law applied across the streamlines for steady, inviscid, incompressible flow is obtained as

$$\boxed{p + \rho \int \frac{V^2}{\mathcal{R}} \, dn + \gamma z = \text{constant across the streamline}} \tag{3.10}$$

V3.2 Free vortex

EXAMPLE 3.3   Pressure Variation Normal to a Streamline

GIVEN Shown in Figs. E3.3a,b are two flow fields with circular streamlines. The velocity distributions are

$$V(r) = (V_0/r_0)r \quad \text{for case } (a)$$

and

$$V(r) = \frac{(V_0 r_0)}{r} \quad \text{for case } (b)$$

where V_0 is the velocity at $r = r_0$.

FIND Determine the pressure distributions, $p = p(r)$, for each, given that $p = p_0$ at $r = r_0$.

SOLUTION

We assume the flows are steady, inviscid, and incompressible with streamlines in the horizontal plane ($dz/dn = 0$). Because the streamlines are circles, the coordinate n points in a direction opposite of that of the radial coordinate, $\partial/\partial n = -\partial/\partial r$, and the radius of curvature is given by $\mathcal{R} = r$. Hence, Eq. 3.9 becomes

$$\frac{\partial p}{\partial r} = \frac{\rho V^2}{r}$$

For case (a) this gives

$$\frac{\partial p}{\partial r} = \rho (V_0/r_0)^2 r$$

whereas for case (b) it gives

$$\frac{\partial p}{\partial r} = \frac{\rho (V_0 r_0)^2}{r^3}$$

For either case the pressure increases as r increases since $\partial p/\partial r > 0$. Integration of these equations with respect to r,

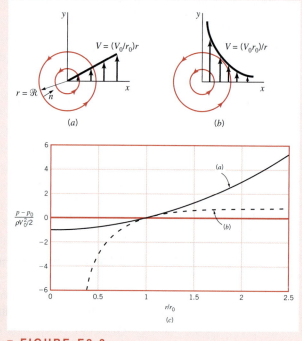

■ **FIGURE E3.3**

starting with a known pressure $p = p_0$ at $r = r_0$, gives

$$p - p_0 = (\rho V_0^2/2)[(r/r_0)^2 - 1] \tag{Ans}$$

for case (a) and

$$p - p_0 = (\rho V_0^2/2)[1 - (r_0/r)^2] \tag{Ans}$$

for case (b). These pressure distributions are shown in Fig. E3.3c.

COMMENT The pressure distributions needed to balance the centrifugal accelerations in cases (*a*) and (*b*) are not the same because the velocity distributions are different. In fact, for case (*a*) the pressure increases without bound as $r \to \infty$, whereas for case (*b*) the pressure approaches a finite value as $r \to \infty$. The streamline patterns are the same for each case, however.

Physically, case (*a*) represents rigid body rotation (as obtained in a can of water on a turntable after it has been "spun up") and case (*b*) represents a free vortex (an approximation to a tornado or the swirl of water in a drain, the "bathtub vortex").

3.4 Physical Interpretation

An alternate but equivalent form of the Bernoulli equation is obtained by dividing each term of Eq. 3.6 by the specific weight, γ, to obtain

$$\frac{p}{\gamma} + \frac{V^2}{2g} + z = \text{constant on a streamline} \tag{3.11}$$

Each of the terms in this equation has the units of length and represents a certain type of head.

The elevation term, z, is related to the potential energy of the particle and is called the *elevation head.* The pressure term, p/γ, is called the ***pressure head*** and represents the height of a column of the fluid that is needed to produce the pressure, p. The velocity term, $V^2/2g$, is the ***velocity head*** and represents the vertical distance needed for the fluid to fall freely (neglecting friction) if it is to reach velocity V from rest. The Bernoulli equation states that the sum of the pressure head, the velocity head, and the elevation head is constant along a streamline.

EXAMPLE 3.4   Kinetic, Potential, and Pressure Energy

GIVEN Consider the flow of water from the syringe shown in Fig. E3.4. A force applied to the plunger will produce a pressure greater than atmospheric at point (1) within the syringe. The water flows from the needle, point (2), with relatively high velocity and coasts up to point (3) at the top of its trajectory.

FIND Discuss the energy of the fluid at points (1), (2), and (3) using the Bernoulli equation.

SOLUTION

If the assumptions (steady, inviscid, incompressible flow) of the Bernoulli equation are approximately valid, it then follows that the flow can be explained in terms of the partition of the total energy of the water. According to Eq. 3.11 the sum of the three types of energy (kinetic, potential, and pressure) or heads (velocity, elevation, and pressure) must remain constant. The following table indicates the relative magnitude of each of these energies at the three points shown in Fig. E3.4.

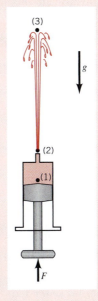

■ FIGURE E3.4

| | Energy Type | | |
Point	Kinetic $\rho V^2/2$	Potential γz	Pressure p
1	Small	Zero	Large
2	Large	Small	Zero
3	Zero	Large	Zero

(Ans)

The motion results in (or is due to) a change in the magnitude of each type of energy as the fluid flows from one location to another. An alternate way to con-sider this flow is as follows. The pressure gradient between (1) and (2) produces an acceleration to eject the water from the needle. Gravity acting on the particle between (2) and (3) produces a deceleration to cause the water to come to a momentary stop at the top of its flight.

COMMENTS   If friction (viscous) effects were important, there would be an energy loss between (1) and (3) and for the given p_1 the water would not be able to reach the height indicated in Fig. E3.4. Such friction may arise in the needle (see Chapter 8, pipe flow) or between the water stream and the surrounding air (see Chapter 9, external flow).

F l u i d s   i n   t h e   N e w s

Armed with a water jet for hunting Archerfish, known for their ability to shoot down insects resting on foliage, are like submarine water pistols. With their snout sticking out of the water, they eject a high-speed water jet at their prey, knocking it onto the water surface where they snare it for their meal. The barrel of their water pistol is formed by placing their tongue against a groove in the roof of their mouth to form a tube. By snapping shut their gills, water is forced through the tube and directed with the tip of their tongue.

The archerfish can produce a *pressure head* within their gills large enough so that the jet can reach 2 to 3 m. However, it is accurate to only about 1 m. Recent research has shown that archerfish are very adept at calculating where their prey will fall. Within 100 milliseconds (a reaction time twice as fast as a human's), the fish has extracted all the information needed to predict the point where the prey will hit the water. Without further visual cues it charges directly to that point. (See Problem 3.18.) ■

When a fluid particle travels along a curved path, a net force directed toward the center of curvature is required. Under the assumptions valid for Eq. 3.10 this force may be gravity, pressure, or a combination of both. In many instances the streamlines are nearly straight ($\mathcal{R} = \infty$) so that centrifugal effects are negligible and the pressure variation across the streamlines is merely hydrostatic (because of gravity alone), even though the fluid is in motion.

EXAMPLE 3.5   Pressure Variation in a Flowing Stream

GIVEN   Consider the inviscid, incompressible, steady flow shown in Fig. E3.5. From section A to B the streamlines are straight, whereas from C to D they follow circular paths.

FIND   Describe the pressure variation between

(a)   points (1) and (2)

(b)   and points (3) and (4).

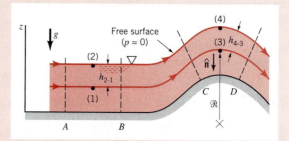

■ **FIGURE E3.5**

SOLUTION

(a)   With the assumptions given in the problem statement and the fact that $\mathcal{R} = \infty$ for the portion from A to B, Eq. 3.10 becomes

$$p + \gamma z = \text{constant}$$

The constant can be determined by evaluating the known variables at the two locations using $p_2 = 0$ (gage), $z_1 = 0$, and

$z_2 = h_{2-1}$ to give

$$p_1 = p_2 + \gamma(z_2 - z_1) = p_2 + \gamma h_{2-1} \qquad \text{(Ans)}$$

COMMENT Note that since the radius of curvature of the streamline is infinite, the pressure variation in the vertical direction is the same as if the fluid were stationary.

(b) If we apply Eq. 3.10 between points (3) and (4) we obtain (using $dn = -dz$)

$$p_4 + \rho \int_{z_3}^{z_4} \frac{V^2}{\mathcal{R}} (-dz) + \gamma z_4 = p_3 + \gamma z_3$$

With $p_4 = 0$ and $z_4 - z_3 = h_{4-3}$ this becomes

$$p_3 = \gamma h_{4-3} - \rho \int_{z_3}^{z_4} \frac{V^2}{\mathcal{R}} \, dz \qquad \text{(Ans)}$$

COMMENT To evaluate the integral we must know the variation of V and $\mathcal{R}$ with z. Even without this detailed information we note that the integral has a positive value. Thus, the pressure at (3) is less than the hydrostatic value, γh_{4-3}, by an amount equal to $\rho \int_{z_3}^{z_4} (V^2/\mathcal{R}) \, dz$. This lower pressure, caused by the curved streamline, is necessary in order to accelerate (centrifugal acceleration) the fluid around the curved path.

3.5   Static, Stagnation, Dynamic, and Total Pressure

Each term of the Bernoulli equation, Eq. 3.6, has the dimensions of force per unit area—psi, lb/ft^2, N/m^2. The first term, p, is the actual thermodynamic pressure of the fluid as it flows. To measure its value, one could move along with the fluid, thus being "static" relative to the moving fluid. Hence, it is normally termed the ***static pressure.*** Another way to measure the static pressure would be to drill a hole in a flat surface and fasten a piezometer tube as indicated by the location of point (3) in Fig. 3.4.

The third term in Eq. 3.5, γz, is termed the *hydrostatic pressure,* in obvious regard to the hydrostatic pressure variation discussed in Chapter 2. It is not actually a pressure, but does represent the change in pressure possible due to potential energy variations of the fluid as a result of elevation changes.

V3.3 Stagnation point flow

The second term in the Bernoulli equation, $\rho V^2/2$, is termed the ***dynamic pressure.*** Its interpretation can be seen in Fig. 3.4 by considering the pressure at the end of a small tube inserted into the flow and pointing upstream. After the initial transient motion has died out, the liquid will fill the tube to a height of H as shown. The fluid in the tube, including that at its tip, (2), will be stationary. That is, $V_2 = 0$, or point (2) is a ***stagnation point.***

If we apply the Bernoulli equation between points (1) and (2), using $V_2 = 0$ and assuming that $z_1 = z_2$, we find that

$$p_2 = p_1 + \tfrac{1}{2} \rho V_1^2$$

Hence, the pressure at the stagnation point, termed the ***stagnation pressure,*** is greater than the static pressure, p_1, by an amount $\rho V_1^2/2$, the dynamic pressure. It can be shown

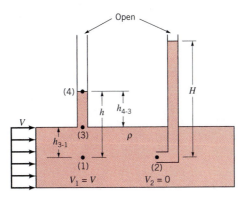

■ **FIGURE 3.4**   Measurement of static and stagnation pressures.

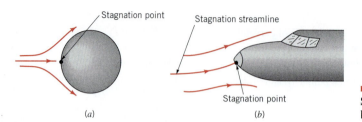

■ **FIGURE 3.5**
Stagnation points on bodies in flowing fluids.

(a)   (b)

Stagnation point

Stagnation streamline

Stagnation point

that there is a stagnation point on any stationary body that is placed into a flowing fluid (see Fig. 3.5).

F l u i d s   i n   t h e   N e w s

Pressurized eyes Our eyes need a certain amount of internal pressure in order to work properly, with the normal range being between 10 and 20 mm of mercury. The pressure is determined by a balance between the fluid entering and leaving the eye. If the pressure is above the normal level, damage may occur to the optic nerve where it leaves the eye, leading to a loss of the visual field termed glaucoma. Measurement of the pressure within the eye can be done by several different noninvasive types of instru- ments, all of which measure the slight deformation of the eyeball when a force is put on it. Some methods use a physical probe that makes contact with the front of the eye, applies a known force, and measures the deformation. One noncontact method uses a calibrated "puff" of air that is blown against the eye. The *stagna- tion pressure* resulting from the air blowing against the eyeball causes a slight deformation, the magnitude of which is corre- lated with the pressure within the eyeball. (See Problem 3.11.) ■

The sum of the static pressure, hydrostatic pressure, and dynamic pressure is termed the **total pressure,** p_T. The Bernoulli equation is a statement that the total pressure remains constant along a streamline. That is,

$$p + \tfrac{1}{2}\rho V^2 + \gamma z = p_T = \text{constant along a streamline} \qquad \textbf{(3.12)}$$

V3.4 Airspeed indicator

Knowledge of the values of the static and stagnation pressures in a fluid implies that the fluid speed can be calculated. This is the principle on which the **Pitot-static tube** is based. As shown in Fig. 3.6, two concentric tubes are attached to two pressure gages. The center tube measures the stagnation pressure at its open tip. If elevation changes are negligible,

$$p_3 = p + \tfrac{1}{2}\rho V^2$$

where p and V are the pressure and velocity of the fluid upstream of point (2). The outer tube is made with several small holes at an appropriate distance from the tip so that they measure the static pressure. If the elevation difference between (1) and (4) is negligible, then

$$p_4 = p_1 = p$$

These two equations can be rearranged to give

$$V = \sqrt{2(p_3 - p_4)/\rho} \qquad \textbf{(3.13)}$$

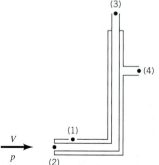

(3)

(4)

(1)

V

p

(2)

■ **FIGURE 3.6**   **The Pitot-static tube.**

EXAMPLE 3.6   Pitot-Static Tube

GIVEN An airplane flies 100 mi/hr at an elevation of 10,000 ft in a standard atmosphere as shown in Fig. E3.6a.

FIND Determine the pressure at point (1) far ahead of the airplane, the pressure at the stagnation point on the nose of the airplane, point (2), and the pressure difference indicated by a Pitot-static probe attached to the fuselage.

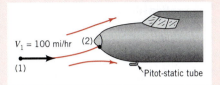

■ **FIGURE E3.6*a***

SOLUTION

From Table C.1 we find that the static pressure at the altitude given is

$$p_1 = 1456 \text{ lb/ft}^2 \text{ (abs)} = 10.11 \text{ psia} \qquad \textbf{(Ans)}$$

Also, the density is $\rho = 0.001756$ slug/ft^3.

If the flow is steady, inviscid, and incompressible and elevation changes are neglected, Eq. 3.6 becomes

$$p_2 = p_1 + \frac{\rho V_1^2}{2}$$

With $V_1 = 100$ mi/hr $= 146.7$ ft/s and $V_2 = 0$ (since the coordinate system is fixed to the airplane) we obtain

$$p_2 = 1456 \text{ lb/ft}^2 + (0.001756 \text{ slugs/ft}^3)(146.7 \text{ ft/s})^2/2$$
$$= (1456 + 18.9) \text{ lb/ft}^2 \text{ (abs)}$$

Hence, in terms of gage pressure

$$p_2 = 18.9 \text{ lb/ft}^2 = 0.1313 \text{ psi} \qquad \textbf{(Ans)}$$

Thus, the pressure difference indicated by the Pitot-static tube is

$$p_2 - p_1 = \frac{\rho V_1^2}{2} = 0.1313 \text{ psi} \qquad \textbf{(Ans)}$$

COMMENTS Note that it is very easy to obtain incorrect results by using improper units. Do not add lb/in.^{2} and lb/ft^2. Note that (slug/ft^3)(ft^2/s^2) = (slug·ft/s^2)/(ft^2) = lb/ft^2.

It was assumed that the flow is incompressible—the density remains constant from (1) to (2). However, because $\rho = p/RT$, a change in pressure (or temperature) will cause a change in density. For this relatively low speed, the ratio of the absolute pressures is nearly unity [i.e., $p_1/p_2 = (10.11$ psia)/(10.11 + 0.1313 psia) = 0.987] so that the density change is negligible. However, by repeating the calculations for various values of the speed, V_1, the results shown in Fig. E3.6b are obtained. Clearly at the 500 to 600 mph speeds normally flown by commercial airliners, the pressure ratio is such that density changes are important. In such situations it is necessary to use compressible flow concepts to obtain accurate results.

■ **FIGURE E3.6*b***

F l u i d s   i n   t h e   N e w s

Bugged and plugged Pitot tubes Although a *Pitot tube* is a simple device for measuring aircraft airspeed, many airplane accidents have been caused by inaccurate Pitot tube readings. Most of these accidents are the result of having one or more of the holes blocked and, therefore, not indicating the correct pressure (speed). Usually this is discovered during takeoff when time to resolve the issue is short. The two most common causes for such a blockage are either that the pilot (or ground crew) has forgotten to remove the protective Pitot tube cover, or that insects have built their nest

within the tube where the standard visual check cannot detect it. The most serious accident (in terms of number of fatalities) caused by a blocked Pitot tube involved a Boeing 757 and occurred shortly after takeoff from Puerto Plata in the Dominican Republic. The incorrect airspeed data was automatically fed to the computer causing the autopilot to change the angle of attack and the engine power. The flight crew became confused by the false indications, the aircraft stalled, and then plunged into the Caribbean Sea killing all aboard. (See Problem 3.12.) ■

3.6 Examples of Use of the Bernoulli Equation

Between any two points, (1) and (2), on a streamline in steady, inviscid, incompressible flow the Bernoulli equation (Eq. 3.6) can be applied in the form

$$p_1 + \tfrac{1}{2}\rho V_1^2 + \gamma z_1 = p_2 + \tfrac{1}{2}\rho V_2^2 + \gamma z_2 \tag{3.14}$$

The use of this equation is discussed in this section.

3.6.1 Free Jets

Consider flow of a liquid from a large reservoir as is shown in Fig. 3.7. A jet of liquid of diameter d flows from the nozzle with velocity V. Application of Eq. 3.14 between points (1) and (2) on the streamline shown gives

$$\gamma h = \tfrac{1}{2}\rho V^2$$

We have used the facts that $z_1 = h$, $z_2 = 0$, the reservoir is large ($V_1 = 0$), open to the atmosphere ($p_1 = 0$ gage), and the fluid leaves as a *"free jet"* ($p_2 = 0$). Thus, we obtain

$$V = \sqrt{2\frac{\gamma h}{\rho}} = \sqrt{2gh} \tag{3.15}$$

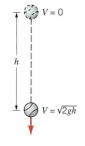
$V = 0$

h

$V = \sqrt{2gh}$

Be careful when applying Eq. 3.15. It is valid only if the reservoir is large enough so that the initial fluid speed is negligible ($V_1 = 0$), the beginning and ending pressures are the same ($p_1 = p_2$), and viscous effects are absent. Note that as shown in the figure in the margin, a ball starting from rest will also obtain a speed of $V = \sqrt{2gh}$ after it has dropped through a distance h, provided friction is negligible.

Once outside the nozzle, the stream continues to fall as a free jet with zero pressure throughout ($p_5 = 0$) and as seen by applying Eq. 3.14 between points (1) and (5), the speed increases according to

$$V = \sqrt{2g(h + H)}$$

where H is the distance the fluid has fallen outside the nozzle.

Equation 3.15 could also be obtained by writing the Bernoulli equation between points (3) and (4) using the fact that $z_4 = 0$, $z_3 = \ell$. Also, $V_3 = 0$ since it is far from the nozzle, and from hydrostatics, $p_3 = \gamma(h - \ell)$.

If the exit of the tank shown in Fig. 3.7 is not a smooth, well-contoured nozzle, the diameter of the jet, d_j, will be less than the diameter of the hole, d_h. This phenomenon, called a *vena contracta* effect, is a result of the inability of the fluid to turn a sharp 90° corner.

V3.5 Flow from a tank

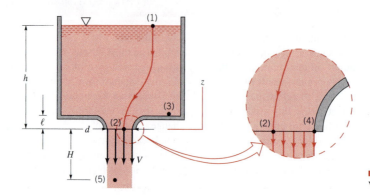

■ **FIGURE 3.7**
Vertical flow from a tank.

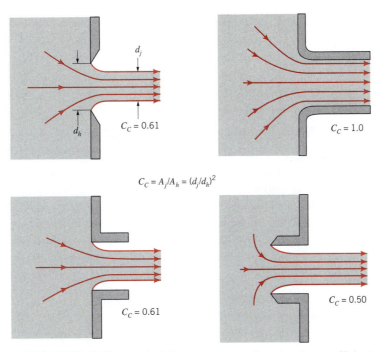

$$C_C = A_j/A_h = (d_j/d_h)^2$$

■ **FIGURE 3.8** **Typical flow patterns and contraction coefficients for various round exit configurations.**

Figure 3.8 shows typical values of the experimentally obtained *contraction coefficient,* $C_c = A_j/A_h$, where A_j and A_h are the areas of the jet at the vena contracta and the area of the hole, respectively.

3.6.2   Confined Flows

In many cases the fluid is physically constrained within a device so that its pressure cannot be prescribed a priori as was done for the free jet examples given earlier. For these situations it is necessary to use the concept of conservation of mass (the continuity equation) along with the Bernoulli equation.

Consider a fluid flowing through a fixed volume (such as a tank) that has one inlet and one outlet as shown in Fig. 3.9. If the flow is steady so that there is no additional accumulation of fluid within the volume, the rate at which the fluid flows into the volume must equal the rate at which it flows out of the volume (otherwise mass would not be conserved).

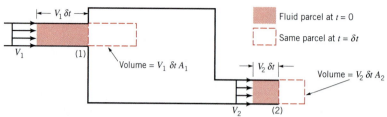

■ **FIGURE 3.9** **Steady flow into and out of a tank.**

The *mass flowrate* from an outlet, $\dot{m}$ (slugs/s or kg/s), is given by $\dot{m} = \rho Q$, where Q (ft³/s or m³/s) is the **volume flowrate.** If the outlet area is A and the fluid flows across this area (normal to the area) with an average velocity V, then the volume of the fluid crossing this area in a time interval δt is $VA\,\delta t$, equal to that in a volume of length $V\,\delta t$ and cross-sectional area A (see Fig. 3.9). Hence, the volume flowrate (volume per unit time) is $Q = VA$. Thus, $\dot{m} = \rho VA$. To conserve mass, the inflow rate must equal the outflow rate. If the inlet is designated as (1) and the outlet as (2), it follows that $\dot{m}_1 = \dot{m}_2$. Thus, conservation of mass requires

$$\rho_1 A_1 V_1 = \rho_2 A_2 V_2$$

If the density remains constant, then $\rho_1 = \rho_2$ and the above becomes the **continuity equation** for incompressible flow

$$A_1 V_1 = A_2 V_2, \text{ or } Q_1 = Q_2 \tag{3.16}$$

EXAMPLE 3.7 | Flow from a Tank—Gravity Driven

GIVEN A stream of water of diameter $d = 0.1$ m flows steadily from a tank of diameter $D = 1.0$ m as shown in Fig. E3.7a.

FIND Determine the flow rate, Q, needed from the inflow pipe. The water depth is constant at $h = 2.0$ m.

SOLUTION

For steady, inviscid, incompressible flow the Bernoulli equation applied between points (1) and (2) is

$$p_1 + \tfrac{1}{2}\rho V_1^2 + \gamma z_1 = p_2 + \tfrac{1}{2}\rho V_2^2 + \gamma z_2 \tag{1}$$

With the assumptions that $p_1 = p_2 = 0$, $z_1 = h$, and $z_2 = 0$, Eq. 1 becomes

$$\tfrac{1}{2}V_1^2 + gh = \tfrac{1}{2}V_2^2 \tag{2}$$

Although the water level remains constant ($h = $ constant), there is an average velocity, V_1, across section (1) because of the flow from the tank. From Eq. 3.16 for steady incompressible flow, conservation of mass requires $Q_1 = Q_2$, where $Q = AV$. Thus, $A_1 V_1 = A_2 V_2$, or

$$\frac{\pi}{4}D^2 V_1 = \frac{\pi}{4}d^2 V_2$$

Hence,

$$V_1 = \left(\frac{d}{D}\right)^2 V_2 \tag{3}$$

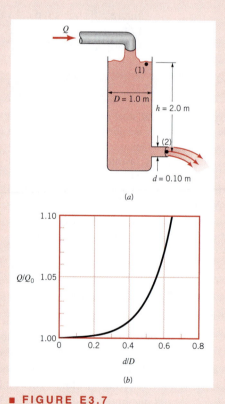

(a)

(b)

■ **FIGURE E3.7**

Equations 1 and 3 can be combined to give

$$V_2 = \sqrt{\frac{2gh}{1 - (d/D)^4}} = \sqrt{\frac{2(9.81\,\text{m/s}^2)(2.0\,\text{m})}{1 - (0.1\,\text{m}/1\,\text{m})^4}}$$

$$= 6.26\,\text{m/s}$$

Thus,

$$Q = A_1V_1 = A_2V_2 = \frac{\pi}{4}(0.1\,\text{m})^2(6.26\,\text{m/s})$$

$$= 0.0492\,\text{m}^3/\text{s} \qquad \textbf{(Ans)}$$

COMMENT In this example we have not neglected the kinetic energy of the water in the tank ($V_1 \neq 0$). If the tank

diameter is large compared to the jet diameter ($D \gg d$), Eq. 3 indicates that $V_1 \ll V_2$ and the assumption that $V_1 \approx 0$ would be reasonable. The error associated with this assumption can be seen by calculating the ratio of the flowrate assuming $V_1 \neq 0$, denoted Q, to that assuming $V_1 = 0$, denoted Q_0. This ratio, written as

$$\frac{Q}{Q_0} = \frac{V_2}{V_2|_{D=\infty}} = \frac{\sqrt{2gh/[1 - (d/D)^4]}}{\sqrt{2gh}} = \frac{1}{\sqrt{1 - (d/D)^4}}$$

is plotted in Fig. E3.7b. With $0 < d/D < 0.4$ it follows that $1 < Q/Q_0 \lesssim 1.01$, and the error in assuming $V_1 = 0$ is less than 1%. Thus, it is often reasonable to assume $V_1 = 0$.

The fact that a kinetic energy change is often accompanied by a change in pressure is shown by Example 3.8.

EXAMPLE 3.8   Flow from a Tank—Pressure Driven

GIVEN Air flows steadily from a tank, through a hose of diameter $D = 0.03$ m, and exits to the atmosphere from a nozzle of diameter $d = 0.01$ m as shown in Fig. E3.8a. The pressure in the tank remains constant at 3.0 kPa (gage), and the atmospheric conditions are standard temperature and pressure.

FIND Determine

(a) the flowrate and

(b) the pressure in the hose.

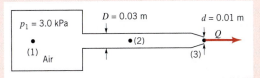

■ **FIGURE E3.8a**

SOLUTION

(a) If the flow is assumed steady, inviscid, and incompressible, we can apply the Bernoulli equation along the streamline from (1) to (2) to (3) as

$$p_1 + \tfrac{1}{2}\rho V_1^2 + \gamma z_1 = p_2 + \tfrac{1}{2}\rho V_2^2 + \gamma z_2$$
$$= p_3 + \tfrac{1}{2}\rho V_3^2 + \gamma z_3$$

With the assumption that $z_1 = z_2 = z_3$ (horizontal hose), $V_1 = 0$ (large tank), and $p_3 = 0$ (free jet) this becomes

$$V_3 = \sqrt{\frac{2p_1}{\rho}}$$

and

$$p_2 = p_1 - \tfrac{1}{2}\rho V_2^2 \qquad \textbf{(1)}$$

The density of the air in the tank is obtained from the perfect gas law, using standard absolute pressure and temperature, as

$$\rho = \frac{p_1}{RT_1}$$

$$= [(3.0 + 101)\,\text{kN/m}^2]$$

$$\times \frac{10^3\,\text{N/kN}}{(286.9\,\text{N}\cdot\text{m/kg}\cdot\text{K})(15 + 273)\text{K}}$$

$$= 1.26\,\text{kg/m}^3$$

Thus, we find that

$$V_3 = \sqrt{\frac{2(3.0 \times 10^3\,\text{N/m}^2)}{1.26\,\text{kg/m}^3}} = 69.0\,\text{m/s}$$

or

$$Q = A_3V_3 = \frac{\pi}{4}d^2V_3 = \frac{\pi}{4}(0.01\,\text{m})^2(69.0\,\text{m/s})$$

$$= 0.00542\,\text{m}^3/\text{s} \qquad \textbf{(Ans)}$$

COMMENT Note that the value of V_3 is determined strictly by the value of p_1 (and the assumptions involved in the Bernoulli equation), independent of the "shape" of the nozzle. The pressure head within the tank, $p_1/\gamma = (3.0\,\text{kPa})/(9.81\,\text{m/s}^2)(1.26\,\text{kg/m}^3) = 243$ m, is converted to the velocity head at the exit, $V_2^2/2g = (69.0\,\text{m/s})^2/(2 \times 9.81\,\text{m/s}^2) = 243$ m. Although we used gage pressure in the Bernoulli equation ($p_3 = 0$), we had to use absolute pressure in the perfect gas law when calculating the density.

(b) The pressure within the hose can be obtained from Eq. 1 and the continuity equation (Eq. 3.16)

$$A_2V_2 = A_3V_3$$

Hence,

$$V_2 = A_3V_3/A_2 = \left(\frac{d}{D}\right)^2 V_3$$

$$= \left(\frac{0.01 \text{ m}}{0.03 \text{ m}}\right)^2 (69.0 \text{ m/s}) = 7.67 \text{ m/s}$$

and from Eq. 1

$$p_2 = 3.0 \times 10^3 \text{ N/m}^2 - \tfrac{1}{2}(1.26 \text{ kg/m}^3)(7.67 \text{ m/s})^2$$

$$= (3000 - 37.1)\text{N/m}^2 = 2963 \text{ N/m}^2 \qquad \textbf{(Ans)}$$

COMMENTS In the absence of viscous effects, the pressure throughout the hose is constant and equal to p_2. Physically, decreases in pressure from p_1 to p_2 to p_3 accelerate the air and increase its kinetic energy from zero in the tank to an intermediate value in the hose and finally to its maximum value at the nozzle exit. Since the air velocity in the nozzle exit is nine times that in the hose, most of the pressure drop occurs across the nozzle ($p_1 = 3000 \text{ N/m}^2$, $p_2 = 2963 \text{ N/m}^2$ and $p_3 = 0$).

Because the pressure change from (1) to (3) is not too great [that is, in terms of absolute pressure $(p_1 - p_3)/p_1 = 3.0/101 = 0.03$], it follows from the perfect gas law that the density change is also not significant. Hence, the incompressibility assumption is reasonable for this problem. If the tank pressure were considerably larger or if viscous effects were important, the aforementioned results would be incorrect.

By repeating the calculations for various nozzle diameters, d, the results shown in Figs. E3.8b,c are obtained. The flowrate increases as the nozzle is opened (i.e., larger d). Note that if the nozzle diameter is the same as that of the hose ($d = 0.03$ m), the pressure throughout the hose is atmospheric (zero gage).

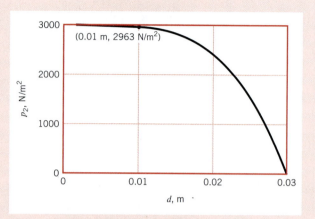

■ **FIGURE E3.8b**

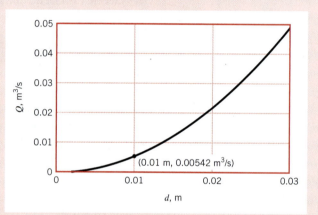

■ **FIGURE E3.8c**

In many situations the combined effects of kinetic energy, pressure, and gravity are important. Example 3.9 illustrates this.

<table>
</table>

EXAMPLE 3.9 Flow in a Variable Area Pipe

GIVEN Water flows through a pipe reducer as is shown in Fig. E3.9. The static pressures at (1) and (2) are measured by the inverted U-tube manometer containing oil of specific gravity, SG, less than one.

FIND Determine the manometer reading, h.

SOLUTION

With the assumptions of steady, inviscid, incompressible flow, the Bernoulli equation can be written as

$$p_1 + \tfrac{1}{2}\rho V_1^2 + \gamma z_1 = p_2 + \tfrac{1}{2}\rho V_2^2 + \gamma z_2$$

The continuity equation (Eq. 3.16) provides a second relationship between V_1 and V_2 if we assume the velocity profiles are uniform at those two locations and the fluid is incompressible:

$$Q = A_1 V_1 = A_2 V_2$$

By combining these two equations we obtain

$$p_1 - p_2 = \gamma(z_2 - z_1) + \tfrac{1}{2}\rho V_2^2[1 - (A_2/A_1)^2] \quad \text{(1)}$$

This pressure difference is measured by the manometer and can be determined using the pressure-depth ideas developed in Chapter 2. Thus,

$$p_1 - \gamma(z_2 - z_1) - \gamma\ell - \gamma h + SG\,\gamma h + \gamma\ell = p_2$$

or

$$p_1 - p_2 = \gamma(z_2 - z_1) + (1 - SG)\gamma h \quad \text{(2)}$$

As discussed in Chapter 2, this pressure difference is neither merely γh nor $\gamma(h + z_1 - z_2)$.

Equations 1 and 2 can be combined to give the desired result as follows

$$(1 - SG)\gamma h = \tfrac{1}{2}\rho V_2^2\left[1 - \left(\frac{A_2}{A_1}\right)^2\right]$$

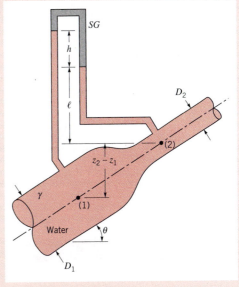

■ **FIGURE E3.9**

or since $V_2 = Q/A_2$

$$h = (Q/A_2)^2\,\frac{1 - (A_2/A_1)^2}{2g(1 - SG)} \quad \text{(Ans)}$$

COMMENTS The difference in elevation, $z_1 - z_2$, was not needed because the change in the elevation term in the Bernoulli equation exactly cancels the elevation term in the manometer equation. However, the pressure difference, $p_1 - p_2$, depends on the angle θ because of the elevation, $z_1 - z_2$, in Eq. 1. Thus, for a given flowrate, the pressure difference, $p_1 - p_2$, as measured by a pressure gage, would vary with θ, but the manometer reading, h, would be independent of θ.

V3.6 Venturi channel

In general, an increase in velocity is accompanied by a decrease in pressure. If the differences in velocity are considerable, the differences in pressure can also be considerable. For flows of liquids, this may result in cavitation, a potentially dangerous situation that results when the liquid pressure is reduced to the vapor pressure and the liquid "boils."

One way to produce cavitation in a flowing liquid is noted from the Bernoulli equation. If the fluid velocity is increased (for example, by a reduction in flow area as shown in Fig. 3.10) the pressure will decrease. This pressure decrease (needed to accelerate the fluid through the constriction) can be large enough so that the pressure in the liquid is reduced to its vapor pressure.

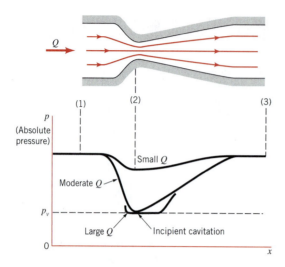

■ **FIGURE 3.10** Pressure variation and cavitation in a variable area pipe.

EXAMPLE 3.10 Siphon and Cavitation

GIVEN Water at 60° F is siphoned from a large tank through a constant diameter hose as shown in Fig. E3.10a. The end of the siphon is 5 ft below the bottom of the tank. Atmospheric pressure is 14.7 psia.

FIND Determine the maximum height of the hill, H, over which the water can be siphoned without cavitation occurring.

SOLUTION

If the flow is steady, inviscid, and incompressible, we can apply the Bernoulli equation along the streamline from (1) to (2) to (3) as follows

$$p_1 + \tfrac{1}{2}\rho V_1^2 + \gamma z_1 = p_2 + \tfrac{1}{2}\rho V_2^2 + \gamma z_2$$
$$= p_3 + \tfrac{1}{2}\rho V_3^2 + \gamma z_3 \qquad (1)$$

With the tank bottom as the datum, we have $z_1 = 15$ ft, $z_2 = H$, and $z_3 = -5$ ft. Also, $V_1 = 0$ (large tank), $p_1 = 0$ (open tank), $p_3 = 0$ (free jet), and from the continuity equation $A_2V_2 = A_3V_3$, or because the hose is constant diameter, $V_2 = V_3$. Thus, the speed of the fluid in the hose is determined from Eq. 1 to be

$$V_3 = \sqrt{2g(z_1 - z_3)} = \sqrt{2(32.2 \text{ ft/s}^2)[15 - (-5)]\text{ft}}$$
$$= 35.9 \text{ ft/s} = V_2$$

Use of Eq. 1 between points (1) and (2) then gives the pressure p_2 at the top of the hill as

$$p_2 = p_1 + \tfrac{1}{2}\rho V_1^2 + \gamma z_1 - \tfrac{1}{2}\rho V_2^2 - \gamma z_2$$
$$= \gamma(z_1 - z_2) - \tfrac{1}{2}\rho V_2^2 \qquad (2)$$

From Table B.1, the vapor pressure of water at 60° F is 0.256 psia. Hence, for incipient cavitation the lowest pressure

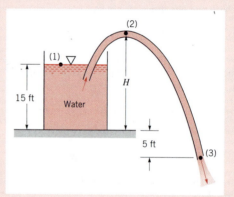

■ **FIGURE E3.10a**

in the system will be $p = 0.256$ psia. Careful consideration of Eq. 2 and Fig. E3.10a will show that this lowest pressure will occur at the top of the hill. Because we have used gage pressure at point (1) ($p_1 = 0$), we must use gage pressure at point (2) also. Thus, $p_2 = 0.256 - 14.7 = -14.4$ psi and Eq. 2 gives

$$(-14.4 \text{ lb/in.}^2)(144 \text{ in.}^2/\text{ft}^2)$$
$$= (62.4 \text{ lb/ft}^3)(15 - H)\text{ft} - \tfrac{1}{2}(1.94 \text{ slugs/ft}^3)(35.9 \text{ ft/s})^2$$

or

$$H = 28.2 \text{ ft} \qquad \textbf{(Ans)}$$

For larger values of H, vapor bubbles will form at point (2) and the siphon action may stop.

COMMENTS Note that we could have used absolute pressure throughout ($p_2 = 0.256$ psia and $p_1 = 14.7$ psia) and obtained the same result. The lower the elevation of point (3), the larger the flowrate and, therefore, the smaller the value of H allowed.

We could also have used the Bernoulli equation between (2) and (3), with $V_2 = V_3$, to obtain the same value of H. In this case it would not have been necessary to determine V_2 by use of the Bernoulli equation between (1) and (3).

The aforementioned results are independent of the diameter and length of the hose (provided viscous effects are not important). Proper design of the hose (or pipe) is needed to ensure that it will not collapse due to the large pressure difference (vacuum) between the inside and the outside of the hose.

By using the fluid properties listed in Table 1.5 and repeating the calculations for various fluids, the results shown in Fig. E3.10b are obtained. The value of H is a function of both the specific weight of the fluid, γ, and its vapor pressure, p_v.

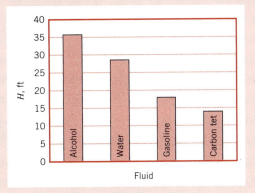

■ **FIGURE E3.10b**

3.6.3 Flowrate Measurement

Many types of devices using principles involved in the Bernoulli equation have been developed to measure fluid velocities and flowrates.

An effective way to measure the flowrate through a pipe is to place some type of restriction within the pipe as shown in Fig. 3.11 and to measure the pressure difference between the low-velocity, high-pressure upstream section (1) and the high-velocity, low-pressure downstream section (2). Three commonly used types of *flowmeters* are illustrated: the *orifice meter,* the *nozzle meter,* and the *Venturi meter.* The operation of each is based on the same physical principles—an increase in velocity causes a decrease in pressure.

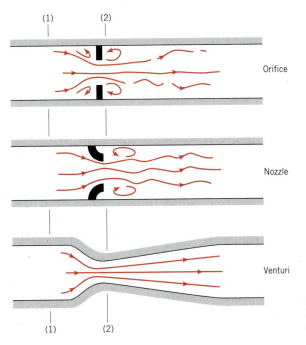

■ **FIGURE 3.11** Typical devices for measuring flowrate in pipes.

We assume the flow is horizontal ($z_1 = z_2$), steady, inviscid, and incompressible between points (1) and (2). The Bernoulli equation becomes

$$p_1 + \tfrac{1}{2}\rho V_1^2 = p_2 + \tfrac{1}{2}\rho V_2^2$$

In addition, the continuity equation (Eq. 3.16) can be written as

$$Q = A_1 V_1 = A_2 V_2$$

where A_2 is the small ($A_2 < A_1$) flow area at section (2). Combining these two equations results in the following theoretical flowrate

$$Q = A_2\sqrt{\frac{2(p_1 - p_2)}{\rho[1 - (A_2/A_1)^2]}} \qquad (3.17)$$

The actual measured flowrate, Q_{actual}, will be smaller than this theoretical result because of various differences between the "real world" and the assumptions used in the derivation of Eq. 3.17. These differences (which are quite consistent and may be as small as 1 to 2% or as large as 40% depending on the geometry used) are discussed in Chapter 8. Note that as shown in the figure in the margin, the flowrate is proportional to the square root of the pressure difference.

(margin figure) $Q \sim \sqrt{\Delta p}$; $\Delta p = p_1 - p_2$

EXAMPLE 3.11 | Venturi Meter

GIVEN Kerosene ($SG = 0.85$) flows through the Venturi meter shown in Fig. E3.11a with flowrates between 0.005 and 0.050 m³/s.

FIND Determine the range in pressure difference, $p_1 - p_2$, needed to measure these flowrates.

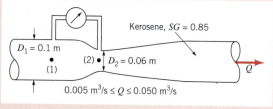

Kerosene, $SG = 0.85$

$D_1 = 0.1$ m

(2) ● $D_2 = 0.06$ m

(1)

Q

$0.005 \text{ m}^3/\text{s} \le Q \le 0.050 \text{ m}^3/\text{s}$

■ **FIGURE E3.11a**

SOLUTION

If the flow is assumed to be steady, inviscid, and incompressible, the relationship between flowrate and pressure is given by Eq. 3.17. This can be rearranged to give

$$p_1 - p_2 = \frac{Q^2\rho[1 - (A_2/A_1)^2]}{2A_2^2}$$

With a density of the flowing fluid of

$$\rho = SG\,\rho_{H_2O} = 0.85(1000 \text{ kg/m}^3) = 850 \text{ kg/m}^3$$

and the area ratio

$$A_2/A_1 = (D_2/D_1)^2 = (0.06 \text{ m}/0.10 \text{ m})^2 = 0.36$$

the pressure difference for the smallest flowrate is

$$p_1 - p_2 = (0.005 \text{ m}^3/\text{s})^2(850 \text{ kg/m}^3)\frac{(1 - 0.36^2)}{2\left[(\pi/4)(0.06 \text{ m})^2\right]^2}$$

$$= 1160 \text{ N/m}^2 = 1.16 \text{ kPa}$$

Likewise, the pressure difference for the largest flowrate is

$$p_1 - p_2 = (0.05)^2(850)\frac{(1 - 0.36^2)}{2[(\pi/4)(0.06)^2]^2}$$

$$= 1.16 \times 10^2 \text{ N/m}^2 = 116 \text{ kPa}$$

Thus,

$$1.16 \text{ kPa} \le p_1 - p_2 \le 116 \text{ kPa} \qquad \textbf{(Ans)}$$

These values represent pressure differences for inviscid, steady, incompressible conditions. The ideal results presented here are independent of the particular flowmeter geometry—an orifice, nozzle, or Venturi meter (see Fig. 3.11).

COMMENT Equation 3.17 shows that the pressure difference varies as the square of the flowrate. Hence, as indicated by the numerical results and shown in Fig. E3.11b, a 10-fold increase in flowrate requires a 100-fold increase in pressure difference. This nonlinear relationship can cause difficulties when measuring flowrates over a wide range of values. Such measurements would require pressure transducers with a wide range of operation. An alternative is to use two flowmeters in parallel—one for the larger and one for the smaller flowrate ranges.

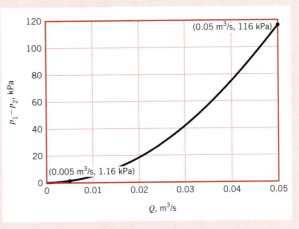

■ FIGURE E3.11b

Other flowmeters based on the Bernoulli equation are used to measure flowrates in open channels such as flumes and irrigation ditches. The *sluice gate* as shown in Fig. 3.12 is an example.

We apply the Bernoulli and continuity equations between points on the free surfaces at (1) and (2) to give

$$p_1 + \tfrac{1}{2}\rho V_1^2 + \gamma z_1 = p_2 + \tfrac{1}{2}\rho V_2^2 + \gamma z_2$$

and

$$Q = A_1 V_1 = b V_1 z_1 = A_2 V_2 = b V_2 z_2$$

With the fact that $p_1 = p_2 = 0$, these equations can be combined to give the flowrate as

$$Q = z_2 b \sqrt{\frac{2g(z_1 - z_2)}{1 - (z_2/z_1)^2}} \tag{3.18}$$

The downstream depth, z_2, not the gate opening, a, was used to obtain the result of Eq. 3.18 since a vena contracta results with a contraction coefficient, $C_c = z_2/a$, less than 1. Typically C_c is approximately 0.61 over the depth ratio range of $0 < a/z_1 < 0.2$. For larger values of a/z_1 the value of C_c increases rapidly.

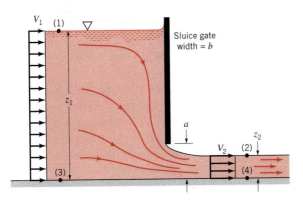

■ FIGURE 3.12   Sluice gate geometry.

EXAMPLE 3.12 Sluice Gate

GIVEN Water flows under the sluice gate shown in Fig. E3.12a.

FIND Determine the approximate flowrate per unit width of the channel.

SOLUTION

Under the assumptions of steady, inviscid, incompressible flow, we can apply Eq. 3.18 to obtain Q/b, the flowrate per unit width, as

$$\frac{Q}{b} = z_2 \sqrt{\frac{2g(z_1 - z_2)}{1 - (z_2/z_1)^2}}$$

In this instance, $z_1 = 5.0$ m and $a = 0.80$ m so the ratio $a/z_1 = 0.16 < 0.20$, and we can assume that the contraction coefficient is approximately $C_c = 0.61$. Thus, $z_2 = C_c a = 0.61$ (0.80 m) = 0.488 m and we obtain the flowrate

$$\frac{Q}{b} = (0.488 \text{ m}) \sqrt{\frac{2(9.81 \text{ m/s}^2)(5.0 \text{ m} - 0.488 \text{ m})}{1 - (0.488 \text{ m}/5.0 \text{ m})^2}}$$

$$= 4.61 \text{ m}^2/\text{s} \qquad \textbf{(Ans)}$$

COMMENTS If we consider $z_1 \gg z_2$ and neglect the kinetic energy of the upstream fluid, we would have

$$\frac{Q}{b} = z_2 \sqrt{2gz_1} = 0.488 \text{ m} \sqrt{2(9.81 \text{ m/s}^2)(5.0 \text{ m})}$$

$$= 4.83 \text{ m}^2/\text{s}$$

In this case the difference in Q with or without including V_1 is not too significant because the depth ratio is fairly large $(z_1/z_2 = 5.0/0.488 = 10.2)$. Thus, it is often reasonable to neglect the kinetic energy upstream from the gate compared to that downstream of it.

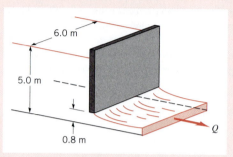

■ **FIGURE E3.12a**

By repeating the calculations for various flow depths, z_1, the results shown in Fig. E3.12b are obtained. Note that the flowrate is not directly proportional to the flow depth. Thus, for example, if during flood conditions the upstream depth doubled from $z_1 = 5$ m to $z_1 = 10$ m, the flowrate per unit width of the channel would not double, but would increase only from 4.61 m^2/s to 6.67 m^2/s.

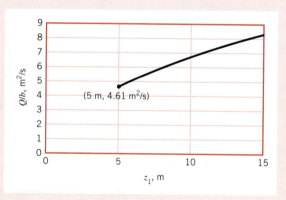

■ **FIGURE E3.12b**

3.7 The Energy Line and the Hydraulic Grade Line

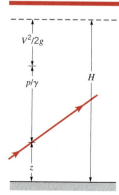

A useful interpretation of the Bernoulli equation can be obtained through the use of the concepts of the *hydraulic grade line* (HGL) and the *energy line* (EL). These ideas represent a geometrical interpretation of a flow.

For steady, inviscid, incompressible flow the Bernoulli equation states that the sum of the pressure head, the velocity head, and the elevation head is constant along a streamline. This constant, called the *total head*, H is shown in the figure in the margin.

$$\frac{p}{\gamma} + \frac{V^2}{2g} + z = \text{constant on a streamline} = H \qquad \textbf{(3.19)}$$

The energy line is a line that represents the total head available to the fluid. As shown in Fig. 3.13, elevation of the energy line can be obtained by measuring the stagnation pressure

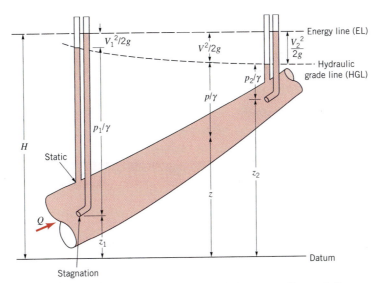

■ FIGURE 3.13 **Representation of the energy line and the hydraulic grade line.**

with a Pitot tube. The stagnation point at the end of the Pitot tube provides a measurement of the total head (or energy) of the flow. The static pressure tap connected to the piezometer tube shown, on the other hand, measures the sum of the pressure head and the elevation head, $p/\gamma + z$. This sum is often called the *piezometric head*.

A Pitot tube at another location in the flow will measure the same total head, as is shown in Fig. 3.13. The elevation head, velocity head, and pressure head may vary along the streamline, however.

The locus of elevations provided by a series of Pitot tubes is termed the energy line, EL. That provided by a series of piezometer taps is termed the hydraulic grade line, HGL. Under the assumptions of the Bernoulli equation, the energy line is horizontal. If the fluid velocity changes along the streamline, the hydraulic grade line will not be horizontal.

The energy line and hydraulic grade line for flow from a large tank are shown in Fig. 3.14. If the flow is steady, incompressible, and inviscid, the energy line is horizontal and at the elevation of the liquid in the tank. The hydraulic grade line lies a distance of one velocity head, $V^2/2g$, below the energy line.

The distance from the pipe to the hydraulic grade line indicates the pressure within the pipe as is shown in Fig. 3.15. If the pipe lies below the hydraulic grade line, the pressure within the pipe is positive (above atmospheric). If the pipe lies above the hydraulic grade line, the pressure is negative (below atmospheric).

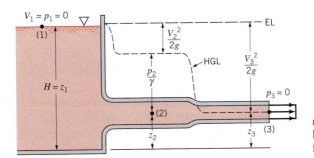

■ FIGURE 3.14 **The energy line and hydraulic grade line for flow from a tank.**

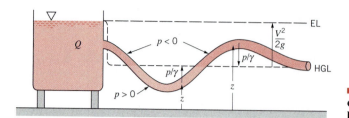

■ **FIGURE 3.15** Use of the energy line and the hydraulic grade line.

EXAMPLE 3.13 | Energy Line and Hydraulic Grade Line

GIVEN Water is siphoned from the tank shown in Fig. E3.13 through a hose of constant diameter. A small hole is found in the hose at location (1) as indicated.

FIND When the siphon is used, will water leak out of the hose or will air leak into the hose, thereby possibly causing the siphon to malfunction?

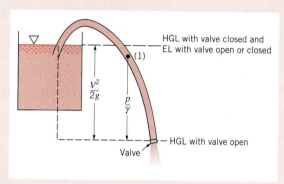

■ **FIGURE E3.13**

SOLUTION

Whether air will leak into or water will leak out of the hose depends on whether the pressure within the hose at (1) is less than or greater than atmospheric. Which happens can be determined easily by using the energy line and hydraulic grade line concepts. With the assumption of steady, incompressible, inviscid flow it follows that the total head is constant—thus, the energy line is horizontal.

Because the hose diameter is constant, it follows from the continuity equation (AV = constant) that the water velocity in the hose is constant throughout. Thus, the hydraulic grade line is a constant distance, $V^2/2g$, below the energy line as shown in Fig. E3.13. Because the pressure at the end of the hose is atmospheric, it follows that the hydraulic grade line is at the same elevation as the end of the hose outlet. The fluid within the hose at any point above the hydraulic grade line will be at less than atmospheric pressure.

Thus, air will leak into the hose through the hole at point (1). **(Ans)**

COMMENT In practice, viscous effects may be quite important, making this simple analysis (horizontal energy line) incorrect. However, if the hose is "not too small diameter," "not too long," the fluid "not too viscous," and the flowrate "not too large," the aforementioned result may be very accurate. If any of these assumptions are relaxed, a more detailed analysis is required (see Chapter 8). If the end of the hose were closed so the flowrate were zero, the hydraulic grade line would coincide with the energy line ($V^2/2g = 0$ throughout), the pressure at (1) would be greater than atmospheric, and water would leak through the hole at (1).

3.8 Restrictions on the Use of the Bernoulli Equation

One of the main assumptions in deriving the Bernoulli equation is that the fluid is incompressible. Although this is reasonable for most liquid flows, it can, in certain instances, introduce considerable errors for gases.

In the previous section we saw that the stagnation pressure is greater than the static pressure by an amount $\rho V^2/2$, provided that the density remains constant. If this dynamic pressure is not too large compared with the static pressure, the density change between two points is not very large and the flow can be considered incompressible. However, because the dynamic pressure varies as V^2, the error associated with the assumption that a fluid is incompressible increases with the square of the velocity of the fluid.

A "rule of thumb" is that the flow of a perfect gas may be considered as incompressible provided the Mach number is less than about 0.3. The Mach number, $Ma = V/c$, is the ratio of the fluid speed, V, to the speed of sound in the fluid, c. In standard air ($T_1 = 59°$ F, $c_1 = \sqrt{kRT_1} = 1117$ ft/s) this corresponds to a speed of $V_1 = c_1 Ma_1 = 0.3\,(1117$ ft/s$) = 335$ ft/s $= 228$ mi/hr. At higher speeds, compressibility may become important.

Another restriction of the Bernoulli equation (Eq. 3.6) is the assumption that the flow is steady. For such flows, on a given streamline the velocity is a function of only s, the location along the streamline. That is, along a streamline $V = V(s)$. For unsteady flows the velocity is also a function of time, so that along a streamline $V = V(s, t)$. Thus, when taking the time derivative of the velocity to obtain the streamwise acceleration, we obtain $a_s = \partial V/\partial t + V\,\partial V/\partial s$ rather than just $a_s = V\,\partial V/\partial s$ as is true for steady flow. The unsteady term, $\partial V/\partial t$, does not allow the equation of motion to be integrated easily (as was done to obtain the Bernoulli equation) unless additional assumptions are made.

Another restriction on the Bernoulli equation is that the flow is inviscid. Recall that the Bernoulli equation is actually a first integral of Newton's second law along a streamline. This general integration was possible because, in the absence of viscous effects, the fluid system considered was a conservative system. The total energy of the system remains constant. If viscous effects are important, the system is nonconservative and energy losses occur. A more detailed analysis is needed for these cases. Such material is presented in Chapter 8.

The final basic restriction on use of the Bernoulli equation is that there are no mechanical devices (pumps or turbines) in the system between the two points along the streamline for which the equation is applied. These devices represent sources or sinks of energy. Since the Bernoulli equation is actually one form of the energy equation, it must be altered to include pumps or turbines, if these are present. The inclusion of pumps and turbines is covered in Chapters 5 and 11.

3.9 Chapter Summary and Study Guide

steady flow
streamline
Bernoulli equation
elevation head
pressure head
velocity head
static pressure
dynamic pressure
stagnation point
stagnation pressure
total pressure
Pitot-static tube
free jet
volume flowrate
continuity equation
flowmeter
hydraulic
grade line
energy line

In this chapter, several aspects of the steady flow of an inviscid, incompressible fluid are discussed. Newton's second law, $F = m\mathbf{a}$, is applied to flows for which the only important forces are those due to pressure and gravity (weight)—viscous effects are assumed negligible. The result is the often-used Bernoulli equation, which provides a simple relationship among pressure, elevation, and velocity variations along a streamline. A similar but less often used equation is also obtained to describe the variations in these parameters normal to a streamline.

The concept of a stagnation point and the corresponding stagnation pressure is introduced, as are the concepts of static, dynamic, and total pressure and their related heads.

Several applications of the Bernoulli equation are discussed. In some flow situations, such as the use of a Pitot-static tube to measure fluid velocity or the flow of a liquid as a free jet from a tank, a Bernoulli equation alone is sufficient for the analysis. In other instances, such as confined flows in tubes and flowmeters, it is necessary to use both the Bernoulli equation and the continuity equation, which is a statement of the fact that mass is conserved as fluid flows.

The following checklist provides a study guide for this chapter. When your study of the entire chapter and end-of-chapter exercises has been completed, you should be able to

- write out meanings of the terms listed here in the margin and understand each of the related concepts. These terms are particularly important and are set in color and bold type in the text.

- explain the origin of the pressure, elevation, and velocity terms in the Bernoulli equation and how they are related to Newton's second law of motion.
- apply the Bernoulli equation to simple flow situations, including Pitot-static tubes, free jet flows, confined flows, and flowmeters.
- use the concept of conservation of mass (the continuity equation) in conjunction with the Bernoulli equation to solve simple flow problems.
- apply Newton's second law across streamlines for appropriate steady, inviscid, incompressible flows.
- use the concepts of pressure, elevation, velocity, and total heads to solve various flow problems.
- explain and use the concepts of static, stagnation, dynamic, and total pressures.
- use the energy line and the hydraulic grade line concepts to solve various flow problems.

Review Problems

Go to Appendix F for a set of review problems with answers. Detailed solutions can be found in *Student Solution Manual for a Brief Introduction to Fluid Mechanics,* by Young, et al. (© 2006 John Wiley and Sons, Inc.).

Problems

Note: Unless otherwise indicated use the values of fluid properties found in the tables on the inside of the front cover. Problems designated with an (*) are intended to be solved with the aid of a programmable calculator or a computer. Problems designated with a (†) are "open-ended" problems and require critical thinking in that to work them one must make various assumptions and provide the necessary data. There is not a unique answer to these problems.

Answers to the even-numbered problems are listed at the end of the book. Access to the videos that accompany problems can be obtained through the book's web site, www.wiley.com/college/young. The lab-type problems and FE problems can also be accessed on this web site.

Section 3.2   F = ma Along a Streamline

3.1   Water flows steadily through the variable area horizontal pipe shown in Fig. P3.1. The centerline velocity is given by $V = 10(1+x)\hat{\imath}$ ft/s, where x is in feet. Viscous effects are neglected. **(a)** Determine the pressure gradient, $\partial p/\partial x$ (as a function of x), needed to produce this flow. **(b)** If the pressure at section (1) is 50 psi, determine the pressure at (2) by (i) integration of the pressure gradient obtained in **(a)** and (ii) application of the Bernoulli equation.

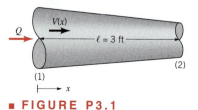

■ **FIGURE P3.1**

3.2   Repeat Problem 3.1 if the pipe is vertical with the flow upward.

3.3   An incompressible fluid flows steadily past a circular cylinder as shown in Fig. P3.3 (see **Video V3.3** also). The fluid velocity along the dividing streamline $(-\infty \leq x \leq -a)$ is found to be $V = V_0(1 - a^2/x^2)$, where a is the radius of the cylinder and V_0 is the upstream velocity. **(a)** Determine the pressure gradient along this streamline. **(b)** If the upstream pressure is p_0, integrate the pressure gradient to obtain the pressure $p(x)$ for $-\infty \leq x \leq -a$. **(c)** Show from the result of part **(b)** that the pressure at the stagnation point $(x = -a)$ is $p_0 + \rho V_0^2/2$, as expected from the Bernoulli equation.

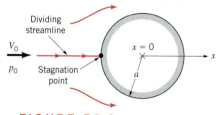

■ **FIGURE P3.3**

3.4   What pressure gradient along the streamline, dp/ds, is required to accelerate water in a horizontal pipe at a rate of 10 m/s²?

3.5   What pressure gradient along the streamline, dp/ds, is required to accelerate water upward in a vertical pipe at a rate of 30 ft/s²? What is the answer if the flow is downward?

3.6   The Bernoulli equation is valid for steady, inviscid, incompressible flows with constant acceleration of gravity. Consider flow on a planet where the acceleration of gravity varies with height so that $g = g_0 - cz$, where g_0 and c are constants. Integrate "$\mathbf{F} = m\mathbf{a}$" along a streamline to obtain the equivalent of the Bernoulli equation for this flow.

Section 3.3 **F = ma Normal to a Streamline**

3.7   Water in a container and air in a tornado flow in horizontal circular streamlines of radius r and speed V as shown in **Video V3.2** and Fig. P3.7. Determine the radial pressure gradient, $\partial p/\partial r$, needed for the following situations: **(a)** The fluid is water with $r = 3$ in. and $V = 0.8$ ft/s and **(b)** the fluid is air with $r = 300$ ft and $V = 200$ mph.

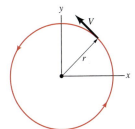

■ **FIGURE P3.7**

3.8   Water flows around the vertical two-dimensional bend with circular streamlines and constant velocity as shown in Fig. P3.8. If the pressure is 40 kPa at point (1), determine the pressures at points (2) and (3). Assume that the velocity profile is uniform as indicated.

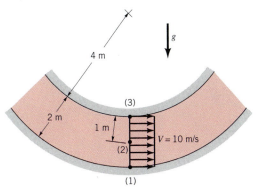

■ **FIGURE P3.8**

†3.9   Air flows smoothly past your face as you ride your bike, but bugs and particles or dust pelt your face and get into your eyes. Explain why this is so.

Section 3.5 **Static, Stagnation, Dynamic, and Total Pressure**

3.10   A person holds her hand out of an open car window while the car drives through still air at 65 mph. Under standard atmospheric conditions, what is the maximum pressure on her hand? What would be the maximum pressure if the "car" were an Indy 500 racer traveling 200 mph?

3.11   (See Fluids in the News article titled "Pressurized eyes," Section 3.5.) Determine the air velocity needed to produce a stagnation pressure equal to 10 mm of mercury.

3.12   (See Fluids in the News article titled "Bugged and plugged Pitot tubes," Section 3.5.) An airplane's Pitot tube (see **Video V3.4**) used to indicate airspeed is partially plugged by an insect nest so that it measures 60% of the stagnation pressure rather than the actual stagnation pressure. If the airspeed indicator indicates that the plane is flying 150 mph, what is the actual airspeed?

3.13   Carbon tetrachloride flows in a pipe of variable diameter with negligible viscous effects. At point A in the pipe the pressure and velocity are 20 psi and 30 ft/s, respectively. At location B the pressure and velocity are 23 psi and 14 ft/s. Which point is at the higher elevation and by how much?

3.14   (See Fluids in the News article titled "Incorrect raindrop shape," Section 3.2.) The speed, V, at which a raindrop falls is a function of its diameter, D, as shown in Fig. P3.14. For what size raindrop will the stagnation pressure be equal to half the internal pressure caused by surface tension? Use the fact that due to surface tension effects the pressure inside a drop is $\Delta p = 4\sigma/D$ greater than the surrounding pressure, where σ is the surface tension.

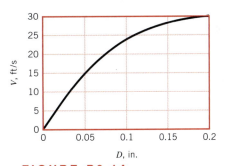

■ **FIGURE P3.14**

3.15   An inviscid fluid flows steadily along the stagnation streamline shown in Fig. P3.15 and **Video V3.3**, starting with speed V_0 far upstream of the object. Upon leaving the stagnation point, point (1), the fluid speed along the surface of the object is assumed to be given by $V = 2V_0 \sin\theta$, where θ is the angle indicated. At what angular position, θ_2, should a hole be drilled to give a pressure difference of $p_1 - p_2 = \rho V_0^2/2$? Gravity is negligible.

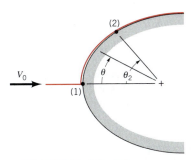

■ **FIGURE P3.15**

3.16   When an airplane is flying 200 mph at 5000-ft altitude in a standard atmosphere, the air velocity at a certain point on the wing is 273 mph relative to the airplane. What suction pressure is developed on the wing at that point? What is the pressure at the leading edge (a stagnation point) of the wing?

Section 3.6.1 Free Jets

3.17 Several holes are punched into a tin can as shown in Fig. P3.17. (See **Video V3.5**). Which of the figures represents the variation of the water velocity as it leaves the holes? Justify your choice.

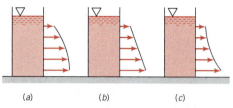

(a)   (b)   (c)

■ **FIGURE P3.17**

3.18 (See Fluids in the News article titled "Armed with a water jet for hunting," Section 3.4.) Determine the pressure needed in the gills of an archerfish if it can shoot a jet of water 1 m vertically upward. Assume steady, inviscid flow.

3.19 A rotameter is a volumetric flowmeter that consists of a tapered glass tube that contains a float as indicated in Fig. P3.19 and **Video V8.6.** The scale reading on the rotameter shown is directly proportional to the volumetric flowrate. With a scale reading of 2.6 the water bubbles up approximately 3 in. How far will it bubble up if the scale reading is 5.0?

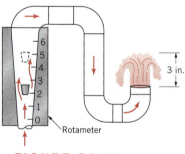

■ **FIGURE P3.19**

3.20 Water flowing from a pipe or a tank is acted upon by gravity and follows a curved trajectory as shown in Fig. P3.20 and **Videos V3.5** and **V4.3.** A simple flowmeter can be constructed as shown in Fig P3.20. A point gage mounted a distance L from the end of the horizontal pipe is adjusted to indicate that the top of the water stream is distance x below the outlet of the pipe. Show that the flowrate from this pipe of diameter D is given by $Q = \pi D^2 L\, g^{1/2} / (2^{5/2}\, x^{1/2})$.

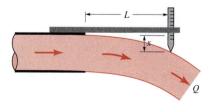

■ **FIGURE P3.20**

3.21 An inviscid, incompressible liquid flows steadily from the large pressurized tank shown in Fig. P3.21. The velocity at the exit is 40 ft/s. Determine the specific gravity of the liquid in the tank.

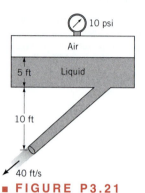

■ **FIGURE P3.21**

3.22 Streams of water from two tanks impinge upon each other as shown in Fig. P3.22. If viscous effects are negligible and point A is a stagnation point, determine the height h.

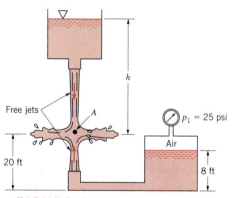

■ **FIGURE P3.22**

Section 3.6.2 Confined Flows (also see Problems 3.60 and 3.62)

3.23 A fire hose nozzle has a diameter of $1\frac{1}{8}$ in. According to some fire codes, the nozzle must be capable of delivering at least 300 gal/min. If the nozzle is attached to a 3-in.-diameter hose, what pressure must be maintained just upstream of the nozzle to deliver this flowrate?

3.24 Water flows into the sink shown in Fig. P3.24 and **Video V5.1** at a rate of 2 gal/min. If the drain is closed, the water will eventually flow through the overflow drain holes rather than over the edge of the sink. How many 0.4-in.-diameter drain holes are needed to ensure that the water does not overflow the sink? Neglect viscous effects.

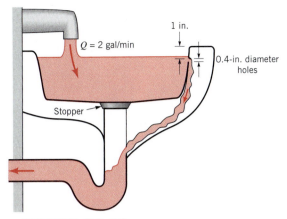

■ **FIGURE P3.24**

3.25 Air is drawn into a wind tunnel used for testing automobiles as shown in Fig. P3.25. **(a)** Determine the manometer reading, h, when the velocity in the test section is 60 mph. Note that there is a 1-in. column of oil on the water in the manometer. **(b)** Determine the difference between the stagnation pressure on the front of the automobile and the pressure in the test section.

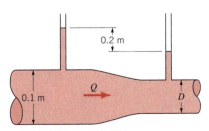

■ **FIGURE P3.25**

3.26 Water flows through the pipe contraction shown in Fig. P3.26. For the given 0.2-m difference in the manometer level, determine the flowrate as a function of the diameter of the small pipe, D.

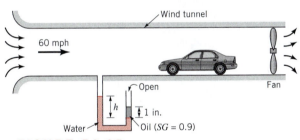

■ **FIGURE P3.26**

3.27 Water flows through the pipe contraction shown in Fig. P3.27. For the given 0.2-m difference in the manometer level, determine the flowrate as a function of the diameter of the small pipe, D.

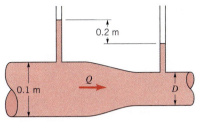

■ **FIGURE P3.27**

3.28 Water flows steadily with negligible viscous effects through the pipe shown in Fig. P3.28. Determine the diameter, D, of the pipe at the outlet (a free jet) if the velocity there is 20 ft/s.

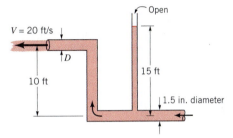

■ **FIGURE P3.28**

3.29 The circular stream of water from a faucet is observed to taper from a diameter of 20 to 10 mm in a distance of 50 cm. Determine the flowrate.

3.30 Water is siphoned from the tank shown in Fig. P3.30. The water barometer indicates a reading of 30.2 ft. Determine the maximum value of h allowed without cavitation occurring. Note that the pressure of the vapor in the closed end of the barometer equals the vapor pressure.

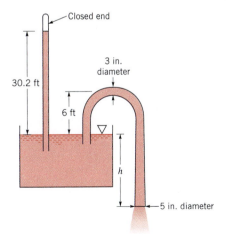

■ **FIGURE P3.30**

3.31 An inviscid fluid flows steadily through the contraction shown in Fig. P3.31. Derive an expression for the fluid velocity at (2) in terms of D_1, D_2, ρ, ρ_m, and h if the flow is assumed incompressible.

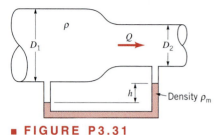

■ **FIGURE P3.31**

3.32   A 50-mm-diameter plastic tube is used to siphon water from the large tank shown in Fig. P3.32. If the pressure on the outside of the tube is more than 30 kPa greater than the pressure within the tube, the tube will collapse and the siphon will stop. If viscous effects are negligible, determine the minimum value of h allowed without the siphon stopping.

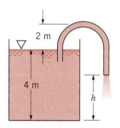

■ **FIGURE P3.32**

3.33   Carbon dioxide flows at a rate of 1.5 ft³/s from a 3-in. pipe in which the pressure and temperature are 20 psi (gage) and 120° F, respectively, into a 1.5-in. pipe. If viscous effects are neglected and incompressible conditions are assumed, determine the pressure in the smaller pipe.

3.34   Oil of specific gravity 0.83 flows in the pipe shown in Fig. P3.34. If viscous effects are neglected, what is the flowrate?

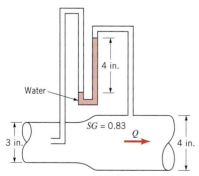

■ **FIGURE P3.34**

3.35   Water flows steadily downward through the pipe shown in Fig. P3.35. Viscous effects are negligible, and the

pressure gage indicates the pressure is zero at point (1). Determine the flowrate and the pressure at point (2).

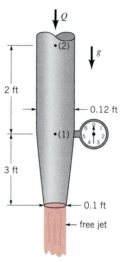

■ **FIGURE P3.35**

3.36   For the pipe enlargement shown in Fig. P3.36, the pressures at sections (1) and (2) are 56.3 and 58.2 psi, respectively. Determine the weight flowrate (lb/s) of the gasoline in the pipe. Assume steady, inviscid, incompressible flow.

■ **FIGURE P3.36**

3.37   Water is pumped from a lake through an 8-in. pipe at a rate of 10 ft³/s. If viscous effects are negligible, what is the pressure in the suction pipe (the pipe between the lake and the pump) at an elevation 6 ft above the lake?

3.38   Water flows steadily with negligible viscous effects through the pipe shown in Fig. 3.38. It is known that the 4-in.-diameter section of thin-walled tubing will collapse if the pressure within it becomes less than 10 psi below atmospheric pressure. Determine the maximum value that h can have without causing collapse of the tubing.

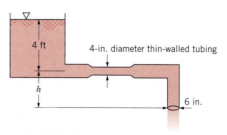

■ **FIGURE P3.38**

3.39 The vent on the tank shown in Fig. P3.39 is closed and the tank pressurized to increase the flowrate. What pressure, p_1, is needed to produce twice the flowarte of that when the vent is open?

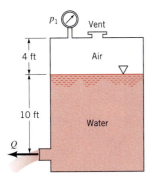

■ **FIGURE P3.39**

3.40 Water is siphoned from the tank shown in Fig. P3.40. Determine the flowrate from the tank and the pressure at points (1), (2), and (3) if viscous effects are negligible.

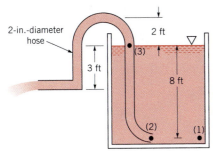

■ **FIGURE P3.40**

3.41 Redo Problem 3.40 if a 1-in.-diameter nozzle is placed at the end of the tube.

3.42 The specific gravity of the manometer fluid shown in Fig. P3.42 is 1.07. Determine the volume flowrate, Q, if the flow is inviscid and incompressible and the flowing fluid is (a) water, (b) gasoline, or (c) air at standard conditions.

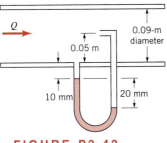

■ **FIGURE P3.42**

3.43 JP-4 fuel ($SG = 0.77$) flows through the Venturi meter shown in Fig. P3.43 with a velocity of 15 ft/s in the 6-in. pipe. If viscous effects are negligible, determine the elevation, h, of the fuel in the open tube connected to the throat of the Venturi meter.

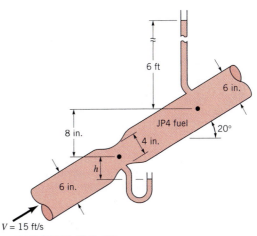

■ **FIGURE P3.43**

3.44 Oil flows through the system shown in Fig. P3.44 with negligible losses. Determine the flowrate.

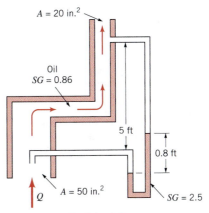

■ **FIGURE P3.44**

3.45 Air flows steadily through a converging–diverging rectangular channel of constant width as shown in Fig. P3.45 and **Video V3.6.** The height of the channel at the exit and the exit velocity are H_0 and V_0, respectively. The channel is to be shaped so that the distance, d, that water is drawn up into tubes attached to static pressure taps along the channel wall is linear with distance along the channel. That is, $d = (d_{max}/L)x$, where L is the channel length and d_{max} is the maximum water depth (at the minimum channel height; $x = L$). Determine the height, $H(x)$, as a function of x and the other important parameters.

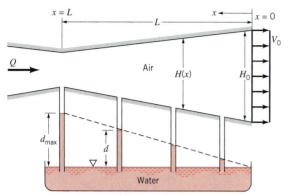

■ **FIGURE P3.45**

3.46 Water flows through the hrizontal branching pipe shown in Fig. P3.46. at a rate of $10 \text{ft}^3/\text{s}$. If viscous effects are negligible, determine the water speed at section (2), the pressure at section (3), and the flowrate at section (4).

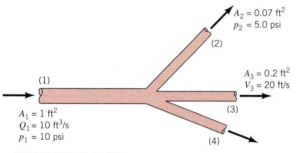

■ **FIGURE P3.46**

3.47 An air cushion vehicle is supported by forcing air into the chamber created by a skirt around the periphery of the vehicle as shown in Fig. P3.47. The air escapes through the 3-in. clearance between the lower end of the skirt and the ground (or water). Assume the vehicle weights 10,000 lb and is essentially rectangular in shape, 30 by 50 ft. The volume of the chamber is large enough so that the kinetic energy of the air within the chamber is negligible. Determine the flowrate, Q, needed to support the vehicle. If the ground clearance were reduced to 2 in., what flowrate would be needed? If the vehicle weight were reduced to 5000 lb and the ground clearance maintained at 3 in., what flowrate would be needed?

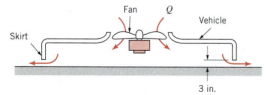

■ **FIGURE P3.47**

3.48 A conical plug is used to regulate the air flow from the pipe shown in Fig. P3.48. The air leaves the edge of the cone

with a uniform thickness of 0.02 m. If viscous effects are negligible and the flowrate is $0.50 \text{ m}^3/\text{s}$, determine the pressure within the pipe.

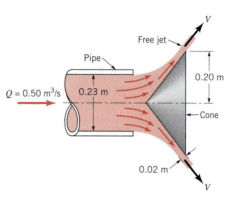

■ **FIGURE P3.48**

***3.49** The surface area, A, of the pond shown in Fig. P3.49 varies with the water depth, h, as shown in the table. At time $t = 0$ a valve is opened and the pond is allowed to drain through a pipe of diameter D. If viscous effects are negligible and quasi-steady conditions are assumed, plot the water depth as a function of time from when the valve is opened ($t = 0$) until the pond is drained for pipe diameters of $D = 0.5, 1.0, 1.5, 2.0, 2.5$, and 3.0 ft. Assume $h = 18$ ft at $t = 0$.

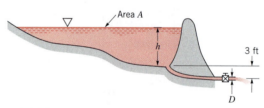

■ **FIGURE P3.49**

h (ft)	A [acres (1 acre = 43,560 ft²)]
0	0
2	0.3
4	0.5
6	0.8
8	0.9
10	1.1
12	1.5
14	1.8
16	2.4
18	2.8

3.50 Water flows over the spillway shown in Fig. P3.50. If the velocity is uniform at sections (1) and (2) and viscous effects are negligible, determine the flowrate per unit width of the spillway.

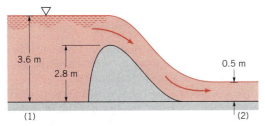

■ **FIGURE P3.50**

3.51 Water flows in a rectangular channel that is 0.5 m wide as shown in Fig. P3.51. The upstream depth is 70 mm. The water surface rises 40 mm as it passes over a portion where the channel bottom rises 10 mm. If viscous effects are negligible, what is the flowrate?

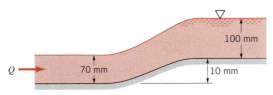

■ **FIGURE P3.51**

Section 3.6.3 Flowrate Measurement (also see Lab Problems 3.61 and 3.63)

3.52 Air flows through the device shown in Fig. P3.52 and **Video V3.6.** If the flowrate is large enough, the pressure within the constriction will be low enough to draw the water up into the tube. Determine the flowrate, Q, and the pressure needed at section (1) to draw the water into section (2). Neglect compressibility and viscous effects.

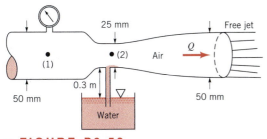

■ **FIGURE P3.52**

3.53 Determine the flowrate through the Venturi meter shown in Fig. P3.53 if ideal conditions exist.

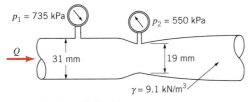

■ **FIGURE P3.53**

3.54 Air (assumed frictionless and incompressible) flows steadily through the device shown in Fig. P3.54. The exit velocity is 100 ft/s, and the differential pressure across the nozzle is 6 lb/ft^2. **(a)** Determine the reading, H, for the water-filled manometer attached to the Pitot tube. **(b)** Determine the diameter, d, of the nozzle.

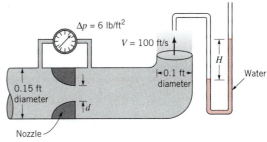

■ **FIGURE P3.54**

3.55 Water flows under the sluice gate shown in Fig. P3.55. Determine the flowrate if the gate is 8 ft wide.

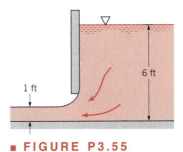

■ **FIGURE P3.55**

Section 3.7 The Energy Line and the Hydraulic Grade Line

3.56 Water flows in a 0.15-m-diameter vertical pipe at a rate of 0.2 m^3/s and a pressure of 200 kPa at an elevation of 25 m. Determine the velocity head and pressure head at elevations of 20 and 55 m.

3.57 Draw the energy line and the hydraulic grade line for the flow of Problem 3.38.

3.58 Draw the energy line and hydraulic grade line for the flow shown in Problem 3.40.

3.59 Draw the energy line and hydraulic grade line for the flow shown in Problem 3.41.

Lab Problems

3.60   This problem involves pressure distribution between two parallel circular plates. To proceed with this problem, go to the book's web site, www.wiley.com/college/young.

3.61   This problem involves calibration of a nozzle-type flowmeter. To proceed with this problem, go to the book's web site, www.wiley.com/college/young.

3.62   This problem involves the pressure distribution in a two-dimensional channel. To proceed with this problem, go to the book's web site, www.wiley.com/college/young.

3.63   This problem involves determination of the flow-rate under a sluice gate as a function of the water depth. To proceed with this problem, go to the book's web site, www.wiley.com/college/young.

FE Exam Problems

Sample FE (Fundamentals of Engineering) exam questions for fluid mechanics are provided on the book's web site, www.wiley.com/college/young.

CHAPTER 4

Fluid Kinematics

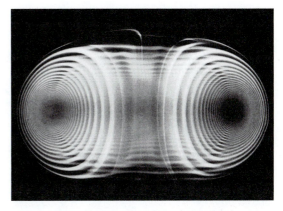

A vortex ring: The complex, three-dimensional structure of a smoke ring is indicated in this cross-sectional view. (Smoke in air.) (Photograph courtesy of R. H. Magarvey and C. S. MacLatchy, Ref. 3.)

LEARNING OBJECTIVES

After completing this chapter, you should be able to:

- discuss the differences between the Eulerian and Lagrangian description of fluid motion.
- identify various flow characteristics based on the velocity field.
- determine the streamline pattern and acceleration field given a velocity field.
- discuss the differences between a system and control volume.
- apply the Reynolds transport theorem and the material derivative.

In this chapter we will discuss various aspects of fluid motion without being concerned with the actual forces necessary to produce the motion. That is, we will consider the *kinematics* of the motion—the velocity and acceleration of the fluid, and the description and visualization of its motion.

4.1   The Velocity Field

The infinitesimal particles of a fluid are tightly packed together (as is implied by the continuum assumption). Thus, at a given instant in time, a description of any fluid property (such as density, pressure, velocity, and acceleration) may be given as a function of the fluid's location. This representation of fluid parameters as functions of the spatial coordinates is termed a *field representation* of the flow. Of course, the specific field representation may be different at different times, so that to describe a fluid flow we must determine the various parameters not only as a function of the spatial coordinates (x, y, z, for example) but also as a function of time, t.

One of the most important fluid variables is the *velocity field,*

$$\mathbf{V} = u(x, y, z, t)\hat{\mathbf{i}} + v(x, y, z, t)\hat{\mathbf{j}} + w(x, y, z, t)\hat{\mathbf{k}}$$

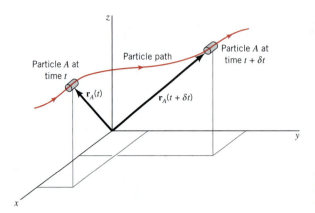

■ **FIGURE 4.1** **Particle location in terms of its position vector.**

V4.1 Velocity field

where u, v, and w are the x, y, and z components of the velocity vector. By definition, the velocity of a particle is the time rate of change of the position vector for that particle. As is illustrated in Fig. 4.1, the position of particle A relative to the coordinate system is given by its *position vector*, $\mathbf{r}_A$, which (if the particle is moving) is a function of time. The time derivative of this position gives the *velocity* of the particle, $d\mathbf{r}_A/dt = \mathbf{V}_A$.

F l u i d s   i n   t h e   N e w s

Follow those particles Superimpose two photographs of a bouncing ball taken a short time apart and draw an arrow between the two images of the ball. This arrow represents an approximation of the velocity (displacement/time) of the ball. The particle image velocimeter (PIV) uses this technique to provide the instantaneous *velocity field* for a given cross section of a flow. The flow being studied is seeded with numerous micron-sized particles, which are small enough to follow the flow, yet big enough to reflect enough light to be captured by the camera. The flow is illuminated with a light sheet from a double-pulsed laser. A digital camera captures both light pulses on the same image frame, allowing the movement of the particles to be tracked. By using appropriate computer software to carry out a pixel-by-pixel interrogation of the double image, it is possible to track the motion of the particles and determine the two components of velocity in the given cross section of the flow. By using two cameras in a stereoscopic arrangement it is possible to determine all three components of velocity. (See Problem 4.7.) ■

EXAMPLE 4.1   Velocity Field Representation

GIVEN A velocity field is given by $\mathbf{V} = (V_0/\ell)(x\hat{\mathbf{i}} - y\hat{\mathbf{j}})$, where V_0 and ℓ are constants.

FIND At what location in the flow field is the speed equal to V_0? Make a sketch of the velocity field in the first quadrant ($x \geq 0$, $y \geq 0$) by drawing arrows representing the fluid velocity at representative locations.

SOLUTION

The x, y, and z components of the velocity are given by $u = V_0 x/\ell$, $v = -V_0 y/\ell$, and $w = 0$ so that the fluid speed, V, is

$$V = (u^2 + v^2 + w^2)^{1/2} = \frac{V_0}{\ell}(x^2 + y^2)^{1/2} \qquad (1)$$

The speed is $V = V_0$ at any location on the circle of radius ℓ centered at the origin $[(x^2 + y^2)^{1/2} = \ell]$ as shown in Fig. E4.1a. **(Ans)**

The direction of the fluid velocity relative to the x axis is given in terms of $\theta = \arctan(v/u)$ as shown in Fig. E4.1b. For this flow

$$\tan\theta = \frac{v}{u} = \frac{-V_0 y/\ell}{V_0 x/\ell} = \frac{-y}{x}$$

Thus, along the x axis ($y = 0$) we see that $\tan\theta = 0$, so that $\theta = 0°$ or $\theta = 180°$. Similarly, along the y axis ($x = 0$) we obtain $\tan \theta = \pm \infty$ so that $\theta = 90°$ or $\theta = 270°$. Also, for $y = 0$ we find $\mathbf{V} = (V_0 x/\ell)\hat{\mathbf{i}}$, while for $x = 0$ we have $\mathbf{V} = (-V_0 y/\ell)\hat{\mathbf{j}}$, indicating (if $V_0 > 0$) that the flow is directed toward the origin along the y axis and away from the origin along the x axis as shown in Fig. E4.1a. By determining $\mathbf{V}$ and θ for other locations in the $x - y$ plane, the velocity field can be sketched as shown in Fig. E4.1. For example, on the line $y = x$ the velocity is at a $-45°$ angle relative to the x axis ($\tan \theta = v/u = -y/x = -1$).

COMMENT   At the origin $x = y = 0$ so that $\mathbf{V} = 0$. This point is a stagnation point. The farther from the origin the fluid is, the faster it is flowing (as seen from Eq. 1). By careful consideration of the velocity field, it is possible to determine considerable information about the flow.

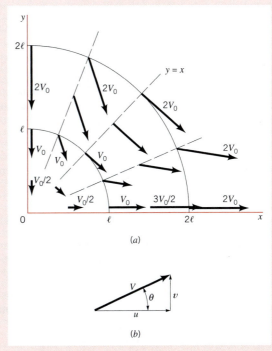

(a)

(b)

■ **FIGURE E4.1**

4.1.1   Eulerian and Lagrangian Flow Descriptions

There are two general approaches in analyzing fluid mechanics problems. The first method, called the *Eulerian method,* uses the field concept introduced earlier. In this case, the fluid motion is given by completely prescribing the necessary properties (pressure, density, velocity, etc.) as functions of space and time. From this method we obtain information about the flow in terms of what happens at fixed points in space as the fluid flows past those points.

The second method, called the *Lagrangian method,* involves following individual fluid particles as they move about and determining how the fluid properties associated with these particles change as a function of time. That is, the fluid particles are "tagged" or identified, and their properties are determined as they move.

The difference between the two methods of analyzing fluid problems can be seen in the example of smoke discharging from a chimney, as is shown in Fig. 4.2. In the Eulerian

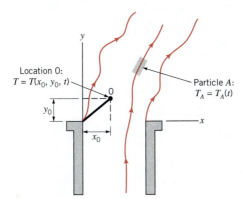

■ **FIGURE 4.2**   Eulerian and Lagrangian description of temperature of a flowing fluid.

method one may attach a temperature-measuring device to the top of the chimney (point 0) and record the temperature at that point as a function of time. That is, $T = T(x_0, y_0, z_0, t)$. The use of numerous temperature-measuring devices fixed at various locations would provide the temperature field, $T = T(x, y, z, t)$.

In the Lagrangian method, one would attach the temperature-measuring device to a particular fluid particle (particle A) and record that particle's temperature as it moves about. Thus, one would obtain that particle's temperature as a function of time, $T_A = T_A(t)$. The use of many such measuring devices moving with various fluid particles would provide the temperature of these fluid particles as a function of time. In fluid mechanics it is usually easier to use the Eulerian method to describe a flow.

Example 4.1 provides an Eulerian description of the flow. For a Lagrangian description we would need to determine the velocity as a function of time for each particle as it flows along from one point to another.

4.1.2 One-, Two-, and Three-Dimensional Flows

V4.2 Flow past a wing

In almost any flow situation, the velocity field actually contains all three velocity components (u, v, and w, for example). In many situations the ***three-dimensional flow*** characteristics are important in terms of the physical effects they produce. A feel for the three-dimensional structure of such flows can be obtained by studying Fig. 4.3, which is a photograph of the flow past a model airfoil.

In many situations one of the velocity components may be small (in some sense) relative to the two other components. In situations of this kind it may be reasonable to neglect the smaller component and assume ***two-dimensional flow.*** That is, $\mathbf{V} = u\hat{\mathbf{i}} + v\hat{\mathbf{j}}$, where u and v are functions of x and y (and possibly time, t).

It is sometimes possible to further simplify a flow analysis by assuming that two of the velocity components are negligible, leaving the velocity field to be approximated as a ***one-dimensional flow*** field. That is, $\mathbf{V} = u\hat{\mathbf{i}}.$ There are many flow fields for which the one-dimensional flow assumption provides a reasonable approximation. There are also many flow situations for which use of a one-dimensional flow field assumption will give completely erroneous results.

4.1.3 Steady and Unsteady Flows

V4.3 Flow types

For ***steady flow*** the velocity at a given point in space does not vary with time, $\partial \mathbf{V}/\partial t = 0$. In reality, almost all flows are unsteady in some sense. That is, the velocity does vary with

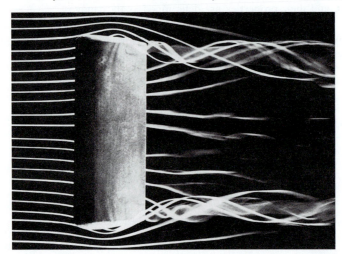

■ **FIGURE 4.3**

Flow visualization of the complex three-dimensional flow past a model airfoil. (Photograph by M. R. Head)

time. An example of a nonperiodic, *unsteady flow* is that produced by turning off a faucet to stop the flow of water. In other flows the unsteady effects may be periodic, occurring time after time in basically the same manner. The periodic injection of the air–gasoline mixture into the cylinder of an automobile engine is such an example.

In many situations the unsteady character of a flow is quite random. That is, there is no repeatable sequence or regular variation to the unsteadiness. This behavior occurs in *turbulent flow* and is absent from *laminar flow*. The "smooth" flow of highly viscous syrup onto a pancake represents a "deterministic" laminar flow. It is quite different from the turbulent flow observed in the "irregular" splashing of water from a faucet onto the sink below it. The "irregular" gustiness of the wind represents another random turbulent flow.

V4.4 Jupiter red spot

F l u i d s   i n   t h e   N e w s

New pulsed liquid-jet scalpel High-speed liquid-jet cutters are used for cutting a wide variety of materials such as leather goods, jigsaw puzzles, plastic, ceramic, and metal. Typically, compressed air is used to produce a continuous stream of water that is ejected from a tiny nozzle. As this stream impacts the material to be cut, a high pressure (the stagnation pressure) is produced on the surface of the material, thereby cutting the material. Such liquid-jet cutters work well in air, but are difficult to control if the jet must pass through a liquid as often happens in surgery. Researchers have developed a new pulsed jet cutting tool that may allow surgeons to perform microsurgery on tissues that are immersed in water. Rather than using a steady water jet, the system uses *unsteady flow*. A high-energy electrical discharge inside the nozzle momentarily raises the temperature of the microjet to approximately 10,000 °C. This creates a rapidly expanding vapor bubble in the nozzle and expels a tiny fluid jet from the nozzle. Each electrical discharge creates a single, brief jet, which makes a small cut in the material. ■

4.1.4   Streamlines, Streaklines, and Pathlines

A *streamline* is a line that is everywhere tangent to the velocity field. If the flow is steady, nothing at a fixed point (including the velocity direction) changes with time, so the streamlines are fixed lines in space.

For two-dimensional flows the slope of the streamline, dy/dx, must be equal to the tangent of the angle that the velocity vector makes with the x axis or, as shown in the figure in the margin,

$$\frac{dy}{dx} = \frac{v}{u} \tag{4.1}$$

If the velocity field is known as a function of x and y (and t if the flow is unsteady), this equation can be integrated to give the equation of the streamlines.

V4.5 Streamlines

EXAMPLE 4.2   **Streamlines for a Given Velocity Field**

GIVEN Consider the two-dimensional steady flow discussed in Example 4.1, $\mathbf{V} = (V_0/\ell)(x\hat{\mathbf{i}} - y\hat{\mathbf{j}})$.

FIND Determine the streamlines for this flow.

SOLUTION_____

Since

$$u = (V_0/\ell)x \text{ and } v = -(V_0/\ell)y \tag{1}$$

it follows that streamlines are given by solution of the equation

$$\frac{dy}{dx} = \frac{v}{u} = \frac{-(V_0/\ell)y}{(V_0/\ell)x} = -\frac{y}{x}$$

in which variables can be separated and the equation integrated to give

$$\int \frac{dy}{y} = -\int \frac{dx}{x}$$

or

$$\ln y = -\ln x + \text{constant}$$

Thus, along the streamline

$$xy = C, \qquad \text{where } C \text{ is a constant} \qquad \textbf{(Ans)}$$

By using different values of the constant C, we can plot various lines in the $x - y$ plane—the streamlines. The streamlines in the first quadrant are plotted in Fig. E4.2. A comparison of this figure with Fig. E4.1a illustrates the fact that streamlines are lines tangent to the velocity field.

COMMENT Note that a flow is not completely specified by the shape of the streamlines alone. For example, the stream-lines for the flow with $V_0/\ell = 10$ have the same shape as those for the flow with $V_0/\ell = -10$. However, the direction of the flow is opposite for these two cases. The arrows in Fig. E4.2

representing the flow direction are correct for $V_0/\ell = 10$ since, from Eq. 1, $u = 10x$ and $v = -10y$. That is, the flow is "down and to the right." For $V_0/\ell = -10$ the arrows are reversed. The flow is "up and to the left."

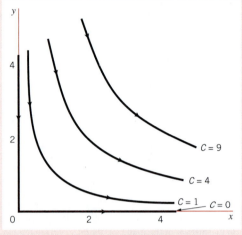

■ **FIGURE E4.2**

A *streakline* consists of all particles in a flow that have previously passed through a common point. Streaklines are more of a laboratory tool than an analytical tool. They can be obtained by taking instantaneous photographs of marked particles that all passed through a given location in the flow field at some earlier time. Such a line can be produced by continuously injecting marked fluid (neutrally buoyant smoke in air, or dye in water) at a given location (Ref. 1).

A *pathline* is the line traced out by a given particle as it flows from one point to another. The pathline is a Lagrangian concept that can be produced in the laboratory by marking a fluid particle (dyeing a small fluid element) and taking a time exposure photograph of its motion.

V4.6 Pathlines

Pathlines, streamlines, and streaklines are the same for steady flows. For unsteady flows none of these three types of lines need be the same (Ref. 2). Often one sees pictures of "streamlines" made visible by the injection of smoke or dye into a flow as is shown in Fig. 4.3. Actually, such pictures show streaklines rather than streamlines. However, for steady flows the two are identical; only the nomenclature is used incorrectly.

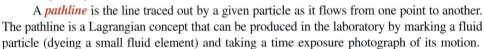

F l u i d s   i n   t h e   N e w s

Winds on Earth and Mars The wind has considerable temporal and spatial variation. For example, the lowest monthly average wind speed in Chicago (dubbed "the Windy City," though there are several windier cities) occurs in August (8.2 mph); the highest occurs in April (11.9 mph). The wind speed variation throughout the day has a shorter time scale. The normally calm morning wind increases during the day because solar heating produces a buoyancy-driven vertical motion that mixes the faster moving air aloft with the slower moving air near the ground. When the sun sets, the vertical motion subsides and the winds become calm. The quickest variations in the wind speed are those irregular, turbulent gusts that are nearly always present (i.e., "The wind is 20 mph with gusts to 35"). Spacecraft data have shown that the wind on Mars has similar characteristics. These martian winds are usually fairly light, rarely more than 15 mph, although during dust storms they can reach 300 mph. Recent photos from spacecraft orbiting Mars show swirling, corkscrew paths in the dust on the Martian surface (a visualization of surface *pathlines*), indicating that "dust devil" type flows occur on Mars just as they do on Earth. (See Problem 4.5.) ■

EXAMPLE 4.3   Comparison of Streamlines, Pathlines, and Streaklines

GIVEN Water flowing from the oscillating slit shown in Fig. E4.3a produces a velocity field given by $\mathbf{V} = u_0 \sin[\omega(t - y/v_0)]\hat{\mathbf{i}} + v_0\hat{\mathbf{j}}$, where u_0, v_0, and ω are constants. Thus, the y component of velocity remains constant ($v = v_0$) and the x component of velocity at $y = 0$ coincides with the velocity of the oscillating sprinkler head [$u = u_0 \sin(\omega t)$ at $y = 0$].

FIND **(a)** Determine the streamline that passes through the origin at $t = 0$; at $t = \pi/2\omega$.

(b) Determine the pathline of the particle that was at the origin at $t = 0$; at $t = \pi/2\omega$.

(c) Discuss the shape of the streakline that passes through the origin.

SOLUTION

(a) Since $u = u_0 \sin [\omega(t - y/v_0)]$ and $v = v_0$ it follows from Eq. 4.1 that streamlines are given by the solution of

$$\frac{dy}{dx} = \frac{v}{u} = \frac{v_0}{u_0 \sin[\omega(t - y/v_0)]}$$

in which the variables can be separated and the equation integrated (for any given time t) to give

$$u_0 \int \sin\left[\omega\left(t - \frac{y}{v_0}\right)\right]dy = v_0 \int dx,$$

or

$$u_0(v_0/\omega) \cos\left[\omega\left(t - \frac{y}{v_0}\right)\right] = v_0 x + C \quad \textbf{(1)}$$

where C is a constant. For the streamline at $t = 0$ that passes through the origin ($x = y = 0$), the value of C is obtained from Eq. 1 as $C = u_0 v_0/\omega$. Hence, the equation for this streamline is

$$x = \frac{u_0}{\omega}\left[\cos\left(\frac{\omega y}{v_0}\right) - 1\right] \quad \textbf{(2)} \quad \textbf{(Ans)}$$

Similarly, for the streamline at $t = \pi/2\omega$ that passes through the origin, Eq. 1 gives $C = 0$. Thus, the equation for this streamline is

$$x = \frac{u_0}{\omega} \cos\left[\omega\left(\frac{\pi}{2\omega} - \frac{y}{v_0}\right)\right] = \frac{u_0}{\omega} \cos\left(\frac{\pi}{2} - \frac{\omega y}{v_0}\right)$$

or

$$x = \frac{u_0}{\omega} \sin\left(\frac{\omega y}{v_0}\right) \quad \textbf{(3)} \quad \textbf{(Ans)}$$

COMMENT These two streamlines, plotted in Figure E4.3b, are not the same because the flow is unsteady. For

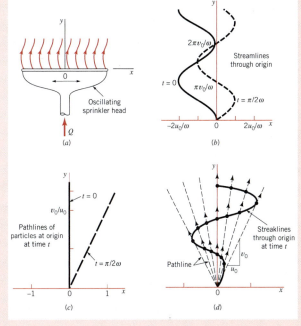

■ **FIGURE E4.3**

example, at the origin ($x = y = 0$) the velocity is $\mathbf{V} = v_0\hat{\mathbf{j}}$ at $t = 0$ and $\mathbf{V} = u_0\hat{\mathbf{i}} + v_0\hat{\mathbf{j}}$ at $t = \pi/2\omega$. Thus, the angle of the streamline passing through the origin changes with time. Similarly, the shape of the entire streamline is a function of time.

(b) The pathline of a particle (the location of the particle as a function of time) can be obtained from the velocity field and the definition of the velocity. Since $u = dx/dt$ and $v = dy/dt$ we obtain

$$\frac{dx}{dt} = u_0 \sin\left[\omega\left(t - \frac{y}{v_0}\right)\right] \quad \text{and} \quad \frac{dy}{dt} = v_0$$

The y equation can be integrated (since $v_0 =$ constant) to give the y coordinate of the pathline as

$$y = v_0 t + C_1 \quad \textbf{(4)}$$

where C_1 is a constant. With this known $y = y(t)$ dependence, the x equation for the pathline becomes

$$\frac{dx}{dt} = u_0 \sin\left[\omega\left(t - \frac{v_0 t + C_1}{v_0}\right)\right] = -u_0 \sin\left(\frac{C_1\omega}{v_0}\right)$$

This can be integrated to give the x component of the pathline as

$$x = -\left[u_0 \sin\left(\frac{C_1\omega}{v_0}\right)\right]t + C_2 \quad \textbf{(5)}$$

where C_2 is a constant. For the particle that was at the origin ($x = y = 0$) at time $t = 0$, Eqs. 4 and 5 give $C_1 = C_2 = 0$. Thus, the pathline is

$$x = 0 \quad \text{and} \quad y = v_0 t \qquad \text{(6) \quad (Ans)}$$

Similarly, for the particle that was at the origin at $t = \pi/2\omega$, Eqs. 4 and 5 give $C_1 = -\pi v_0/2\omega$ and $C_2 = -\pi v_0/2\omega$. Thus, the pathline for this particle is

$$x = u_0\left(t - \frac{\pi}{2\omega}\right) \quad \text{and} \quad y = v_0\left(t - \frac{\pi}{2\omega}\right) \qquad \text{(7)}$$

The pathline can be drawn by plotting the locus of $x(t)$, $y(t)$ values for $t \geq 0$ or by eliminating the parameter t from Eq. 7 to give

$$y = \frac{v_0}{u_0}x \qquad \text{(8) \quad (Ans)}$$

COMMENT The pathlines given by Eqs. 6 and 8, shown in Fig. E4.3c, are straight lines from the origin (rays). The pathlines and streamlines do not coincide because the flow is unsteady.

(c) The streakline through the origin at time $t = 0$ is the locus of particles at $t = 0$ that previously ($t < 0$) passed through the origin. The general shape of the streaklines can be seen as follows. Each particle that flows through the origin travels in a straight line (pathlines are rays from the origin), the slope of which lies between $\pm v_0/u_0$ as shown in Fig. E4.3d. Particles passing through the origin at different times are located on different rays from the origin and at different distances from the origin. The net result is that a stream of dye continually injected at the origin (a streakline) would have the shape shown in Fig. E4.3d. Because of the unsteadiness, the streakline will vary with time, although it will always have the oscillating, sinuous character shown. Similar streaklines are given by the stream of water from a garden hose nozzle that oscillates back and forth in a direction normal to the axis of the nozzle.

COMMENT In this example streamlines, pathlines, and streaklines do not coincide. If the flow were steady, all of these lines would be the same.

4.2   The Acceleration Field

To apply Newton's second law ($\mathbf{F} = m\mathbf{a}$) we must be able to describe the particle acceleration in an appropriate fashion. For the infrequently used Lagrangian method, we describe the fluid acceleration just as is done in solid body dynamics—$\mathbf{a} = \mathbf{a}(t)$ for each particle. For the Eulerian description we describe the *acceleration field* as a function of position and time without actually following any particular particle. This is analogous to describing the flow in terms of the velocity field, $\mathbf{V} = \mathbf{V}(x, y, z, t)$, rather than the velocity for particular particles.

4.2.1   The Material Derivative

Consider a fluid particle moving along its pathline as is shown in Fig. 4.4. In general, the particle's velocity, denoted $\mathbf{V}_A$ for particle A, is a function of its location and the time.

$$\mathbf{V}_A = \mathbf{V}_A(\mathbf{r}_A, t) = \mathbf{V}_A[x_A(t), y_A(t), z_A(t), t]$$

where $x_A = x_A(t)$, $y_A = y_A(t)$, and $z_A = z_A(t)$, define the location of the moving particle. By definition, the acceleration of a particle is the time rate of change of its velocity. Thus, we use the chain rule of differentiation to obtain the acceleration of particle A, $\mathbf{a}_A$, as

$$\mathbf{a}_A(t) = \frac{d\mathbf{V}_A}{dt} = \frac{\partial \mathbf{V}_A}{\partial t} + \frac{\partial \mathbf{V}_A}{\partial x}\frac{dx_A}{dt} + \frac{\partial \mathbf{V}_A}{\partial y}\frac{dy_A}{dt} + \frac{\partial \mathbf{V}_A}{\partial z}\frac{dz_A}{dt} \qquad \text{(4.2)}$$

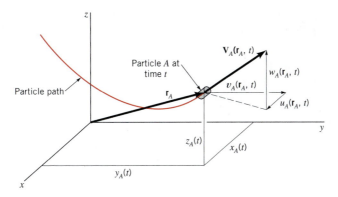

■ **FIGURE 4.4** Velocity and position of particle A at time t.

Using the fact that the particle velocity components are given by $u_A = dx_A/dt$, $v_A = dy_A/dt$, and $w_A = dz_A/dt$, Eq. 4.2 becomes

$$\mathbf{a}_A = \frac{\partial \mathbf{V}_A}{\partial t} + u_A \frac{\partial \mathbf{V}_A}{\partial x} + v_A \frac{\partial \mathbf{V}_A}{\partial y} + w_A \frac{\partial \mathbf{V}_A}{\partial z}$$

Because the equation just described is valid for any particle, we can drop the reference to particle A and obtain the acceleration field from the velocity field as

$$\mathbf{a} = \frac{\partial \mathbf{V}}{\partial t} + u \frac{\partial \mathbf{V}}{\partial x} + v \frac{\partial \mathbf{V}}{\partial y} + w \frac{\partial \mathbf{V}}{\partial z} \tag{4.3}$$

This is a vector result whose scalar components can be written as

$$a_x = \frac{\partial u}{\partial t} + u \frac{\partial u}{\partial x} + v \frac{\partial u}{\partial y} + w \frac{\partial u}{\partial z}$$

$$a_y = \frac{\partial v}{\partial t} + u \frac{\partial v}{\partial x} + v \frac{\partial v}{\partial y} + w \frac{\partial v}{\partial z} \tag{4.4}$$

and

$$a_z = \frac{\partial w}{\partial t} + u \frac{\partial w}{\partial x} + v \frac{\partial w}{\partial y} + w \frac{\partial w}{\partial z}$$

The result given in Eq. 4.4 is often written in shorthand notation as

$$\mathbf{a} = \frac{D\mathbf{V}}{Dt}$$

where the operator

$$\frac{D(\ )}{Dt} \equiv \frac{\partial(\ )}{\partial t} + u \frac{\partial(\ )}{\partial x} + v \frac{\partial(\ )}{\partial y} + w \frac{\partial(\ )}{\partial z} \tag{4.5}$$

is termed the ***material derivative*** or *substantial derivative*. An often-used shorthand notation for the material derivative operator is

$$\frac{D(\ )}{Dt} = \frac{\partial(\ )}{\partial t} + (\mathbf{V} \cdot \nabla)(\ ) \tag{4.6}$$

The dot product of the velocity vector, $\mathbf{V}$, and the gradient operator, $\nabla(\ ) = \partial(\ )/\partial x \,\hat{\mathbf{i}} + \partial(\ )/\partial y \,\hat{\mathbf{j}} + \partial(\ )/\partial z \,\hat{\mathbf{k}}$ (a vector operator), provides a convenient notation for the spatial derivative terms appearing in the Cartesian coordinate representation of the material derivative. Note that the notation $\mathbf{V} \cdot \nabla$ represents the operator $\mathbf{V} \cdot \nabla(\ ) = u\partial(\ )/\partial x + v\partial(\ )/\partial y + w\partial(\ )/\partial z$.

EXAMPLE 4.4 — Acceleration along a Streamline

GIVEN An incompressible, inviscid fluid flows steadily past a sphere of radius R, as shown in Fig. E4.4a. According to a more advanced analysis of the flow, the fluid velocity along streamline A–B is given by

$$\mathbf{V} = u(x)\hat{\mathbf{i}} = V_0\left(1 + \frac{R^3}{x^3}\right)\hat{\mathbf{i}}$$

where V_0 is the upstream velocity far ahead of the sphere.

FIND Determine the acceleration experienced by fluid particles as they flow along this streamline.

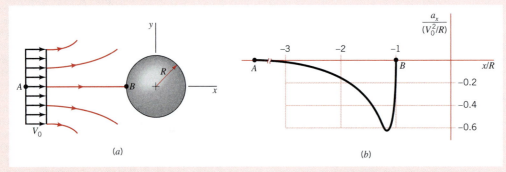

■ **FIGURE E4.4**

SOLUTION

Along streamline A–B there is only one component of velocity ($v = w = 0$) so that from Eq. 4.3

$$\mathbf{a} = \frac{\partial \mathbf{V}}{\partial t} + u\frac{\partial \mathbf{V}}{\partial x} = \left(\frac{\partial u}{\partial t} + u\frac{\partial u}{\partial x}\right)\hat{\mathbf{i}}$$

or

$$a_x = \frac{\partial u}{\partial t} + u\frac{\partial u}{\partial x}, \qquad a_y = 0, \qquad a_z = 0$$

Since the flow is steady, the velocity at a given point in space does not change with time. Thus, $\partial u/\partial t = 0$. With the given velocity distribution along the streamline, the acceleration becomes

$$a_x = u\frac{\partial u}{\partial x} = V_0\left(1 + \frac{R^3}{x^3}\right)V_0[R^3(-3x^{-4})]$$

or

$$a_x = -3(V_0^2/R)\frac{1 + (R/x)^3}{(x/R)^4} \qquad \textbf{(Ans)}$$

velocity of $\mathbf{V} = V_0\hat{\mathbf{i}}$ at $x = -\infty$ to its stagnation point velocity of $\mathbf{V} = 0$ at $x = -R$, the "nose" of the sphere. The variation of a_x along streamline A–B is shown in Fig. E4.4b. It is the same result as is obtained in Example 3.1 by using the streamwise component of the acceleration, $a_x = V\partial V/\partial s$. The maximum deceleration occurs at $x = -1.205R$ and has a value of $a_{x,max} = -0.610V_0^2/R$. Note that this maximum deceleration increases with increasing velocity and decreasing size. As indicated in the following table, typical values of this deceleration can be quite large. For example, the $a_{x,max} = -4.08 \times 10^4$ ft/s² value for a pitched baseball is a deceleration approximately 1500 times that of gravity.

Object	V_0 (ft/s)	R (ft)	$a_{x,max}$ (ft/s²)
Rising weather balloon	1	4.0	−0.153
Soccer ball	20	0.80	−305
Baseball	90	0.121	−4.08 × 10⁴
Golf ball	200	0.070	−3.49 × 10⁵

COMMENT Along streamline A–B ($-\infty \le x \le -a$ and $y = 0$) the acceleration has only an x component and it is negative (a deceleration). Thus, the fluid slows down from its upstream

In general, for fluid particles on streamlines other than A–B, all three components of the acceleration (a_x, a_y, and a_z) will be nonzero.

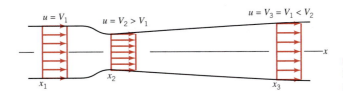

■ **FIGURE 4.5** Uniform, steady flow in a variable area pipe.

4.2.2 Unsteady Effects

As is seen from Eq. 4.5, the material derivative formula contains two types of terms—those involving the time derivative $[\partial(\ )/\partial t]$ and those involving spatial derivatives $[\partial(\ )/\partial x, \partial(\ )/\partial y,$ and $\partial(\ )/\partial z]$. The time derivative portions are denoted as the *local derivative*. They represent effects of the unsteadiness of the flow. If the parameter involved is the acceleration, that portion given by $\partial\mathbf{V}/\partial t$ is termed *local acceleration*. For steady flow the time derivative is zero throughout the flow field $[\partial(\ )/\partial t \equiv 0]$, and the local effect vanishes. Physically, there is no change in flow parameters at a fixed point in space if the flow is steady.

4.2.3 Convective Effects

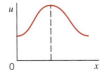

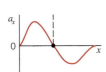

The portion of the material derivative (Eq. 4.5) represented by the spatial derivatives is termed the *convective derivative*. It represents the fact that a flow property associated with a fluid particle may vary because of the motion of the particle from one point in space where the parameter has one value to another point in space where its value is different. This contribution is due to the convection, or motion, of the particle through space in which there is a gradient $[\nabla(\ ) = \partial(\ )/\partial x\,\hat{\mathbf{i}} + \partial(\ )/\partial y\,\hat{\mathbf{j}} + \partial(\ )/\partial z\,\hat{\mathbf{k}}]$ in the parameter value. That portion of the acceleration given by the term $(\mathbf{V}\cdot\nabla)\mathbf{V}$ is termed *convective acceleration.*

Consider flow in a variable area pipe as shown in Fig. 4.5. It is assumed that the flow is steady and one-dimensional with velocity that increases and decreases in the flow direction as indicated. As the fluid flows from section (1) to section (2), its velocity increases from V_1 to V_2. Thus, even though $\partial\mathbf{V}/\partial t = 0$ (i.e., the flow is steady), fluid particles experience an acceleration given by $a_x = u\,\partial u/\partial x$, the convective acceleration. For $x_1 < x < x_2$, it is seen that $\partial u/\partial x > 0$ so that $a_x > 0$—the fluid accelerates. For $x_2 < x < x_3$, it is seen that $\partial u/\partial x < 0$ so that $a_x < 0$—the fluid decelerates. This acceleration and deceleration are shown in the figure in the margin.

EXAMPLE 4.5 | Acceleration from a Given Velocity Field

GIVEN Consider the steady, two-dimensional flow field discussed in Example 4.2.

FIND Determine the acceleration field for this flow.

SOLUTION

In general, the acceleration is given by

$$\mathbf{a} = \frac{D\mathbf{V}}{Dt} = \frac{\partial\mathbf{V}}{\partial t} + (\mathbf{V}\cdot\nabla)(\mathbf{V})$$

$$= \frac{\partial\mathbf{V}}{\partial t} + u\frac{\partial\mathbf{V}}{\partial x} + v\frac{\partial\mathbf{V}}{\partial y} + w\frac{\partial\mathbf{V}}{\partial z} \qquad (1)$$

where the velocity is given by $\mathbf{V} = (V_0/\ell)(x\hat{\mathbf{i}} - y\hat{\mathbf{j}})$ so that $u = (V_0/\ell)x$ and $v = -(V_0/\ell)y$. For steady $[\partial(\ )/\partial t = 0]$, two-dimensional $[w = 0$ and $\partial(\ )/\partial z = 0]$ flow, Eq. 1 becomes

$$\mathbf{a} = u\frac{\partial\mathbf{V}}{\partial x} + v\frac{\partial\mathbf{V}}{\partial y}$$

$$= \left(u\frac{\partial u}{\partial x} + v\frac{\partial u}{\partial y}\right)\hat{\mathbf{i}} + \left(u\frac{\partial v}{\partial x} + v\frac{\partial v}{\partial y}\right)\hat{\mathbf{j}}$$

Hence, for this flow the acceleration is given by

$$\mathbf{a} = \left[\left(\frac{V_0}{\ell}\right)(x)\left(\frac{V_0}{\ell}\right) + \left(\frac{V_0}{\ell}\right)(y)(0)\right]\hat{\mathbf{i}}$$

$$+ \left[\left(\frac{V_0}{\ell}\right)(x)(0) + \left(\frac{-V_0}{\ell}\right)(y)\left(\frac{-V_0}{\ell}\right)\right]\hat{\mathbf{j}}$$

or

$$a_x = \frac{V_0^2 x}{\ell^2}, \qquad a_y = \frac{V_0^2 y}{\ell^2} \qquad \textbf{(Ans)}$$

COMMENTS The fluid experiences an acceleration in both the x and y directions. Since the flow is steady, there is no local acceleration—the fluid velocity at any given point is constant in time. However, there is a convective acceleration due to the change in velocity from one point on the particle's pathline to another. Recall that the velocity is a vector—it has both a magnitude and a direction. In this flow both the fluid speed (magnitude) and the flow direction change with location (see Fig. E4.1a).

For this flow the magnitude of the acceleration is constant on circles centered at the origin, as is seen from the fact that

$$|\mathbf{a}| = (a_x^2 + a_y^2 + a_z^2)^{1/2} = \left(\frac{V_0}{\ell}\right)^2 (x^2 + y^2)^{1/2} \qquad \textbf{(2)}$$

Also, the acceleration vector is oriented at an angle θ from the x axis, where

$$\tan \theta = \frac{a_y}{a_x} = \frac{y}{x}$$

This is the same angle as that formed by a ray from the origin to point (x, y). Thus, the acceleration is directed along rays from the origin and has a magnitude proportional to the distance from the origin. Typical acceleration vectors (from Eq. 2) and velocity vectors (from Example 4.1) are shown in Fig. E4.5 for the flow in the first quadrant. Note that $\mathbf{a}$ and $\mathbf{V}$ are not parallel except along the x and y axes (a fact that is responsible for the curved pathlines of the flow), and that both the acceleration and velocity are zero at the origin ($x = y = 0$). An infinitesimal fluid particle placed precisely at the origin will remain there, but its neighbors (no matter how close they are to the origin) will drift away.

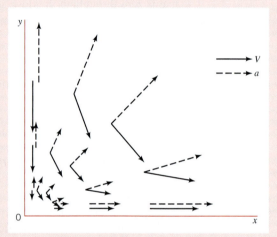

■ **FIGURE E4.5**

4.2.4   Streamline Coordinates

In many flow situations it is convenient to use a coordinate system defined in terms of the streamlines of the flow. An example for steady, two-dimensional flows is illustrated in Fig. 4.6. Such flows can be described in terms of the streamline coordinates involving one coordinate along the streamlines, denoted s, and the second coordinate normal to the streamlines, denoted n. Unit vectors in these two directions are denoted by $\hat{\mathbf{s}}$ and $\hat{\mathbf{n}}$ as shown in Fig. 4.6.

One of the major advantages of using the streamline coordinate system is that the velocity is always tangent to the s direction. That is,

$$\mathbf{V} = V\hat{\mathbf{s}}$$

For steady, two-dimensional flow we can determine the acceleration as (see Eq. 3.1)

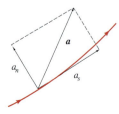

$$\mathbf{a} = V\frac{\partial V}{\partial s}\hat{\mathbf{s}} + \frac{V^2}{\mathcal{R}}\hat{\mathbf{n}} \quad \text{or} \quad a_s = V\frac{\partial V}{\partial s}, \qquad a_n = \frac{V^2}{\mathcal{R}} \qquad \textbf{(4.7)}$$

As shown in the figure in the margin, the first term, $a_s = V\partial V/\partial s$, represents convective acceleration along the streamline, and the second term, $a_n = V^2/\mathcal{R}$, represents centrifugal

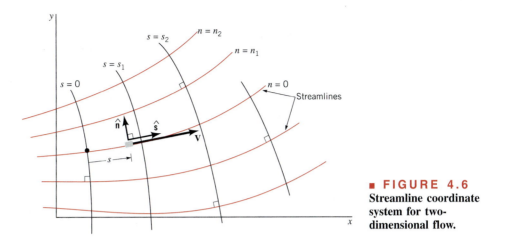

■ FIGURE 4.6
**Streamline coordinate
system for two-
dimensional flow.**

acceleration (one type of convective acceleration) normal to the fluid motion. These forms of acceleration are probably familiar from previous dynamics or physics considerations.

4.3   Control Volume and System Representations

A fluid's behavior is governed by a set of fundamental physical laws, which are approximated by an appropriate set of equations. The application of laws such as the conservation of mass, Newton's laws of motion, and the laws of thermodynamics form the foundation of fluid mechanics analyses. There are various ways that these governing laws can be applied to a fluid, including the system approach and the control volume approach. By definition, a *system* is a collection of matter of fixed identity (always the same atoms or fluid particles), which may move, flow, and interact with its surroundings. A *control volume,* however, is a volume in space (a geometric entity, independent of mass) through which fluid may flow.

We may often be more interested in determining the forces put on a fan, airplane, or automobile by air flowing past the object than we are in the information obtained by following a given portion of the air (a system) as it flows along. For these situations we often use the control volume approach. We identify a specific volume in space (a volume associated with the fan, airplane, or automobile, for example) and analyze the fluid flow within, through, or around that volume.

Examples of control volumes and *control surfaces* (the surface of the control volume) are shown in Fig. 4.7. For case (*a*), fluid flows through a pipe. The fixed control surface consists of the inside surface of the pipe, the outlet end at section (2), and a section across the pipe at (1). Fluid flows across part of the control surface, but not across all of it.

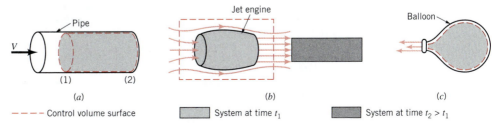

■ FIGURE 4.7   **Typical control volumes: (*a*) fixed control volume, (*b*) fixed or moving control volume, (*c*) deforming control volume.**

Another control volume is the rectangular volume surrounding the jet engine shown in Fig. 4.7b. The air that was within the engine itself at time $t = t_1$ (a system) has passed through the engine and is outside of the control volume at a later time $t = t_2$ as indicated. At this later time other air (a different system) is within the engine.

The deflating balloon shown in Fig. 4.7c provides an example of a deforming control volume. As time increases, the control volume (whose surface is the inner surface of the balloon) decreases in size.

All of the laws governing the motion of a fluid are stated in their basic form in terms of a system approach. For example, "the mass of a system remains constant," or "the time rate of change of momentum of a system is equal to the sum of all the forces acting on the system." Note the word "system," not control volume, in these statements. To use the governing equations in a control volume approach to problem solving, we must rephrase the laws in an appropriate manner. To this end we introduce the Reynolds transport theorem.

4.4 The Reynolds Transport Theorem

We need to describe the laws governing fluid motion using both system concepts (consider a given mass of the fluid) and control volume concepts (consider a given volume). To do this we need an analytical tool to shift from one representation to the other. The *Reynolds transport theorem* provides this tool.

All physical laws are stated in terms of various physical parameters such as velocity, acceleration, mass, temperature, and momentum. Let B represent any of these (or other) fluid parameters and b represent the amount of that parameter per unit mass. That is,

$$B = mb$$

where m is the mass of the portion of fluid of interest. If $B = mV^2/2$, the kinetic energy of the mass, then $b = V^2/2$, the kinetic energy per unit mass. The parameters B and b may be scalars or vectors. Thus, if $\mathbf{B} = m\mathbf{V}$, the momentum of the mass, then $\mathbf{b} = \mathbf{V}$. (The momentum per unit mass is the velocity.) The parameter B is termed an *extensive property*, and the parameter b is termed an *intensive property*.

4.4.1 Derivation of the Reynolds Transport Theorem

A simple version of the Reynolds transport theorem relating system concepts to control volume concepts can be obtained easily for the one-dimensional flow through a fixed control volume as is shown in Fig. 4.8a. We consider the control volume to be that stationary volume

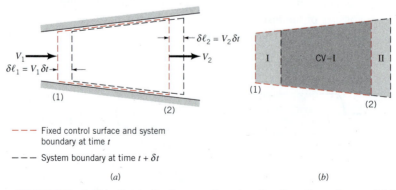

■ **FIGURE 4.8** **Control volume and system for flow through a variable area pipe.**

within the pipe or duct between sections (1) and (2) as indicated. The system that we consider is that fluid occupying the control volume at some initial time t. A short time later, at time $t + \delta t$, the system has moved slightly to the right. The fluid particles that coincided with section (2) of the control surface at time t have moved a distance $\delta \ell_2 = V_2 \, \delta t$ to the right, where V_2 is the velocity of the fluid as it passes section (2). Similarly, the fluid initially at section (1) has moved a distance $\delta \ell_1 = V_1 \delta t$, where V_1 is the fluid velocity at section (1). We assume the fluid flows across sections (1) and (2) in a direction normal to these surfaces and that V_1 and V_2 are constant across sections (1) and (2).

As is shown in Fig. 4.8b, the outflow from the control volume from time t to $t + \delta t$ is denoted as volume II, the inflow as volume I, and the control volume itself as CV. Thus, the system at time t consists of the fluid in section CV ("SYS = CV" at time t), whereas at time $t + \delta t$ the system consists of the same fluid that now occupies sections (CV − I) + II. That is, "SYS = CV − I + II" at time $t + \delta t$. The control volume remains as section CV for all time.

If B is an extensive parameter of the system, then the value of it for the system at time t is the same as that for the control volume B_{cv},

$$B_{sys}(t) = B_{cv}(t)$$

since the system and the fluid within the control volume coincide at this time. Its value at time $t + \delta t$ is

$$B_{sys}(t + \delta t) = B_{cv}(t + \delta t) - B_I(t + \delta t) + B_{II}(t + \delta t)$$

Thus, the change in the amount of B in the system in the time interval δt divided by this time interval is given by

$$\frac{\delta B_{sys}}{\delta t} = \frac{B_{sys}(t + \delta t) - B_{sys}(t)}{\delta t} = \frac{B_{cv}(t + \delta t) - B_I(t + \delta t) + B_{II}(t + \delta t) - B_{sys}(t)}{\delta t}$$

By using the fact that at the initial time t we have $B_{sys}(t) = B_{cv}(t)$, this ungainly expression may be rearranged as follows.

$$\frac{\delta B_{sys}}{\delta t} = \frac{B_{cv}(t + \delta t) - B_{cv}(t)}{\delta t} - \frac{B_1(t + \delta t)}{\delta t} + \frac{B_{II}(t + \delta t)}{\delta t} \tag{4.8}$$

In the limit $\delta t \to 0$, the left-hand side of Eq. 4.8 is equal to the time rate of change of B for the system and is denoted as DB_{sys}/Dt.

In the limit $\delta t \to 0$, the first term on the right-hand side of Eq. 4.8 is seen to be the time rate of change of the amount of B within the control volume

$$\lim_{\delta t \to 0} \frac{B_{cv}(t + \delta t) - B_{cv}(t)}{\delta t} = \frac{\partial B_{cv}}{\partial t} \tag{4.9}$$

The third term on the right-hand side of Eq. 4.8 represents the rate at which the extensive parameter B flows from the control volume, across the control surface. This can be seen from the fact that the amount of B within region II, the outflow region, is its amount per unit volume, ρb, times the volume $\delta V_{II} = A_2 \, \delta \ell_2 = A_2(V_2 \delta t)$. Hence,

$$B_{II}(t + \delta t) = (\rho_2 b_2)(\delta V_{II}) = \rho_2 b_2 A_2 V_2 \, \delta t$$

where b_2 and ρ_2 are the constant values of b and ρ across section (2). Thus, the rate at which this property flows from the control volume, $\dot{B}_{out}$, is given by

$$\dot{B}_{out} = \lim_{\delta t \to 0} \frac{B_{II}(t + \delta t)}{\delta t} = \rho_2 A_2 V_2 b_2 \tag{4.10}$$

Similarly, the inflow of B into the control volume across section (1) during the time interval δt corresponds to that in region I and is given by the amount per unit volume times the volume, $\delta V_I = A_1 \delta \ell_1 = A_1(V_1 \delta t)$. Hence,

$$B_I(t + \delta t) = (\rho_1 b_1)(\delta V_I) = \rho_1 b_1 A_1 V_1 \delta t$$

where b_1 and ρ_1 are the constant values of b and ρ across section (1). Thus, the rate of inflow of property B into the control volume, $\dot{B}_{in}$, is given by

$$\dot{B}_{in} = \lim_{\delta t \to 0} \frac{B_I(t + \delta t)}{\delta t} = \rho_1 A_1 V_1 b_1 \tag{4.11}$$

If we combine Eqs. 4.8, 4.9, 4.10, and 4.11, we see that the relationship between the time rate of change of B for the system and that for the control volume is given by

$$\frac{DB_{sys}}{Dt} = \frac{\partial B_{cv}}{\partial t} + \dot{B}_{out} - \dot{B}_{in} \tag{4.12}$$

or

$$\frac{DB_{sys}}{Dt} = \frac{\partial B_{cv}}{\partial t} + \rho_2 A_2 V_2 b_2 - \rho_1 A_1 V_1 b_1 \tag{4.13}$$

This is a version of the Reynolds transport theorem valid under the restrictive assumptions associated with the flow shown in Fig. 4.8—fixed control volume with one inlet and one outlet having uniform properties (density, velocity, and the parameter b) across the inlet and outlet with the velocity normal to sections (1) and (2). Note that the time rate of change of B for the system (the left-hand side of Eq. 4.13) is not necessarily the same as the rate of change of B within the control volume (the first term on the right-hand side of Eq. 4.13). This is true because the inflow rate ($b_1 \rho_1 V_1 A_1$) and the outflow rate ($b_2 \rho_2 V_2 A_2$) of the property B for the control volume need not be the same.

As indicated in Fig. 4.9, a control volume may contain more than one inlet and one outlet. In such cases a simple extension of Eq. 4.13 gives

$$\boxed{\frac{DB_{sys}}{Dt} = \frac{\partial B_{cv}}{\partial t} + \sum \rho_{out} A_{out} V_{out} b_{out} - \sum \rho_{in} A_{in} V_{in} b_{in}} \tag{4.14}$$

where summations over the inlets (in) and outlets (out) account for all of the flow through the control volume.

Equations 4.13 and 4.14 are simplified versions of the Reynolds transport theorem. A more general form of this theorem, valid for more general flow conditions, is presented in Appendix D.

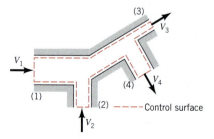

■ **FIGURE 4.9** **Multiple inlet and outlet control volume.**

EXAMPLE 4.6	Time Rate of Change for a System and a Control Volume

GIVEN Consider the flow from the fire extinguisher shown in Fig. E4.6. Let the extensive property of interest be the system mass ($B = m$, the system mass, so that $b = 1$).

FIND Write the appropriate form of the Reynolds transport theorem for this flow.

SOLUTION

We take the control volume to be the fire extinguisher, and the system to be the fluid within it at time $t = 0$. For this case there is no inlet, section (1), across which the fluid flows into the control volume ($A_1 = 0$). There is, however, an outlet, section (2). Thus, the Reynolds transport theorem, Eq. 4.13, along with Eq. 4.9 with $b = 1$ can be written as

$$\frac{Dm_{sys}}{Dt} = \frac{\partial m_{cv}}{\partial t} + \rho_2 A_2 V_2 \qquad \text{(1)} \quad \textbf{(Ans)}$$

COMMENT If we proceed one step further and use the basic law of conservation of mass, we may set the left-hand side of this equation equal to zero (the amount of mass in a system is constant) and rewrite Eq. 1 in the form:

$$\frac{\partial m_{cv}}{\partial t} = -\rho_2 A_2 V_2 \qquad \text{(2)}$$

The physical interpretation of this result is that the rate at which the mass in the tank decreases in time is equal in magnitude but opposite to the rate of flow of mass from the exit, $\rho_2 A_2 V_2$. Note the units for the two terms of Eq. 2 (kg/s or

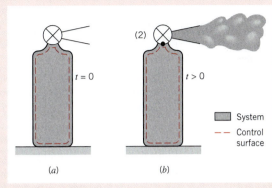

■ **FIGURE E4.6**

slugs/s). Note that if there were both an inlet and an outlet to the control volume shown in Fig. E4.6, Eq. 2 would become

$$\frac{\partial m_{cv}}{\partial t} = \rho_1 A_1 V_1 - \rho_2 A_2 V_2 \qquad \text{(3)}$$

In addition, if the flow were steady, the left-hand side of Eq. 3 would be zero (the amount of mass in the control would be constant in time) and Eq. 3 would become

$$\rho_1 A_1 V_1 = \rho_2 A_2 V_2$$

This is one form of the conservation of mass principle—the mass flowrates into and out of the control volume are equal. Other more general forms are discussed in Chapter 5 and Appendix D.

4.4.2 Selection of a Control Volume

Any volume in space can be considered as a control volume. The ease of solving a given fluid mechanics problem is often very dependent upon the choice of the control volume used. Only by practice can we develop skill at selecting the "best" control volume. None are "wrong," but some are "much better" than others.

Figure 4.10 illustrates three possible control volumes associated with flow through a pipe. If the problem is to determine the pressure at point (1), selection of the control volume (a) is better than that of (b) because point (1) lies on the control surface. Similarly, control volume (a) is better than (c) because the flow is normal to the inlet and exit portions of the control volume. None of these control volumes are wrong—(a) will be easier to use. Proper control

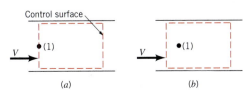
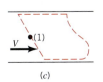

■ **FIGURE 4.10** **Various control volumes for flow through a pipe.**

volume selection will become much clearer in Chapter 5 where the Reynolds transport theorem is used to transform the governing equations from the system formulation into the control volume formulation, and numerous examples using control volume ideas are discussed.

4.5   Chapter Summary and Study Guide

field representation
velocity field
Eulerian method
Lagrangian method
one-, two-, and
  three-
  dimensional
  flow
steady and
  unsteady flow
streamline
streakline
pathline
acceleration field
material derivative
local acceleration
convective
  acceleration
system
control volume
Reynolds transport
  theorem

This chapter considered several fundamental concepts of fluid kinematics. That is, various aspects of fluid motion are discussed without regard to the forces needed to produce this motion. The concepts of a field representation of a flow and the Eulerian and Lagrangian approaches to describing a flow are introduced, as are the concepts of velocity and acceleration fields.

The properties of one-, two-, or three-dimensional flows and steady or unsteady flows are introduced along with the concepts of streamlines, streaklines, and pathlines. Streamlines, which are lines tangent to the velocity field, are identical to streaklines and pathlines if the flow is steady. For unsteady flows, they need not be identical.

As a fluid particle moves about, its properties (i.e., velocity, density, temperature) may change. The rate of change of these properties can be obtained by using the material derivative, which involves both unsteady effects (time rate of change at a fixed location) and convective effects (time rate of change due to the motion of the particle from one location to another).

The concepts of a control volume and a system are introduced, and the Reynolds transport theorem is developed. By using these ideas, the analysis of flows can be carried out using a control volume (a fixed volume through which the fluid flows), whereas the governing principles are stated in terms of a system (a flowing portion of fluid).

The following checklist provides a study guide for this chapter. When your study of the entire chapter and end-of-chapter exercises has been completed you should be able to

- write out the meanings of the terms listed here in the margin and understand each of the related concepts. These terms are particularly important and are set in color and bold type in the text.
- understand the concept of the field representation of a flow and the difference between Eulerian and Lagrangian methods of describing a flow.
- explain the differences among streamlines, streaklines, and pathlines.
- calculate and plot streamlines for flows with given velocity fields.
- use the concept of the material derivative, with its unsteady and convective effects, to determine time rate of change of a fluid property.
- determine the acceleration field for a flow with a given velocity field.
- understand the properties of and differences between a system and a control volume.
- interpret, physically and mathematically, the concepts involved in the Reynolds transport theorem.

References

1. Goldstein, R. J., *Fluid Mechanics Measurements,* Hemisphere, New York, 1983.
2. Homsy, G. M., et al., *Multimedia Fluid Mechanics* CD-ROM, Cambridge University Press, New York, 2000.
3. Magarvey, R. H., and MacLatchy, C. S., The Formation and Structure of Vortex Rings, *Canadian Journal of Physics,* Vol. 42, 1964.

Review Problems

Go to Appendix F for a set of review problems with answers. Detailed solutions can be found in *Student Solution Manual for a Brief Introduction to Fluid Mechanics,* by Young et al. (© 2006 John Wiley and Sons, Inc.).

Problems

Note: Unless otherwise indicated use the values of fluid properties found in the tables on the inside of the front cover. Problems designated with an (*) are intended to be solved with the aid of a programmable calculator or a computer. Problems designated with a (†) are "open-ended" problems and require critical thinking in that to work them one must make various assumptions and provide the necessary data. There is not a unique answer to these problems.

Answers to the even-numbered problems are listed at the end of the book. Access to the videos that accompany problems can be obtained through the book's web site, www.wiley.com/college/young. FE problems can also be accessed on this web site.

Section 4.1 The Velocity Field

4.1 The velocity field of a flow is given by $\mathbf{V} = 10y/(x^2 + y^2)^{1/2}\hat{\mathbf{i}} - 10x/(x^2 + y^2)^{1/2}\hat{\mathbf{j}}$ ft/s, where x and y are in feet. Determine the fluid speed at points along the x axis; along the y axis. What is the angle between the velocity vector and the x axis at points $(x, y) = (5, 0), (5, 5)$ and $(0, 5)$?

4.2 A flow can be visualized by plotting the velocity field as velocity vectors at representative locations in the flow as shown in **Video V4.1** and Fig. E4.1. Consider the velocity field given in polar coordinates by $v_r = -10/r$ and $v_\theta = 10/r$. This flow approximates a fluid swirling into a sink as shown in Fig. P4.2. Plot the velocity field at locations given by $r = 1, 2,$ and 3 with $\theta = 0, 30, 60,$ and $90°$.

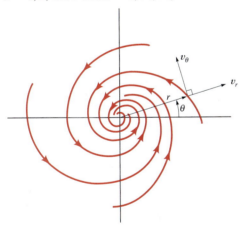

■ **FIGURE P4.2**

4.3 The x and y components of velocity for a two-dimensional flow are $u = 6y$ ft/s and $v = 4$ ft/s, where y is in feet. Determine the equation for the streamlines and sketch representative streamlines in the upper half plane.

4.4 A velocity field is given by $\mathbf{V} = x\hat{\mathbf{i}} + x(x - 1)(y + 1)\hat{\mathbf{j}}$, where u and v are in ft/s and x and y are in feet. Plot the streamline that passes through $x = 0$ and $y = 0$. Compare this streamline with the streakline through the origin.

4.5 (See Fluids in the News article titled "Winds on Earth and Mars," Section 4.1.4.) A 10-ft-diameter dust devil that rotates one revolution per second travels across the Martian surface (in the x-direction) with a speed of 5 ft/s. Plot the pathline etched on the surface by a fluid particle 10 ft from the center of the dust devil for time $0 \le t \le 3$ s. The particle position is given by

the sum of that for a stationary swirl $[x = 10\ \cos(2\pi t), y = 10\ \sin(2\pi t)]$ and that for a uniform velocity $(x = 5t, y = \text{constant})$, where x and y are in feet and t is in seconds.

4.6 In addition to the customary horizontal velocity components of the air in the atmosphere (the "wind"), there often are vertical air currents (thermals) caused by buoyant effects due to uneven heating of the air as indicated in Fig. P4.6. Assume that the velocity field in a certain region is approximated by $u = u_0$, $v = v_0(1 - y/h)$ for $0 < y < h$, and $u = u_0, v = 0$ for $y > h$. Determine the equation for the streamlines and plot the streamline that passes through the origin for values of $u_0/v_0 = 0.5, 1,$ and 2.

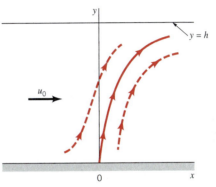

■ **FIGURE P4.6**

4.7 (See Fluids in the News article titled "Follow those particles," Section 4.1.) Two photographs of four particles in a flow past a sphere are superposed as shown in Fig. P4.7. The time interval between the photos is $\Delta t = 0.002$ s. The locations of the particles, as determined from the photos, are shown in the table. **(a)** Determine the fluid velocity for these particles. **(b)** Plot a graph to compare the results of part **(a)** with the theoretical velocity which is given by $V = V_0(1 + a^3/x^3)$, where a is the sphere radius and V_0 is the fluid speed far from the sphere.

Particle	x at $t = 0$ s (ft)	x at $t = 0.002$ s (ft)
1	−0.500	−0.480
2	−0.250	−0.232
3	−0.140	−0.128
4	−0.120	−0.112

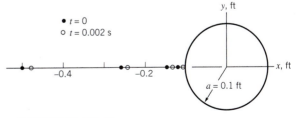

■ **FIGURE P4.7**

Section 4.2 The Acceleration Field

4.8 A three-dimensional velocity field is given by $u = x^2$, $v = -2xy$, and $w = x + y$. Determine the acceleration vector.

†4.9 Estimate the deceleration of a water particle in a rain-drop as it strikes the sidewalk. List all assumptions and show all calculations.

4.10 The velocity of the water in the pipe shown in Fig. P4.10 is given by $V_1 = 0.50t$ m/s and $V_2 = 1.0t$ m/s, where t is in seconds. **(a)** Determine the local acceleration at points (1) and (2). **(b)** Is the average convective acceleration between these two points negative, zero, or positive? Explain.

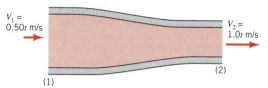

$V_1 = 0.50t$ m/s

$V_2 = 1.0t$ m/s

(2)

(1)

■ **FIGURE P4.10**

4.11 A fluid particle flowing along a stagnation streamline, as shown in **Video V4.5** and Fig. P4.11, slows down as it approaches the stagnation point. Measurements of the dye flow in the video indicate that the location of a particle starting on the stagnation streamline a distance $s = 0.6$ ft upstream of the stagnation point at $t = 0$ is given approximately by $s = 0.6\,e^{-0.5t}$, where t is in seconds and s is in ft. **(a)** Determine the speed of a fluid particle as a function of time, $V_{particle}(t)$, as it flows along the streamline. **(b)** Determine the speed of the fluid as a function of position along the streamline, $V = V(s)$. **(c)** Determine the fluid acceleration along the streamline as a function of position, $a_s = a_s(s)$.

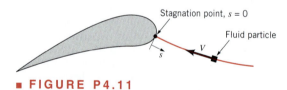

Stagnation point, $s = 0$

Fluid particle

V

s

■ **FIGURE P4.11**

4.12 As a valve is opened, water flows through the diffuser shown in Fig. P4.12 at an increasing flowrate so that the velocity along the centerline is given by $\mathbf{V} = u\hat{\mathbf{i}} = V_0(1 - e^{-ct})(1 - x/\ell)\,\hat{\mathbf{i}}$, where u_0, c, and ℓ are constants. Determine the acceleration as a function of x and t. If $V_0 = 10$ ft/s and $\ell = 5$ ft, what value of c (other than $c = 0$) is needed to make the acceleration zero for any x at $t = 2$ s? Explain how the acceleration can be zero if the flowrate is increasing with time.

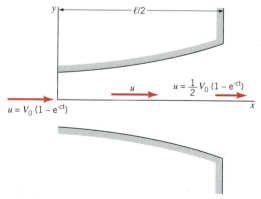

y

$\ell/2$

u

$u = \frac{1}{2} V_0 (1 - e^{-ct})$

x

$u = V_0 (1 - e^{-ct})$

■ **FIGURE P4.12**

4.13 An incompressible fluid flows past a turbine blade as shown in Fig. P4.13a and **Video V4.5.** Far upstream and downstream of the blade the velocity is V_0. Measurements show that the velocity of the fluid along streamline A–F near the blade is as indicated in Fig. P4.13b. Sketch the streamwise component of acceleration, a_s, as a function of distance, s, along the streamline. Discuss the important characteristics of your result.

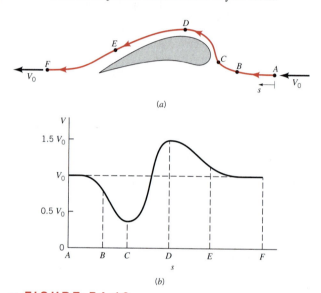

(a)

(b)

■ **FIGURE P4.13**

4.14 A gas flows along the x-axis with a speed $V = 5x$ m/s and a pressure of $p = 10x^2$ N/m², where x is in meters. **(a)** Determine the time rate of change of pressure at the fixed location $x = 1$. **(b)** Determine the time rate of change of pressure for a fluid particle flowing past $x = 1$. **(c)** Explain without using any equations why the answers to parts (a) and (b) are different.

4.15 A hydraulic jump is a rather sudden change in depth of a liquid layer as it flows in an open channel as shown in Fig. P4.15 and **Videos V10.5 and 10.6.** In a relatively short distance (thickness = ℓ) the liquid depth changes from z_1 to z_2, with a corresponding change in velocity from V_1 to V_2. If $V_1 = 5$ m/s, $V_2 = 1$ m/s, and $\ell = 0.2$ m, estimate the average deceleration of the liquid as it flows across the hydraulic jump. How many g's deceleration does this represent?

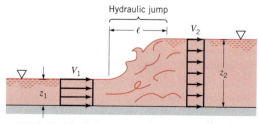

Hydraulic jump

ℓ

V_1

V_2

z_1

z_2

■ **FIGURE P4.15**

4.16 A nozzle is designed to accelerate the fluid from V_1 to V_2 in a linear fashion. That is, $V = ax + b$, where a and b are constants. If the flow is constant with $V_1 = 10$ m/s at $x_1 = 0$ and $V_2 = 25$ m/s at $x_2 = 1$ m, determine local acceleration, convective acceleration, and acceleration of the fluid at points (1) and (2).

4.17 Assume the temperature of the exhaust in an exhaust pipe can be approximated by $T = T_0(1 + ae^{-bx})[1 + c\cos(\omega t)]$, where $T_0 = 100°$ C, $a = 3$, $b = 0.03$ m^{-1}, $c = 0.05$, and $\omega = 100$ rad/s. If the exhaust speed is a constant 3 m/s, determine the time rate of change of temperature of the fluid particles at $x = 0$ and $x = 4$ m when $t = 0$.

4.18 The temperature distribution in a fluid is given by $T = 10x + 5y$, where x and y are the horizontal and vertical coordinates in meters and T is in degrees centigrade. Determine the time rate of change of temperature of a fluid particle traveling **(a)** horizontally with $u = 20$ m/s, $v = 0$ or **(b)** vertically with $u = 0$, $v = 20$ m/s.

4.19 Assume that the streamlines for the wingtip vortices from an airplane (see Fig. P4.19 and **Video V4.2**) can be approximated by circles of radius r and that the speed is $V = K/r$, where K is a constant. Determine the streamline acceleration, a_s, and the normal acceleration, a_n, for this flow.

■ **FIGURE P4.19**

4.20 At the top of its trajectory, the stream of water shown in Fig. P4.20 and **Video V4.3** flows with a horizontal velocity of 1.80 ft/s. The radius of curvature of its streamline at that point is approximately 0.10 ft. Determine the normal component of acceleration at that location.

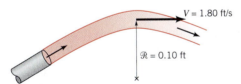

■ **FIGURE P4.20**

4.21 Water flows through the slit at the bottom of a two-dimensional water trough as shown in Fig. P4.21. Throughout most of the trough the flow is approximately radial (along rays from O) with a velocity of $V = c/r$, where r is the radial coordinate and c is a constant. If the velocity is 0.04 m/s when $r = 0.1$ m, determine the acceleration at points A and B.

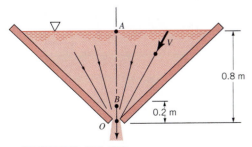

■ **FIGURE P4.21**

4.22 Air flows from a pipe into the region between a circular disk and a cone as shown in Fig. P4.22. The fluid velocity in the gap between the disk and the cone is closely approximated by $V = V_0R^2/r^2$, where R is the radius of the disk, r is the radial coordinate, and V_0 is the fluid velocity at the edge of the disk. Determine the acceleration for $r = 0.5$ and 2 ft if $V_0 = 5$ ft/s and $R = 2$ ft.

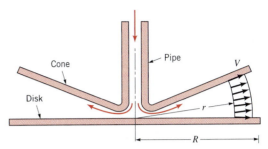

■ **FIGURE P4.22**

Sections 4.3 and 4.4 Control Volume and System Representations and the Reynolds Transport Theorem

4.23 Water flows through a duct of square cross section as shown in Fig. P4.23 with a constant, uniform velocity of $V = 20$ m/s. Consider fluid particles that lie along line A–B at time $t = 0$. Determine the position of these particles, denoted by line A'–B', when $t = 0.20$ s. Use the volume of fluid in the region between lines A–B and A'–B' to determine the flowrate in the duct. Repeat the problem for fluid particles originally along line C–D; along line E–F. Compare your three answers.

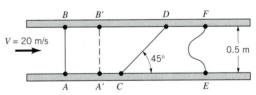

■ **FIGURE P4.23**

4.24 In the region just downstream of a sluice gate, the water may develop a reverse flow region as indicated in Fig. P4.24 and **Video V10.5**. The velocity profile is assumed to consist of two uniform regions, one with velocity $V_a = 10$ fps and the other with $V_b = 3$ fps. Determine the net flowrate of water across the portion of the control surface at section (2) if the channel is 20 ft wide.

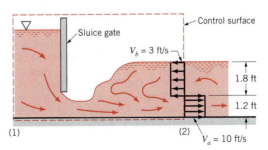

■ **FIGURE P4.24**

4.25 Air enters an elbow with a uniform speed of 10 m/s as shown in Fig. P4.25. At the exit of the elbow the velocity profile is not uniform. In fact, there is a region of separation or reverse flow. The fixed control volume $ABCD$ coincides with the system at time $t = 0$. Make a sketch to indicate (**a**) the system at time $t = 0.01$ s and (**b**) the fluid that has entered and exited the control volume in that time period.

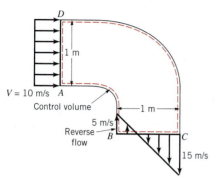

■ **FIGURE P4.25**

4.26 Water flows in the branching pipe shown in Fig. P4.26 with uniform velocity at each inlet and outlet. The fixed control volume indicated coincides with the system at time $t = 20$ s. Make a sketch to indicate (**a**) the boundary of the system at time $t = 20.2$ s, (**b**) the fluid that left the control volume during that 0.2-s interval, and (**c**) the fluid that entered the control volume during that time interval.

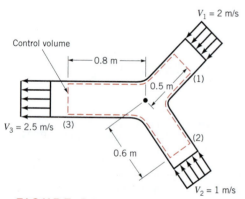

■ **FIGURE P4.26**

4.27 Two plates are pulled in opposite directions with speeds of 1.0 ft/s as shown in Fig. P4.27. The oil between the plates moves with a velocity given by $\mathbf{V} = 10y\hat{\mathbf{i}}$ ft/s, where y is in feet. The fixed control volume $ABCD$ coincides with the system at time $t = 0$. Make a sketch to indicate (**a**) the system at time $t = 0.2$ s and (**b**) the fluid that has entered and exited the control volume in that time period.

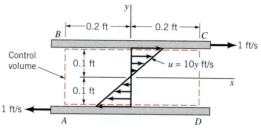

■ **FIGURE P4.27**

4.28 Two liquids with different densities and viscosities fill the gap between parallel plates as shown in Fig. P4.28. The bottom plate is fixed; the top plate moves with a speed of 2 ft/s. The velocity profile consists of two linear segments as indicated. The fixed control volume $ABCD$ coincides with the system at time $t = 0$. Make a sketch to indicate (**a**) the system at time $t = 0.1$ s and (**b**) the fluid that has entered and exited the control volume in that time period.

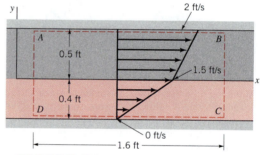

■ **FIGURE P4.28**

4.29 Water enters a 5-ft-wide, 1-ft-deep channel as shown in Fig. P4.29. Across the inlet the water velocity is 6 ft/s in the center portion of the channel and 1 ft/s in the remainder of it. Farther downstream the water flows at a uniform 2-ft/s velocity across the entire channel. The fixed control volume $ABCD$ coincides with the system at time $t = 0$. Make a sketch to indicate (**a**) the system at time $t = 0.5$ s and (**b**) the fluid that has entered and exited the control volume in that time period.

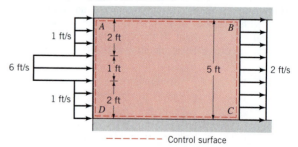

■ **FIGURE P4.29**

FE Exam Problems

Sample FE (Fundamentals of Engineering) exam questions for fluid mechanics are provided on the book's web site, www.wiley.com/college/young.

CHAPTER 5

Finite Control Volume Analysis

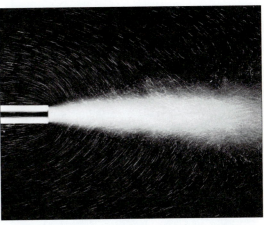

A jet of water injected into stationary water: Upon emerging from the slit at the left, the jet of fluid loses some of its momentum to the surrounding fluid. This causes the jet to slow down and its width to increase (air bubbles in water). (Photograph courtesy of ONERA, France.)

LEARNING OBJECTIVES

After completing this chapter, you should be able to:

- identify an appropriate control volume and draw the corresponding diagram.
- calculate flowrates using the continuity equation.
- calculate forces and torques using the linear momentum and moment-of-momentum equations.
- use the energy equation to account for losses due to friction, as well as effects of pumps and turbines.
- apply the kinetic energy coefficient to nonuniform flows.

Many practical problems in fluid mechanics require analysis of the behavior of the contents of a finite region in space (a control volume). For example, we may be asked to calculate the anchoring force required to hold a jet engine in place during a test. Important questions can be answered readily with finite control volume analyses. The bases of this analysis method are some fundamental principles of physics, namely, conservation of mass, Newton's second law of motion, and the laws of thermodynamics. Thus, as one might expect, the resultant techniques are powerful and applicable to a wide variety of fluid mechanical circumstances that require engineering judgment. Furthermore, the finite control volume formulas are easy to interpret physically and not difficult to use.

5.1   Conservation of Mass—The Continuity Equation

5.1.1   Derivation of the Continuity Equation

A system is defined as a collection of unchanging contents, so the conservation of mass principle for a system is simply stated as time rate of change of the system mass = 0

or

$$\frac{Dm_{sys}}{Dt} = 0 \qquad (5.1)$$

where m_{sys} is the system mass.

Figure 5.1 shows a system and a fixed, nondeforming control volume that are coincident at an instant of time. Since we are considering conservation of mass, we take the extensive property to be mass (i.e., $B = m$ = mass so that $b = B/m = 1$). Thus, the Reynolds transport theorem (Eq. 4.13) allows us to state that

$$\frac{Dm_{sys}}{Dt} = \frac{\partial m_{cv}}{\partial t} + \rho_2 A_2 V_2 - \rho_1 A_1 V_1 \qquad (5.2)$$

or

| time rate of change of the mass of the coincident system | = | time rate of change of the mass of the contents of the coincident control volume | + | net rate of flow of mass through the control surface |

Because the amount of mass in a small volume $d\mathcal{V}$ is $\rho\, d\mathcal{V}$, it follows that the amount of mass in control volume, m_{cv}, can be written as

$$m_{cv} = \int_{cv} \rho\, d\mathcal{V}$$

If the control volume has multiple inlets and outlets, Eq. 5.2 can be modified to account for flow through each of the inlets and outlets to give

$$\frac{Dm_{sys}}{Dt} = \frac{\partial}{\partial t}\int_{cv} \rho\, d\mathcal{V} + \sum \rho_{out} A_{out} V_{out} - \sum \rho_{in} A_{in} V_{in} \qquad (5.3)$$

When a flow is steady, all field properties (i.e., properties at any specified point), including density, remain constant with time and the time rate of change of the mass of the contents of the control volume is zero. That is,

$$\frac{\partial m_{cv}}{\partial t} = 0 \qquad (5.4)$$

so that for steady flow

$$\frac{\partial}{\partial t}\int_{cv} \rho\, d\mathcal{V} = 0$$

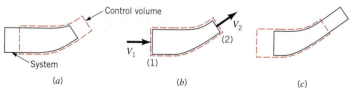

■ **FIGURE 5.1** System and control volume at three different instances of time. (*a*) System and control at time $t - \delta t$. (*b*) System and control volume at t, coincident condition. (*c*) System and control volume at $t + \delta t$.

The control volume expression for *conservation of mass,* commonly called the *continuity equation,* is obtained by combining Eqs. 5.1 and 5.3 to obtain

$$\frac{\partial}{\partial t} \int_{cv} \rho \, dV + \sum \rho_{out} A_{out} V_{out} - \sum \rho_{in} A_{in} V_{in} = 0 \tag{5.5}$$

In words, Eq. 5.5 states that to conserve mass the time rate of change of the mass of the contents of the control volume plus the net rate of mass flow through the control surface must equal zero. Note that Eq. 5.5 is restricted to fixed, nondeforming control volumes having uniform properties across the inlets and outlets, with the velocity normal to the inlet and outlet areas. A more general form of the continuity equation, valid for more general flow conditions, is given in Appendix D.

An often-used expression for the *mass flowrate, $\dot{m}$,* through a section of the control volume having area A is

$$\dot{m} = \rho A V = \rho Q \tag{5.6}$$

where ρ is the fluid density, V is the component of fluid velocity normal to the area A, and $Q = VA$ is the volume flowrate (ft³/s or m³/s). Note the symbols used to denote mass, m (slugs or kg), and mass flowrate, $\dot{m}$ (slugs/s or kg/s).

Often the fluid velocity across a section area A is not uniform. In such cases, the appropriate fluid velocity to use in Eq. 5.6 is the average value of the component of velocity normal to the section area involved. This average value, $\bar{V}$, is defined in Eq. 5.7 and shown in the figure in the margin.

$$\bar{V} = \frac{\int_A \rho V \, dA}{\rho A} \tag{5.7}$$

If the velocity is considered uniformly distributed (one-dimensional flow) over the section area, then $\bar{V} = V$ and the bar notation is not necessary.

5.1.2 Fixed, Nondeforming Control Volume

Several example problems that involve the continuity equation for fixed, nondeforming control volumes (Eq. 5.5) follow.

V5.1 Sink flow

EXAMPLE 5.1   Conservation of Mass—Steady, Incompressible Flow

GIVEN Seawater flows steadily through a simple conical-shaped nozzle at the end of a fire hose as illustrated in Fig. E5.1a. According to local regulations, the nozzle exit velocity must be at least 20 m/s.

FIND Determine the minimum pumping capacity required in m³/s.

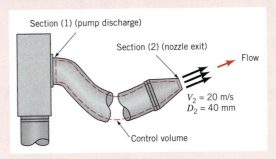

■ **FIGURE E5.1a**

SOLUTION

The pumping capacity sought is the volume flowrate delivered by the fire pump to the hose and nozzle. Since we desire knowledge about the pump discharge flowrate and we have information about the nozzle exit flowrate, we link these two flowrates with the control volume designated with the dashed line in Fig. E5.1a. This control volume contains, at any instant, seawater that is within the hose and nozzle from the pump discharge to the nozzle exit plane.

Equation 5.5 is applied to the contents of this control volume to give

$$\frac{\partial}{\partial t}\int_{cv}\rho\,d\mathcal{V} + \sum\rho_{out}A_{out}V_{out} - \sum\rho_{in}A_{in}V_{in} = 0 \quad (1)$$

with the note: "0 (flow is steady)" over the first term.

The time rate of change of the mass of the contents of this control volume is zero because the flow is steady. Because there is only one inflow [the pump discharge, section (1)] and one outflow [the nozzle exit, section (2)], Eq. (1) becomes

$$\rho_2 A_2 V_2 - \rho_1 A_1 V_1 = 0$$

so that with $\dot{m} = \rho A V$

$$\dot{m}_1 = \dot{m}_2 \quad (2)$$

Because the mass flowrate is equal to the product of fluid density, ρ, and volume flowrate, Q (see Eq. 5.6), we obtain from Eq. 2

$$\rho_2 Q_2 = \rho_1 Q_1 \quad (3)$$

Liquid flow at low speeds, as in this example, may be considered incompressible. Therefore

$$\rho_2 = \rho_1 \quad (4)$$

and from Eqs. 3 and 4

$$Q_2 = Q_1 \quad (5)$$

The pumping capacity is equal to the volume flowrate at the nozzle exit. If, for simplicity, the velocity distribution at the nozzle exit plane, section (2), is considered uniform (one-dimensional), then from Eq. 5

$$Q_1 = Q_2 = V_2 A_2$$

$$= V_2 \frac{\pi}{4} D_2^2 = (20\text{ m/s})\frac{\pi}{4}\left(\frac{40\text{ mm}}{1000\text{ mm/m}}\right)^2$$

$$= 0.0251\text{ m}^3/\text{s} \quad \text{(Ans)}$$

COMMENT   By repeating the calculations for various values of the nozzle exit diameter, D_2, the results shown in Fig. E5.1b are obtained. The flowrate is proportional to the exit area, which varies as the diameter squared. Hence, if the diameter were doubled, the flowrate would increase by a factor of four, provided the exit velocity remained the same.

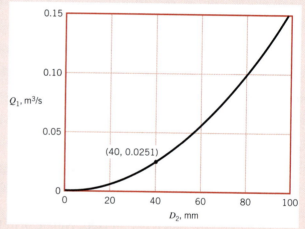

■ **FIGURE E5.1b**

EXAMPLE 5.2   Conservation of Mass—Nonuniform Velocity Profile

GIVEN   Incompressible, laminar water flow develops in a straight pipe having radius R as indicated in Fig. E5.2a. At section (1), the velocity profile is uniform; the velocity is equal to a constant value U and is parallel to the pipe axis everywhere. At section (2), the velocity profile is axisymmetric and parabolic, with zero velocity at the pipe wall and a maximum value of u_{max} at the centerline.

FIND

(a)   How are U and u_{max} related?

(b)   How are the average velocity at section (2), $\bar{V}_2$, and u_{max} related?

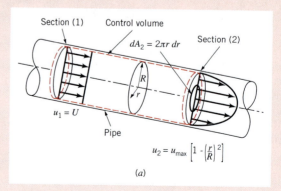

■ **FIGURE E5.2a**

SOLUTION

(a) An appropriate control volume is sketched (dashed lines) in Fig. E5.2a. The application of Eq. 5.5 to the contents of this control volume yields

$$\sum \rho_{out} A_{out} V_{out} - \sum \rho_{in} A_{in} V_{in} = 0 \qquad (1)$$

At the inlet, section (1), the velocity is uniform with $V_1 = U$ so that

$$\sum \rho_{in} A_{in} V_{in} = \rho_1 A_1 U \qquad (2)$$

At the outlet, section (2), the velocity is not uniform. However, the net flowrate through this section is the sum of flows through numerous small washer-shaped areas of size $dA_2 = 2\pi r\, dr$ as shown by the shaded area element in Fig. E5.2b. On each of these infinitesimal areas the fluid velocity is denoted as u_2. Thus, in the limit of infinitesimal area elements, the summation is replaced by an integration and the outflow through section (2) is given by

$$\sum \rho_{out} A_{out} V_{out} = \rho_2 \int_0^R u_2 2\pi r\, dr \qquad (3)$$

By combining Eqs. 1, 2, and 3 we get

$$\rho_2 \int_0^R u_2 2\pi r\, dr - \rho_1 A_1 U = 0 \qquad (4)$$

Since the flow is considered incompressible, $\rho_1 = \rho_2$. The parabolic velocity relationship for flow through section (2) is used in Eq. 4 to yield

$$2\pi u_{max} \int_0^R \left[1 - \left(\frac{r}{R}\right)^2\right] r\, dr - A_1 U = 0 \qquad (5)$$

Integrating, we get from Eq. 5

$$2\pi u_{max} \left(\frac{r^2}{2} - \frac{r^4}{4R^2}\right)_0^R - \pi R^2 U = 0$$

or

$$u_{max} = 2U \qquad \textbf{(Ans)}$$

(b) Since this flow is incompressible, we conclude from Eq. 5.7 that U is the average velocity at all sections of the control volume. Thus, the average velocity at section (2), $\overline{V}_2$, is one-half the maximum velocity, u_{max}, there or

$$\overline{V}_2 = \frac{u_{max}}{2} \qquad \textbf{(Ans)}$$

COMMENT The relationship between the maximum velocity at section (2) and the average velocity is a function of the "shape" of the velocity profile. For the parabolic profile assumed in this example, the average velocity, $u_{max}/2$, is the actual "average" of the maximum velocity at section (2), $u_2 = u_{max}$, and the minimum velocity at that section, $u_2 = 0$. However, as shown in Fig. E5.2c, if the velocity profile is a different shape (non-parabolic), the average velocity is not necessarily one half of the maximum velocity.

(b)

■ **FIGURE E5.2b**

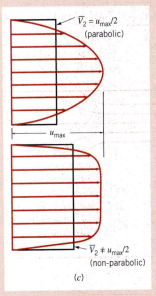

(c)

■ **FIGURE E5.2c**

EXAMPLE 5.3   Conservation of Mass—Unsteady Flow

GIVEN   A bathtub is being filled with water from a faucet. The rate of flow from the faucet is steady at 9 gal/min. The tub volume is approximated by a rectangular space as indicated in Fig. E5.3a.

FIND   Estimate the time rate of change of the depth of water in the tub, $\partial h/\partial t$, in inches per minute at any instant.

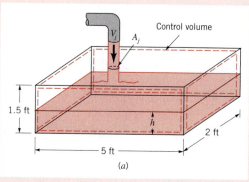

■ **FIGURE E5.3a**

SOLUTION

We use the fixed, nondeforming control volume outlined with a dashed line in Fig. E5.3. This control volume includes in it, at any instant, the water accumulated in the tub, some of the water flowing from the faucet into the tub, and some air. Application of Eqs. 5.5 and 5.6 to these contents of the control volume results in

$$\frac{\partial}{\partial t} \int_{\substack{\text{air} \\ \text{volume}}} \rho_{\text{air}} \, dV_{\text{air}} + \frac{\partial}{\partial t} \int_{\substack{\text{water} \\ \text{volume}}} \rho_{\text{water}} \, dV_{\text{water}}$$

$$- \dot{m}_{\text{water}} + \dot{m}_{\text{air}} = 0 \quad \text{(1)}$$

Recall that the mass, dm, of fluid contained in a small volume dV is $dm = \rho \, dV$. Hence, the two *integrals* in Eq. 1 represent the total amount of air and water in the control volume, and the sum of the first two *terms* is the time rate of change of mass within the control volume.

Note that the time rate of change of air mass and water mass are each not zero. Recognizing, however, that the air mass must be conserved, we know that the time rate of change of the mass of air in the control volume must be equal to the rate of air mass flow out of the control volume. For simplicity, we disregard any water evaporation that occurs. Thus, applying Eqs. 5.5 and 5.6 to the air only and to the water only, we obtain

$$\frac{\partial}{\partial t} \int_{\substack{\text{air} \\ \text{volume}}} \rho_{\text{air}} \, dV_{\text{air}} + \dot{m}_{\text{air}} = 0$$

for air, and

$$\frac{\partial}{\partial t} \int_{\substack{\text{water} \\ \text{volume}}} \rho_{\text{water}} \, dV_{\text{water}} = \dot{m}_{\text{water}} \quad \text{(2)}$$

for water. The volume of water in the control volume is given by

$$\int_{\substack{\text{water} \\ \text{volume}}} \rho_{\text{water}} \, dV_{\text{water}} = \rho_{\text{water}} \left[h(2 \text{ ft})(5 \text{ ft}) \right.$$

$$\left. + (1.5 \text{ ft} - h)A_j \right] \quad \text{(3)}$$

where A_j is the cross-sectional area of the water flowing from the faucet into the tub. Combining Eqs. 2 and 3, we obtain

$$\rho_{\text{water}} \left(10 \text{ ft}^2 - A_j \right) \frac{\partial h}{\partial t} = \dot{m}_{\text{water}}$$

and, thus, since $\dot{m} = \rho Q$,

$$\frac{\partial h}{\partial t} = \frac{Q_{\text{water}}}{(10 \text{ ft}^2 - A_j)}$$

For $A_j \ll 10 \text{ ft}^2$ we can conclude that

$$\frac{\partial h}{\partial t} = \frac{Q_{\text{water}}}{(10 \text{ ft}^2)}$$

or

$$\frac{\partial h}{\partial t} = \frac{(9 \text{ gal/min})(12 \text{ in./ft})}{(7.48 \text{ gal/ft}^3)(10 \text{ ft}^2)} = 1.44 \text{ in./min} \quad \text{(Ans)}$$

COMMENT   By repeating the calculations for the same flowrate but with various water jet diameters, D_j, the results shown in Fig. E5.3b are obtained. With the flowrate held constant, the value of $\partial h/\partial t$ is nearly independent of the jet diameter for values of the diameter less than about 10 in.

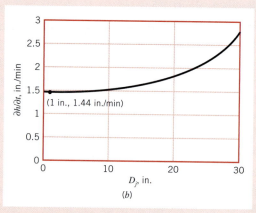

■ **FIGURE E5.3b**

The preceding example problems illustrate that when the flow is steady, the time rate of change of the mass of the contents of the control volume is zero and the net amount of mass flowrate, $\dot{m}$, through the control surface is therefore also zero

$$\sum \dot{m}_{\text{out}} - \sum \dot{m}_{\text{in}} = 0 \tag{5.8}$$

If the steady flow is also incompressible, the net amount of volume flowrate, Q, through the control surface is also zero

$$\sum Q_{\text{out}} - \sum Q_{\text{in}} = 0 \tag{5.9}$$

When the flow is unsteady, the instantaneous time rate of change of the mass of the contents of the control volume is not necessarily zero.

For steady flow involving only one stream of a specific fluid flowing through the control volume at sections (1) and (2),

$$\dot{m} = \rho_1 A_1 V_1 = \rho_2 A_2 V_2 \tag{5.10}$$

V5.2 Shop vac filter

and for incompressible flow,

$$Q = A_1 V_1 = A_2 V_2 \tag{5.11}$$

F l u i d s   i n   t h e   N e w s

New 1.6 GPF standards Toilets account for approximately 40% of all indoor household water use. To conserve water, the new standard is 1.6 gallons of water per flush (gpf). Old toilets use up to 7 gpf; those manufactured after 1980 use 3.5 gpf. Neither is considered a low-flush toilet. A typical 3.2 person household in which each person flushes a 7-gpf toilet 4 times a day uses 32,700 gallons of water each year; with a 3.5-gpf toilet the amount is reduced to 16,400 gallons. Clearly the new 1.6-gpf toilets will save even more water. However, designing a toilet that flushes properly with such a small amount of water is not simple. Today there are two basic types involved: those that are gravity powered and those that are pressure powered. Gravity toilets (typical of most currently in use) have rather long cycle times. The water starts flowing under the action of gravity and the swirling vortex motion initiates the siphon action which builds to a point of discharge. In the newer pressure-assisted models, the *flowrate* is large but the cycle time is short and the amount of water used is relatively small. (See Problem 5.13.) ∎

5.1.3   Moving, Nondeforming Control Volume

When a moving control volume is used, the velocity relative to the moving control volume (relative velocity) is an important flow field variable. The relative velocity, **W,** is the fluid velocity seen by an observer moving with the control volume. The control volume velocity, $\mathbf{V}_{\text{cv}}$, is the velocity of the control volume as seen from a fixed coordinate system. The absolute velocity, **V,** is the fluid velocity seen by a stationary observer in a fixed coordinate system. As indicated by the figure in the margin, these velocities are related to each other by the vector equation

$$\mathbf{V} = \mathbf{W} + \mathbf{V}_{\text{cv}} \tag{5.12}$$

The control volume expression for conservation of the mass (the continuity equation) for a moving, nondeforming control volume is the same as that for a stationary control volume, provided the absolute velocity is replaced by the relative velocity. Thus,

$$\frac{\partial}{\partial t} \int_{\text{cv}} \rho \, d\forall + \sum \rho_{\text{out}} A_{\text{out}} W_{\text{out}} - \sum \rho_{\text{in}} A_{\text{in}} W_{\text{in}} = 0 \tag{5.13}$$

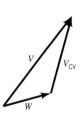

Note that Eq. 5.13 is restricted to control volumes having uniform properties across the inlets and outlets, with the velocity normal to the inlet and outlet areas.

EXAMPLE 5.4 Conservation of Mass—Relative Velocity

GIVEN Water enters a rotating lawn sprinkler through its base at the steady rate of 1000 ml/s as sketched in Fig. E5.4. The exit area of each of the two nozzles is 30 mm^2.

FIND Determine the average speed of the water leaving the nozzle, relative to the nozzle, if

(a) the rotary sprinkler head is stationary,

(b) the sprinkler head rotates at 600 rpm, and

(c) the sprinkler head accelerates from 0 to 600 rpm.

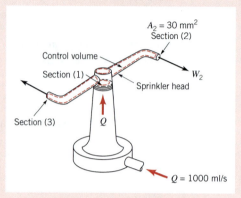

■ **FIGURE E5.4**

SOLUTION

(a) We specify a control volume that contains the water in the rotary sprinkler head at any instant. This control volume is non-deforming, but it moves (rotates) with the sprinkler head. The application of Eq. 5.13 to the contents of this control volume for situation **(a)**, **(b)**, or **(c)** of the problem results in the same expression, namely

$$\underbrace{\frac{\partial}{\partial t}\int_{cv}\rho\,dV}_{\substack{0\text{ flow is steady or the}\\ \text{control volume is filled with}\\ \text{an incompressible fluid}}} + \sum \rho_{\text{out}}A_{\text{out}}W_{\text{out}} - \sum \rho_{\text{in}}A_{\text{in}}W_{\text{in}} = 0 \quad (1)$$

The time rate of change of the mass of water in the control volume is zero because the flow is steady and the control volume is filled with water.

Because there is only one inflow [at the base of the rotating arm, section (1)] and two outflows [the two nozzles at the tips of the arm, sections (2) and (3), each have the same area and fluid velocity], Eq. 1 becomes

$$\rho_2 A_2 W_2 + \rho_3 A_3 W_3 - \rho_1 A_1 W_1 = 0 \quad (2)$$

Hence, for incompressible flow with $\rho_1 = \rho_2 = \rho_3$, Eq. 2 becomes

$$A_2 W_2 + A_3 W_3 - A_1 W_1 = 0$$

With $Q = A_1 W_1$, $A_2 = A_3$, and $W_2 = W_3$ it follows that

$$W_2 = \frac{Q}{2A_2}$$

or

$$W_2 = \frac{(1000 \text{ ml/s})(0.001 \text{ m}^3/\text{liter})(10^6 \text{ mm}^2/\text{m}^2)}{(1000 \text{ ml/liter})(2)(30 \text{ mm}^2)}$$

$$= 16.7 \text{ m/s} \quad \textbf{(Ans)}$$

(b), (c) The value of W_2 is independent of the speed of rotation of the sprinkler head and represents the average velocity of the water exiting from each nozzle with respect to the nozzle for cases **(a)**, **(b)**, and **(c)**.

COMMENT The velocity of water discharging from each nozzle, when viewed from a stationary reference (i.e., V_2), will vary as the rotation speed of the sprinkler head varies since from Eq. 5.12,

$$V_2 = W_2 - U$$

where $U = \omega R$ is the speed of the nozzle and ω and R are the angular velocity and radius of the sprinkler head, respectively.

5.2 Newton's Second Law—The Linear Momentum and Moment-of-Momentum Equations

5.2.1 Derivation of the Linear Momentum Equation

Newton's second law of motion for a system is

$$\begin{array}{c}\text{time rate of change of the}\\ \text{linear momentum of the system}\end{array} = \begin{array}{c}\text{sum of external forces}\\ \text{acting on the system}\end{array}$$

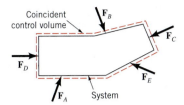

■ **FIGURE 5.2** External forces acting on system and coincident control volume.

Because momentum is mass times velocity, the momentum of a small particle of mass $\rho d V$ is $\mathbf{V}\rho d V$. Hence, the momentum of the entire system, $\int_{sys} \mathbf{V}\rho dV$, is the sum (integral) of the momentum of each small volume element in the system and Newton's law becomes

$$\frac{D}{Dt}\int_{sys}\mathbf{V}\rho\,dV = \sum\mathbf{F}_{sys} \tag{5.14}$$

V5.3 Smokestack plume momentum

When a control volume is coincident with a system at an instant of time, the forces acting on the system and the forces acting on the contents of the coincident control volume (see Fig. 5.2) are instantaneously identical, that is,

$$\sum\mathbf{F}_{sys} = \sum\mathbf{F}_{\substack{\text{contents of the}\\\text{coincident control volume}}} \tag{5.15}$$

Furthermore, for a system and the contents of a coincident control volume that is fixed and nondeforming, the Reynolds transport theorem [Eq. 4.14 with b set equal to the velocity (i.e., momentum per unit mass), and B_{sys} being the system momentum] allows us to conclude that

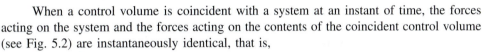

$$\frac{D}{Dt}\int_{sys}\mathbf{V}\rho\,dV = \frac{\partial}{\partial t}\int_{cv}\mathbf{V}\rho\,dV + \sum\mathbf{V}_{out}\rho_{out}A_{out}V_{out} - \sum\mathbf{V}_{in}\rho_{in}A_{in}V_{in} \tag{5.16}$$

Note that the integral $\int_{cv}\mathbf{V}\rho\,dV$ represents the amount of momentum within the control volume. In words, Eq. 5.16 can be written as

time rate of change of the linear momentum of the system	=	time rate of change of the linear momentum of the contents of the control volume	+	net rate of flow of linear momentum through the control surface

Equation 5.16 states that the time rate of change of system linear momentum is expressed as the sum of the two control volume quantities: the time rate of change of the *linear momentum of the contents of the control volume* and the net rate of *linear momentum flow through the control surface*. As particles of mass move into or out of a control volume through the control surface, they carry linear momentum in or out. Thus, linear momentum flow should seem no more unusual than mass flow.

By combining Eqs. 5.14, 5.15, and 5.16 we obtain the following mathematical statement of Newton's second law of motion:

$$\frac{\partial}{\partial t}\int_{cv}\mathbf{V}\rho dV + \sum\mathbf{V}_{out}\rho_{out}A_{out}V_{out} - \sum\mathbf{V}_{in}\rho_{in}A_{in}V_{in} = \sum\mathbf{F}_{\substack{\text{contents of the}\\\text{control volume}}}$$

If the flow is steady, the velocity (and, therefore, the momentum) of the fluid occupying a small volume element dV within the control volume is constant in time. Hence, the total amount of momentum within the *control volume*, $\int_{cv}\mathbf{V}\rho\,dV$, is also constant in time and the time derivative term in the aforementioned equation is zero. The momentum problems considered in this text all involve steady flow. Thus, for steady flow

$$\sum \mathbf{V}_{\text{out}} \rho_{\text{out}} A_{\text{out}} V_{\text{out}} - \sum \mathbf{V}_{\text{in}} \rho_{\text{in}} A_{\text{in}} V_{\text{in}} = \sum \mathbf{F}_{\substack{\text{contents of the} \\ \text{control volume}}} \qquad \textbf{(5.17)}$$

Equation 5.17 is the *linear momentum equation.* The form presented here is restricted to steady flows through fixed, nondeforming control volumes having uniform properties across the inlets and outlets, with the velocity normal to the inlet and outlet areas. A more general form of the linear momentum equation, valid for more general flow conditions, is given in Appendix D.

The forces involved in Eq. 5.17 are body and surface forces that act on what is contained in the control volume. The only body force we consider in this chapter is the one associated with the action of gravity. We experience this body force as weight. The surface forces are basically exerted on the contents of the control volume by material just outside the control volume in contact with material just inside the control volume. For example, a wall in contact with fluid can exert a reaction surface force on the fluid it bounds. Similarly, fluid just outside the control volume can push on fluid just inside the control volume at a common interface, usually an opening in the control surface through which fluid flow occurs.

5.2.2   Application of the Linear Momentum Equation

V5.4 Force due to a water jet

The linear momentum equation for the inertial control volume is a *vector equation* (Eq. 5.17). In engineering applications, components of this vector equation resolved along orthogonal coordinates, for example, x, y, and z (rectangular coordinate system) or r, θ, and x (cylindrical coordinate system), will normally be used. A simple example involving one-dimensional, steady, incompressible flow is considered first.

EXAMPLE 5.5   Linear Momentum—Change in Flow Direction

GIVEN   As shown in Fig. E5.5a, a horizontal jet of water exits a nozzle with a uniform speed of $V_1 = 10$ ft/s, strikes a vane, and is turned through an angle θ.

FIND   Determine the anchoring force needed to hold the vane stationary if gravity and viscous effects are negligible.

SOLUTION

We select a control volume that includes the vane and a portion of the water (see Figs. E5.5b, c) and apply the linear momentum equation to this fixed control volume. The only portions of the control surface across which fluid flows are section (1) (the entrance) and section (2) (the exit). Hence, the x and z components of Eq. 5.17 become

$$u_2 \rho A_2 V_2 - u_1 \rho A_1 V_1 = \sum F_x \qquad (1)$$

and

$$w_2 \rho A_2 V_2 - w_1 \rho A_1 V_1 = \sum F_z \qquad (2)$$

where $\mathbf{V} = u\hat{\mathbf{i}} + w\hat{\mathbf{k}}$, and $\sum F_x$ and $\sum F_z$ are the net x and z components of force acting on the contents of the control volume.

Depending on the particular flow situation being considered and the coordinate system chosen, the x and z components of velocity, u and w, can be positive, negative, or zero. In this example the flow is in the positive directions at both the inlet and the outlet.

The water enters and leaves the control volume as a free jet at atmospheric pressure. Hence, there is atmospheric pressure surrounding the entire control volume, and the net pressure force on the control volume surface is zero. If we neglect the weight of the water and vane, the only forces applied to the control volume contents are the horizontal and vertical components of the anchoring force, F_{Ax} and F_{Az}, respectively.

With negligible gravity and viscous effects, and since $p_1 = p_2$, the speed of the fluid remains constant so that $V_1 = V_2 = 10$ ft/s

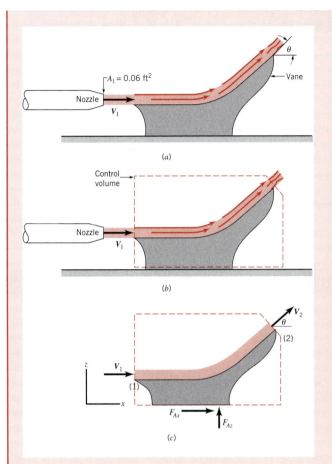

(a)

(b)

(c)

■ **FIGURE E5.5**

(see the Bernoulli equation, Eq. 3.6). Hence, at section 1, $u_1 = V_1, w_1 = 0$, and at section 2, $u_2 = V_1 \cos\theta, w_2 = V_1 \sin\theta$.

By using this information, Eqs. 1 and 2 can be written as

$$V_1 \cos\theta \, \rho A_2 V_1 - V_1 \rho A_1 V_1 = F_{Ax} \qquad (3)$$

and

$$V_1 \sin\theta \, \rho A_2 V_1 - 0 \, \rho A_1 V_1 = F_{Az} \qquad (4)$$

Equations 3 and 4 can be simplified by using conservation of mass, which states that for this incompressible flow $A_1 V_1 = A_2 V_2$, or $A_1 = A_2$ since $V_1 = V_2$. Thus

$$F_{Ax} = -\rho A_1 V_1^2 + \rho A_1 V_1^2 \cos\theta = -\rho A_1 V_1^2 (1 - \cos\theta) \quad (5)$$

and

$$F_{Az} = \rho A_1 V_1^2 \sin\theta \qquad (6)$$

With the given data we obtain

$$F_{Ax} = -(1.94 \text{ slugs/ft}^3)(0.06 \text{ ft}^2)(10 \text{ ft/s})^2(1 - \cos\theta)$$
$$= -11.64(1 - \cos\theta) \text{ slugs·ft/s}^2$$
$$= -11.64(1 - \cos\theta) \text{ lb} \qquad \textbf{(Ans)}$$

and

$$F_{Az} = (1.94 \text{ slugs/ft}^3)(0.06 \text{ ft}^2)(10 \text{ ft/s})^2 \sin\theta$$
$$= 11.64 \sin\theta \text{ lb} \qquad \textbf{(Ans)}$$

COMMENTS The values of F_{Ax} and F_{Az} as a function of θ are shown in Fig. E5.5d. Note that if $\theta = 0$ (i.e., the vane does not turn the water), the anchoring force is zero. The inviscid fluid merely slides along the vane without putting any force on it. If $\theta = 90°$, then $F_{Ax} = -11.64$ lb and $F_{Az} = -11.64$ lb. It is necessary to push on the vane (and, hence, for the vane to push on the water) to the left (F_{Ax} is negative) and up in order to change the direction of flow of the water from horizontal to vertical. A momentum change requires a force. If $\theta = 180°$, the water jet is turned back on itself. This requires no vertical force ($F_{Az} = 0$), but the horizontal force ($F_{Ax} = -23.3$ lb) is two times that required if $\theta = 90°$. This force must eliminate the incoming fluid momentum and create the outgoing momentum.

Note that the anchoring force (Eqs. 5, 6) can be written in terms of the mass flowrate, $\dot{m} = \rho A_1 V_1$, as

$$F_{Ax} = -\dot{m} V_1 (1 - \cos\theta)$$

and

$$F_{Az} = \dot{m} V_1 \sin\theta$$

In this example the anchoring force is needed to produce the nonzero net momentum flowrate (mass flowrate times the change in x or z component of velocity) across the control surface.

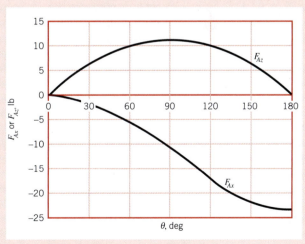

■ **FIGURE E5.5d**

EXAMPLE 5.6   Linear Momentum—Weight, Pressure, and Change in Speed

GIVEN   As shown in Fig. E5.6a, water flows through a nozzle attached to the end of a laboratory sink faucet with a flowrate of 0.6 liters/s. The nozzle inlet and exit diameters are 16 and 5 mm, respectively, and the nozzle axis is vertical. The mass of the nozzle is 0.1 kg, and the mass of the water in the nozzle is 3×10^{-3} kg. The pressure at section (1) is 464 kPa.

FIND   Determine the anchoring force required to hold the nozzle in place.

SOLUTION

The anchoring force sought is the reaction between the faucet and nozzle threads. This force can be obtained by application of the linear momentum equation, Eq. 5.17, to an appropriate control volume.

$$\sum \mathbf{V}_{out} \rho_{out} A_{out} V_{out} - \sum \mathbf{V}_{in} \rho_{in} A_{in} V_{in}$$

$$= \sum \mathbf{F}_{\substack{\text{contents of the} \\ \text{control volume}}} \quad \textbf{(1)}$$

We select a control volume that includes the entire nozzle and the water contained in the nozzle at an instant, as is indicated in Figs. E5.6a and E5.6b. All of the vertical forces acting on the contents of this control volume are identified in Fig. E5.6b. The action of atmospheric pressure cancels out in every direction and is not shown. Gage pressure forces do not cancel out in the vertical direction and are shown. Application of the vertical or z direction component of Eq. 1 to the contents of this control volume leads to

$$w_2 \rho A_2 V_2 - w_1 \rho A_1 V_1$$
$$= F_A - \mathcal{W}_n - p_1 A_1 - \mathcal{W}_w + p_2 A_2 \quad \textbf{(2)}$$

where w is the z direction component of fluid velocity, and the various parameters are identified in the figure.

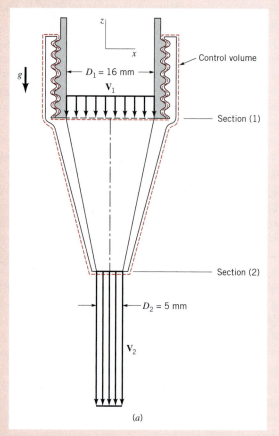

■ **FIGURE E5.6a**

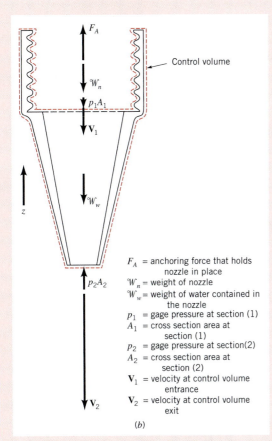

= anchoring force that holds nozzle in place
F_A = anchoring force that holds nozzle in place
$\mathcal{W}_n$ = weight of nozzle
$\mathcal{W}_w$ = weight of water contained in the nozzle
p_1 = gage pressure at section (1)
A_1 = cross section area at section (1)
p_2 = gage pressure at section(2)
A_2 = cross section area at section (2)
$\mathbf{V}_1$ = velocity at control volume entrance
$\mathbf{V}_2$ = velocity at control volume exit

(b)

■ **FIGURE E5.6b**

Note that the positive direction is considered "up" for the forces. We will use this same sign convention for the fluid velocity, w, in Eq. 2. Hence, because the flow is "down" at section (1), it follows that $w_1 = -V_1$. Similarly, at section (2), $w_2 = -V_2$ so that Eq. 2 can be written as

$$(-V_2)\dot{m}_2 - (-V_1)\dot{m}_1$$
$$= F_A - \mathcal{W}_n - p_1A_1 - \mathcal{W}_w + p_2A_2 \qquad (3)$$

where $\dot{m} = \rho AV$ is the mass flowrate.

Solving Eq. 3 for the anchoring force, F_A, we obtain

$$F_A = \dot{m}_1V_1 - \dot{m}_2V_2 + \mathcal{W}_n + p_1A_1 + \mathcal{W}_w - p_2A_2 \qquad (4)$$

From the conservation of mass equation, Eq. 5.10, we obtain

$$\dot{m}_1 = \dot{m}_2 = \dot{m} \qquad (5)$$

which when combined with Eq. 4 gives

$$F_A = \dot{m}(V_1 - V_2) + \mathcal{W}_n + p_1A_1 + \mathcal{W}_w - p_2A_2 \qquad (6)$$

It is instructive to note how the anchoring force is affected by the different actions involved. As expected, the nozzle weight, $\mathcal{W}_n$, the water weight, $\mathcal{W}_w$, and gage pressure force at

section (1), p_1A_1, all increase the anchoring force, while the gage pressure force at section (2), p_2A_2, acts to decrease the anchoring force. The change in the vertical momentum flowrate, $\dot{m}(V_1 - V_2)$, will, in this instance, decrease the anchoring force because this change is negative ($V_2 > V_1$).

To complete this example we use quantities given in the problem statement to quantify the terms on the right-hand side of Eq. 6. From Eq. 5.6,

$$\dot{m} = \rho V_1 A_1 = \rho Q$$
$$= (999 \text{ kg/m}^3)(0.6 \text{ liter/s})(10^{-3} \text{ m}^3/\text{liter})$$
$$= 0.599 \text{ kg/s} \qquad (7)$$

and

$$V_1 = \frac{Q}{A_1} = \frac{Q}{\pi(D_1^2/4)}$$
$$= \frac{(0.6 \text{ liter/s})(10^{-3} \text{ m}^3/\text{liter})}{\pi(16 \text{ mm})^2/4(1000^2 \text{ mm}^2/\text{m}^2)} = 2.98 \text{ m/s} \qquad (8)$$

Also from Eq. 5.6,

$$V_2 = \frac{Q}{A_2} = \frac{Q}{\pi(D_2^2/4)}$$
$$= \frac{(0.6 \text{ liter/s})(10^{-3} \text{ m}^3/\text{liter})}{\pi(5 \text{ mm})^2/4(1000^2 \text{ mm}^2/\text{m}^2)} = 30.6 \text{ m/s} \qquad (9)$$

The weight of the nozzle, $\mathcal{W}_n$, can be obtained from the nozzle mass, m_n, with

$$\mathcal{W}_n = m_ng = (0.1 \text{ kg})(9.81 \text{ m/s}^2) = 0.981 \text{ N} \qquad (10)$$

Similarly, the weight of the water in the control volume, $\mathcal{W}_w$, can be obtained from the mass of the water, m_w, as

$$\mathcal{W}_w = m_wg = 3 \times 10^{-3}\text{kg} (9.81 \text{ m/s}^2)$$
$$= 2.94 \times 10^{-2}\text{N} \qquad (11)$$

The gage pressure at section (2), p_2, is zero since, as discussed in Section 3.6.1, when a subsonic flow discharges to the atmosphere as in the present situation, the discharge pressure is essentially atmospheric. The anchoring force, F_A, can now be determined from Eqs. 6 through 11 with

$$F_A = (0.599 \text{ kg/s})(2.98 \text{ m/s} - 30.6 \text{ m/s}) + 0.981 \text{ N}$$
$$+ (464 \text{ kPa})(1000 \text{ Pa/kPa}) \frac{\pi(16 \text{ mm})^2}{4(1000^2 \text{ mm}^2/\text{m}^2)}$$
$$+ 0.0294 \text{ N} - 0$$

or

$$F_A = -16.5 \text{ N} + 0.981 \text{ N} + 93.3 \text{ N}$$
$$+ 0.0294 \text{ N} = 77.8 \text{ N} \qquad \textbf{(Ans)}$$

Because the anchoring force, F_A, is positive, it acts upward in the z direction. The nozzle would be pushed off the pipe if it were not fastened securely.

COMMENT The control volume selected earlier to solve problems such as this is not unique. The following is an alternate solution that involves two other control volumes—one containing only the nozzle and the other containing only the water in the nozzle. These control volumes are shown in Figs. E5.6c and E5.6d along with the vertical forces acting on the contents of each control volume. The new force involved, R_z, represents the interaction between the water and the conical inside surface of the nozzle. It includes the net pressure and viscous forces at this interface.

Application of Eq. 5.17 to the contents of the control volume of Fig. E5.6c leads to

$$F_A = \mathcal{W}_n + R_z - p_{atm}(A_1 - A_2) \qquad (12)$$

The term $p_{atm}(A_1 - A_2)$ is the resultant force from the atmospheric pressure acting upon the exterior surface of the nozzle (i.e., that portion of the surface of the nozzle that is not in contact with the water). Recall that the pressure force on a curved surface (such as the exterior surface of the nozzle) is equal to the pressure times the projection of the surface area on a plane perpendicular to the axis of the nozzle. The projection of this area on a plane perpendicular to the z direction is $A_1 - A_2$. The effect of the atmospheric pressure on the internal area (between the nozzle and the water) is already included in R_z, which represents the net force on this area.

Similarly, for the control volume of Fig. E5.6d we obtain

$$R_z = \dot{m}(V_1 - V_2) + \mathcal{W}_w$$
$$+ (p_1 + p_{atm})A_1 - (p_2 + p_{atm})A_2 \qquad (13)$$

where p_1 and p_2 are gage pressures. From Eq. 13 it is clear that the value of R_z depends on the value of the atmospheric pressure, p_{atm}, since $A_1 \neq A_2$. That is, we must use absolute pressure, not gage pressure, to obtain the correct value of R_z.

By combining Eqs. 12 and 13 we obtain the same result as before (Eq. 6) for F_A:

$$F_A = \dot{m}(V_1 - V_2) + \mathcal{W}_n + p_1 A_1 + \mathcal{W}_w - p_2 A_2$$

Note that although the force between the fluid and the nozzle wall, R_z, is a function of p_{atm}, the anchoring force, F_A, is not. That is, we were correct in using gage pressure when solving for F_A by means of the original control volume shown in Fig. E5.6b.

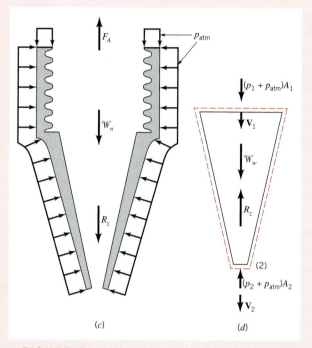

(c)   (d)

■ **FIGURE E5.6** c & d

Several important generalities about the application of the linear momentum equation (Eq. 5.17) are apparent in the example just considered.

1. Linear momentum is directional; it can have components in as many as three orthogonal coordinate directions. Furthermore, along any one coordinate, the linear momentum of a fluid particle can be in the positive or negative direction and thus be considered as a positive or a negative quantity. In Example 5.6, only the linear momentum in the z direction was considered (all of it was in the negative z direction).

2. The time rate of change of the linear momentum of the contents of a nondeforming *control volume* is zero for steady flow. The time rate of change of the linear momentum of a *system*, however, is generally not zero, even for steady flow. This is due to the fact that the momentum of fluid particles making up the system may change as they

flow through the control volume. That is, the particles may speed up, slow down, or change their direction of motion, even though the flow is steady (i.e., nothing changes with time at any given location). The momentum problems considered in this text all involve steady flow.

3. If the control surface is selected so that it is perpendicular to the flow where fluid enters or leaves the control volume, the surface force exerted at these locations by fluid outside the control volume on fluid inside will be due to pressure. Furthermore, when subsonic flow exits from a control volume into the atmosphere, atmospheric pressure prevails at the exit cross section. In Example 5.6, the flow was subsonic and so we set the exit flow pressure at the atmospheric level. The continuity equation (Eq. 5.10) allowed us to evaluate the fluid flow velocities V_1 and V_2 at sections (1) and (2).

4. Forces due to atmospheric pressure acting on the control surface may need consideration as indicated by Eq. 13 for the reaction force between the nozzle and the fluid. When calculating the anchoring force, F_A, forces due to atmospheric pressure on the control surface cancel each other (for example, after combining Eqs. 12 and 13 the atmospheric pressure forces are no longer involved) and gage pressures may be used.

5. The external forces have an algebraic sign, positive if the force is in the assigned positive coordinate direction and negative otherwise.

6. Only external forces acting on the contents of the control volume are considered in the linear momentum equation (Eq. 5.17). If the fluid alone is included in a control volume, reaction forces between the fluid and the surface or surfaces in contact with the fluid [wetted surface(s)] will need to be in Eq. 5.17. If the fluid and the wetted surface or surfaces are within the control volume, the reaction forces between fluid and wetted surface(s) do not appear in the linear momentum equation (Eq. 5.17) because they are internal, not external forces. The anchoring force that holds the wetted surface(s) in place is an external force, however, and must therefore be in Eq. 5.17.

7. The force required to anchor an object will generally exist in response to surface pressure and/or shear forces acting on the control surface, to a change in linear momentum flow through the control volume containing the object, and to the weight of the object and the fluid contained in the control volume. In Example 5.6 the nozzle anchoring force was required mainly because of pressure forces and partly because of a change in linear momentum flow associated with accelerating the fluid in the nozzle. The weight of the water and the nozzle contained in the control volume influenced the size of the anchoring force only slightly.

F l u i d s   i n   t h e   N e w s

Motorized surfboard When Bob Montgomery, a former professional surfer, started to design his motorized surfboard (called a jet board), he discovered that there were many engineering challenges to the design. The idea is to provide surfing to anyone, no matter where they live, near or far from the ocean. The rider stands on the device like a surfboard and steers it like a surfboard by shifting his/her body weight. A new, sleek, compact 45-horsepower engine and pump was designed to fit within the surfboard hull. Thrust is produced in response to the change in *linear momentum* of the water stream as it enters through the inlet passage and exits through an appropriately designed nozzle. Some of the fluid dynamic problems associated with designing the craft included one-way valves so that water does not get into the engine (at both the intake or exhaust ports), buoyancy, hydrodynamic lift, drag, thrust, and hull stability. (See Problem 5.23.) ■

EXAMPLE 5.7   Linear Momentum—Pressure and Change in Flow Direction

GIVEN Water flows through a horizontal, 180° pipe bend as illustrated in Fig. E5.7a. The flow cross-sectional area is constant at a value of 0.1 ft² through the bend. The flow velocity everywhere in the bend is axial and 50 ft/s. The absolute pressures at the entrance and exit of the bend are 30 and 24 psia, respectively.

FIND Calculate the horizontal (x and y) components of the anchoring force required to hold the bend in place.

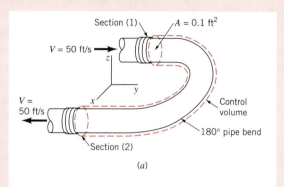

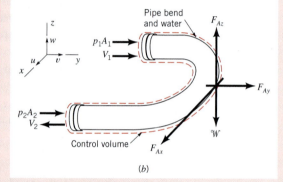

■ **FIGURE E5.7** *a & b*

SOLUTION

Since we want to evaluate components of the anchoring force to hold the pipe bend in place, an appropriate control volume (see dashed line in Fig. E5.7a) contains the bend and the water in the bend at an instant. The horizontal forces acting on the contents of this control volume are identified in Fig. E5.7b. Note that the weight of the water is vertical (in the negative z direction) and does not contribute to the x and y components of the anchoring force. All of the horizontal normal and tangential forces exerted on the fluid and the pipe bend are resolved and combined into the two resultant components, F_{Ax} and F_{Ay}. These two forces act on the control volume contents, and thus for the x direction, Eq. 5.17 leads to

$$u_2 \rho A_2 V_2 - u_1 \rho A_1 V_1 = F_{Ax} \tag{1}$$

where

$$V = u\hat{\mathbf{i}} + v\hat{\mathbf{j}} + w\hat{\mathbf{k}}$$

At sections (1) and (2), the flow is in the y direction and therefore $u = 0$ at both cross sections. There is no x direction momentum flow into or out of the control volume, and we conclude from Eq. 1 that

$$F_{Ax} = 0. \tag{Ans}$$

For the y direction, we get from Eq. 5.17

$$v_2 \rho A_2 V_2 - v_1 \rho A_1 V_1 = F_{Ay} + p_1 A_1 + p_2 A_2 \tag{2}$$

The y components of velocity, v_1 and v_2, at the inlet and outlet sections of the control volume are $v_1 = V_1 = V$ and $v_2 = -V_2 = -V$, respectively, where $V = V_1 = V_2 = 50$ ft/s. Hence, Eq. 2 becomes

$$-V_2 \rho A_2 V_2 - V_1 \rho A_1 V_1 = F_{Ay} + p_1 A_1 + p_2 A_2 \tag{3}$$

Note that the y component of velocity is positive at section (1) but is negative at section (2). From the continuity equation (Eq. 5.10), we get

$$\dot{m} = \rho A_2 V_2 = \rho A_1 V_1 \tag{4}$$

and thus Eq. 3 can be written as

$$-\dot{m}(V_1 + V_2) = F_{Ay} + p_1 A_1 + p_2 A_2 \tag{5}$$

Solving Eq. 5 for F_{Ay} we obtain

$$F_{Ay} = -\dot{m}(V_1 + V_2) - p_1 A_1 - p_2 A_2 \tag{6}$$

From the given data we can calculate $\dot{m}$, the mass flowrate, from Eq. 4 as

$$\dot{m} = \rho_1 A_1 V_1 = (1.94 \text{ slugs/ft}^3)(0.1 \text{ ft}^2)(50 \text{ ft/s})$$
$$= 9.70 \text{ slugs/s}$$

For determining the anchoring force, F_{Ay}, the effects of atmospheric pressure cancel and thus gage pressures for p_1 and p_2 are appropriate. By substituting numerical values of variables into Eq. 6 and using the fact that 1 lb = 1 slug · ft/s², we get

$$F_{Ay} = -(9.70 \text{ slugs/s})(50 \text{ ft/s} + 50 \text{ ft/s})$$
$$-(30 \text{ psia} - 14.7 \text{ psia})(144 \text{ in.}^2/\text{ft}^2)(0.1 \text{ ft}^2)$$
$$-(24 \text{ psia} - 14.7 \text{ psia})(144 \text{ in.}^2/\text{ft}^2)(0.1 \text{ ft}^2)$$
$$F_{Ay} = -970 \text{ lb} - 220 \text{ lb} - 134 \text{ lb} = -1324 \text{ lb} \tag{Ans}$$

The negative sign for F_{Ay} is interpreted as meaning that the y component of the anchoring force is actually in the negative y direction, not the positive y direction as originally indicated in Fig. E5.7b.

COMMENT As with Example 5.6, the anchoring force for the pipe bend is independent of the atmospheric pressure. However, the force that the bend puts on the fluid inside of it, R_y, depends on the atmospheric pressure. We can see this by using a control volume that surrounds only the fluid within the bend as shown in Fig. E5.7c. Application of the momentum equation to this situation gives

$$R_y = -\dot{m}(V_1 + V_2) - p_1 A_1 - p_2 A_2$$

where p_1 and p_2 must be in terms of absolute pressure because the force between the fluid and the pipe wall, R_y, is the complete pressure effect (i.e., absolute pressure).

Thus, we obtain

$$
\begin{aligned}
R_y = &-(9.70 \text{ slugs/s})(50 \text{ ft/s} + 50 \text{ ft/s}) \\
&-(30 \text{ psia})(144 \text{ in.}^2/\text{ft}^2)(0.1 \text{ ft}^2) \\
&-(24 \text{ psia})(144 \text{ in.}^2/\text{ft}^2)(0.1 \text{ ft}^2) \\
= &-1748 \text{ lb}
\end{aligned}
\qquad (7)
$$

We can use the control volume that includes just the pipe bend (without the fluid inside it) as shown in Fig. E5.7d to determine F_{Ay}, the anchoring force component in the y direction necessary to hold the bend stationary. The y component of the momentum equation applied to this control volume gives

$$F_{Ay} = R_y + p_{\text{atm}}(A_1 + A_2) \qquad (8)$$

where R_y is given by Eq. 7. The $p_{\text{atm}}(A_1 + A_2)$ term represents the net pressure force on the outside portion of the control volume. Recall that the pressure force on the inside of the bend is accounted for by R_y. By combining Eqs. 7 and 8 and using the fact that $p_{\text{atm}} = 14.7 \text{ lb/in.}^2 (144 \text{ in.}^2/\text{ft}^2) = 2117 \text{ lb/ft}^2$, we obtain.

$$
\begin{aligned}
F_{Ay} = &-1748 \text{ lb} + 2117 \text{ lb/ft}^2(0.1 \text{ ft}^2 + 0.1 \text{ ft}^2) \\
= &-1324 \text{ lb}
\end{aligned}
$$

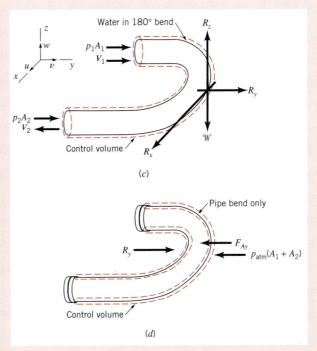

■ **FIGURE E5.7** *c & d*

in agreement with the original answer obtained using the control volume of Fig. E5.7b.

F l u i d s   i n   t h e   N e w s

Bow thrusters In the past, large ships required the use of tugboats for precise maneuvering, especially when docking. Nowadays, most large ships (and many moderate to small ones as well) are equipped with bow thrusters to help steer in close quarters. The units consist of a mechanism (usually a ducted propeller mounted at right angles to the fore/aft axis of the ship) that takes water from one side of the bow and ejects it as a water jet on the other side. The *momentum flux* of this jet produces a starboard or port force on the ship for maneu-

vering. Sometimes a second unit is installed in the stern. Initially used in the bows of ferries, these versatile control devices have became popular in offshore oil servicing boats, fishing vessels, and larger ocean-going craft. They permit unassisted maneuvering alongside of oilrigs, vessels, loading platforms, fishing nets, and docks. They also provide precise control at slow speeds through locks, narrow channels, and bridges, where the rudder becomes very ineffective. (See Problem 5.30.) ■

From Examples 5.5, 5.6, and 5.7, we see that changes in flow speed and/or direction result in a reaction force. Other types of problems that can be solved with the linear momentum equation (Eq. 5.17) are illustrated in the following examples.

EXAMPLE 5.8	Linear Momentum—Weight, Pressure, Friction, and Nonuniform Velocity Profile

GIVEN Assume that the flow of Example 5.2 is vertically upward as shown by Fig. E5.8.

FIND Develop an expression for the fluid pressure drop that occurs between section (1) and section (2).

SOLUTION

A control volume (see dashed lines in Fig. E5.2) that includes only fluid from section (1) to section (2) is selected. Forces acting on the fluid in this control volume are identified in Fig. E5.8. The application of the axial component of Eq. 5.17 to the fluid in this control volume results in

$$\sum w_{\text{out}} \rho A_{\text{out}} V_{\text{out}} - \sum w_{\text{in}} \rho A_{\text{in}} V_{\text{in}}$$
$$= p_1 A_1 - R_z - \mathcal{W} - p_2 A_2 \qquad (1)$$

where R_z is the resultant force of the wetted pipe wall on the fluid, $\mathcal{W}$ is the weight of the fluid in the pipe between sections (1) and (2), p_1 and p_2 are the pressures at the inlet and outlet, respectively, and w is the vertical (axial) component of velocity. Because the velocity is uniform across section (1), it follows that the momentum flux across the inlet is simply

$$\sum w_{\text{in}} \rho A_{\text{in}} V_{\text{in}} = w_1 \rho A_1 w_1 = w_1 \dot{m}_1 \qquad (2)$$

At the outlet, section (2), the velocity is not uniform. However, the net momentum flux across this section, $\sum w_{\text{out}} \rho A_{\text{out}} V_{\text{out}}$, is the sum of the momentum fluxes through numerous small washer-shaped areas of size $dA_2 = 2\pi r\, dr$. On each of these infinitesimal areas the fluid velocity is denoted as w_2. Thus, in the limit of infinitesimal area elements, the summation is replaced by an integration, and by using the parabolic velocity profile from Example 5.2, $w_2 = 2w_1[1 - (r/R)^2]$, the momentum outflow through section (2) is given by

$$\sum w_{\text{out}} \rho A_{\text{out}} V_{\text{out}} = \int_{A_2} w_2 \rho w_2\, dA_2 = \rho \int_0^R w_2^2\, 2\pi r\, dr$$
$$= 2\pi\rho \int_0^R (2w_1)^2 \left[1 - \left(\frac{r}{R}\right)^2\right]^2 r\, dr$$

or

$$\sum w_{\text{out}} \rho A_{\text{out}} V_{\text{out}} = 4\pi\rho w_1^2 \frac{R^2}{3} \qquad (3)$$

Combining Eqs. 1, 2, and 3 we obtain

$$\tfrac{4}{3} w_1^2 \rho\pi R^2 - w_1^2 \rho\pi R^2 = p_1 A_1 - R_z - \mathcal{W} - p_2 A_2 \qquad (4)$$

Solving Eq. 4 for the pressure drop from section (1) to section (2), $p_1 - p_2$, we obtain

$$p_1 - p_2 = \frac{\rho w_1^2}{3} + \frac{R_z}{A_1} + \frac{\mathcal{W}}{A_1} \qquad \textbf{(Ans)}$$

COMMENT We see that the drop in pressure from section (1) to section (2) occurs because of the following:

1. The change in momentum flow between the two sections associated with going from a uniform velocity profile to

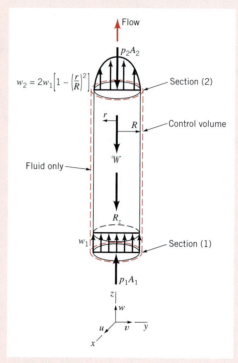

■ **FIGURE E5.8**

a parabolic velocity profile. As shown in Eq. 1, the momentum flux at section (1) is $w_1\dot{m}_1$. From Eq. 3, the momentum flux at section (2) can be written as $4\pi\rho w_1^2 R^2/3 = (4/3)w_1\dot{m}_1$. Thus, even though the mass flowrates are equal at section (1) and section (2), the momentum flowrates at these two sections are not equal. The momentum flux associated with a nonuniform velocity profile is always greater than that for a uniform velocity profile carrying the same mass flowrate.

2. Pipe wall friction R_z.

3. The weight of the water column; a hydrostatic pressure effect.

If the velocity profiles had been identically parabolic at sections (1) and (2), the momentum flowrate at each section would have been identical, a condition we call "fully developed" flow. Then, the pressure drop, $p_1 - p_2$, would be due only to pipe wall friction and the weight of the water column. If in addition to being fully developed the flow involved negligible weight effects (for example, horizontal flow of liquids or the flow of gases in any direction), the drop in pressure between any two sections, $p_1 - p_2$, would be a result of pipe wall friction only.

EXAMPLE 5.9   Linear Momentum—Nonuniform Pressure

GIVEN A sluice gate across a channel of width b is shown in the closed and open positions in Figs. E5.9a and E5.9b.

FIND Is the anchoring force required to hold the gate in place larger when the gate is closed or when it is open?

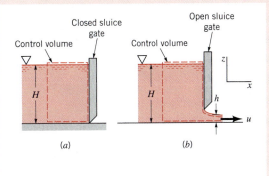

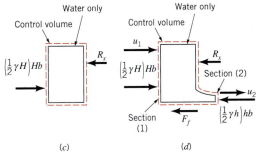

■ **FIGURE E5.9**

SOLUTION

We will answer this question by comparing expressions for the horizontal reaction force, R_x, between the gate and the water when the gate is closed and when the gate is open. The control volume used in each case is indicated with dashed lines in Figs. E5.9a and E5.9b.

When the gate is closed, the horizontal forces acting on the contents of the control volume are identified in Fig. E5.9c. Application of Eq. 5.17 to the contents of this control volume yields

$$\sum u_{out}\rho A_{out}V_{out} - \sum u_{in}\rho A_{in}V_{in} = \tfrac{1}{2}\gamma H^2 b - R_x \quad (1)$$

(with 0 and 0 (no flow) over the two summation terms)

Note that the hydrostatic pressure force, $\gamma H^2 b/2$, is used. From Eq. 1, the force exerted on the water by the gate (which is equal to the force necessary to hold the gate stationary) is

$$R_x = \tfrac{1}{2}\gamma H^2 b \quad (2)$$

which is equal in magnitude to the hydrostatic force exerted on the gate by the water.

When the gate is open, the horizontal forces acting on the contents of the control volume are shown in Fig. E5.9d. Application of Eq. 5.17 to the contents of this control volume leads to

$$\sum u_{out}\rho A_{out}V_{out} - \sum u_{in}\rho A_{in}V_{in}$$
$$= \tfrac{1}{2}\gamma H^2 b - R_x - \tfrac{1}{2}\gamma h^2 b - F_f \quad (3)$$

Note that we have assumed that the pressure distribution is hydrostatic in the water at sections (1) and (2). (See Section 3.4.) Also, the frictional force between the channel bottom and the water is specified as F_f. With the assumption of uniform velocity distributions

$$\sum u_{out}\rho A_{out}V_{out} - \sum u_{in}\rho A_{in}V_{in} = u_2\rho h b u_2 - u_1\rho H b u_1 \quad (4)$$

Equations 3 and 4 can be combined to form

$$\rho u_2^2 h b - \rho u_1^2 H b = \tfrac{1}{2}\gamma H^2 b - R_x - \tfrac{1}{2}\gamma h^2 b - F_f \quad (5)$$

Solving Eq. 5 for the reaction force, R_x, we obtain

$$R_x = \tfrac{1}{2}\gamma H^2 b - \tfrac{1}{2}\gamma h^2 b - F_f - \rho u_2^2 h b + \rho u_1^2 H b \quad (6)$$

By using the continuity equation, $\dot{m} = \rho b H u_1 = \rho b h u_2$, Eq. (6) can be rewritten as

$$R_x = \tfrac{1}{2}\gamma H^2 b - \tfrac{1}{2}\gamma h^2 b - F_f - \dot{m}(u_2 - u_1) \quad (7)$$

Hence, since $u_2 > u_1$, by comparing the expressions for R_x (Eqs. 2 and 7) we conclude that the reaction force between the gate and the water (and therefore the anchoring force required to hold the gate in place) is smaller when the gate is open than when it is closed. **(Ans)**

It should be clear from the preceding examples that fluid flows can lead to a reaction force in the following ways:

1. Linear momentum flow variation in direction and/or magnitude.
2. Fluid pressure forces.
3. Fluid friction forces.
4. Fluid weight.

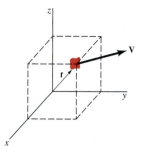

■ **FIGURE 5.3** Inertial coordinate system.

The selection of a control volume is an important matter. An appropriate control volume can make a problem solution straightforward.

5.2.3   Derivation of the Moment-of-Momentum Equation

In many engineering problems, the moment of a force with respect to an axis, namely *torque,* is important. Newton's second law of motion has already led to a useful relationship between forces and linear momentum flow (see Eq. 5.17). The linear momentum equation can also be used to solve problems involving torques. However, by forming the moment of the linear momentum and the resultant force associated with each particle of fluid with respect to a point in an inertial coordinate system, we obtain a *moment-of-momentum equation* that relates *torques* and *angular momentum flow* for the contents of a control volume.

We define **r** as the position vector from the origin of the inertial coordinate system to the fluid particle (Fig. 5.3) and form the moment of each side of Eq. 5.14 with respect to the origin of an inertial coordinate system. The result is

$$\frac{D}{Dt}\int_{\text{sys}} (\mathbf{r} \times \mathbf{V})\rho \, d\mathcal{V} = \sum (\mathbf{r} \times \mathbf{F})_{\text{sys}} \tag{5.18}$$

or

$$\begin{array}{l}\text{the time rate of change of the}\\ \text{moment-of-momentum of the system}\end{array} = \begin{array}{l}\text{sum of the external torques}\\ \text{acting on the system}\end{array}$$

For a control volume that is instantaneously coincident with the system, the torques acting on the system and on the control volume contents will be identical:

$$\sum (\mathbf{r} \times \mathbf{F})_{\text{sys}} = \sum (\mathbf{r} \times \mathbf{F})_{\text{cv}} \tag{5.19}$$

Recall from Section 5.2.1 that $\int_{\text{sys}} \mathbf{V}\rho \, d\mathcal{V}$ and $\int_{\text{cv}} \mathbf{V}\rho \, d\mathcal{V}$ represent the momentum of the system and the momentum of the contents of the control volume, respectively. Similarly, the terms $\int_{\text{sys}}(\mathbf{r} \times \mathbf{V})\rho \, d\mathcal{V}$ and $\int_{\text{cv}}(\mathbf{r} \times \mathbf{V})\rho \, d\mathcal{V}$ represent the moment-of-momentum of the system and the moment-of-momentum of the contents of the control volume, respectively. Hence, we can use the Reynolds transport theorem (Eq. 4.14) with B set equal to the moment-of-momentum and b equal to $(\mathbf{r} \times \mathbf{V})$, the moment-of-momentum per unit mass, to obtain

$$\frac{D}{Dt}\int_{\text{sys}} (\mathbf{r} \times \mathbf{V})\rho \, d\mathcal{V} = \frac{\partial}{\partial t}\int_{\text{cv}} (\mathbf{r} \times \mathbf{V})\rho \, d\mathcal{V} + \sum (\mathbf{r} \times \mathbf{V})_{\text{out}}\,\rho_{\text{out}}A_{\text{out}}V_{\text{out}}$$
$$- \sum (\mathbf{r} \times \mathbf{V})_{\text{in}}\,\rho_{\text{in}}A_{\text{in}}V_{\text{in}} \tag{5.20}$$

In words, Eq. 5.20 can be written as

time rate of change of the moment-of-momentum of the system	=	time rate of change of the moment-of-momentum of the contents of the control volume	+	net rate of flow of the moment-of-momentum through the control surface

By combining Eqs. 5.18, 5.19, and 5.20 we obtain the following mathematical statement of the moment-of-momentum equation for a fixed control volume.

$$\frac{\partial}{\partial t} \int_{cv} (\mathbf{r} \times \mathbf{V})\rho \, dV + \sum (\mathbf{r} \times \mathbf{V})_{out} \rho_{out} A_{out} V_{out}$$
$$- \sum (\mathbf{r} \times \mathbf{V})_{in} \rho_{in} A_{in} V_{in} = \sum (\mathbf{r} \times \mathbf{F})_{\substack{\text{contents of the} \\ \text{control volume}}}$$

If the flow is steady, the time derivative term in the aforementioned equation is zero. Moment-of-momentum problems considered in this text all involve steady flow. Thus, for steady flow

$$\sum (\mathbf{r} \times \mathbf{V})_{out} \rho_{out} A_{out} V_{out} - \sum (\mathbf{r} \times \mathbf{V})_{in} \rho_{in} A_{in} V_{in} = \sum (\mathbf{r} \times \mathbf{F})_{\substack{\text{contents of the} \\ \text{control volume}}} \quad \textbf{(5.21)}$$

Equation 5.21 is the ***moment-of-momentum equation.*** The form presented here is restricted to steady flows through fixed, nondeforming control volumes having uniform properties across the inlets and outlets, with the velocity normal to the inlet and outlet areas. A more general form of the moment-of-momentum equation, valid for more general flow conditions, is given in Appendix D.

Equation 5.21 is a vector equation, which can be written in terms of its components in the radial, r, tangential, θ, and axial, z, directions. For applications involved in this text, we will need to consider only the axial component of this equation.

5.2.4   Application of the Moment-of-Momentum Equation

V5.5 Rotating lawn sprinkler

Consider the rotating sprinkler sketched in Fig. 5.4. Because the direction and magnitude of the flow through the sprinkler from the inlet [section (1)] to the outlet [section (2)] of the arm change, the water exerts a torque on the sprinkler head, causing it to tend to rotate or to actually rotate in the direction shown, much like a turbine rotor. In applying the

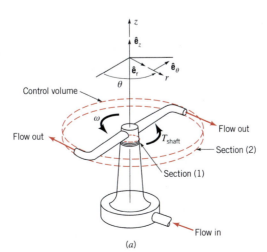

(a)

■ **FIGURE 5.4**   (*a*) **Rotary water sprinkler.**

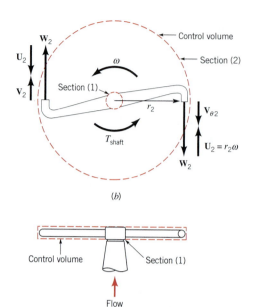

(b)

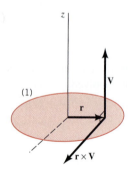

(c)

■ **FIGURE 5.4** continued
(b) **Rotary water sprinkler, plan view.**
(c) **Rotary water sprinkler, side view.**

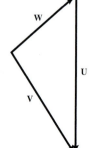

moment-of-momentum equation (Eq. 5.21) to this flow situation, we elect to use the fixed and nondeforming control volume shown in Fig. 5.4. The disk-shaped control volume contains within its boundaries the spinning or stationary sprinkler head and the portion of the water flowing through the sprinkler contained in the control volume at an instant. The control surface cuts through the sprinkler head's solid material so that the shaft torque that resists motion can be clearly identified. When the sprinkler is rotating, the flow field in the stationary control volume is cyclical and unsteady, but steady in the mean. We proceed to use the axial component of the moment-of-momentum equation (Eq. 5.21) to analyze this flow.

Water enters the control volume axially through the hollow stem of the sprinkler at section (1). At this portion of the control surface, the component of $\mathbf{r} \times \mathbf{V}$ resolved along the axis of rotation is zero because, as indicated in the figure in the margin, $\mathbf{r} \times \mathbf{V}$ and the axis of rotation (the z axis) are perpendicular. Thus, there is no axial moment-of-momentum flow in at section (1). Water leaves the control volume through each of the two nozzle openings at section (2). For the exiting flow, the *magnitude* of the axial component of $\mathbf{r} \times \mathbf{V}$ is $r_2 V_{\theta 2}$, where r_2 is the radius from the axis of rotation to the nozzle centerline and $V_{\theta 2}$ is the value of the tangential component of the velocity of the flow exiting each nozzle as observed from a frame of reference attached to the fixed and nondeforming control volume. The fluid velocity measured relative to a fixed control surface is an absolute velocity, $\mathbf{V}$. The velocity of the nozzle exit flow as viewed from the nozzle is called the relative velocity, $\mathbf{W}$. As indicated by the figure in the margin, the absolute and relative velocities, $\mathbf{V}$ and $\mathbf{W}$, are related by the vector relationship

$$\mathbf{V} = \mathbf{W} + \mathbf{U} \tag{5.22}$$

where $\mathbf{U}$ is the velocity of the moving nozzle as measured relative to the fixed control surface.

The algebraic sign to assign the axial component of $\mathbf{r} \times \mathbf{V}$ can be ascertained by using the right-hand rule. The positive direction along the axis of rotation is the direction in which the thumb of the right hand points when it is extended and the remaining

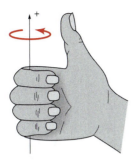

■ **FIGURE 5.5** **Right-hand rule convention.**

fingers are curled around the rotation axis in the positive direction of rotation as illustrated in Fig. 5.5. The direction of the axial component of $\mathbf{r} \times \mathbf{V}$ is similarly ascertained by noting the direction of the cross product of the radius from the axis of rotation, $r\hat{\mathbf{e}}_r$, and the tangential component of absolute velocity, $V_\theta \hat{\mathbf{e}}_\theta$. Thus, for the sprinkler of Fig. 5.4, we can state that

$$\left[\sum (\mathbf{r} \times \mathbf{V})_{\text{out}} \rho_{\text{out}} A_{\text{out}} V_{\text{out}} - \sum (\mathbf{r} \times \mathbf{V})_{\text{in}} \rho_{\text{in}} A_{\text{in}} V_{\text{in}} \right]_{\text{axial}} = (-r_2 V_{\theta 2})\dot{m} \quad (5.23)$$

where, because of mass conservation, $\dot{m}$ is the total mass flowrate through both nozzles. As was demonstrated in Example 5.4, the mass flowrate is the same whether the sprinkler rotates or not. The correct algebraic sign of the axial component of $\mathbf{r} \times \mathbf{V}$ can be easily remembered in the following way: if $\mathbf{V}_\theta$ and $\mathbf{U}$ are in the same direction, use $+$; if $\mathbf{V}_\theta$ and $\mathbf{U}$ are in opposite directions, use $-$. For the sprinkler of Fig. 5.4

$$\sum \left[(\mathbf{r} \times \mathbf{F})_{\substack{\text{contents of the} \\ \text{control volume}}} \right]_{\text{axial}} = T_{\text{shaft}} \quad (5.24)$$

Note that we have entered T_{shaft} as a positive quantity in Eq. 5.24. This is equivalent to assuming that T_{shaft} is in the same direction as rotation.

For the sprinkler of Fig. 5.4, the axial component of the moment-of-momentum equation (Eq. 5.21) is, from Eqs. 5.23 and 5.24,

$$-r_2 V_{\theta 2}\dot{m} = T_{\text{shaft}} \quad (5.25)$$

We interpret T_{shaft} being a negative quantity from Eq. 5.25 to mean that the shaft torque actually opposes the rotation of the sprinkler arms. The shaft torque, T_{shaft}, opposes rotation in all turbine devices.

We could evaluate the ***shaft power,*** $\dot{W}_{\text{shaft}}$, associated with ***shaft torque,*** T_{shaft}, by forming the product T_{shaft} and the rotational speed of the shaft, ω. Thus, from Eq. 5.25 we get

$$\dot{W}_{\text{shaft}} = T_{\text{shaft}}\omega = -r_2 V_{\theta 2}\dot{m}\omega \quad (5.26)$$

Since $r_2\omega$ is the speed, U, of each sprinkler nozzle, we can also state Eq. 5.26 in the form

$$\dot{W}_{\text{shaft}} = -U_2 V_{\theta 2}\dot{m} \quad (5.27)$$

Shaft work per unit mass, w_{shaft}, is equal to $\dot{W}_{\text{shaft}}/\dot{m}$. Dividing Eq. 5.27 by the mass flowrate, $\dot{m}$, we obtain

$$w_{\text{shaft}} = -U_2 V_{\theta 2} \quad (5.28)$$

V5.6 Impulse-type lawn sprinkler

Negative shaft work as in Eqs. 5.26, 5.27, and 5.28 is work out of the control volume, that is, work done by the fluid on the rotor and thus its shaft.

EXAMPLE 5.10	Moment-of-Momentum—Torque

GIVEN Water enters a rotating lawn sprinkler through its base at the steady rate of 1000 ml/s as sketched in Fig. E5.10a. The exit area of each of the two nozzles is 30 mm^2 and the flow leaving each nozzle is in the tangential direction. The radius from the axis of rotation to the centerline of each nozzle is 200 mm.

FIND **(a)** Determine the resisting torque required to hold the sprinkler head stationary.

(b) Determine the resisting torque associated with the sprinkler rotating with a constant speed of 500 rev/min.

(c) Determine the speed of the sprinkler if no resisting torque is applied.

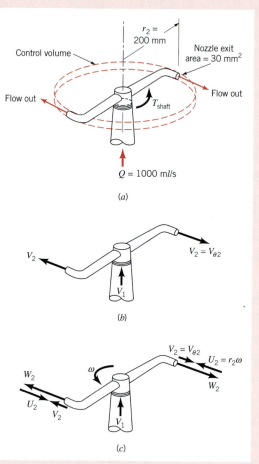

■ **FIGURE E5.10**

SOLUTION

To solve parts **(a)**, **(b)**, and **(c)** of this example we can use the same fixed and nondeforming, disk-shaped control volume illustrated in Fig. 5.4. As indicated in Fig. E5.10a, the only axial torque considered is the one resisting motion, T_{shaft}.

(a) When the sprinkler head is held stationary as specified in part **(a)** of this example problem, the velocities of the fluid entering and leaving the control volume are shown in Fig. E5.10b. Equation 5.25 applies to the contents of this control volume. Thus,

$$T_{shaft} = -r_2 V_{\theta 2} \dot{m} \qquad (1)$$

Since the control volume is fixed and nondeforming and the flow exiting from each nozzle is tangential,

$$V_{\theta 2} = V_2 \qquad (2)$$

Equations 1 and 2 give

$$T_{shaft} = -r_2 V_2 \dot{m} \qquad (3)$$

In Example 5.4, we ascertained that $V_2 = 16.7$ m/s. Thus, from Eq. 3 with

$$\dot{m} = Q\rho = \frac{(1000 \text{ ml/s})(10^{-3} \text{ m}^3/\text{liter})(999 \text{ kg/m}^3)}{(1000 \text{ ml/liter})}$$

$$= 0.999 \text{ kg/s}$$

we obtain

$$T_{shaft} = -\frac{(200 \text{ mm})(16.7 \text{ m/s})(0.999 \text{ kg/s})[1 (\text{N/kg})/(\text{m/s}^2)]}{(1000 \text{ mm/m})}$$

or

$$T_{shaft} = -3.34 \text{ N} \cdot \text{m} \qquad \textbf{(Ans)}$$

(b) When the sprinkler is rotating at a constant speed of 500 rpm, the flow field in the control volume is unsteady but cyclical. Thus, the flow field is steady in the mean. The velocities of the flow entering and leaving the control volume are as indi-

cated in Fig. E5.10c. The absolute velocity of the fluid leaving each nozzle, V_2, is from Eq. 5.22,

$$V_2 = W_2 - U_2 \qquad (4)$$

where

$$W_2 = 16.7 \text{ m/s}$$

as determined in Example 5.4. The speed of the nozzle, U_2, is obtained from

$$U_2 = r_2 \omega \qquad (5)$$

Application of the axial component of the moment-of-momentum equation (Eq. 5.25) leads again to Eq. 3. From Eqs. 4 and 5,

$$V_2 = 16.7 \text{ m/s} - r_2 \omega$$

$$= 16.7 \text{ m/s} - \frac{(200 \text{ mm})(500 \text{ rev/min})(2\pi \text{ rad/rev})}{(1000 \text{ mm/m})(60 \text{ s/min})}$$

or

$$V_2 = 16.7 \text{ m/s} - 10.5 \text{ m/s} = 6.2 \text{ m/s}$$

Thus, using Eq. 3, with $\dot{m} = 0.999$ kg/s (as calculated previously), we get

$$T_{\text{shaft}} = -\frac{(200 \text{ mm})(6.2 \text{ m/s}) \, 0.999 \text{ kg/s} \, [1 \, (\text{N/kg})/(\text{m/s}^2)]}{(1000 \text{ mm/m})}$$

or

$$T_{\text{shaft}} = -1.24 \text{ N} \cdot \text{m} \qquad \textbf{(Ans)}$$

COMMENT   Note that the resisting torque associated with sprinkler head rotation is much less than the resisting torque that is required to hold the sprinkler stationary.

(c)   When no resisting torque is applied to the rotating sprinkler head, a maximum constant speed of rotation will occur as demonstrated below. Application of Eqs. 3, 4, and 5 to the contents of the control volume results in

$$T_{\text{shaft}} = -r_2(W_2 - r_2\omega)\dot{m} \qquad \textbf{(6)}$$

For no resisting torque, Eq. 6 yields

$$0 = -r_2(W_2 - r_2\omega)\dot{m}$$

Thus,

$$\omega = \frac{W_2}{r_2} \qquad \textbf{(7)}$$

In Example 5.4, we learned that the relative velocity of the fluid leaving each nozzle, W_2, is the same regardless of the speed of rotation of the sprinkler head, ω, as long as the mass flowrate of the fluid, $\dot{m}$, remains constant. Thus, by using Eq. 7 we obtain

$$\omega = \frac{W_2}{r_2} = \frac{(16.7 \text{ m/s})(1000 \text{ mm/m})}{(200 \text{ mm})} = 83.5 \text{ rad/s}$$

or

$$\omega = \frac{(83.5 \text{ rad/s})(60 \text{ s/min})}{2\pi \text{ rad/rev}} = 797 \text{ rpm} \qquad \textbf{(Ans)}$$

For this condition ($T_{\text{shaft}} = 0$), the water both enters and leaves the control volume with zero angular momentum.

COMMENT   By repeating the calculations for various values of the angular velocity, ω, the results shown in Fig. E5.10d are obtained. It is seen that the magnitude of the resisting torque associated with rotation is less than the torque required to hold the rotor stationary. Even in the absence of a resisting torque, the rotor maximum speed is finite.

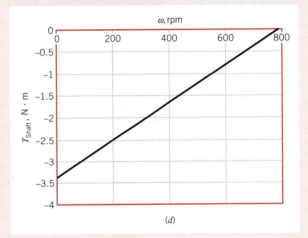

■ **FIGURE E5.10d**

When the moment-of-momentum equation (Eq. 5.21) is applied to a more general, one-dimensional flow through a rotating machine, we obtain

$$T_{\text{shaft}} = -\dot{m}_{\text{in}}(\pm r_{\text{in}} V_{\theta\text{in}}) + \dot{m}_{\text{out}}(\pm r_{\text{out}} V_{\theta\text{out}}) \tag{5.29}$$

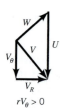

$rV_\theta > 0$

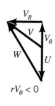

$rV_\theta < 0$

by applying the same kind of analysis used with the sprinkler of Fig. 5.4. Whether "+" or "−" is used with the rV_θ product depends on the direction of $(\mathbf{r} \times \mathbf{V})_{\text{axial}}$. A simple way to determine the sign of the rV_θ product is to compare the direction of the V_θ and the blade speed, U. If V_θ and U are in the same direction, then the rV_θ product is positive. If V_θ and U are in opposite directions, the rV_θ product is negative. These two situations are shown in the figure in the margin. The sign of the shaft torque is "+" if T_{shaft} is in the same direction along the axis of rotation as ω, and "−" otherwise.

The shaft power, $\dot{W}_{\text{shaft}}$, is related to shaft torque, T_{shaft}, by

$$\dot{W}_{\text{shaft}} = T_{\text{shaft}}\omega \tag{5.30}$$

Thus, using Eq. 5.29 and 5.30 with a "+" for T_{shaft} in Eq. 5.29, we obtain

$$\dot{W}_{\text{shaft}} = -\dot{m}_{\text{in}}(\pm r_{\text{in}}\omega V_{\theta\text{in}}) + \dot{m}_{\text{out}}(\pm r_{\text{out}}\omega V_{\theta\text{out}}) \tag{5.31}$$

or since $r\omega = U$

$$\dot{W}_{\text{shaft}} = -\dot{m}_{\text{in}}(\pm U_{\text{in}} V_{\theta\text{in}}) + \dot{m}_{\text{out}}(\pm U_{\text{out}} V_{\theta\text{out}}) \tag{5.32}$$

The "+" is used for the UV_θ product when U and V_θ are in the same direction; the "−" is used when U and V_θ are in opposite directions. Also, since $+ T_{\text{shaft}}$ was used to obtain Eq. 5.32, when $\dot{W}_{\text{shaft}}$ is positive, power is into the control volume (e.g., pump), and when $\dot{W}_{\text{shaft}}$ is negative, power is out of the control volume (e.g., turbine).

The shaft work per unit mass, $w_{\text{shaft}} = \dot{W}_{\text{shaft}}/\dot{m}$, can be obtained from the shaft power, $\dot{W}_{\text{shaft}}$, by dividing Eq. 5.32 by the mass flowrate, $\dot{m}$. By conservation of mass,

$$\dot{m} = \dot{m}_{\text{in}} = \dot{m}_{\text{out}}$$

From Eq. 5.32, we obtain

$$w_{\text{shaft}} = -(\pm U_{\text{in}} V_{\theta\text{in}}) + (\pm U_{\text{out}} V_{\theta\text{out}}) \tag{5.33}$$

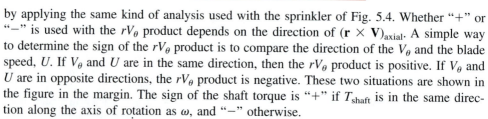

EXAMPLE 5.11   Moment-of-Momentum—Power

GIVEN An air fan has a bladed rotor of 12-in. outside diameter and 10-in. inside diameter as illustrated in Fig. E5.11a. The height of each rotor blade is constant at 1 in. from blade inlet to outlet. The flowrate is steady, on a time-average basis, at 230 ft³/min and the absolute velocity of the air at blade inlet, $\mathbf{V}_1$, is radial. The blade discharge angle is 30° from the tangential direction. The rotor rotates at a constant speed of 1725 rpm.

FIND Estimate the power required to run the fan.

SOLUTION

We select a fixed and nondeforming control volume that includes the rotating blades and the fluid within the blade row at an instant, as shown with a dashed line in Fig. E5.11a. The flow within this control volume is cyclical, but steady in the mean.

The only torque we consider is the driving shaft torque, T_{shaft}. This torque is provided by a motor. We assume that the entering and leaving flows are each represented by uniformly distributed velocities and flow properties. Since shaft power is

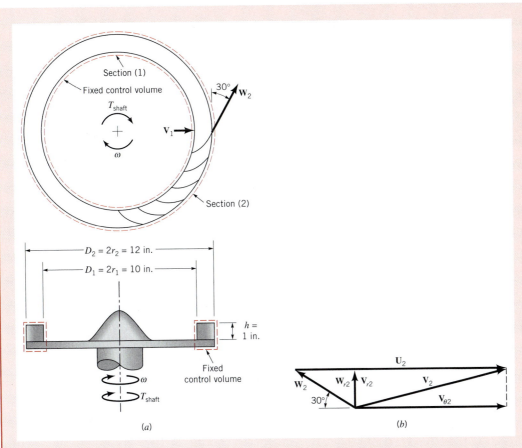

■ **FIGURE E5.11**

sought, Eq. 5.32 is appropriate. Application of Eq. 5.32 to the contents of the control volume in Fig. E5.11 gives

$$\dot{W}_{shaft} = -\dot{m}_1(\pm U_1 V_{\theta1}) + \dot{m}_2(\pm U_2 V_{\theta2}) \qquad (1)$$

$$0\ (\mathbf{V}_1\ \text{is radial})$$

From Eq. 1 we see that to calculate fan power, we need mass flowrate, $\dot{m}$, rotor exit blade velocity, U_2, and fluid tangential velocity at blade exit, $V_{\theta2}$. The mass flowrate, $\dot{m}$, is easily obtained from Eq. 5.6 as

$$\dot{m} = \rho Q = \frac{(2.38 \times 10^{-3}\ \text{slug/ft}^3)(230\ \text{ft}^3/\text{min})}{(60\ \text{s/min})}$$

$$= 0.00912\ \text{slug/s} \qquad (2)$$

The rotor exit blade speed, U_2, is

$$U_2 = r_2\omega = \frac{(6\ \text{in.})(1725\ \text{rpm})(2\pi\ \text{rad/rev})}{(12\ \text{in./ft})(60\ \text{s/min})}$$

$$= 90.3\ \text{ft/s} \qquad (3)$$

To determine the fluid tangential speed at the fan rotor exit, $V_{\theta2}$, we use Eq. 5.22 to get

$$\mathbf{V}_2 = \mathbf{W}_2 + \mathbf{U}_2 \qquad (4)$$

The vector addition of Eq. 4 is shown in the form of a "velocity triangle" in Fig. E5.11b. From Fig. E5.11b, we can see that

$$V_{\theta2} = U_2 - W_2 \cos 30° \qquad (5)$$

To solve Eq. 5 for $V_{\theta2}$ we need a value of W_2, in addition to the value of U_2 already determined (Eq. 3). To get W_2, we recognize that

$$W_2 \sin 30° = V_{r2} \qquad (6)$$

where V_{r2} is the radial component of either $\mathbf{W}_2$ or $\mathbf{V}_2$. Also, using Eq. 5.6, we obtain

$$\dot{m} = \rho A_2 V_{r2} \qquad (7)$$

or since

$$A_2 = 2\pi r_2 h \qquad (8)$$

where h is the blade height, Eqs. 7 and 8 combine to form

$$\dot{m} = \rho 2\pi r_2 h V_{r2} \qquad (9)$$

Taking Eqs. 6 and 9 together we get

$$W_2 = \frac{\dot{m}}{\rho 2\pi r_2 h \sin 30°} \qquad (10)$$

Substituting known values into Eq. 10, we obtain

$$W_2 = \frac{(0.00912 \text{ slugs/s})(12 \text{ in./ft})(12 \text{ in./ft})}{(2.38 \times 10^{-3} \text{ slugs/ft}^3)2\pi(6 \text{ in.})(1 \text{ in.}) \sin 30°}$$

$$= 29.3 \text{ ft/s}$$

By using this value of W_2 in Eq. 5 we get

$$V_{\theta 2} = U_2 - W_2 \cos 30°$$

$$= 90.3 \text{ ft/s} - (29.3 \text{ ft/s})(0.866) = 64.9 \text{ ft/s}$$

Equation 1 can now be used to obtain

$$\dot{W}_{\text{shaft}} = \dot{m} U_2 V_{\theta 2} = \frac{(0.00912 \text{ slug/s})(90.3 \text{ ft/s})(64.9 \text{ ft/s})}{[1 (\text{slug} \cdot \text{ft/s}^2)/\text{lb}][550(\text{ft} \cdot \text{lb})/(\text{hp} \cdot \text{s})]}$$

or

$$\dot{W}_{\text{shaft}} = 0.0972 \text{ hp} \qquad \textbf{(Ans)}$$

COMMENT Note that the "+" was used with the $U_2 V_{\theta 2}$ product because U_2 and $V_{\theta 2}$ are in the same direction. This result, 0.0972 hp, is the power that needs to be delivered through the fan shaft for the given conditions. Ideally, all of this power would go into the flowing air. However, because of fluid friction, only some of this power will produce a useful effect (e.g., pressure rise) in the air. How much useful effect depends on the efficiency of the energy transfer between the fan blades and the fluid.

5.3 First Law of Thermodynamics—The Energy Equation

5.3.1 Derivation of the Energy Equation

The *first law of thermodynamics* for a system is, in words

time rate of increase of the total stored energy of the system	=	net time rate of energy addition by heat transfer into the system	+	net time rate of energy addition by work transfer into the system

In symbolic form, this statement is

$$\frac{D}{Dt} \int_{\text{sys}} e\rho \, d\mathcal{V} = \left(\sum \dot{Q}_{\text{in}} - \sum \dot{Q}_{\text{out}} \right)_{\text{sys}} + \left(\sum \dot{W}_{\text{in}} - \sum \dot{W}_{\text{out}} \right)_{\text{sys}}$$

or

$$\frac{D}{Dt} \int_{\text{sys}} e\rho \, d\mathcal{V} = (\dot{Q}_{\substack{\text{net} \\ \text{in}}} + \dot{W}_{\substack{\text{net} \\ \text{in}}})_{\text{sys}} \qquad \textbf{(5.34)}$$

Some of these variables deserve a brief explanation before proceeding further. The total stored energy per unit mass for each particle in the system, e, is related to the internal energy per unit mass, $\check{u}$, the kinetic energy per unit mass, $V^2/2$, and the potential energy per unit mass, gz, by the equation

$$e = \check{u} + \frac{V^2}{2} + gz \qquad \textbf{(5.35)}$$

Thus, $\int_{\text{sys}} e\rho \, d\mathcal{V}$ is the total stored energy of the system.

The net *heat transfer rate* into the system is denoted with $\dot{Q}_{\text{net in}}$, and the net rate of work transfer into the system is labeled $\dot{W}_{\text{net in}}$. Heat transfer and work transfer are considered "+" going into the system and "−" coming out.

For the control volume that is coincident with the system at an instant of time

$$(\dot{Q}_{\substack{\text{net} \\ \text{in}}} + \dot{W}_{\substack{\text{net} \\ \text{in}}})_{\text{sys}} = (\dot{Q}_{\substack{\text{net} \\ \text{in}}} + \dot{W}_{\substack{\text{net} \\ \text{in}}})_{\substack{\text{coincident} \\ \text{control volume}}} \qquad \textbf{(5.36)}$$

Furthermore, for the system and the contents of the coincident control volume that is fixed and nondeforming, the Reynolds transport theorem (Eq. 4.14 with the parameter B set equal to the total stored energy and, therefore, the parameter b set equal to e) allows us to conclude that

$$\frac{D}{Dt}\int_{sys} e\rho \, d\mathcal{V} = \frac{\partial}{\partial t}\int_{cv} e\rho \, d\mathcal{V} + \sum e_{out}\rho_{out}A_{out}V_{out} - \sum e_{in}\rho_{in}A_{in}V_{in} \qquad (5.37)$$

or, in words,

the time rate of increase of the total stored energy of the system	=	the time rate of increase of the total stored energy of the contents of the control volume	+	the net rate of flow of the total stored energy out of the control volume through the control surface

Combining Eqs. 5.34, 5.36, and 5.37 we get the control volume formula for the first law of thermodynamics:

$$\frac{\partial}{\partial t}\int_{cv} e\rho \, d\mathcal{V} + \sum e_{out}\rho_{out}A_{out}V_{out} - \sum e_{in}\rho_{in}A_{in}V_{in} = (\dot{Q}_{\underset{in}{net}} + \dot{W}_{\underset{in}{net}})_{cv} \qquad (5.38)$$

The heat transfer rate, $\dot{Q}$, represents all of the ways in which energy is exchanged between the control volume contents and surroundings because of a temperature difference. Thus, radiation, conduction, and/or convection are possible. Heat transfer into the control volume is considered positive, heat transfer out is negative. In many engineering applications, the process is *adiabatic*; the heat transfer rate, $\dot{Q}$, is zero. The net heat transfer rate, $\dot{Q}_{net\,in}$, can also be zero when $\sum \dot{Q}_{in} - \sum \dot{Q}_{out} = 0$.

The work transfer rate, $\dot{W}$, also called *power*, is positive when work is done on the contents of the control volume by the surroundings. Otherwise, it is considered negative.

In many instances, work is transferred across the control surface by a moving shaft. In rotary devices such as turbines, fans, and propellers, a rotating shaft transfers work across that portion of the control surface that slices through the shaft. Since work is the dot product of force and related displacement, rate of work (or power) is the dot product of force and related displacement per unit time. For a rotating shaft, the power transfer, $\dot{W}_{shaft}$, is related to the shaft torque that causes the rotation, T_{shaft}, and the angular velocity of the shaft, ω, by the relationship

$$\dot{W}_{shaft} = T_{shaft}\omega$$

When the control surface cuts through the shaft material, the shaft torque is exerted by shaft material at the control surface. To allow for consideration of problems involving more than one shaft we use the notation

$$\dot{W}_{\underset{net\,in}{shaft}} = \sum \dot{W}_{\underset{in}{shaft}} - \sum \dot{W}_{\underset{out}{shaft}} \qquad (5.39)$$

Work transfer can also occur at the control surface when the force associated with the fluid pressure (a normal stress) acts over a distance. For example, as shown in Fig. 5.6, the uniform pressure force acting on the contents of the control surface across the inlet is $F_{normal\,stress\,in} = p_{in}A_{in}$. This force acts inward, in the same direction as the inlet flow. Hence, the power transfer (force times velocity) associated with this inlet flow is

$$\dot{W}_{normal\,stress\,in} = F_{normal\,stress\,in}V_{in} = p_{in}A_{in}V_{in}$$

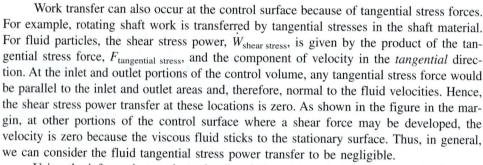

■ **FIGURE 5.6** **Inlet and outlet pressure forces.**

Note that $\dot{W}_{\text{normal stress in}} > 0$; the inlet pressure force transfers power *to* the control volume contents. As shown in Fig. 5.6, the pressure force also acts inward at the control volume outlet. Recall that pressure is a compressive normal stress, independent of the direction of the velocity. Thus, at the outlet the pressure force acts inward and the velocity is outward. That is, the pressure force and velocity are oppositely directed so that the power transfer associated with the outlet flow is negative and given by

$$\dot{W}_{\text{normal stress out}} = -F_{\text{normal stress out}} V_{\text{out}} = -p_{\text{out}} A_{\text{out}} V_{\text{out}}$$

The outlet pressure force transfers power *from* the control volume to its surroundings. Thus, the net power transfer due to fluid normal stress, $\dot{W}_{\text{normal stress}}$, is

$$\dot{W}_{\text{normal stress}} = p_{\text{in}} A_{\text{in}} V_{\text{in}} - p_{\text{out}} A_{\text{out}} V_{\text{out}} \tag{5.40}$$

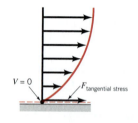

Work transfer can also occur at the control surface because of tangential stress forces. For example, rotating shaft work is transferred by tangential stresses in the shaft material. For fluid particles, the shear stress power, $\dot{W}_{\text{shear stress}}$, is given by the product of the tangential stress force, $F_{\text{tangential stress}}$, and the component of velocity in the *tangential* direction. At the inlet and outlet portions of the control volume, any tangential stress force would be parallel to the inlet and outlet areas and, therefore, normal to the fluid velocities. Hence, the shear stress power transfer at these locations is zero. As shown in the figure in the margin, at other portions of the control surface where a shear force may be developed, the velocity is zero because the viscous fluid sticks to the stationary surface. Thus, in general, we can consider the fluid tangential stress power transfer to be negligible.

Using the information we have developed about power, we can express the first law of thermodynamics for the contents of a control volume by combining Eqs. 5.38, 5.39, and 5.40 to obtain

$$\frac{\partial}{\partial t} \int_{\text{cv}} e\rho \, d\mathcal{V} + \sum e_{\text{out}} \rho_{\text{out}} A_{\text{out}} V_{\text{out}} - \sum e_{\text{in}} \rho_{\text{in}} A_{\text{in}} V_{\text{in}}$$

$$= \dot{Q}_{\substack{\text{net} \\ \text{in}}} + \dot{W}_{\substack{\text{shaft} \\ \text{net in}}} + \sum p_{\text{in}} A_{\text{in}} V_{\text{in}} - \sum p_{\text{out}} A_{\text{out}} V_{\text{out}} \tag{5.41}$$

When the equation for total stored energy (Eq. 5.35) is considered with Eq. 5.41, we obtain the *energy equation*:

$$\frac{\partial}{\partial t} \int_{\text{cv}} e\rho \, d\mathcal{V} + \sum \left(\check{u} + \frac{p}{\rho} + \frac{V^2}{2} + gz \right)_{\text{out}} \rho_{\text{out}} A_{\text{out}} V_{\text{out}}$$

$$- \sum \left(\check{u} + \frac{p}{\rho} + \frac{V^2}{2} + gz \right)_{\text{in}} \rho_{\text{in}} A_{\text{in}} V_{\text{in}} = \dot{Q}_{\substack{\text{net} \\ \text{in}}} + \dot{W}_{\substack{\text{shaft} \\ \text{net in}}} \tag{5.42}$$

Equation 5.42 is restricted to flows through fixed, nondeforming control volumes having uniform properties across the inlets and outlets, with the velocity normal to the inlet and outlet areas. A more general form of the energy equation, valid for more general flow conditions, is given in Appendix D.

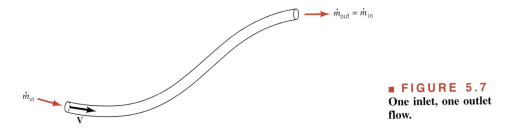

■ **FIGURE 5.7**
One inlet, one outlet flow.

5.3.2   Application of the Energy Equation

For many applications, Eq. 5.42 can be simplified. For example, the term $\partial/\partial t \int_{cv} e\rho \, dV$ represents the time rate of change of the total stored energy, e, of the contents of the control volume. This term is zero when the flow is steady. This term is also zero in the mean when the flow is steady in the mean (cyclical). Furthermore, for many applications there is only one stream entering and leaving the control volume, and the properties are all assumed to be uniformly distributed over the flow cross-sectional areas involved. For such flows, as shown in Fig. 5.7, the continuity equation (Eq. 5.10) gives

$$\rho_{in} A_{in} V_{in} = \rho_{out} A_{out} V_{out} = \dot{m} \tag{5.43}$$

and Eq. 5.42 can be simplified to form

$$\dot{m}\left[\check{u}_{out} - \check{u}_{in} + \left(\frac{p}{\rho}\right)_{out} - \left(\frac{p}{\rho}\right)_{in} + \frac{V_{out}^2 - V_{in}^2}{2} + g(z_{out} - z_{in}) \right] = \dot{Q}_{\substack{net \\ in}} + \dot{W}_{\substack{shaft \\ net\ in}} \tag{5.44}$$

We call Eq. 5.44 the *one-dimensional energy equation for steady flow*. Note that Eq. 5.44 is valid for incompressible and compressible flows. Often, the fluid property called *enthalpy*, $\check{h}$, where

$$\check{h} = \check{u} + \frac{p}{\rho} \tag{5.45}$$

is used in Eq. 5.44. With enthalpy, the one-dimensional energy equation for steady flow (Eq. 5.44) is

$$\dot{m}\left[\check{h}_{out} - \check{h}_{in} + \frac{V_{out}^2 - V_{in}^2}{2} + g(z_{out} - z_{in}) \right] = \dot{Q}_{\substack{net \\ in}} + \dot{W}_{\substack{shaft \\ net\ in}} \tag{5.46}$$

Equation 5.46 is often used for solving compressible flow problems.

EXAMPLE 5.12   **Energy—Pump Power**

GIVEN A pump delivers water at a steady rate of 300 gal/min as shown in Fig. E5.12. Just upstream of the pump [section (1)] where the pipe diameter is 3.5 in., the pressure is 18 psi. Just downstream of the pump [section (2)] where the pipe diameter is 1 in., the pressure is 60 psi. The change in water elevation across the pump is zero. The rise in internal energy of water,

$\check{u}_2 - \check{u}_1$, associated with a temperature rise across the pump is 3000 ft·lb/slug. The pumping process is considered to be adiabatic.

FIND Determine the power (hp) required by the pump.

SOLUTION

We include in our control volume the water contained in the pump between its entrance and exit sections. Application of Eq. 5.44 to the contents of this control volume on a time-average basis yields

0 (no elevation change)

$$\dot{m}\left[\breve{u}_2 - \breve{u}_1 + \left(\frac{p}{\rho}\right)_2 - \left(\frac{p}{\rho}\right)_1 + \frac{V_2^2 - V_1^2}{2} + g(z_2 - z_1) \right]$$

0 (adiabatic flow)

$$= \dot{Q}_{\substack{\text{net} \\ \text{in}}} + \dot{W}_{\substack{\text{shaft} \\ \text{net in}}} \quad (1)$$

We can solve directly for the power required by the pump, $\dot{W}_{\text{shaft net in}}$, from Eq. 1, after we first determine the mass flowrate, $\dot{m}$, the speed of flow into the pump, V_1, and the speed of the flow out of the pump, V_2. All other quantities in Eq. 1 are given in the problem statement. From Eq. 5.6, we get

$$\dot{m} = \rho Q = \frac{(1.94 \text{ slugs/ft}^3)(300 \text{ gal/min})}{(7.48 \text{ gal/ft}^3)(60 \text{ s/min})}$$

$$= 1.30 \text{ slugs/s} \quad (2)$$

Also from Eq. 5.6,

$$V = \frac{Q}{A} = \frac{Q}{\pi D^2/4}$$

so

$$V_1 = \frac{Q}{A_1} = \frac{(300 \text{ gal/min})4(12 \text{ in./ft})^2}{(7.48 \text{ gal/ft}^3)(60 \text{ s/min})\pi(3.5 \text{ in.})^2}$$

$$= 10.0 \text{ ft/s} \quad (3)$$

and

$$V_2 = \frac{Q}{A_2} = \frac{(300 \text{ gal/min})4(12 \text{ in./ft})^2}{(7.48 \text{ gal/ft}^3)(60 \text{ s/min})\pi(1 \text{ in.})^2}$$

$$= 123 \text{ ft/s} \quad (4)$$

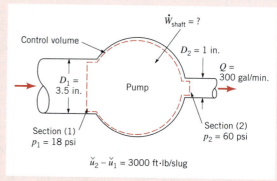

■ **FIGURE E5.12**

Substituting the values of Eqs. 2, 3, and 4 and values from the problem statement into Eq. 1 we obtain

$$\dot{W}_{\substack{\text{shaft} \\ \text{net in}}} = (1.30 \text{ slugs/s})\bigg[(3000 \text{ ft} \cdot \text{lb/slug})$$

$$+ \frac{(60 \text{ psi})(144 \text{ in.}^2/\text{ft}^2)}{(1.94 \text{ slugs/ft}^3)}$$

$$- \frac{(18 \text{ psi})(144 \text{ in.}^2/\text{ft}^2)}{(1.94 \text{ slugs/ft}^3)}$$

$$+ \frac{(123 \text{ ft/s})^2 - (10.0 \text{ ft/s})^2}{2[1 (\text{slug} \cdot \text{ft})/(\text{lb} \cdot \text{s}^2)]}\bigg]$$

$$\times \frac{1}{[550(\text{ft} \cdot \text{lb/s})/\text{hp}]} = 32.2 \text{ hp} \quad \textbf{(Ans)}$$

COMMENT   Of the total 32.2 hp, internal energy change accounts for 7.09 hp, the pressure rise accounts for 7.37 hp, and the kinetic energy increase accounts for 17.8 hp.

EXAMPLE 5.13   Energy—Turbine Power per Unit Mass of Flow

GIVEN   Steam enters a turbine with a velocity of 30 m/s and enthalpy, h_1, of 3348 kJ/kg (see Fig. E5.13). The steam leaves the turbine as a mixture of vapor and liquid having a velocity of 60 m/s and an enthalpy of 2550 kJ/kg. The flow through the turbine is adiabatic, and changes in elevation are negligible.

FIND   Determine the work output involved per unit mass of steam through-flow.

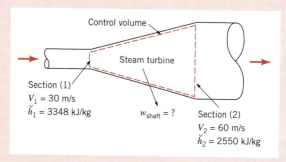

■ **FIGURE E5.13**

SOLUTION

We use a control volume that includes the steam in the turbine from the entrance to the exit as shown in Fig. E5.13. Applying Eq. 5.46 to the steam in this control volume we get

0 (elevation change is negligible)

0 (adiabatic flow)

$$\dot{m}\left[\check{h}_2 - \check{h}_1 + \frac{V_2^2 - V_1^2}{2} + g(z_2 - z_1)\right] = \dot{Q}_{\substack{\text{net} \\ \text{in}}} + \dot{W}_{\substack{\text{shaft} \\ \text{net in}}} \quad (1)$$

The work output per unit mass of steam through-flow, $w_{\text{shaft net in}}$, can be obtained by dividing Eq. 1 by the mass flow rate, $\dot{m}$, to obtain

$$w_{\substack{\text{shaft} \\ \text{net in}}} = \frac{\dot{W}_{\substack{\text{shaft} \\ \text{net in}}}}{\dot{m}} = \check{h}_2 - \check{h}_1 + \frac{V_2^2 - V_1^2}{2} \quad (2)$$

Since $w_{\text{shaft net out}} = -w_{\text{shaft net in}}$, we obtain

$$w_{\substack{\text{shaft} \\ \text{net out}}} = \check{h}_1 - \check{h}_2 + \frac{V_1^2 - V_2^2}{2}$$

or

$$w_{\substack{\text{shaft} \\ \text{net out}}} = 3348 \text{ kJ/kg} - 2550 \text{ kJ/kg}$$

$$+ \frac{[(30 \text{ m/s})^2 - (60 \text{ m/s})^2][1 \text{ J/(N·m)}]}{2[1 (\text{kg·m})/(\text{N·s}^2)](1000 \text{ J/kJ})}$$

Thus,

$$w_{\substack{\text{shaft} \\ \text{net out}}} = 3348 \text{ kJ/kg} - 2550 \text{ kJ/kg} - 1.35 \text{ kJ/kg}$$

$$= 797 \text{ kJ/kg} \qquad \textbf{(Ans)}$$

COMMENT Note that in this particular example, the change in kinetic energy is small in comparison to the difference in enthalpy involved. This is often true in applications involving steam turbines. To determine the power output, $\dot{W}_{\text{shaft}}$, we must know the mass flowrate, $\dot{m}$.

5.3.3 Comparison of the Energy Equation with the Bernoulli Equation

For steady, incompressible flow with zero shaft power, the energy equation (Eq. 5.44) becomes

$$\dot{m}\left[\check{u}_{\text{out}} - \check{u}_{\text{in}} + \frac{p_{\text{out}}}{\rho} - \frac{p_{\text{in}}}{\rho} + \frac{V_{\text{out}}^2 - V_{\text{in}}^2}{2} + g(z_{\text{out}} - z_{\text{in}})\right] = \dot{Q}_{\substack{\text{net} \\ \text{in}}} \qquad \textbf{(5.47)}$$

Dividing Eq. 5.47 by the mass flowrate, $\dot{m}$, and rearranging terms we obtain

$$\frac{p_{\text{out}}}{\rho} + \frac{V_{\text{out}}^2}{2} + gz_{\text{out}} = \frac{p_{\text{in}}}{\rho} + \frac{V_{\text{in}}^2}{2} + gz_{\text{in}} - (\check{u}_{\text{out}} - \check{u}_{\text{in}} - q_{\substack{\text{net} \\ \text{in}}}) \qquad \textbf{(5.48)}$$

where

$$q_{\substack{\text{net} \\ \text{in}}} = \frac{\dot{Q}_{\text{net in}}}{\dot{m}}$$

is the heat transfer rate per mass flowrate, or heat transfer per unit mass. Note that Eq. 5.48 involves energy per unit mass and is applicable to one-dimensional flow of a single stream of fluid between two sections or flow along a streamline between two sections.

   If the steady, incompressible flow we are considering also involves negligible viscous effects (frictionless flow), then the Bernoulli equation, Eq. 3.6, can be used to describe what happens between two sections in the flow as

$$p_{\text{out}} + \frac{\rho V_{\text{out}}^2}{2} + \gamma z_{\text{out}} = p_{\text{in}} + \frac{\rho V_{\text{in}}^2}{2} + \gamma z_{\text{in}} \qquad \textbf{(5.49)}$$

where $\gamma = \rho g$ is the specific weight of the fluid. To get Eq. 5.49 in terms of energy per unit mass, so that it can be compared directly with Eq. 5.48, we divide Eq. 5.49 by density, ρ, and obtain

$$\frac{p_{out}}{\rho} + \frac{V_{out}^2}{2} + gz_{out} = \frac{p_{in}}{\rho} + \frac{V_{in}^2}{2} + gz_{in} \tag{5.50}$$

A comparison of Eqs. 5.48 and 5.50 prompts us to conclude that

$$\check{u}_{out} - \check{u}_{in} - q_{\substack{net \\ in}} = 0 \tag{5.51}$$

when the steady, incompressible flow is frictionless. For steady, incompressible flow with friction, we learn from experience that

$$\check{u}_{out} - \check{u}_{in} - q_{\substack{net \\ in}} > 0 \tag{5.52}$$

In Eqs. 5.48 and 5.50, we consider the combination of variables

$$\frac{p}{\rho} + \frac{V^2}{2} + gz$$

as equal to *useful* or *available energy*. Thus, from inspection of Eqs. 5.48 and 5.50, we can conclude that $\check{u}_{out} - \check{u}_{in} - q_{net\ in}$ represents the **loss** of useful or available energy that occurs in an incompressible fluid flow because of friction. In equation form we have

$$\check{u}_{out} - \check{u}_{in} - q_{\substack{net \\ in}} = loss \tag{5.53}$$

For a frictionless flow, Eqs. 5.48 and 5.50 tell us that loss equals zero.

It is often convenient to express Eq. 5.48 in terms of loss as

$$\frac{p_{out}}{\rho} + \frac{V_{out}^2}{2} + gz_{out} = \frac{p_{in}}{\rho} + \frac{V_{in}^2}{2} + gz_{in} - loss \tag{5.54}$$

V5.7 Energy transfer

An example of the application of Eq. 5.54 follows.

EXAMPLE 5.14   Energy—Effect of Loss of Available Energy

GIVEN As shown in Fig. E5.14a, air flows from a room through two different vent configurations: a cylindrical hole in the wall having a diameter of 120 mm and the same diameter cylindrical hole in the wall but with a well-rounded entrance. The room pressure is held constant at 1.0 kPa above atmospheric pressure. Both vents exhaust into the atmosphere. As discussed in Section 8.4.2, the loss in available energy associated with flow through the cylindrical vent from the room to the vent exit is $0.5V_2^2/2$ where V_2 is the uniformly distributed exit velocity of air. The loss in available energy associated with flow through the rounded entrance vent from the room to the vent exit is $0.05V_2^2/2$, where V_2 is the uniformly distributed exit velocity of air.

FIND Compare the volume flowrates associated with the two different vent configurations.

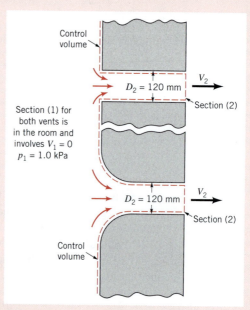

Section (1) for both vents is in the room and involves $V_1 = 0$ $p_1 = 1.0$ kPa

■ **FIGURE E5.14a**

SOLUTION

We use the control volume for each vent sketched in Fig. E5.14a. What is sought is the flowrate, $Q = A_2V_2$, where A_2 is the vent exit cross-sectional area, and V_2 is the uniformly distributed exit velocity. For both vents, application of Eq. 5.54 leads to

$$\frac{p_2}{\rho} + \frac{V_2^2}{2} + g\cancel{z_2}^{\;0 \text{ (no elevation change)}} = \frac{p_1}{\rho} + \cancel{\frac{V_1^2}{2}}^{\;0\,(V_1 \approx 0)} + g\cancel{z_1} - {}_1\text{loss}_2 \tag{1}$$

where ${}_1\text{loss}_2$ is the loss between sections (1) and (2). Solving Eq. 1 for V_2 we get

$$V_2 = \sqrt{2\left[\left(\frac{p_1 - p_2}{\rho}\right) - {}_1\text{loss}_2\right]} \tag{2}$$

Since

$${}_1\text{loss}_2 = K_L \frac{V_2^2}{2} \tag{3}$$

where K_L is the loss coefficient ($K_L = 0.5$ and 0.05 for the two vent configurations involved), we can combine Eqs. 2 and 3 to get

$$V_2 = \sqrt{2\left[\left(\frac{p_1 - p_2}{\rho}\right) - K_L \frac{V_2^2}{2}\right]} \tag{4}$$

Solving Eq. 4 for V_2 we obtain

$$V_2 = \sqrt{\frac{p_1 - p_2}{\rho[(1 + K_L)/2]}} \tag{5}$$

Therefore, for flowrate, Q, we obtain

$$Q = A_2V_2 = \frac{\pi D_2^2}{4}\sqrt{\frac{p_1 - p_2}{\rho[(1 + K_L)/2]}} \tag{6}$$

For the rounded entrance cylindrical vent, Eq. 6 gives

$$Q = \frac{\pi(120 \text{ mm})^2}{4(1000 \text{ mm/m})^2}$$
$$\times \sqrt{\frac{(1.0 \text{ kPa})(1000 \text{ Pa/kPa})[1(\text{N/m}^2)/(\text{Pa})]}{(1.23 \text{ kg/m}^3)[(1 + 0.05)/2][1(\text{N}\cdot\text{s}^2)/(\text{kg}\cdot\text{m})]}}$$

or

$$Q = 0.445 \text{ m}^3/\text{s} \tag{Ans}$$

For the cylindrical vent, Eq. 6 gives us

$$Q = \frac{\pi(120 \text{ mm})^2}{4(1000 \text{ mm/m})^2}$$
$$\times \sqrt{\frac{(1.0 \text{ kPa})(1000 \text{ Pa/kPa})[1(\text{N/m}^2)/(\text{Pa})]}{(1.23 \text{ kg/m}^3)[(1 + 0.5)/2][1(\text{N}\cdot\text{s}^2)/(\text{kg}\cdot\text{m})]}}$$

or

$$Q = 0.372 \text{ m}^3/\text{s} \tag{Ans}$$

COMMENT   By repeating the calculations for various values of the loss coefficient, K_L, the results shown in Fig. E5.14b are obtained. Note that the rounded entrance vent allows the passage of more air than does the cylindrical vent because the loss associated with the rounded entrance vent is less than that for the cylindrical one. For this flow the pressure drop, $p_1 - p_2$, has two purposes: (1) overcome the loss associated with the flow, and (2) produce the kinetic energy at the exit. Even if there were no loss (i.e., $K_L = 0$) a pressure drop would be needed to accelerate the fluid through the vent.

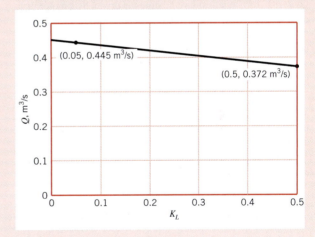

■ **FIGURE E5.14***b*

An important group of fluid mechanics problems involves one-dimensional, incompressible, steady flow with friction and shaft work. Included in this category are constant density flows through pumps, blowers, fans, and turbines. For this kind of flow, Eq. 5.44 becomes

$$\dot{m}\left[\breve{u}_{\text{out}} - \breve{u}_{\text{in}} + \frac{p_{\text{out}}}{\rho} - \frac{p_{\text{in}}}{\rho} + \frac{V_{\text{out}}^2 - V_{\text{in}}^2}{2} + g(z_{\text{out}} - z_{\text{in}})\right] = \dot{Q}_{\substack{\text{net} \\ \text{in}}} + \dot{W}_{\substack{\text{shaft} \\ \text{net in}}} \tag{5.55}$$

We divide Eq. 5.55 by the mass flowrate, use the facts that loss $= \breve{u}_{\text{out}} - \breve{u}_{\text{in}} - q_{\text{net in}}$ and work per unit mass $= w_{\text{shaft net in}} = \dot{W}_{\text{shaft net in}}/\dot{m}$ to obtain

$$\boxed{\frac{p_{\text{out}}}{\rho} + \frac{V_{\text{out}}^2}{2} + gz_{\text{out}} = \frac{p_{\text{in}}}{\rho} + \frac{V_{\text{in}}^2}{2} + gz_{\text{in}} + w_{\substack{\text{shaft} \\ \text{net in}}} - \text{loss}} \tag{5.56}$$

This is a form of the energy equation for steady flow that is often used for incompressible flow problems. It is sometimes called the *mechanical energy* equation or the *extended Bernoulli* equation. Note that Eq. 5.56 involves energy per unit mass (ft·lb/slug = ft²/s² or N·m/kg = m²/s²).

If Eq. 5.56 is divided by the acceleration of gravity, g, we get

$$\frac{p_{out}}{\gamma} + \frac{V_{out}^2}{2g} + z_{out} = \frac{p_{in}}{\gamma} + \frac{V_{in}^2}{2g} + z_{in} + h_s - h_L \qquad (5.57)$$

where

$$h_s = w_{shaft\ net\ in}/g = \frac{\dot{W}_{shaft\ net\ in}}{\dot{m}g} = \frac{\dot{W}_{shaft\ net\ in}}{\gamma Q} \qquad (5.58)$$

V5.8 Water plant aerator

and $h_L = $ loss/g. Equation 5.57 involves *energy per unit weight* (ft·lb/lb = ft or N·m/N = m). Section 3.7 introduced the notion of "head," which is energy per unit weight. Units of length (e.g., ft, m) are used to quantify the amount of head involved. If a turbine is in the control volume, the notation $h_s = -h_T$ (with $h_T > 0$) is sometimes used, particularly in the field of hydraulics. For a pump in the control volume, $h_s = h_P$. The quantity h_T is termed the **turbine head** and h_P is the **pump head**. The loss term, h_L, is often referred to as **head loss**.

EXAMPLE 5.15 | Energy—Fan Work and Efficiency

GIVEN An axial-flow ventilating fan driven by a motor that delivers 0.4 kW of power to the fan blades produces a 0.6-m-diameter axial stream of air having a speed of 12 m/s. The flow upstream of the fan involves negligible speed.

FIND Determine how much of the work to the air actually produces a useful effect, that is, a rise in available energy, and estimate the fluid mechanical efficiency of this fan.

SOLUTION

We select a fixed and nondeforming control volume as is illustrated in Fig. E5.15. The application of Eq. 5.56 to the contents of this control volume leads to

$$w_{shaft\ net\ in} - loss = \left(\frac{p_2}{\rho} + \frac{V_2^2}{2} + gz_2\right) - \left(\frac{p_1}{\rho} + \frac{V_1^2}{2} + gz_1\right) \quad (1)$$

0 (atmospheric pressures cancel) 0 ($V_1 \approx 0$) 0 (no elevation change)

where $w_{shaft\ net\ in} - $ loss is the amount of work added to the air that produces a useful effect. Equation 1 leads to

$$w_{shaft\ net\ in} - loss = \frac{V_2^2}{2} = \frac{(12\ m/s)^2}{2[1(kg·m)/(N·s^2)]}$$
$$= 72.0\ N·m/kg \qquad (2) \quad \textbf{(Ans)}$$

A reasonable estimate of *efficiency*, η, would be the ratio of amount of work that produces a useful effect, Eq. 2, to the amount of work delivered to the fan blades. That is

$$\eta = \frac{w_{shaft\ net\ in} - loss}{w_{shaft\ net\ in}} \qquad (3)$$

To calculate the efficiency, we need a value of $w_{shaft\ net\ in}$, which is related to the power delivered to the blades, $\dot{W}_{shaft\ net\ in}$. We note that

$$w_{shaft\ net\ in} = \frac{\dot{W}_{shaft\ net\ in}}{\dot{m}} \qquad (4)$$

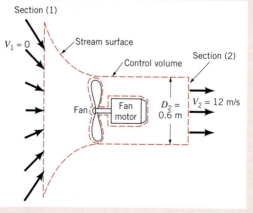

■ FIGURE E5.15

where the mass flowrate, $\dot{m}$, is (from Eq. 5.6)

$$\dot{m} = \rho A V = \rho \frac{\pi D_2^2}{4} V_2 \qquad (5)$$

For fluid density, ρ, we use 1.23 kg/m³ (standard air) and, thus, from Eqs. 4 and 5 we obtain

$$w_{\substack{\text{shaft} \\ \text{net in}}} = \frac{\dot{W}_{\substack{\text{shaft} \\ \text{net in}}}}{(\rho \pi D_2^2/4) V_2}$$

$$= \frac{(0.4 \text{ kW})[1000 \text{ (N·m)/(s·kW)}]}{(1.23 \text{ kg/m}^3)[(\pi)(0.6 \text{ m})^2/4](12 \text{ m/s})}$$

or

$$w_{\substack{\text{shaft} \\ \text{net in}}} = 95.8 \text{ N·m/kg} \qquad (6)$$

From Eqs. 2, 3, and 6 we obtain

$$\eta = \frac{72.0 \text{ N·m/kg}}{95.8 \text{ N·m/kg}} = 0.752 \qquad \text{(Ans)}$$

COMMENT   Note that only 75% of the power that was delivered to the air resulted in a useful effect, and, thus, 25% of the shaft power is lost to air friction.

F l u i d s   i n   t h e   N e w s

Curtain of air An air curtain is produced by blowing air through a long rectangular nozzle to produce a high-velocity sheet of air, or a "curtain of air." This air curtain is typically directed over a doorway or opening as a replacement for a conventional door. The air curtain can be used for such things as keeping warm air from infiltrating dedicated cold spaces, preventing dust and other contaminants from entering a clean environment, and even just keeping insects out of the workplace, still allowing people to enter or exit. A disadvantage over conventional doors is the added *power requirements* to operate the air curtain, although the advantages can outweigh the disadvantage for various industrial applications. New applications for current air curtain designs continue to be developed. For example, the use of air curtains as a means of road tunnel fire security is currently being investigated. In such an application, the air curtain would act to isolate a portion of the tunnel where fire has broken out and not allow smoke and fumes to infiltrate the entire tunnel system. (See Problem 5.70.) ■

EXAMPLE 5.16   Energy—Head Loss and Power Loss

GIVEN The pump shown in Fig. E5.16a adds 10 horsepower to the water as it pumps water from the lower lake to the upper lake. The elevation difference between the lake surfaces is 30 ft and the head loss is 15 ft.

FIND Determine

(a) the flowrate and

(b) the power loss associated with this flow.

SOLUTION

(a) The energy equation (Eq. 5.57) for this flow is

$$\frac{p_A}{\gamma} + \frac{V_A^2}{2g} + z_A = \frac{p_B}{\gamma} + \frac{V_B^2}{2g} + z_B + h_s - h_L \qquad (1)$$

where points A and B (corresponding to "out" and "in" in Eq. 5.57) are located on the lake surfaces. Thus, $p_A = p_B = 0$ and $V_A = V_B = 0$ so that Eq. 1 becomes

$$h_s = h_L + z_A - z_B \qquad (2)$$

where $z_A = 30$ ft, $z_B = 0$, and $h_L = 15$ ft. The pump head is obtained from Eq. 5.58 as

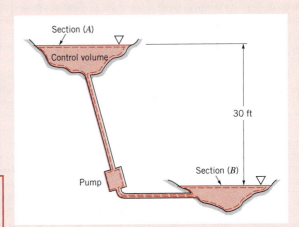

■ **FIGURE E5.16a**

$$h_s = \dot{W}_{\text{shaft net in}}/\gamma Q$$
$$= (10 \text{ hp})(550 \text{ ft·lb/s/hp})/(62.4 \text{ lb/ft}^3) Q$$
$$= 88.1/Q$$

where h_s is in ft when Q is in ft³/s.

Hence, from Eq. 2,

$$88.1/Q = 15 \text{ ft} + 30 \text{ ft}$$

or

$$Q = 1.96 \text{ ft}^3/\text{s} \qquad \textbf{(Ans)}$$

COMMENT   Note that in this example the purpose of the pump is to lift the water (a 30-ft head) and overcome the head loss (a 15-ft head); it does not, overall, alter the water's pressure or velocity.

(b)   The power lost due to friction can be obtained from Eq. 5.58 as

$$\dot{W}_{\text{loss}} = \gamma Q h_L = (62.4 \text{ lb/ft}^3)(1.96 \text{ ft}^3/\text{s})(15 \text{ ft})$$
$$= 1830 \text{ ft·lb/s } (1 \text{ hp}/550 \text{ ft·lb/s})$$
$$= 3.33 \text{ hp} \qquad \textbf{(Ans)}$$

COMMENTS   The remaining 10 hp − 3.33 hp = 6.67 hp that the pump adds to the water is used to lift the water from the lower to the upper lake. This energy is not "lost," but it is stored as potential energy.

By repeating the calculations for various head losses, h_L, the results shown in Fig. E5.16b are obtained. Note that as the head loss increases, the flowrate decreases because an increasing portion of the 10 hp supplied by the pump is lost and, therefore, not available to lift the fluid to the higher elevation.

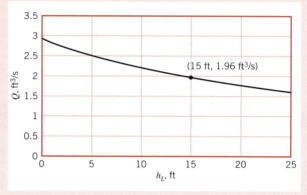

■ **FIGURE E5.16***b*

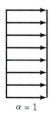

$\alpha = 1$

$\alpha > 1$

5.3.4   Application of the Energy Equation to Nonuniform Flows

The forms of the energy equation discussed in Sections 5.3.1 through 5.3.3 are applicable to one-dimensional flows, flows that are approximated with uniform velocity distributions where fluid crosses the control surface.

If the velocity profile at any section where flow crosses the control surface is not uniform, the kinetic energy term in the energy equation must be slightly modified. For example, for nonuniform velocity profiles, the energy equation on an energy per unit mass basis (Eq. 5.56) is

$$\frac{p_{\text{out}}}{\rho} + \frac{\alpha_{out}\overline{V}_{\text{out}}^2}{2} + gz_{\text{out}} = \frac{p_{\text{in}}}{\rho} + \frac{\alpha_{\text{in}}\overline{V}_{\text{in}}^2}{2} + gz_{\text{in}} + w_{\text{shaft}} - \text{loss} \qquad \textbf{(5.59)}$$
$$\text{net in}$$

where α is the *kinetic energy coefficient* and $\overline{V}$ is the average velocity (see Eq. 5.7). It can be shown that for any velocity profile, $\alpha \geq 1$, with $\alpha = 1$ only for uniform flow. This property is indicated by the figure in the margin of the previous page. However, for many practical applications $\alpha \approx 1.0$.

EXAMPLE 5.17   Energy—Effect of Nonuniform Velocity Profile

GIVEN   The small fan shown in Fig. E5.17 moves air at a mass flowrate of 0.1 kg/min. Upstream of the fan, the pipe diameter is 60 mm, the flow is laminar, the velocity distribution is parabolic, and the kinetic energy coefficient, α_1, is equal to 2.0. Downstream of the fan, the pipe diameter is 30 mm, the flow is turbulent, the velocity profile is quite uniform, and the kinetic energy coefficient, α_2, is equal to 1.08. The rise in static pressure across the fan is 0.1 kPa, and the fan motor draws 0.14 W.

FIND   Compare the value of loss calculated

(a)   assuming uniform velocity distributions, and

(b)   considering actual velocity distributions.

SOLUTION

Application of Eq. 5.59 to the contents of the control volume shown in Fig. E5.17 leads to

$$\frac{p_2}{\rho} + \frac{\alpha_2 \overline{V}_2^2}{2} + \cancel{gz_2} \overset{0 \text{ (change in } gz \text{ is negligible)}}{=} \frac{p_1}{\rho} + \frac{\alpha_1 \overline{V}_1^2}{2} + \cancel{gz_1}$$
$$- \text{loss} + w_{\substack{\text{shaft} \\ \text{net in}}} \qquad (1)$$

or solving Eq. 1 for loss we get

$$\text{loss} = w_{\substack{\text{shaft} \\ \text{net in}}} - \left(\frac{p_2 - p_1}{\rho}\right) + \frac{\alpha_1 \overline{V}_1^2}{2} - \frac{\alpha_2 \overline{V}_2^2}{2} \qquad (2)$$

To proceed further, we need values of $w_{\text{shaft net in}}$, $\overline{V}_1$, and $\overline{V}_2$. These quantities can be obtained as follows. For shaft work

$$w_{\substack{\text{shaft} \\ \text{net in}}} = \frac{\text{power to fan motor}}{\dot{m}}$$

or

$$w_{\substack{\text{shaft} \\ \text{net in}}} = \frac{(0.14 \text{ W})([1 \text{ (N·m/s)/W}]}{0.1 \text{ kg/min}}(60 \text{ s/min})$$

so that

$$w_{\substack{\text{shaft} \\ \text{net in}}} = 84.0 \text{ N·m/kg} \qquad (3)$$

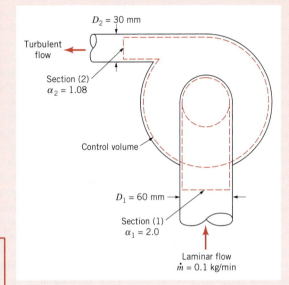

■ FIGURE E5.17

For the average velocity at section (1), $\overline{V}_1$, from Eq. 5.10 we obtain

$$\overline{V}_1 = \frac{\dot{m}}{\rho A_1}$$

$$= \frac{\dot{m}}{\rho(\pi D_1^2/4)}$$

$$= \frac{(0.1 \text{ kg/min})(1 \text{min/60 s})(1000 \text{ mm/m})^2}{(1.23 \text{ kg/m}^3)[\pi(60 \text{ mm})^2/4]}$$

$$= 0.479 \text{ m/s} \qquad (4)$$

For the average velocity at section (2), $\overline{V}_2$,

$$\overline{V}_2 = \frac{(0.1 \text{ kg/min})(1 \text{ min/60 s})(1000 \text{ mm/m})^2}{(1.23 \text{ kg/m}^3)[\pi(30 \text{ mm})^2/4]}$$

$$= 1.92 \text{ m/s} \qquad (5)$$

(a)   For the assumed uniform velocity profiles ($\alpha_1 = \alpha_2 = 1.0$), Eq. 2 yields

$$\text{loss} = w_{\substack{\text{shaft} \\ \text{net in}}} - \left(\frac{p_2 - p_1}{\rho}\right) + \frac{\overline{V}_1^2}{2} - \frac{\overline{V}_2^2}{2} \qquad (6)$$

Using Eqs. 3, 4, and 5 and the pressure rise given in the problem statement, Eq. 6 gives

$$\text{loss} = 84.0\,\frac{\text{N·m}}{\text{kg}} - \frac{(0.1\text{ kPa})(1000\text{ Pa/kPa})(1\text{ N/m}^2/\text{Pa})}{1.23\text{ kg/m}^3}$$
$$+ \frac{(0.479\text{ m/s})^2}{2[1\text{ (kg·m)/(N·s}^2)]} - \frac{(1.92\text{ m/s})^2}{2[1\text{ (kg·m)/(N·s}^2)]}$$

or

$$\text{loss} = 84.0\text{ N·m/kg} - 81.3\text{ N·m/kg}$$
$$+ 0.115\text{ N·m/kg} - 1.84\text{ N·m/kg}$$
$$= 0.975\text{ N·m/kg} \qquad \textbf{(Ans)}$$

(b) For the actual velocity profiles ($\alpha_1 = 2$, $\alpha_2 = 1.08$), Eq. 1 gives

$$\text{loss} = w_{\substack{\text{shaft} \\ \text{net in}}} - \left(\frac{p_2 - p_1}{\rho}\right) + \alpha_1\frac{\overline{V}_1^2}{2} - \alpha_2\frac{\overline{V}_2^2}{2} \qquad (7)$$

If we use Eqs. 3, 4, and 5 and the given pressure rise, Eq. 7 yields

$$\text{loss} = 84\text{ N·m/kg} - \frac{(0.1\text{ kPa})(1000\text{ Pa/kPa})(1\text{ N/m}^2/\text{Pa})}{1.23\text{ kg/m}^3}$$
$$+ \frac{2(0.479\text{ m/s})^2}{2[1\text{ (kg·m)/(N·s}^2)]} - \frac{1.08(1.92\text{ m/s})^2}{2[1\text{ (kg·m)/(N·s}^2)]}$$

or

$$\text{loss} = 84.0\text{ N·m/kg} - 81.3\text{ N·m/kg}$$
$$+ 0.230\text{ N·m/kg} - 1.99\text{ N·m/kg}$$
$$= 0.940\text{ N·m/kg} \qquad \textbf{(Ans)}$$

COMMENT The difference in loss calculated assuming uniform velocity profiles and actual velocity profiles is not large compared to $w_{\text{shaft net in}}$ for this fluid flow situation.

5.4   Chapter Summary and Study Guide

conservation
  of mass
continuity equation
mass flowrate
linear momentum
  equation
moment-of-
  momentum
  equation
shaft power
shaft torque
first law of
  thermo-
  dynamics
heat transfer rate
energy equation
loss
turbine and
  pump head
head loss
kinetic energy
  coefficient

In this chapter the flow of a fluid is analyzed by using the principles of conservation of mass, momentum, and energy as applied to control volumes. The Reynolds transport theorem is used to convert basic system-orientated conservation laws into the corresponding control volume formulation.

The continuity equation, a statement of the fact that mass is conserved, is obtained in a form that can be applied to any flows—steady or unsteady, incompressible or compressible. Simplified forms of the continuity equation are used to solve problems dealing with mass and volume flowrates.

The linear momentum equation, a form of Newton's second law of motion applicable to flow through a control volume, is obtained and used to solve flow problems dealing with surface and body forces acting on the fluid and its surroundings. The net flux of momentum through the control surface is directly related to the net force exerted on the contents of the control volume.

The moment-of-momentum equation, which involves the relationship between torque and angular momentum flowrate, is obtained and used to solve flow problems dealing with turbines (that remove energy from a fluid) and pumps (that supply energy to a fluid).

The steady-state energy equation, obtained from the first law of thermodynamics, is written in several forms. The first (Eq. 5.44) involves terms such as the mass flowrate, internal energy per unit mass, heat transfer rate, and shaft work rate. The second form (Eq. 5.56 or 5.57) is termed the mechanical energy equation or the extended Bernoulli equation. It consists of the Bernoulli equation with extra terms that account for losses due to friction in the flow, as well as terms accounting for the existence of pumps or turbines in the flow.

The following checklist provides a study guide for this chapter. When your study of the entire chapter and end-of-chapter exercises has been completed you should be able to

■ write out meanings of the terms listed here in the margin and understand each of the related concepts. These terms are particularly important and are set in color and bold type in the text.

- select an appropriate control volume for a given problem and draw an accurately labeled control volume diagram.
- use the continuity equation and a control volume to solve problems involving mass or volume flowrate.
- use the linear momentum equation and a control volume, in conjunction with the continuity equation as necessary, to solve problems involving forces and momentum flowrate.
- use the moment-of-momentum equation to solve problems involving torque and angular momentum flowrate.
- use the energy equation, in one of its appropriate forms, to solve problems involving losses due to friction (head loss) and energy input by pumps or removal by turbines.
- use the kinetic energy coefficient in the energy equation when solving problems involving nonuniform flows.

Review Problems

Go to Appendix F for a set of review problems with answers. Detailed solutions can be found in *Student Solution Manual for a Brief Introduction to Fluid Mechanics,* by Young et al. (© 2006 John Wiley and Sons, Inc.).

Problems

Note: Unless otherwise indicated, use the values of fluid properties found in the tables on the inside of the front cover. Problems designated with an (*) are intended to be solved with the aid of a programmable calculator or a computer. Problems designated with a (†) are "open-ended" problems and require critical thinking in that to work them one must make various assumptions and provide the necessary data. There is not a unique answer to these problems.

Answers to the even-numbered problems are listed at the end of the book. Access to the videos that accompany problems can be obtained through the book's web site, www.wiley.com/college/young. The lab-type problems and FE problems can also be accessed on this web site.

Section 5.1 Conservation of Mass—Uniform Flow

5.1   Water enters a cylindrical tank through two pipes at rates of 250 and 100 gal/min (see Fig. P5.1). If the level of the water in the tank remains constant, calculate the average velocity of the flow leaving the tank through an 8-in. inside-diameter pipe.

5.2   A hydroelectric turbine passes 2 million gal/min through its blades. If the average velocity of the flow in the circular cross-section conduit leading to the turbine is not to exceed 30 ft/s, determine the minimum allowable diameter of the conduit.

5.3   Air flows steadily between two cross sections in a long, straight section of 0.25-m inside-diameter pipe. The static temperature and pressure at each section are indicated in Fig. P5.3. If the average air velocity at section (2) is 320 m/s, determine the average air velocity at section (1).

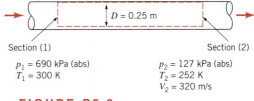

Section (1)   Section (2)
p_1 = 690 kPa (abs) p_2 = 127 kPa (abs)
T_1 = 300 K T_2 = 252 K
V_2 = 320 m/s

■ FIGURE P5.3

5.4   The wind blows through a 7×10-ft garage door with a speed of 2 ft/s as shown in Fig. P5.4. Determine the average speed, V, of the air through the two 3×4-ft windows.

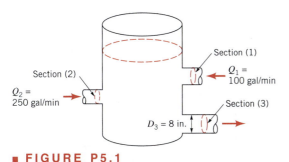

Section (1)

$Q_1 =$ 100 gal/min

Section (2)

$Q_2 =$ 250 gal/min

Section (3)

D_3 = 8 in.

■ FIGURE P5.1

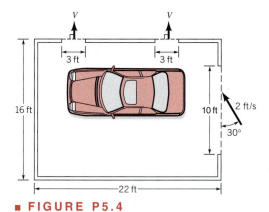

■ FIGURE P5.4

5.5 Water flows out through a set of thin, closely spaced blades as shown in Fig. P5.5 with a speed of $V = 10$ ft/s around the entire circumference of the outlet. Determine the mass flowrate through the inlet pipe.

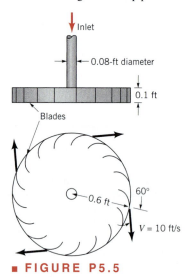

■ FIGURE P5.5

5.6 Water flows into a sink as shown in **Video V5.1** and Fig. P5.6 at a rate of 2 gallons per minute. Determine the average

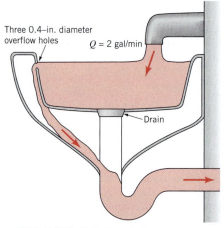

■ FIGURE P5.6

velocity through each of the three 0.4-in.-diameter overflow holes if the drain is closed and the water level in the sink remains constant.

5.7 An evaporative cooling tower (see Fig. P5.7) is used to cool water from 110 to 80°F. Water enters the tower at a rate of 250,000 lb/hr. Dry air (no water vapor) flows into the tower at a rate of 151,000 lb/hr. If the rate of wet air flow out of the tower is 156,900 lb/hr, determine the rate of water evaporation in lb/hr and the rate of cooled water flow in lb/hr.

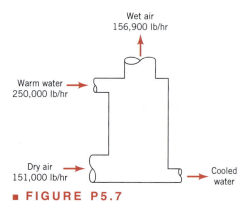

■ FIGURE P5.7

5.8 Water flows into a rain gutter on a house as shown in Fig. P5.8 and in **Video V10.3** at a rate of 0.0040 ft³/s per foot of length of the gutter. At the beginning of the gutter ($x = 0$), the water depth is zero. **(a)** If the water flows with a velocity of 1.0 ft/s throughout the entire gutter, determine an equation for the water depth, h, as a function of location, x. **(b)** At what location will the gutter overflow?

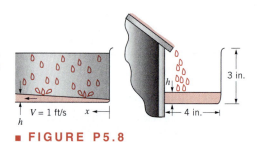

■ FIGURE P5.8

5.9 It takes you 1 min to fill your car's fuel tank with 8.8 gal of gasoline. What is the average velocity of the gasoline leaving the 0.60-in.-diameter nozzle at this pump?

5.10 Various types of attachments can be used with the shop vac shown in **Video V5.2.** Two such attachments are shown in Fig. P5.10—a nozzle and a brush. The flowrate is 1 ft³/s. **(a)** Determine the average velocity through the nozzle entrance, V_n. **(b)** Assume the air enters the brush attachment in a radial direction all around the brush with a velocity profile that varies linearly from 0 to V_b along the length of the bristles as shown in the figure. Determine the value of V_b.

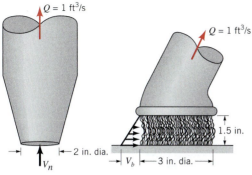

■ **FIGURE P5.10**

5.11 An appropriate turbulent pipe flow velocity profile is

$$\mathbf{V} = u_c\left(\frac{R - r}{R}\right)^{1/n}\hat{\mathbf{i}}$$

where u_c = centerline velocity, r = local radius, R = pipe radius, and $\hat{\mathbf{i}}$ = unit vector along pipe centerline. Determine the ratio of average velocity, $\bar{u}$, to centerline velocity, u_c, for **(a)** $n = 5$, **(b)** $n = 6$, **(c)** $n = 7$, **(d)** $n = 8$, **(e)** $n = 9$, **(f)** $n = 10$.

5.12 A water jet pump (see Fig. P5.12) involves a jet cross-sectional area of 0.01 m², and a jet velocity of 30 m/s. The jet is surrounded by entrained water. The total cross-sectional area associated with the jet and entrained streams is 0.075 m². These two fluid streams leave the pump thoroughly mixed with an average velocity of 6 m/s through a cross-sectional area of 0.075 m². Determine the pumping rate (i.e., the entrained fluid flowrate) involved in liters/s.

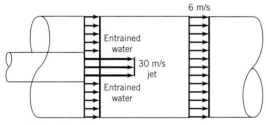

■ **FIGURE P5.12**

Section 5.1 Conservation of Mass—Unsteady Flow

5.13 (See Fluids in the News article titled "New 1.6 gpf standards," Section 5.1.2.) When a toilet is flushed, the water depth, h, in the tank as a function of time, t, is as given in the table. The size of the rectangular tank is 19 in. by 7.5 in. **(a)** Determine the volume of water used per flush, gpf. **(b)** Plot the flowrate for $0 \le t \le 6$ s.

t (s)	h (in.)
0	5.70
0.5	5.33
1.0	4.80
2.0	3.45
3.0	2.40
4.0	1.50
5.0	0.75
6.0	0

5.14 The Hoover Dam (see **Video V2.3**) backs up the Colorado River and creates Lake Meade, which is approximately 115 miles long and has a surface area of approximately 225 square miles. If during flood conditions the Colorado River flows into the lake at a rate of 45,000 cfs and the outflow from the dam is 8000 cfs, how many feet per 24-hr day will the lake level rise?

Section 5.2.1 Linear Momentum—Uniform Flow
(also see Lab Problems 5.74, 5.75, 5.76, and 5.77)

5.15 Exhaust (assumed to have the properties of standard air) leaves the 4-ft-diameter chimney shown in **Video V5.3** and Fig. P5.15 with a speed of 6 ft/s. Because of the wind, after a few diameters downstream the exhaust flows in a horizontal direction with the speed of the wind, 15 ft/s. Determine the horizontal component of the force that the blowing wind puts on the exhaust gases.

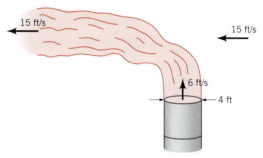

■ **FIGURE P5.15**

5.16 Water enters the horizontal, circular cross-sectional, sudden contraction nozzle sketched in Fig. P5.16 at section (1) with a uniformly distributed velocity of 25 ft/s and a pressure of 75 psi. The water exits from the nozzle into the atmosphere at section (2) where the uniformly distributed velocity is 100 ft/s. Determine the axial component of the anchoring force required to hold the contraction in place.

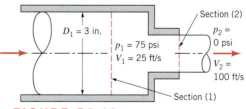

■ **FIGURE P5.16**

5.17 Water flows as two free jets from the tee attached to the pipe shown in Fig. P5.17. The exit speed is 15 m/s. If viscous effects and gravity are negligible, determine the x and y components of the force that the pipe exerts on the tee.

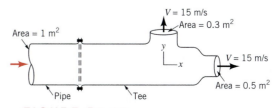

■ **FIGURE P5.17**

5.18 A nozzle is attached to a vertical pipe and discharges water into the atmosphere as shown in Fig. P5.18. When the discharge is 0.1 m³/s, the gage pressure at the flange is 40 kPa. Determine the vertical component of the anchoring force required to hold the nozzle in place. The nozzle has a weight of 200 N, and the volume of water in the nozzle is 0.012 m³. Is the anchoring force directed upward or downward?

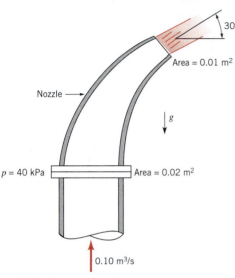

■ **FIGURE P5.18**

5.19 A converging elbow (see Fig. P5.19) turns water through an angle of 135° in a vertical plane. The flow cross-sectional diameter is 400 mm at the elbow inlet, section (1), and 200 mm at the elbow outlet, section (2). The elbow flow passage volume is 0.2 m³ between sections (1) and (2). The water volume flowrate is 0.4 m³/s, and the elbow inlet and outlet pressures are 150 and 90 kPa. The elbow mass is 12 kg. Calculate the horizontal (x direction) and vertical (z direction) anchoring forces required to hold the elbow in place.

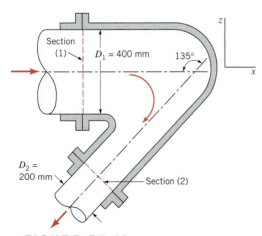

■ **FIGURE P5.19**

5.20 A circular plate having a diameter of 300 mm is held perpendicular to an axisymmetric horizontal jet of air having a velocity of 40 m/s and a diameter of 80 mm as shown in Fig. P5.20. A

hole at the center of the plate results in a discharge jet of air having a velocity of 40 m/s and a diameter of 20 mm. Determine the horizontal component of force required to hold the plate stationary.

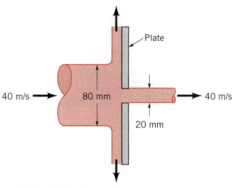

■ **FIGURE P5.20**

5.21 (See Fluids in the News article titled "Where the plume goes," Section 5.2.2.) Air flows into the jet engine shown in Fig. P5.21 at a rate of 9 slugs/s and a speed of 300 ft/s. Upon landing, the engine exhaust exits through the reverse thrust mechanism with a speed of 900 ft/s in the direction indicated. Determine the reverse thrust applied by the engine to the airplane. Assume that the inlet and exit pressures are atmospheric and that the mass flowrate of fuel is negligible compared to the air flowrate through the engine.

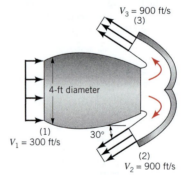

■ **FIGURE P5.21**

5.22 Water flows steadily into and out of a tank that sits on frictionless wheels as shown in Fig. P5.22. Determine the diameter D so that the tank remains motionless if $F = 0$.

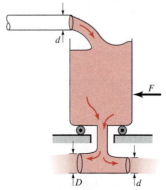

■ **FIGURE P5.22**

5.23 (See Fluids in the News article titled "Motorized surfboard," Section 5.2.2.) The thrust to propel the powered surfboard shown in Fig. P5.23 (or a jet ski; see **Video V 9.7**) is a result of water pumped through the board that exits as a high-speed 2.75-in.-diameter jet. Determine the flowrate and the velocity of the exiting jet if the thrust is to be 300 lb. Neglect the momentum of the water entering the pump.

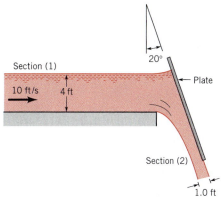

■ FIGURE P5.25

5.26 A vertical, circular cross-sectional jet of air strikes a conical deflector as indicated in Fig. P5.26. A vertical anchoring force of 0.1 N is required to hold the deflector in place. Determine the mass (kg) of the deflector. The magnitude of velocity of the air remains constant.

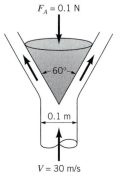

■ FIGURE P5.26

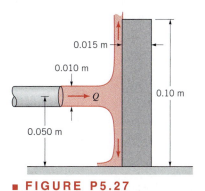

■ FIGURE P5.23

5.24 Determine the magnitude of the horizontal component of the anchoring force per unit width required to hold in place the sluice gate shown in Fig. P5.24. Compare this result with the size of the horizontal component of the anchoring force required to hold in place the sluice gate when it is closed and the depth of water upstream is 6 ft.

5.27 A 10-mm-diameter jet of water is deflected by a homogeneous rectangular block ($15 \times 200 \times 100$ mm) that weighs 6 N as shown in **Video V5.4** and Fig. P5.27. Determine the minimum volume flowrate needed to tip the block.

■ FIGURE P5.24

5.25 Water flows from a two-dimensional open channel and is diverted by an inclined plate as illustrated in Fig. P5.25. When the velocity at section (1) is 10 ft/s, what horizontal force (per unit width) is required to hold the plate in position? At section (1) the pressure distribution is hydrostatic, and the fluid acts as a free jet at section (2). Neglect friction.

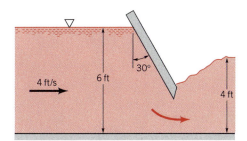

■ FIGURE P5.27

5.28 Air flows into the atmosphere from a nozzle and strikes a vertical plate as shown in Fig. P5.28. A horizontal force of 9 N is required to hold the plate in place. Determine the reading on the pressure gage. Assume the flow to be incompressible and frictionless.

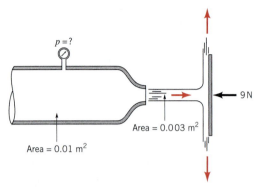

■ **FIGURE P5.28**

5.29 Two water jets of equal size and speed strike each other as shown in Fig. P5.29. Determine the speed, V, and direction, θ, of the resulting combined jet. Gravity is negligible.

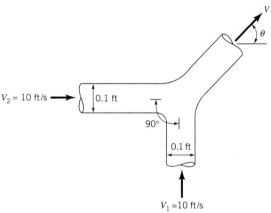

■ **FIGURE P5.29**

5.30 (See Fluids in the News article titled "Bow thrusters," Section 5.2.2.) The bow thruster on the boat shown in Fig. P5.30 is used to turn the boat. The thruster produces a 1-m-diameter jet of water with a velocity of 10 m/s. Determine the force produced by the thruster. Assume that the inlet and outlet pressures are zero and that the momentum of the water entering the thruster is negligible.

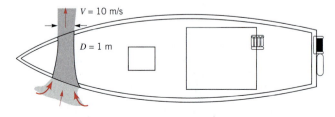

■ **FIGURE P5.30**

5.31 Assuming frictionless, incompressible, one-dimensional flow of water through the horizontal tee connection sketched in Fig. P5.31, estimate values of the x and y components of the force exerted by the tee on the water. Each pipe has an inside diameter of 1 m.

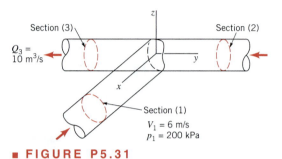

■ **FIGURE P5.31**

5.32 A vertical jet of water leaves a nozzle at a speed of 10 m/s and a diameter of 20 mm. It suspends a plate having a mass of 1.5 kg as indicated in Fig. P5.32. What is the vertical distance h?

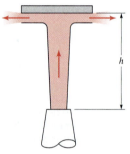

■ **FIGURE P5.32**

†5.33 A table tennis ball "balances" on a jet of air as shown in **Video V3.1** and in Fig. P5.33. Explain why the ball sits at the height it does and why the ball does not "roll off" the jet.

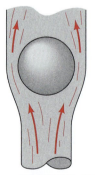

■ **FIGURE P5.33**

5.34 Air discharges from a 2-in.-diameter nozzle and strikes a curved vane, which is in a vertical plane as shown in Fig. P5.34. A stagnation tube connected to a water U-tube manometer is located in the free air jet. Determine the horizontal component of the force that the air jet exerts on the vane. Neglect the weight of the air and all friction.

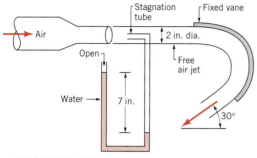

■ **FIGURE P5.34**

Section 5.2.1 Linear Momentum—Nonuniform Flow

5.35 Water is sprayed radially outward over 180° as indicated in Fig. P5.35. The jet sheet is in the horizontal plane. If the jet velocity at the nozzle exit is 30 ft/s, determine the direction and magnitude of the resultant horizontal anchoring force required to hold the nozzle in place.

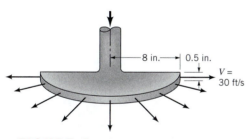

■ **FIGURE P5.35**

5.36 A sheet of water of uniform thickness ($h = 0.01$ m) flows from the device shown in Fig. P5.36. The water enters

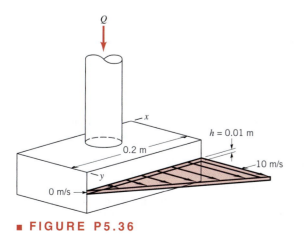

■ **FIGURE P5.36**

vertically through the inlet pipe and exits horizontally with a speed that varies linearly from 0 to 10 m/s along the 0.2-m length of the slit. Determine the y component of the anchoring force necessary to hold this device stationary.

5.37 A variable mesh screen produces a linear and axisymmetric velocity profile as indicated in Fig. P5.37 in the air flow through a 2-ft-diameter circular cross-sectional duct. The static pressures upstream and downstream of the screen are 0.2 and 0.15 psi and are uniformly distributed over the flow cross-sectional-area. Neglecting the force exerted by the duct wall on the flowing air, calculate the screen drag force.

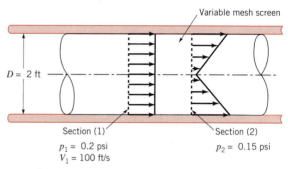

■ **FIGURE P5.37**

Section 5.2.3 Derivation of the Moment-of-Momentum Equation

5.38 Five liters per second of water enter the rotor shown in **Video V5.5** and Fig. P5.38 along the axis of rotation. The cross-sectional area of each of the three nozzle exits normal to the relative velocity is 18 mm². How large is the resisting torque required to hold the rotor stationary if (a) $\theta = 0°$, (b) $\theta = 30°$, and (c) $\theta = 60°$?

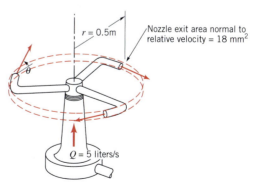

■ **FIGURE P5.38**

5.39 Five liters per second of water enter the rotor shown in **Video V5.5** and Fig. P5.39 along the axis of rotation. The cross-sectional area of each of the three nozzle exits normal to the relative velocity is 18 mm². How fast will the rotor spin steadily if the resisting torque is reduced to zero and (a) $\theta = 0°$, (b) $\theta = 30°$, or (c) $\theta = 60°$?

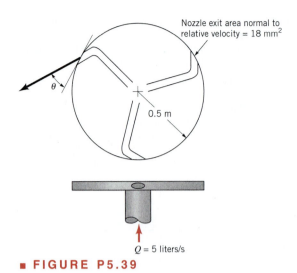

■ **FIGURE P5.39**

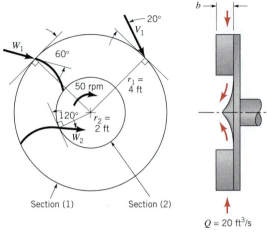

■ **FIGURE P5.41**

5.40 An inward flow radial turbine (see Fig. P5.40) involves a nozzle angle, α_1, of 60° and an inlet rotor tip speed, U_1, of 6 m/s. The ratio of rotor inlet to outlet diameters is 2.0. The absolute velocity leaving the rotor at section (2) is radial with a magnitude of 12 m/s. Determine the energy transfer per unit of mass of fluid flowing through this turbine if the fluid is **(a)** air or **(b)** water.

5.42 A water turbine with radial flow has the dimensions shown in Fig. P5.42. The absolute entering velocity is 15 m/s, and it makes an angle of 30° with the tangent to the rotor. The absolute exit velocity is directed radically inward. The angular speed of the rotor is 30 rpm. Find the power delivered to the shaft of the turbine.

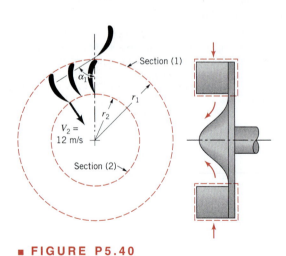

■ **FIGURE P5.40**

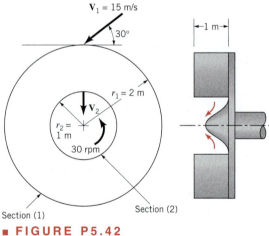

■ **FIGURE P5.42**

5.41 A water turbine wheel rotates at the rate of 50 rpm in the direction shown in Fig. P5.41. The inner radius, r_2, of the blade row is 2 ft, and the outer radius, r_1, is 4 ft. The absolute velocity vector at the turbine rotor entrance makes an angle of 20° with the tangential direction. The inlet blade angle is 60° relative to the tangential direction. The blade outlet angle is 120°. The flowrate is 20 ft³/s. For the flow tangent to the rotor blade surface at inlet and outlet, determine an appropriate constant blade height, b, and the corresponding power available at the rotor shaft.

5.43 A fan (see Fig. P5.43) has a bladed rotor of 12-in. outside diameter and 5-in. inside diameter and runs at 1725 rpm. The width of each rotor blade is 1 in. from blade inlet to outlet. The volume flowrate is steady at 230 ft³/min, and the absolute velocity of the air at blade inlet, V_1, is purely radial. The blade discharge angle is 30° measured with respect to the tangential direction at the outside diameter of the rotor. **(a)** What would be a reasonable blade inlet angle (measured with respect to the tangential direction at the inside diameter of the rotor)? **(b)** Find the power required to run the fan.

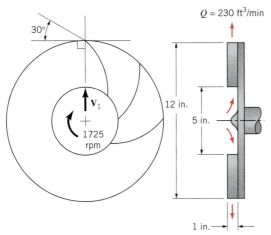

FIGURE P5.43

row is axial viewed from the stationary casing. Is this device a turbine or a pump? Estimate the amount of power transferred to or from the fluid.

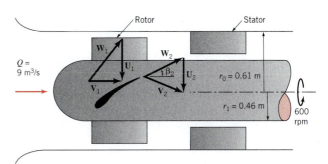

FIGURE P5.45

5.44 An axial flow gasoline pump (see Fig. P5.44) consists of a rotating row of blades (rotor) followed downstream by a stationary row of blades (stator). The gasoline enters the rotor axially (without any angular momentum) with an absolute velocity of 3 m/s. The rotor blade inlet and exit angles are 60° and 45° from axial directions. The pump annulus passage cross-sectional area is constant. Consider the flow as being tangent to the blades involved. Sketch velocity triangles for flow just upstream and downstream of the rotor and just downstream of the stator where the flow is axial. How much energy is added to each kilogram of gasoline?

5.46 By using velocity triangles for flow upstream (1) and downstream (2) of a turbomachine rotor, prove that the shaft work in per unit mass flowing through the rotor is

$$w_{\substack{\text{shaft} \\ \text{net in}}} = \frac{V_2^2 - V_1^2 + U_2^2 - U_1^2 + W_1^2 - W_2^2}{2}$$

where V is absolute flow velocity magnitude, W is relative flow velocity magnitude, and U is blade speed.

5.47 An axial-flow turbomachine rotor involves the upstream (1) and downstream (2) velocity triangles shown in Fig. P5.47. Is this turbomachine a turbine or a fan? Sketch an appropriate blade section and determine the energy transferred per unit mass of fluid.

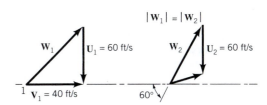

FIGURE P5.47

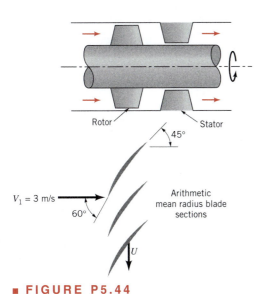

FIGURE P5.44

5.45 The single-stage, axial-flow turbomachine shown in Fig. P5.45 involves water flow at a volumetric flowrate of 9 m³/s. The rotor revolves at 600 rpm. The inner and outer radii of the annular flow path through the stage are 0.46 m and 0.61 m, and $\beta_2 = 30°$. The flow entering the rotor row and leaving the stator

5.48 Water enters an axial-flow turbine rotor with an absolute velocity tangential component, V_θ, of 15 ft/s. The corresponding blade velocity, U, is 50 ft/s. The water leaves the rotor blade row with no angular momentum. If the stagnation pressure drop across the turbine is 12 psi, determine the hydraulic efficiency of the turbine.

5.49 (See Fluids in the News article titled "Tailless helicopters" Section 5.2.4.) Exhaust gas from a tailless helicopter turbojet engine flows through the three 1-ft-diameter rotor blade nozzles shown in Fig. P5.49 at a rate of 500 ft³/s. Determine the angular velocity of the rotor if the torque on the rotor is negligible.

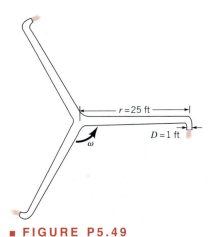

■ FIGURE P5.49

Section 5.3 First Law of Thermodynamics—The Energy Equation

5.50   A 100-ft-wide river with a flowrate of 2400 ft³/s flows over a rock pile as shown in Fig. P5.50. Determine the direction of flow and the head loss associated with the flow across the rock pile.

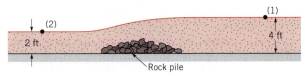

■ FIGURE P5.50

5.51   Air flows past an object in a 2-m-diameter pipe and exits as a free jet as shown in Fig. P5.51. The velocity and pressure upstream are uniform at 10 m/s and 50 N/m², respectively. At the pipe exit the velocity is nonuniform as indicated. The shear stress along the pipe wall is negligible. **(a)** Determine the head loss associated with a particle as it flows from the uniform velocity upstream of the object to a location in the wake at the exit plane of the pipe. **(b)** Determine the force that the air puts on the object.

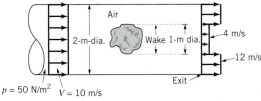

■ FIGURE P5.51

†5.52   Explain how, in terms of the loss of available energy involved, a home water faucet valve works to vary the flow through a lawn hose from the shutoff condition to maximum flow.

5.53   A water siphon having a constant inside diameter of 3 in. is arranged as shown in Fig. P5.53. If the friction loss between A and B is $0.6V^2/2$, where V is the velocity of flow in the siphon, determine the flowrate involved.

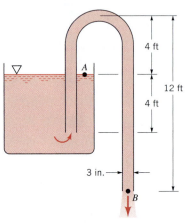

■ FIGURE P5.53

5.54   Water flows through a valve (see Fig. P5.54) with a weight flowrate, $\dot{m}g$, of 1000 lb/s. The pressure just upstream of the valve is 90 psi, and the pressure drop across the valve is 5 psi. The inside diameters of the valve inlet and exit pipes are 12 and 24 in. If the flow through the valve occurs in a horizontal plane, determine the loss in available energy across the valve.

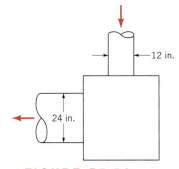

■ FIGURE P5.54

5.55   Water flows steadily from one location to another in the inclined pipe shown in Fig. P5.55. At one section, the

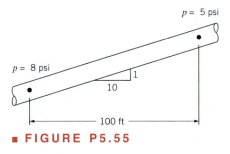

■ FIGURE P5.55

static pressure is 8 psi. At the other section, the static pressure is 5 psi. Which way is the water flowing? Explain.

5.56 (See Fluids in the News article titled "Smart shocks," Section 5.3.3.) A 200-lb force applied to the end of the piston of the shock absorber shown in Fig. P5.56 causes the two ends of the shock absorber to move toward each other with a speed of 5 ft/s. Determine the head loss associated with the flow of the oil through the channel. Neglect gravity and any friction force between the piston and cylinder walls.

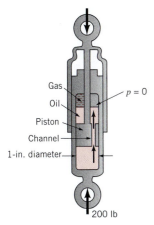

■ **FIGURE P5.56**

Section 5.3 The Energy Equation Involving a Pump or Turbine

5.57 What is the maximum possible power output of the hydroelectric turbine shown in Fig. P5.57?

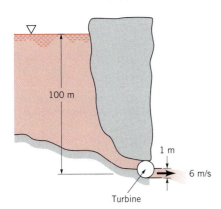

■ **FIGURE P5.57**

5.58 A hydroelectric turbine passes 4 million gal/min across a head of 100 ft of water. What is the maximum amount of power output possible? Why will the actual amount be less?

5.59 A hydraulic turbine is provided with 4.25 m³/s of water at 415 kPa. A vacuum gage in the turbine discharge 3 m be-

low the turbine inlet centerline reads 250 mm Hg vacuum. If the turbine shaft output power is 1100 kW, calculate the power loss through the turbine. The supply and discharge pipe inside diameters are identically 800 mm.

5.60 Water is supplied at 150 ft³/s and 60 psi to a hydraulic turbine through a 3-ft inside-diameter inlet pipe as indicated in Fig. P5.60. The turbine discharge pipe has a 4-ft inside diameter. The static pressure at section (2), 10 ft below the turbine inlet, is 10 in. Hg vacuum. If the turbine develops 2500 hp, determine the rate of loss of available energy between sections (1) and (2).

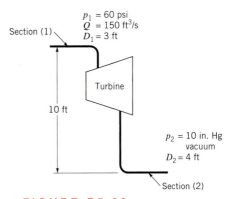

■ **FIGURE P5.60**

5.61 Water is pumped from a tank, point (1), to the top of a water plant aerator, point (2), as shown in **Video V5.8** and Fig. P5.61, at a rate of 3.0 ft³/s. **(a)** Determine the power that the pump adds to the water if the head loss from (1) to (2) where $V_2 = 0$ is 4 ft. **(b)** Determine the head loss from (2) to the bottom of the aerator column, point (3), if the average velocity at (3) is $V_3 = 2$ ft/s.

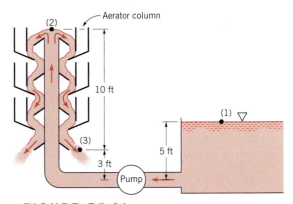

■ **FIGURE P5.61**

5.62 Water is pumped from the tank shown in Fig. P5.62a. The head loss is known to be 1.2 $V^2/2g$, where V is the average velocity in the pipe. According to the pump manufacturer, the relationship between the pump head and the flowrate is as shown in Fig. P5.62b: $h_p = 20 - 2000\ Q^2$, where h_p is in meters and Q is in m³/s. Determine the flowrate, Q.

loss in available energy associated with this flow? If this same amount of loss is associated with pumping the fluid from the lower lake to the higher one at the same flowrate, estimate the amount of pumping power required.

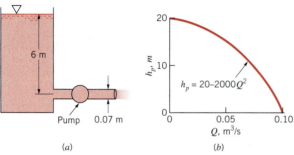

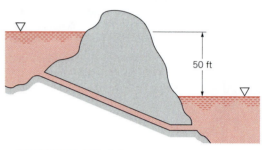

(a)   (b)

■ **FIGURE P5.62**

■ **FIGURE P5.64**

5.63 A pump transfers water from the upper reservoir to the lower one as shown in Fig. P5.63a. The difference in elevation between the two reservoirs is 100 ft. The friction head loss in the piping is given by $K_L \bar{V}^2/2g$, where $\bar{V}$ is the average fluid velocity in the pipe and K_L is the loss coefficient, which is considered constant. The relation between the head added to the water by the pump and the flowrate, Q, through the pump is given in Fig. P5.63b. If $K_L = 20$, and the pipe diameter is 4 in., what is the flowrate through the pump?

5.65 The turbine shown in Fig. P5.65 develops 100 hp when the flowrate of water is 20 ft³/s. If all losses are negligible, determine **(a)** the elevation h, **(b)** the pressure difference across the turbine, and **(c)** the flowrate expected if the turbine were removed.

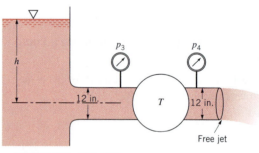

■ **FIGURE P5.65**

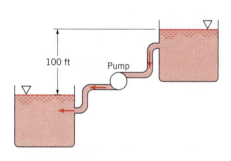

(a)

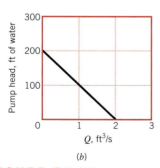

(b)

■ **FIGURE P5.63**

5.66 Gasoline ($SG = 0.68$) flows through a pump at 0.12 m³/s as indicated in Fig. P5.66. The loss between sections

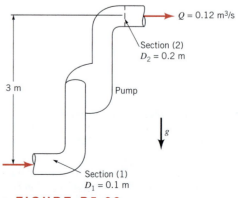

■ **FIGURE P5.66**

5.64 Water flows by gravity from one lake to another as sketched in Fig. P5.64 at the steady rate of 100 gpm. What is the

(1) and (2) is equal to $0.3V_1^2/2$. What will the difference in pressures between sections (1) and (2) be if 20 kW is delivered by the pump to the fluid?

5.67 Oil ($SG = 0.88$) flows in an inclined pipe at a rate of 5 ft³/s as shown in Fig. P5.67. If the differential reading in the mercury manometer is 3 ft, calculate the power that the pump supplies to the oil if head losses are negligible.

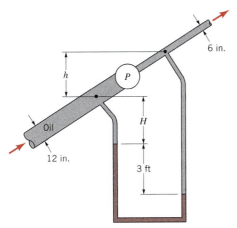

■ **FIGURE P5.67**

5.68 Water is to be moved from one large reservoir to another at a higher elevation as indicated in Fig. P5.68. The loss in available energy associated with 2.5 ft³/s being pumped from sections (1) to (2) is $61\overline{V}^2/2$ ft²/s², where $\overline{V}$ is the average velocity of water in the 8-in. inside-diameter piping involved. Determine the amount of shaft power required.

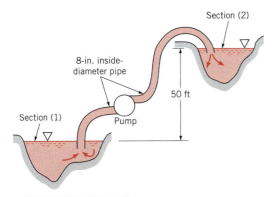

■ **FIGURE P5.68**

5.69 The pumper truck shown in Fig. P5.69 is to deliver 1.5 ft³/s to a maximum elevation of 60 ft above the hydrant. The pressure at the 4-in.-diameter outlet of the hydrant is 10 psi. If head losses are negligibly small, determine the power that the pump must add to the water.

■ **FIGURE P5.69**

5.70 (See Fluids in the News article titled "Curtain of air," Section 5.3.3.) The fan shown in Fig. P5.70 produces an air curtain to separate a loading dock from a cold storage room. The air curtain is a jet of air 10 ft wide, 0.5-ft thick moving with speed $V = 30$ ft/s. The loss associated with this flow is $loss = K_L V^2/2$, where $K_L = 5$. How much power must the fan supply to the air to produce this flow?

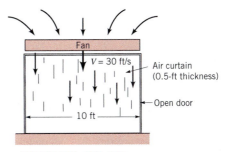

■ **FIGURE P5.70**

Section 5.3 The Energy and Linear Momentum Equations

5.71 Water flows steadily down the inclined pipe as indicated in Fig. P5.71. Determine the following: **(a)** the difference

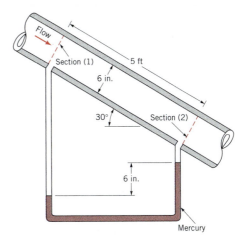

■ **FIGURE P5.71**

in pressure $p_1 - p_2$, **(b)** the loss between sections (1) and (2), and **(c)** the net axial force exerted by the pipe wall on the flowing water between sections (1) and (2).

5.72 Water flows through a 2-ft-diameter pipe arranged horizontally in a circular arc as shown in Fig. P5.72. If the pipe discharges to the atmosphere ($p = 14.7$ psia), determine the x and y components of the resultant force needed to hold the piping between sections (1) and (2) stationary. The steady flowrate is 3000 ft³/min. The loss in pressure due to fluid friction between sections (1) and (2) is 25 psi.

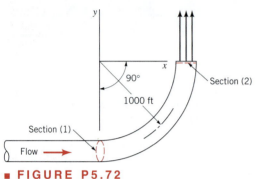

■ FIGURE P5.72

5.73 When fluid flows through an abrupt expansion as indicated in Fig. P5.73, the loss in available energy across the expansion, loss_{ex}, is often expressed as

$$\text{loss}_{ex} = \left(1 - \frac{A_1}{A_2}\right)^2 \frac{V_1^2}{2}$$

where A_1 is the cross-sectional area upstream of expansion, A_2 is the cross-sectional area downstream of expansion, and V_1 is the velocity of flow upstream of expansion. Derive this relationship.

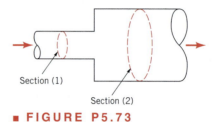

■ FIGURE P5.73

Lab Problems

5.74 This problem involves the force that a jet of air exerts on a flat plate as the air is deflected by the plate. To proceed with this problem, go to the book's web site, www.wiley.com/college/young.

5.75 This problem involves the pressure distribution produced on a flat plate that deflects a jet of air. To proceed with this problem, go to the book's web site, www.wiley.com/college/young.

5.76 This problem involves the force that a jet of water exerts on a vane when the vane turns the jet through a given angle. To proceed with this problem, go to the book's web site, www.wiley.com/college/young.

5.77 This problem involves the force needed to hold a pipe elbow stationary. To proceed with this problem, go to the book's web site, www.wiley.com/college/young.

FE Exam Problems

Sample FE (Fundamentals of Engineering) exam questions for fluid mechanics are provided on the book's web site, www.wiley.com/college/young.

CHAPTER 6

Differential Analysis of Fluid Flow

Flow past an inclined plate: The streamlines of a viscous fluid flowing slowly past a two-dimensional object placed between two closely spaced plates (a Hele–Shaw cell) approximate inviscid, irrotational (potential) flow. (Dye in water between glass plates spaced 1 mm apart) (Photography courtesy of D. H. Peregrine.)

LEARNING OBJECTIVES

After completing this chapter, you should be able to:

- determine various kinematic elements of the flow given the velocity field.
- explain the conditions necessary for a velocity field to satisfy the continuity equation.
- apply the concepts of stream function and velocity potential.
- characterize simple potential flow fields.
- analyze certain types of flows using the Navier–Stokes equations.

The previous chapter focused attention on the use of finite control volumes for the solution of a variety of fluid mechanics problems. This approach is very practical and useful, as it does not generally require a detailed knowledge of the pressure and velocity variations within the control volume. Typically, we found that only conditions on the surface of the control volume entered the problem, and thus, problems could be solved without a detailed knowledge of the flow field. Unfortunately, many situations arise in which the details of the flow are important and the finite control volume approach will not yield the desired information. For example, we may need to know how the velocity varies over the cross section of a pipe or how the pressure and shear stress vary along the surface of an airplane wing.

In these circumstances we need to develop relationships that apply at a point, or at least in a very small region (infinitesimal volume), within a given flow field. This approach, which involves an *infinitesimal control volume,* as distinguished from a finite control volume, is commonly referred to as *differential analysis,* since (as we will soon discover) the governing equations are differential equations.

We begin our introduction to differential analysis by reviewing and extending some of the ideas associated with fluid kinematics that were introduced in Chapter 4. With this background the remainder of the chapter will be devoted to the derivation of the basic differential equations (which will be based on the principle of conservation of mass and Newton's second law of motion) and to some applications.

6.1 Fluid Element Kinematics

In this section we will be concerned with the mathematical description of the motion of fluid elements moving in a flow field. A small fluid element in the shape of a cube, which is initially in one position, will move to another position during a short time interval δt as illustrated in Fig. 6.1. Because of the generally complex velocity variation within the field, we expect the element to not only translate from one position, but to also have its volume changed (linear deformation), to rotate, and to undergo a change in shape (angular deformation). Although these movements and deformations occur simultaneously, we can consider each one separately as illustrated in Fig. 6.1. Because element motion and deformation are intimately related to the velocity and variation of velocity throughout the flow field, we will briefly review the manner in which velocity and acceleration fields can be described.

6.1.1 Velocity and Acceleration Fields Revisited

As discussed in detail in Section 4.1, the velocity field can be described by specifying the velocity **V** at all points, and at all times, within the flow field of interest. Thus, in terms of rectangular coordinates, the notation **V** (x, y, z, t) means that the velocity of a fluid particle depends on where it is located within the flow field (as determined by its coordinates x, y, and z) and when it occupies the particular point (as determined by the time, t). As pointed out in Section 4.1.1, this method of describing fluid motion is called the Eulerian method. It is also convenient to express the velocity in terms of three rectangular components so that

$$\mathbf{V} = u\hat{\mathbf{i}} + v\hat{\mathbf{j}} + w\hat{\mathbf{k}} \tag{6.1}$$

where u, v, and w are the velocity components in the x, y, and z directions, respectively, and $\hat{\mathbf{i}}$, $\hat{\mathbf{j}}$, and $\hat{\mathbf{k}}$ are the corresponding unit vectors. Of course, each of these components will, in general, be a function of x, y, z, and t. One of the goals of differential analysis is to determine how these velocity components specifically depend on x, y, z, and t for a particular problem.

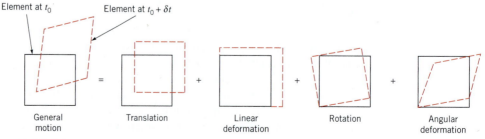

Element at t_0 Element at $t_0 + \delta t$

| General motion | Translation | Linear deformation | Rotation | Angular deformation |

■ **FIGURE 6.1** **Types of motion and deformation for a fluid element.**

With this description of the velocity field it was also shown in Section 4.2.1 that the acceleration of a fluid particle can be expressed as

$$\mathbf{a} = \frac{\partial \mathbf{V}}{\partial t} + u \frac{\partial \mathbf{V}}{\partial x} + v \frac{\partial \mathbf{V}}{\partial y} + w \frac{\partial \mathbf{V}}{\partial z} \tag{6.2}$$

and in component form:

$$a_x = \frac{\partial u}{\partial t} + u \frac{\partial u}{\partial x} + v \frac{\partial u}{\partial y} + w \frac{\partial u}{\partial z} \tag{6.3a}$$

$$a_y = \frac{\partial v}{\partial t} + u \frac{\partial v}{\partial x} + v \frac{\partial v}{\partial y} + w \frac{\partial v}{\partial z} \tag{6.3b}$$

$$a_z = \frac{\partial w}{\partial t} + u \frac{\partial w}{\partial x} + v \frac{\partial w}{\partial y} + w \frac{\partial w}{\partial z} \tag{6.3c}$$

The acceleration is also concisely expressed as

$$\mathbf{a} = \frac{D\mathbf{V}}{Dt} \tag{6.4}$$

where the operator

$$\frac{D(\ )}{Dt} = \frac{\partial(\ )}{\partial t} + u \frac{\partial(\ )}{\partial x} + v \frac{\partial(\ )}{\partial y} + w \frac{\partial(\ )}{\partial z} \tag{6.5}$$

is termed the *material derivative,* or *substantial derivative.* In vector notation

$$\frac{D(\ )}{Dt} = \frac{\partial(\ )}{\partial t} + (\mathbf{V} \cdot \mathbf{\nabla})(\ ) \tag{6.6}$$

where the gradient operator, $\mathbf{\nabla}(\ )$, is

$$\mathbf{\nabla}(\ ) = \frac{\partial(\ )}{\partial x}\hat{\mathbf{i}} + \frac{\partial(\ )}{\partial y}\hat{\mathbf{j}} + \frac{\partial(\ )}{\partial z}\hat{\mathbf{k}} \tag{6.7}$$

which was introduced in Chapter 2. As shown in the following sections, the motion and deformation of a fluid element depend on the velocity field. The relationship between the motion and the forces causing the motion depends on the acceleration field.

6.1.2 Linear Motion and Deformation

The simplest type of motion that a fluid element can undergo is translation, as illustrated in Fig. 6.2. In a small time interval δt a particle located at point O will move to point O' as is

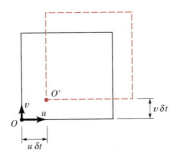

■ **FIGURE 6.2** **Translation of a fluid element.**

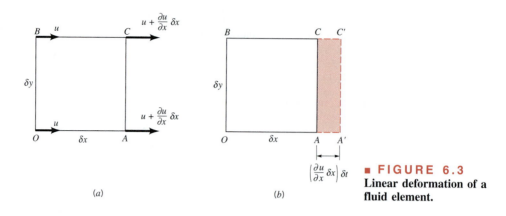

■ **FIGURE 6.3**

Linear deformation of a fluid element.

illustrated in Fig. 6.2. If all points in the element have the same velocity (which is only true if there are no velocity gradients), then the element will simply translate from one position to another. However, because of the presence of velocity gradients, the element will generally be deformed and rotated as it moves. For example, consider the effect of a single velocity gradient $\partial u/\partial x$ on a small cube having sides δx, δy, and δz. As is shown in Fig. 6.3a, if the x component of velocity of O and B is u, then at nearby points A and C the x component of the velocity can be expressed as $u + (\partial u/\partial x)\, \delta x$. This difference in velocity causes a "stretching" of the volume element by an amount $(\partial u/\partial x)(\delta x)(\delta t)$ during the short time interval δt in which line OA stretches to OA' and BC to BC' (Fig. 6.3b). The corresponding change in the original volume, $\delta V = \delta x\, \delta y\, \delta z$, would be

$$\text{Change in } \delta V = \left(\frac{\partial u}{\partial x}\, \delta x \right)(\delta y\, \delta z)(\delta t)$$

and the *rate* at which the volume δV is changing *per unit volume* due to the gradient $\partial u/\partial x$ is

$$\frac{1}{\delta V}\frac{d(\delta V)}{dt} = \lim_{\delta t \to 0}\left[\frac{(\partial u/\partial x)\, \delta t}{\delta t} \right] = \frac{\partial u}{\partial x} \tag{6.8}$$

If velocity gradients $\partial v/\partial y$ and $\partial w/\partial z$ are also present, then using a similar analysis it follows that, in the general case,

$$\frac{1}{\delta V}\frac{d(\delta V)}{dt} = \frac{\partial u}{\partial x} + \frac{\partial v}{\partial y} + \frac{\partial w}{\partial z} = \nabla \cdot \mathbf{V} \tag{6.9}$$

This rate of change of the volume per unit volume is called the ***volumetric dilatation rate.*** Thus, we see that the volume of a fluid may change as the element moves from one location to another in the flow field. However, for an *incompressible fluid* the volumetric dilatation rate is zero, as the element volume cannot change without a change in fluid density (the element mass must be conserved). Variations in the velocity in the direction of the velocity, as represented by the derivatives $\partial u/\partial x$, $\partial v/\partial y$, and $\partial w/\partial z$, simply cause a *linear deformation* of the element in the sense that the shape of the element does not change. Cross derivatives, such as $\partial u/\partial y$ and $\partial v/\partial x$, will cause the element to rotate and generally to undergo an *angular deformation,* which changes the shape of the element.

6.1.3 Angular Motion and Deformation

For simplicity we will consider motion in the x–y plane, but the results can be readily extended to the more general case. The velocity variation that causes rotation and angular deformation is illustrated in Fig. 6.4a. In a short time interval δt the line segments OA and

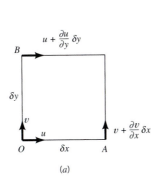

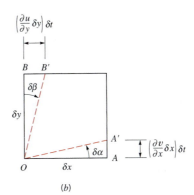

(a)   (b)

■ **FIGURE 6.4**
**Angular motion and
deformation of a fluid
element.**

V6.1 Shear
deformation

OB will rotate through the angles $\delta\alpha$ and $\delta\beta$ to the new positions OA' and OB' as shown in Fig. 6.4b. The angular velocity of line OA, ω_{OA}, is

$$\omega_{OA} = \lim_{\delta t \to 0} \frac{\delta\alpha}{\delta t}$$

For small angles

$$\tan \delta\alpha \approx \delta\alpha = \frac{(\partial v/\partial x)\, \delta x\, \delta t}{\delta x} = \frac{\partial v}{\partial x} \delta t \qquad (6.10)$$

so that

$$\omega_{OA} = \lim_{\delta t \to 0} \left[\frac{(\partial v/\partial x)\, \delta t}{\delta t} \right] = \frac{\partial v}{\partial x}$$

Note that if $\partial v/\partial x$ is positive, ω_{OA} will be counterclockwise. Similarly, the angular velocity of the line OB is

$$\omega_{OB} = \lim_{\delta t \to 0} \frac{\delta\beta}{\delta t}$$

and

$$\tan \delta\beta \approx \delta\beta = \frac{(\partial u/\partial y)\, \delta y\, \delta t}{\delta y} = \frac{\partial u}{\partial y} \delta t \qquad (6.11)$$

so that

$$\omega_{OB} = \lim_{\delta t \to 0} \left[\frac{(\partial u/\partial y)\, \delta t}{\delta t} \right] = \frac{\partial u}{\partial y}$$

In this instance if $\partial u/\partial y$ is positive, ω_{OB} will be clockwise. The *rotation*, ω_z, of the element about the z axis is defined as the average of the angular velocities ω_{OA} and ω_{OB} of the two mutually perpendicular lines OA and OB.[1] Thus, if counterclockwise rotation is considered to be positive, it follows that

$$\omega_z = \frac{1}{2}\left(\frac{\partial v}{\partial x} - \frac{\partial u}{\partial y} \right) \qquad (6.12)$$

[1]With this definition ω_z can also be interpreted to be the angular velocity of the bisector of the angle between the lines OA and OB.

Rotation of the fluid element about the other two coordinate axes can be obtained in a similar manner with the result that for rotation about the x axis

$$\omega_x = \frac{1}{2}\left(\frac{\partial w}{\partial y} - \frac{\partial v}{\partial z}\right) \tag{6.13}$$

and for rotation about the y axis

$$\omega_y = \frac{1}{2}\left(\frac{\partial u}{\partial z} - \frac{\partial w}{\partial x}\right) \tag{6.14}$$

The three components ω_x, ω_y, and ω_z can be combined to give the rotation vector, $\boldsymbol{\omega}$, in the form

$$\boldsymbol{\omega} = \omega_x\hat{\mathbf{i}} + \omega_y\hat{\mathbf{j}} + \omega_z\hat{\mathbf{k}} \tag{6.15}$$

An examination of this result reveals that $\boldsymbol{\omega}$ is equal to one-half the curl of the velocity vector. That is,

$$\boldsymbol{\omega} = \tfrac{1}{2} \operatorname{curl} \mathbf{V} = \tfrac{1}{2}\boldsymbol{\nabla} \times \mathbf{V} \tag{6.16}$$

since by definition of the vector operator $\boldsymbol{\nabla} \times \mathbf{V}$

$$\frac{1}{2}\boldsymbol{\nabla} \times \mathbf{V} = \frac{1}{2}\begin{vmatrix} \hat{\mathbf{i}} & \hat{\mathbf{j}} & \hat{\mathbf{k}} \\ \dfrac{\partial}{\partial x} & \dfrac{\partial}{\partial y} & \dfrac{\partial}{\partial z} \\ u & v & w \end{vmatrix}$$

$$= \frac{1}{2}\left(\frac{\partial w}{\partial y} - \frac{\partial v}{\partial z}\right)\hat{\mathbf{i}} + \frac{1}{2}\left(\frac{\partial u}{\partial z} - \frac{\partial w}{\partial x}\right)\hat{\mathbf{j}} + \frac{1}{2}\left(\frac{\partial v}{\partial x} - \frac{\partial u}{\partial y}\right)\hat{\mathbf{k}}$$

The *vorticity*, $\boldsymbol{\zeta}$, is defined as a vector that is twice the rotation vector; that is,

$$\boldsymbol{\zeta} = 2\,\boldsymbol{\omega} = \boldsymbol{\nabla} \times \mathbf{V} \tag{6.17}$$

The use of the vorticity to describe the rotational characteristics of the fluid simply eliminates the $\left(\frac{1}{2}\right)$ factor associated with the rotation vector.

We observe from Eq. 6.12 that the fluid element will rotate about the z axis as an *undeformed* block (i.e., $\omega_{OA} = -\omega_{OB}$) only when $\partial u/\partial y = -\partial v/\partial x$. Otherwise, the rotation will be associated with an angular deformation. We also note from Eq. 6.12 that when $\partial u/\partial y = \partial v/\partial x$ the rotation around the z axis is zero. More generally if $\boldsymbol{\nabla} \times \mathbf{V} = 0$, then the rotation (and the vorticity) is zero, and flow fields for which this condition applies are termed *irrotational.* We will find in Section 6.4 that the condition of irrotationality often greatly simplifies the analysis of complex flow fields. However, it is probably not immediately obvious why some flow fields would be irrotational, and we will need to examine this concept more fully in Section 6.4.

EXAMPLE 6.1 Vorticity

GIVEN For a certain two-dimensional flow field the velocity is given by the equation

$$\mathbf{V} = (x^2 - y^2)\hat{\mathbf{i}} - 2xy\hat{\mathbf{j}}$$

FIND Is this flow irrotational?

SOLUTION

For an irrotational flow the rotation vector, $\boldsymbol{\omega}$, having the components given by Eqs. 6.12, 6.13, and 6.14 must be zero. For the prescribed velocity field

$$u = x^2 - y^2 \qquad v = -2xy \qquad w = 0$$

and therefore

$$\omega_x = \frac{1}{2}\left(\frac{\partial w}{\partial y} - \frac{\partial v}{\partial z}\right) = 0$$

$$\omega_y = \frac{1}{2}\left(\frac{\partial u}{\partial z} - \frac{\partial w}{\partial x}\right) = 0$$

$$\omega_z = \frac{1}{2}\left(\frac{\partial v}{\partial x} - \frac{\partial u}{\partial y}\right) = \frac{1}{2}\left[(-2y) - (-2y)\right] = 0$$

Thus, the flow is irrotational. **(Ans)**

COMMENTS It is to be noted that for a two-dimensional flow field (where the flow is in the x–y plane) ω_x and ω_y will always be zero, as by definition of two-dimensional flow u and v are not functions of z, and w is zero. In this instance the condition for irrotationality simply becomes $\omega_z = 0$ or $\partial v/\partial x = \partial u/\partial y$.

The streamlines for the steady, two-dimensional flow of this example are shown in Fig. E6.1. (Information about how to calculate streamlines for a given velocity field is given in Sections 4.1.1 and 6.2.3.) It is noted that all of the streamlines (except for the one through the origin) are curved. However, because the flow is irrotational, there is no rotation of the fluid elements. That is, lines OA and OB of Fig. 6.4 rotate with the same speed but in opposite directions.

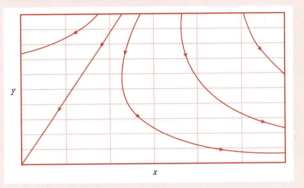

■ **FIGURE E6.1**

In addition to the rotation associated with the derivatives $\partial u/\partial y$ and $\partial v/\partial x$, it is observed from Fig. 6.4b that these derivatives can cause the fluid element to undergo an *angular deformation,* which results in a change in shape of the element. The change in the original right angle formed by the lines OA and OB is termed the shearing strain, $\delta\gamma$, and from Fig. 6.4b

$$\delta\gamma = \delta\alpha + \delta\beta$$

where $\delta\gamma$ is considered to be positive if the original right angle is decreasing. The rate of change of $\delta\gamma$ is called the *rate of shearing strain* or the *rate of angular deformation* and is commonly denoted with the symbol $\dot{\gamma}$. Angles $\delta\alpha$ and $\delta\beta$ are related to the velocity gradients through Eqs. 6.10 and 6.11 so that

$$\dot{\gamma} = \lim_{\delta t \to 0}\frac{\delta\gamma}{\delta t} = \lim_{\delta t \to 0}\left[\frac{(\partial v/\partial x)\,\delta t + (\partial u/\partial y)\,\delta t}{\delta t}\right]$$

and, therefore,

$$\dot{\gamma} = \frac{\partial v}{\partial x} + \frac{\partial u}{\partial y} \tag{6.18}$$

As we will learn in Section 6.7, the rate of angular deformation is related to a corresponding shearing stress, which causes the fluid element to change in shape. From Eq. 6.18 we note that if $\partial u/\partial y = -\partial v/\partial x$, the rate of angular deformation is zero, and this condition corresponds to the case in which the element is simply rotating as an undeformed block (Eq. 6.12). In the remainder of this chapter we will see how the various kinematical relationships developed in this section play an important role in the development and subsequent analysis of the differential equations that govern fluid motion.

6.2 Conservation of Mass

As discussed in Section 5.2, conservation of mass requires that the mass, M, of a system remain constant as the system moves through the flow field. In equation form this principle is expressed as

$$\frac{DM_{\text{sys}}}{Dt} = 0$$

We found it convenient to use the control volume approach for fluid flow problems, with the control volume representation of the conservation of mass written as

$$\frac{\partial}{\partial t} \int_{\text{cv}} \rho \, d\Psi + \sum \rho_{\text{out}} A_{\text{out}} V_{\text{out}} - \sum \rho_{\text{in}} A_{\text{in}} V_{\text{in}} = 0 \qquad \textbf{(6.19)}$$

where this equation (commonly called the *continuity equation*) can be applied to a finite control volume (cv). The first term on the left side of Eq. 6.19 represents the rate at which the mass within the control volume is increasing, and the other terms represent the net rate at which mass is flowing out through the control surface (rate of mass outflow − rate of mass inflow). To obtain the differential form of the continuity equation, Eq. 6.19 is applied to an infinitesimal control volume.

6.2.1 Differential Form of Continuity Equation

We will take as our control volume the small, stationary cubical element shown in Fig. 6.5a. At the center of the element the fluid density is ρ and the velocity has components u, v, and w. Since the element is small the volume integral in Eq. 6.19 can be expressed as

$$\frac{\partial}{\partial t} \int_{\text{cv}} \rho \, d\Psi \approx \frac{\partial \rho}{\partial t} \delta x \, \delta y \, \delta z \qquad \textbf{(6.20)}$$

The rate of mass flow through the surfaces of the element can be obtained by considering the flow in each of the coordinate directions separately. For example, in Fig. 6.5b flow in the x direction is depicted. If we let ρu represent the x component of the mass rate of flow per unit area at the center of the element, then on the right face

$$\rho u \big|_{x+(\delta x/2)} = \rho u + \frac{\partial(\rho u)}{\partial x} \frac{\delta x}{2} \qquad \textbf{(6.21)}$$

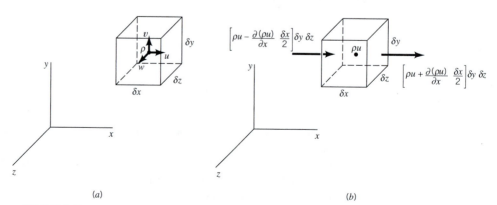

(a)   *(b)*

■ **FIGURE 6.5** **A differential element for the development of conservation of mass equation.**

and on the left face

$$\rho u\big|_{x-(\delta x/2)} = \rho u - \frac{\partial(\rho u)}{\partial x}\frac{\delta x}{2} \tag{6.22}$$

Note that we are really using a Taylor series expansion of ρu and neglecting higher order terms such as $(\delta x)^2$, $(\delta x)^3$, and so on. When the right-hand sides of Eqs. 6.21 and 6.22 are multiplied by the area $\delta y\,\delta z$, the rate at which mass is crossing the right and left sides of the element is obtained, as illustrated in Fig. 6.5b. When these two expressions are combined, the net rate of mass flowing from the element through the two surfaces can be expressed as

$$\begin{aligned}
\text{Net rate of mass} &= \left[\rho u + \frac{\partial(\rho u)}{\partial x}\frac{\delta x}{2}\right]\delta y\,\delta z \\
\text{outflow in } x \text{ direction} & \\
&\quad - \left[\rho u - \frac{\partial(\rho u)}{\partial x}\frac{\delta x}{2}\right]\delta y\,\delta z = \frac{\partial(\rho u)}{\partial x}\delta x\,\delta y\,\delta z
\end{aligned} \tag{6.23}$$

For simplicity, only flow in the x direction has been considered in Fig. 6.5b, but, in general, there will also be flow in the y and z directions. An analysis similar to the one used for flow in the x direction shows that

$$\begin{aligned}
\text{Net rate of mass} &= \frac{\partial(\rho v)}{\partial y}\delta x\,\delta y\,\delta z \\
\text{outflow in } y \text{ direction}
\end{aligned} \tag{6.24}$$

and

$$\begin{aligned}
\text{Net rate of mass} &= \frac{\partial(\rho w)}{\partial z}\delta x\,\delta y\,\delta z \\
\text{outflow in } z \text{ direction}
\end{aligned} \tag{6.25}$$

Thus,

$$\begin{aligned}
\text{Net rate of} &= \left[\frac{\partial(\rho u)}{\partial x} + \frac{\partial(\rho v)}{\partial y} + \frac{\partial(\rho w)}{\partial z}\right]\delta x\,\delta y\,\delta z \\
\text{mass outflow}
\end{aligned} \tag{6.26}$$

From Eqs. 6.19, 6.20, and 6.26 it now follows that the differential equation for the conservation of mass is

$$\boxed{\frac{\partial\rho}{\partial t} + \frac{\partial(\rho u)}{\partial x} + \frac{\partial(\rho v)}{\partial y} + \frac{\partial(\rho w)}{\partial z} = 0} \tag{6.27}$$

As mentioned previously, this equation is also commonly referred to as the continuity equation.

The continuity equation is one of the fundamental equations of fluid mechanics and, as expressed in Eq. 6.27, is valid for steady or unsteady flow, and compressible or incompressible fluids. In vector notation Eq. 6.27 can be written as

$$\frac{\partial\rho}{\partial t} + \nabla \cdot \rho\mathbf{V} = 0 \tag{6.28}$$

Two special cases are of particular interest. For *steady* flow of *compressible* fluids

$$\nabla \cdot \rho\mathbf{V} = 0$$

or

$$\frac{\partial(\rho u)}{\partial x} + \frac{\partial(\rho v)}{\partial y} + \frac{\partial(\rho w)}{\partial z} = 0 \tag{6.29}$$

This follows since by definition ρ is not a function of time for steady flow, but could be a function of position. For *incompressible* fluids the fluid density, ρ, is a constant throughout the flow field so that Eq. 6.28 becomes

$$\nabla \cdot \mathbf{V} = 0 \tag{6.30}$$

or

$$\frac{\partial u}{\partial x} + \frac{\partial v}{\partial y} + \frac{\partial w}{\partial z} = 0 \tag{6.31}$$

Equation 6.31 applies to both steady and unsteady flow of incompressible fluids. Note that Eq. 6.31 is the same as that obtained by setting the volumetric dilatation rate (Eq. 6.9) equal to zero. This result should not be surprising since both relationships are based on conservation of mass for incompressible fluids.

6.2.2 Cylindrical Polar Coordinates

For some problems it is more convenient to express the various differential relationships in cylindrical polar coordinates rather than Cartesian coordinates. As is shown in Fig. 6.6, with cylindrical coordinates a point is located by specifying the coordinates r, θ, and z. The coordinate r is the radial distance from the z axis, θ is the angle measured from a line parallel to the x axis (with counterclockwise taken as positive), and z is the coordinate along the z axis. The velocity components, as sketched in Fig. 6.6, are the radial velocity, v_r, the tangential velocity, v_θ, and the axial velocity, v_z. Thus, the velocity at some arbitrary point P can be expressed as

$$\mathbf{V} = v_r \hat{\mathbf{e}}_r + v_\theta \hat{\mathbf{e}}_\theta + v_z \hat{\mathbf{e}}_z \tag{6.32}$$

where $\hat{\mathbf{e}}_r$, $\hat{\mathbf{e}}_\theta$, and $\hat{\mathbf{e}}_z$ are the unit vectors in the r, θ, and z directions, respectively, as illustrated in Fig. 6.6. The use of cylindrical coordinates is particularly convenient when the boundaries of the flow system are cylindrical. Several examples illustrating the use of cylindrical coordinates will be given in succeeding sections in this chapter.

The differential form of the continuity equation in cylindrical coordinates is

$$\frac{\partial \rho}{\partial t} + \frac{1}{r}\frac{\partial (r\rho v_r)}{\partial r} + \frac{1}{r}\frac{\partial (\rho v_\theta)}{\partial \theta} + \frac{\partial (\rho v_z)}{\partial z} = 0 \tag{6.33}$$

This equation can be derived by following the same procedure used in the preceding section. For steady, compressible flow

$$\frac{1}{r}\frac{\partial (r\rho v_r)}{\partial r} + \frac{1}{r}\frac{\partial (\rho v_\theta)}{\partial \theta} + \frac{\partial (\rho v_z)}{\partial z} = 0 \tag{6.34}$$

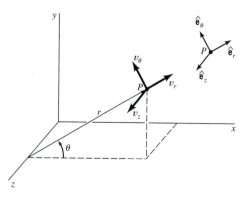

■ **FIGURE 6.6** The representation of velocity components in cylindrical polar coordinates.

For incompressible fluids (for steady or unsteady flow)

$$\frac{1}{r}\frac{\partial(rv_r)}{\partial r} + \frac{1}{r}\frac{\partial v_\theta}{\partial \theta} + \frac{\partial v_z}{\partial z} = 0 \tag{6.35}$$

6.2.3   The Stream Function

Steady, incompressible, plane, two-dimensional flow represents one of the simplest types of flow of practical importance. By plane, two-dimensional flow we mean that there are only two velocity components, such as u and v, when the flow is considered to be in the x–y plane. For this flow the continuity equation, Eq. 6.31, reduces to

$$\frac{\partial u}{\partial x} + \frac{\partial v}{\partial y} = 0 \tag{6.36}$$

We still have two variables, u and v, to deal with, but they must be related in a special way as indicated by Eq. 6.36. This equation suggests that if we define a function $\psi(x, y)$, called the **stream function,** which, as shown by the figure in the margin, relates the velocities as

$$u = \frac{\partial \psi}{\partial y} \qquad v = -\frac{\partial \psi}{\partial x} \tag{6.37}$$

then the continuity equation is identically satisfied. This conclusion can be verified by simply substituting the expressions for u and v into Eq. 6.36 so that

$$\frac{\partial}{\partial x}\left(\frac{\partial \psi}{\partial y}\right) + \frac{\partial}{\partial y}\left(-\frac{\partial \psi}{\partial x}\right) = \frac{\partial^2 \psi}{\partial x\,\partial y} - \frac{\partial^2 \psi}{\partial y\,\partial x} = 0$$

Thus, whenever the velocity components are defined in terms of the stream function we know that conservation of mass will be satisfied. Of course, we still do not know what $\psi(x, y)$ is for a particular problem, but at least we have simplified the analysis by having to determine only one unknown function, $\psi(x, y)$, rather than the two functions, $u(x, y)$ and $v(x, y)$.

Another particular advantage of using the stream function is related to the fact that *lines along which ψ is constant are streamlines.* Recall from Section 4.1.4 that streamlines are lines in the flow field that are everywhere tangent to the velocities, as illustrated in Fig. 6.7. It follows from the definition of the streamline that the slope at any point along a streamline is given by

$$\frac{dy}{dx} = \frac{v}{u}$$

The change in the value of ψ as we move from one point (x, y) to a nearby point $(x + dx, y + dy)$ is given by the relationship:

$$d\psi = \frac{\partial \psi}{\partial x}dx + \frac{\partial \psi}{\partial y}dy = -v\,dx + u\,dy$$

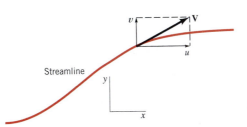

■ **FIGURE 6.7**   **Velocity and velocity components along a streamline.**

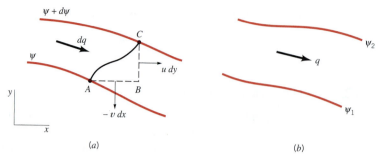

■ **FIGURE 6.8**   **Flow between two streamlines.**

Along a line of constant ψ we have $d\psi = 0$ so that

$$-v\,dx + u\,dy = 0$$

and, therefore, along a line of constant ψ

$$\frac{dy}{dx} = \frac{v}{u}$$

which is the defining equation for a streamline. Thus, if we know the function $\psi(x, y)$ we can plot lines of constant ψ to provide the family of streamlines that are helpful in visualizing the pattern of flow. There are an infinite number of streamlines that make up a particular flow field, since for each constant value assigned to ψ a streamline can be drawn.

The actual numerical value associated with a particular streamline is not of particular significance, but the change in the value of ψ is related to the volume rate of flow. Consider two closely spaced streamlines, as shown in Fig. 6.8a. The lower streamline is designated ψ and the upper one $\psi + d\psi$. Let dq represent the volume rate of flow (per unit width perpendicular to the x–y plane) passing between the two streamlines. Note that flow never crosses streamlines, since by definition the velocity is tangent to the streamline. From conservation of mass we know that the inflow, dq, crossing the arbitrary surface AC of Fig. 6.8a must equal the net outflow through surfaces AB and BC. Thus,

$$dq = u\,dy - v\,dx$$

or in terms of the stream function

$$dq = \frac{\partial \psi}{\partial y}\,dy + \frac{\partial \psi}{\partial x}\,dx \tag{6.38}$$

The right-hand side of Eq. 6.38 is equal to $d\psi$ so that

$$dq = d\psi \tag{6.39}$$

Thus, the volume rate of flow, q, between two streamlines such as ψ_1 and ψ_2 of Fig. 6.8b can be determined by integrating Eq. 6.39 to yield

$$q = \int_{\psi_1}^{\psi_2} d\psi = \psi_2 - \psi_1 \tag{6.40}$$

If the upper streamline, ψ_2, has a value greater than the lower streamline, ψ_1, then q is positive, which indicates that the flow is from left to right. For $\psi_1 > \psi_2$ the flow is from right to left.

In cylindrical coordinates the continuity equation (Eq. 6.35) for incompressible, plane, two-dimensional flow reduces to

$$\frac{1}{r}\frac{\partial(rv_r)}{\partial r} + \frac{1}{r}\frac{\partial v_\theta}{\partial \theta} = 0 \tag{6.41}$$

and as show by the figure in the margin, the velocity components, v_r and v_θ, can be related to the stream function, $\psi(r, \theta)$, through the equations

$$v_r = \frac{1}{r}\frac{\partial \psi}{\partial \theta} \qquad v_\theta = -\frac{\partial \psi}{\partial r} \tag{6.42}$$

Substitution of these expressions for the velocity components into Eq. 6.41 shows that the continuity equation is identically satisfied. The stream function concept can be extended to axisymmetric flows, such as flow in pipes or flow around bodies of revolution, and to two-dimensional compressible flows. However, the concept is not applicable to general three-dimensional flows.

EXAMPLE 6.2   Stream Function

GIVEN The velocity components in a steady, two-dimensional incompressible flow field are

$$u = 2y$$
$$v = 4x$$

FIND

(a) Determine the corresponding stream function and

(b) Show on a sketch several streamlines. Indicate the direction of flow along the streamlines.

SOLUTION

(a) From the definition of the stream function (Eqs. 6.37)

$$u = \frac{\partial \psi}{\partial y} = 2y$$

and

$$v = -\frac{\partial \psi}{\partial x} = 4x$$

The first of these equations can be integrated to give

$$\psi = y^2 + f_1(x)$$

where $f_1(x)$ is an arbitrary function of x. Similarly, from the second equation

$$\psi = -2x^2 + f_2(y)$$

where $f_2(y)$ is an arbitrary function of y. It now follows that in order to satisfy both expressions for the stream function

$$\psi = -2x^2 + y^2 + C \qquad \textbf{(Ans)}$$

where C is an arbitrary constant.

COMMENT Since the velocities are related to the derivatives of the stream function, an arbitrary constant can always be

added to the function, and the value of the constant is actually of no consequence. Usually, for simplicity, we set $C = 0$ so that for this particular example the simplest form for the stream function is

$$\psi = -2x^2 + y^2 \qquad \textbf{(1)} \quad \textbf{(Ans)}$$

Either answer indicated would be acceptable.

(b) Streamlines can now be determined by setting $\psi =$ constant and plotting the resulting curve. With the above expression for ψ (with $C = 0$) the value of ψ at the origin is zero so that the equation of the streamline passing through the origin (the $\psi = 0$ streamline) is

$$0 = -2x^2 + y^2$$

or

$$y = \pm\sqrt{2}x$$

Other streamlines can be obtained by setting ψ equal to various constants. It follows from Eq. 1 that the equations of these streamlines (for $\psi \neq 0$) can be expressed in the form

$$\frac{y^2}{\psi} - \frac{x^2}{\psi/2} = 1$$

which we recognize as the equation of a hyperbola. Thus, the streamlines are a family of hyperbolas with the $\psi = 0$ streamlines as asymptotes. Several of the streamlines are plotted in Fig. E6.2. Since the velocities can be calculated at any point, the direction of flow along a given streamline can be easily deduced. For example, $v = -\partial\psi/\partial x = 4x$ so that $v > 0$ if $x > 0$ and $v < 0$ if $x < 0$. The direction of flow is indicated in Fig. E6.2.

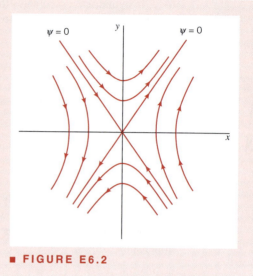

■ **FIGURE E6.2**

6.3 Conservation of Linear Momentum

To develop the differential, linear momentum equations we can start with the linear momentum equation

$$\mathbf{F} = \frac{D\mathbf{P}}{Dt}\bigg|_{\text{sys}} \tag{6.43}$$

where $\mathbf{F}$ is the resultant force acting on a fluid mass, $\mathbf{P}$ is the linear momentum defined as

$$\mathbf{P} = \int_{\text{sys}} \mathbf{V}\, dm$$

and the operator $D(\ )/Dt$ is the material derivative (see Section 4.2.1). In the last chapter it was demonstrated how Eq. 6.43 in the form

$$\Sigma\, \mathbf{F}_{\substack{\text{contents of the} \\ \text{control volume}}} = \frac{\partial}{\partial t}\int_{\text{cv}} \mathbf{V}\rho\, d\forall + \sum \mathbf{V}_{\text{out}}\rho_{\text{out}}A_{\text{out}}V_{\text{out}} - \sum \mathbf{V}_{\text{in}}\rho_{\text{in}}A_{\text{in}}V_{\text{in}} \tag{6.44}$$

could be applied to a finite control volume to solve a variety of flow problems. To obtain the differential form of the linear momentum equation, we can either apply Eq. 6.43 to a differential system, consisting of a mass, δm, or apply Eq. 6.44 to an infinitesimal control volume, $\delta\forall$, which initially bounds the mass δm. It is probably simpler to use the system approach since application of Eq. 6.43 to the differential mass, δm, yields

$$\delta\mathbf{F} = \frac{D(\mathbf{V}\, \delta m)}{Dt}$$

where $\delta\mathbf{F}$ is the resultant force acting on δm. Using this system approach, δm can be treated as a constant so that

$$\delta\mathbf{F} = \delta m\, \frac{D\mathbf{V}}{Dt}$$

But DV/Dt is the acceleration, **a**, of the element. Thus,

$$\delta\mathbf{F} = \delta m \, \mathbf{a} \qquad (6.45)$$

which is simply Newton's second law applied to the mass, δm. This is the same result that would be obtained by applying Eq. 6.44 to an infinitesimal control volume (see Ref. 1). Before we can proceed, it is necessary to examine how the force, $\delta\mathbf{F}$, can be most conveniently expressed.

6.3.1   Description of Forces Acting on Differential Element

In general, two types of forces need to be considered: *surface forces,* which act on the surface of the differential element, and *body forces,* which are distributed throughout the element. For our purpose, the only body force, $\delta\mathbf{F}_b$, of interest is the weight of the element, which can be expressed as

$$\delta\mathbf{F}_b = \delta m \, \mathbf{g} \qquad (6.46)$$

where **g** is the vector representation of the acceleration of gravity. In component form

$$\delta F_{bx} = \delta m \, g_x \qquad (6.47a)$$

$$\delta F_{by} = \delta m \, g_y \qquad (6.47b)$$

$$\delta F_{bz} = \delta m \, g_z \qquad (6.47c)$$

where g_x, g_y, and g_z are the components of the acceleration of gravity vector in the x, y, and z directions, respectively.

Surface forces act on the element as a result of its interaction with its surroundings. At any arbitrary location within a fluid mass, the force acting on a small area, δA, which lies in an arbitrary surface can be represented by $\delta\mathbf{F}_s$, as shown in Fig. 6.9. In general, $\delta\mathbf{F}_s$ will be inclined with respect to the surface. The force $\delta\mathbf{F}_s$ can be resolved into three components, δF_n, δF_1, and δF_2, where δF_n is normal to the area, δA, and δF_1 and δF_2 are parallel to the area and orthogonal to each other. The *normal stress,* σ_n, is defined as

$$\sigma_n = \lim_{\delta A \to 0} \frac{\delta F_n}{\delta A}$$

and the *shearing stresses* are defined as

$$\tau_1 = \lim_{\delta A \to 0} \frac{\delta F_1}{\delta A}$$

and

$$\tau_2 = \lim_{\delta A \to 0} \frac{\delta F_2}{\delta A}$$

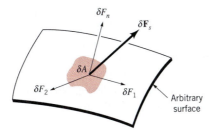

■ **FIGURE 6.9**   Component of force acting on an arbitrary differential area.

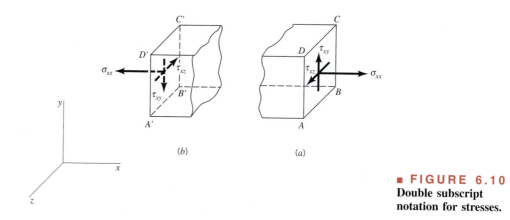

■ **FIGURE 6.10**
Double subscript notation for stresses.

We will use σ for normal stresses and τ for shearing stresses. The intensity of the force per unit area at a point in a body can thus be characterized by a normal stress and two shearing stresses, if the orientation of the area is specified. For purposes of analysis it is usually convenient to reference the area to the coordinate system. For example, for the rectangular coordinate system shown in Fig. 6.10 we choose to consider the stresses acting on planes parallel to the coordinate planes. On the plane $ABCD$ of Fig. 6.10a, which is parallel to the y–z plane, the normal stress is denoted σ_{xx} and the shearing stresses are denoted as τ_{xy} and τ_{xz}. To easily identify the particular stress component we use a double subscript notation. The first subscript indicates the direction of the *normal* to the plane on which the stress acts, and the second subscript indicates the direction of the stress. Thus, normal stresses have repeated subscripts, whereas subscripts for the shearing stresses are always different.

It is also necessary to establish a sign convention for the stresses. We define the positive direction for the stress as the positive coordinate direction on the surfaces for which the outward normal is in the positive coordinate direction. This is the case illustrated in Fig. 6.10a where the outward normal to the area $ABCD$ is in the positive x direction. The positive directions for σ_{xx}, τ_{xy}, and τ_{xz} are as shown in Fig. 6.10a. If the outward normal points in the negative coordinate direction, as in Fig. 6.10b for the area $A'B'C'D'$, then the stresses are considered positive if directed in the negative coordinate directions. Thus, the stresses shown in Fig. 6.10b are considered to be positive when directed as shown. Note that positive normal stresses are tensile stresses; that is, they tend to "stretch" the material.

It should be emphasized that the state of stress at a point in a material is not completely defined by simply three components of a "stress vector." This follows since any particular stress vector depends on the orientation of the plane passing through the point. However, it can be shown that the normal and shearing stresses acting on *any* plane passing through a point can be expressed in terms of the stresses acting on three orthogonal planes passing through the point (Ref. 2).

We now can express the surface forces acting on a small cubical element of fluid in terms of the stresses acting on the faces of the element as shown in Fig. 6.11. It is expected that in general the stresses will vary from point to point within the flow field. Thus, we will express the stresses on the various faces in terms of the corresponding stresses at the center of the element of Fig. 6.11 and their gradients in the coordinate directions. For simplicity, only the forces in the x direction are shown. Note that the stresses must be multiplied by the area on which they act to obtain the force. Summing all these forces in the

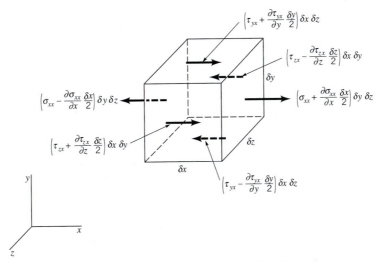

■ FIGURE 6.11 **Surface forces in the x direction acting on a fluid element.**

x direction yields

$$\delta F_{sx} = \left(\frac{\partial \sigma_{xx}}{\partial x} + \frac{\partial \tau_{yx}}{\partial y} + \frac{\partial \tau_{zx}}{\partial z}\right) \delta x \, \delta y \, \delta z \qquad \text{(6.48a)}$$

for the resultant surface force in the x direction. In a similar manner the resultant surface forces in the y and z direction can be obtained and expressed as

$$\delta F_{sy} = \left(\frac{\partial \tau_{xy}}{\partial x} + \frac{\partial \sigma_{yy}}{\partial y} + \frac{\partial \tau_{zy}}{\partial z}\right) \delta x \, \delta y \, \delta z \qquad \text{(6.48b)}$$

and

$$\delta F_{sz} = \left(\frac{\partial \tau_{xz}}{\partial x} + \frac{\partial \tau_{yz}}{\partial y} + \frac{\partial \sigma_{zz}}{\partial z}\right) \delta x \, \delta y \, \delta z \qquad \text{(6.48c)}$$

The resultant surface force can now be expressed as

$$\delta \mathbf{F}_s = \delta F_{sx}\hat{\mathbf{i}} + \delta F_{sy}\hat{\mathbf{j}} + \delta F_{sz}\hat{\mathbf{k}} \qquad \text{(6.49)}$$

and this force combined with the body force, $\delta \mathbf{F}_b$, yields the resultant force, $\delta \mathbf{F}$, acting on the differential mass, δm. That is, $\delta \mathbf{F} = \delta \mathbf{F}_s + \delta \mathbf{F}_b$.

6.3.2 Equations of Motion

The expressions for the body and surface forces can now be used in conjunction with Eq. 6.45 to develop the equations of motion. In component form, Eq. 6.45 can be written as

$$\delta F_x = \delta m \, a_x$$

$$\delta F_y = \delta m \, a_y$$

$$\delta F_z = \delta m \, a_z$$

where $\delta m = \rho \, \delta x \, \delta y \, \delta z$, and the acceleration components are given by Eq. 6.3. It now follows (using Eqs. 6.47 and 6.48 for the forces on the element) that

$$\rho g_x + \frac{\partial \sigma_{xx}}{\partial x} + \frac{\partial \tau_{yx}}{\partial y} + \frac{\partial \tau_{zx}}{\partial z} = \rho \left(\frac{\partial u}{\partial t} + u \frac{\partial u}{\partial x} + v \frac{\partial u}{\partial y} + w \frac{\partial u}{\partial z} \right) \tag{6.50a}$$

$$\rho g_y + \frac{\partial \tau_{xy}}{\partial x} + \frac{\partial \sigma_{yy}}{\partial y} + \frac{\partial \tau_{zy}}{\partial z} = \rho \left(\frac{\partial v}{\partial t} + u \frac{\partial v}{\partial x} + v \frac{\partial v}{\partial y} + w \frac{\partial v}{\partial z} \right) \tag{6.50b}$$

$$\rho g_z + \frac{\partial \tau_{xz}}{\partial x} + \frac{\partial \tau_{yz}}{\partial y} + \frac{\partial \sigma_{zz}}{\partial z} = \rho \left(\frac{\partial w}{\partial t} + u \frac{\partial w}{\partial x} + v \frac{\partial w}{\partial y} + w \frac{\partial w}{\partial z} \right) \tag{6.50c}$$

where the element volume $\delta x \, \delta y \, \delta z$ cancels out.

Equations 6.50 are the general differential equations of motion for a fluid. In fact, they are applicable to any continuum (solid or fluid) in motion or at rest. However, before we can use the equations to solve specific problems, some additional information about the stresses must be obtained. Otherwise, we will have more unknowns (all of the stresses and velocities and the density) than equations. It should not be too surprising that the differential analysis of fluid motion is complicated. We are attempting to describe, in detail, complex fluid motion.

6.4 Inviscid Flow

As discussed in Section 1.6, shearing stresses develop in a moving fluid because of the viscosity of the fluid. We know that for some common fluids, such as air and water, the viscosity is small, and therefore it seems reasonable to assume that under some circumstances we may be able to simply neglect the effect of viscosity (and thus shearing stresses). Flow fields in which the shearing stresses are assumed to be negligible are said to be *inviscid, non-viscous,* or *frictionless.* These terms are used interchangeably. As discussed in Section 2.1, for fluids in which there are no shearing stresses, the normal stress at a point is independent of direction—that is, $\sigma_{xx} = \sigma_{yy} = \sigma_{zz}$. In this instance we define the pressure, p, as the negative of the normal stress so that

$$-p = \sigma_{xx} = \sigma_{yy} = \sigma_{zz}$$

The negative sign is used so that a *compressive* normal stress (which is what we expect in a fluid) will give a *positive* value for p.

In Chapter 3 the inviscid flow concept was used in the development of the Bernoulli equation, and numerous applications of this important equation were considered. In this section we will again consider the Bernoulli equation and will show how it can be derived from the general equations of motion for inviscid flow.

6.4.1 Euler's Equations of Motion

For an inviscid flow in which all the shearing stresses are zero, and the normal stresses are replaced by $-p$, the general equations of motion (Eqs. 6.50) reduce to

$$\rho g_x - \frac{\partial p}{\partial x} = \rho \left(\frac{\partial u}{\partial t} + u \frac{\partial u}{\partial x} + v \frac{\partial u}{\partial y} + w \frac{\partial u}{\partial z} \right) \tag{6.51a}$$

$$\rho g_y - \frac{\partial p}{\partial y} = \rho \left(\frac{\partial v}{\partial t} + u \frac{\partial v}{\partial x} + v \frac{\partial v}{\partial y} + w \frac{\partial v}{\partial z} \right) \tag{6.51b}$$

$$\rho g_z - \frac{\partial p}{\partial z} = \rho \left(\frac{\partial w}{\partial t} + u \frac{\partial w}{\partial x} + v \frac{\partial w}{\partial y} + w \frac{\partial w}{\partial z} \right) \tag{6.51c}$$

These equations are commonly referred to as *Euler's equations of motion,* named in honor of Leonhard Euler, a famous Swiss mathematician who pioneered work on the relationship between pressure and flow. In vector notation Euler's equations can be expressed as

$$\rho \mathbf{g} - \boldsymbol{\nabla} p = \rho \left[\frac{\partial \mathbf{V}}{\partial t} + (\mathbf{V} \cdot \boldsymbol{\nabla}) \mathbf{V} \right] \tag{6.52}$$

Although Eqs. 6.51 are considerably simpler than the general equations of motion, Eqs. 6.50, they are still not amenable to a general analytical solution that would allow us to determine the pressure and velocity at all points within an inviscid flow field. The main difficulty arises from the nonlinear velocity terms (such as $u\, \partial u/\partial x$, $v\, \partial u/\partial y$), which appear in the convective acceleration. Because of these terms, Euler's equations are nonlinear partial differential equations for which we do not have a general method for solving. However, under some circumstances we can use them to obtain useful information about inviscid flow fields. For example, as shown in the following section we can integrate Eq. 6.52 to obtain a relationship (the Bernoulli equation) among elevation, pressure, and velocity along a streamline.

6.4.2 The Bernoulli Equation

In Section 3.2 the Bernoulli equation was derived by a direct application of Newton's second law to a fluid particle moving along a streamline. In this section we will again derive this important equation, starting from Euler's equations. Of course, we should obtain the same result since Euler's equations simply represent a statement of Newton's second law expressed in a general form that is useful for flow problems. We will restrict our attention to steady flow so Euler's equation in vector form becomes

$$\rho \mathbf{g} - \boldsymbol{\nabla} p = \rho (\mathbf{V} \cdot \boldsymbol{\nabla}) \mathbf{V} \tag{6.53}$$

We wish to integrate this differential equation along some arbitrary streamline (Fig. 6.12) and select the coordinate system with the z axis vertical (with "up" being positive) so that the acceleration of gravity vector can be expressed as

$$\mathbf{g} = -g\, \boldsymbol{\nabla} z$$

where g is the magnitude of the acceleration of gravity vector. Also, it will be convenient to use the vector identity

$$(\mathbf{V} \cdot \boldsymbol{\nabla}) \mathbf{V} = \tfrac{1}{2} \boldsymbol{\nabla} (\mathbf{V} \cdot \mathbf{V}) - \mathbf{V} \times (\boldsymbol{\nabla} \times \mathbf{V})$$

Equation 6.53 can now be written in the form

$$-\rho g\, \boldsymbol{\nabla} z - \boldsymbol{\nabla} p = \frac{\rho}{2} \boldsymbol{\nabla} (\mathbf{V} \cdot \mathbf{V}) - \rho (\mathbf{V} \times \boldsymbol{\nabla} \times \mathbf{V})$$

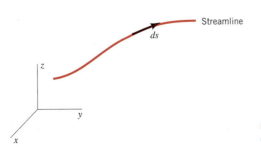

Streamline

ds

z

y

x

■ **FIGURE 6.12** The notation for differential length along a streamline.

and this equation can be rearranged to yield

$$\frac{\nabla p}{\rho} + \frac{1}{2}\nabla(V^2) + g\nabla z = \mathbf{V} \times (\nabla \times \mathbf{V})$$

We next take the dot product of each term with a differential length ds along a streamline (Fig. 6.12). Thus,

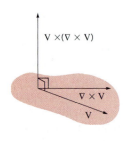

$$\frac{\nabla p}{\rho} \cdot d\mathbf{s} + \frac{1}{2}\nabla(V^2) \cdot d\mathbf{s} + g\nabla z \cdot d\mathbf{s} = [\mathbf{V} \times (\nabla \times \mathbf{V})] \cdot d\mathbf{s} \tag{6.54}$$

Since $d\mathbf{s}$ has a direction along the streamline, the vectors $d\mathbf{s}$ and $\mathbf{V}$ are parallel. However, as shown by the figure in the margin, the vector $\mathbf{V} \times (\nabla \times \mathbf{V})$ is perpendicular to $\mathbf{V}$ (why?), so it follows that

$$[\mathbf{V} \times (\nabla \times \mathbf{V})] \cdot d\mathbf{s} = 0$$

Recall also that the dot product of the gradient of a scalar and a differential length gives the differential change in the scalar in the direction of the differential length. That is, with $d\mathbf{s} = dx\,\hat{\mathbf{i}} + dy\,\hat{\mathbf{j}} + dz\,\hat{\mathbf{k}}$ we can write $\nabla p \cdot d\mathbf{s} = (\partial p/\partial x)\,dx + (\partial p/\partial y)\,dy + (\partial p/\partial z)\,dz = dp$. Thus, Eq. 6.54 becomes

$$\frac{dp}{\rho} + \frac{1}{2}d(V^2) + g\,dz = 0 \tag{6.55}$$

where the change in p, V, and z is along the streamline. Equation 6.55 can now be integrated to give

$$\int \frac{dp}{\rho} + \frac{V^2}{2} + gz = \text{constant} \tag{6.56}$$

which indicates that the sum of the three terms on the left side of the equation must remain a constant along a given streamline. Equation 6.56 is valid for both compressible and incompressible inviscid flows, but for compressible fluids the variation in ρ with p must be specified before the first term in Eq. 6.56 can be evaluated.

For inviscid, incompressible fluids (commonly called **ideal fluids**) Eq. 6.56 can be written as

$$\boxed{\frac{p}{\rho} + \frac{V^2}{2} + gz = \text{constant along a streamline}} \tag{6.57}$$

and this equation is the **Bernoulli equation** used extensively in Chapter 3. It is often convenient to write Eq. 6.57 between two points (1) and (2) along a streamline and to express the equation in the "head" form by dividing each term by g so that

$$\boxed{\frac{p_1}{\gamma} + \frac{V_1^2}{2g} + z_1 = \frac{p_2}{\gamma} + \frac{V_2^2}{2g} + z_2} \tag{6.58}$$

It should be again emphasized that the Bernoulli equation, as expressed by Eqs. 6.57 and 6.58, is restricted to the following:

- inviscid flow
- steady flow
- incompressible flow
- flow along a streamline

You may want to go back and review some of the examples in Chapter 3 that illustrate the use of the Bernoulli equation.

6.4.3   Irrotational Flow

If we make one additional assumption—that the flow is *irrotational*—the analysis of inviscid flow problems is further simplified. Recall from Section 6.1.3 that the rotation of a fluid element is equal to $\frac{1}{2}(\nabla \times \mathbf{V})$, and an irrotational flow field is one for which $\nabla \times \mathbf{V} = 0$. Since the vorticity, ζ, is defined as $\nabla \times \mathbf{V}$, it also follows that in an irrotational flow field the vorticity is zero. The concept of irrotationality may seem to be a rather strange condition for a flow field. Why would a flow field be irrotational? To answer this question we note that if $\frac{1}{2}(\nabla \times \mathbf{V}) = 0$, then each of the components of this vector, as are given by Eqs. 6.12, 6.13, and 6.14, must be equal to zero. Since these components include the various velocity gradients in the flow field, the condition of irrotationality imposes specific relationships among these velocity gradients. For example, for rotation about the z axis to be zero, it follows from Eq. 6.12 that

$$\omega_z = \frac{1}{2}\left(\frac{\partial v}{\partial x} - \frac{\partial u}{\partial y}\right) = 0$$

and, therefore,

$$\frac{\partial v}{\partial x} = \frac{\partial u}{\partial y} \tag{6.59}$$

Similarly from Eqs. 6.13 and 6.14

$$\frac{\partial w}{\partial y} = \frac{\partial v}{\partial z} \tag{6.60}$$

and

$$\frac{\partial u}{\partial z} = \frac{\partial w}{\partial x} \tag{6.61}$$

A general flow field would not satisfy these three equations. However, a uniform flow as is illustrated in Fig. 6.13 does. Since $u = U$ (a constant), $v = 0$, and $w = 0$, it follows that Eqs. 6.59, 6.60, and 6.61 are all satisfied. Therefore, a uniform flow field (in which there are no velocity gradients) is certainly an example of an irrotational flow.

For an inviscid fluid there are no shearing stresses—the only forces acting on a fluid element are its weight and pressure forces. Because the weight acts through the element center of gravity, and the pressure acts in a direction normal to the element surface, neither of these forces can cause the element to rotate. Therefore, for an inviscid fluid, if some part of the flow field is irrotational, the fluid elements emanating from this region will not take on any rotation as they progress through the flow field.

6.4.4   The Bernoulli Equation for Irrotational Flow

In the development of the Bernoulli equation in Section 6.4.2, Eq. 6.54 was integrated along a streamline. This restriction was imposed so the right side of the equation could be set

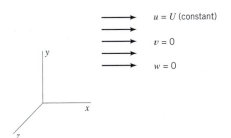

■ **FIGURE 6.13**   **Uniform flow in the x direction.**

equal to zero; that is

$$(\mathbf{V} \times \nabla \times \mathbf{V}) \cdot d\mathbf{s} = 0$$

(since $d\mathbf{s}$ is parallel to $\mathbf{V}$). However, for irrotational flow, $\nabla \times \mathbf{V} = 0$, so the right side of Eq. 6.54 is zero regardless of the direction of $d\mathbf{s}$. We can now follow the same procedure used to obtain Eq. 6.55, where the differential changes dp, $d(V^2)$, and dz can be taken in any direction. Integration of Eq. 6.55 again yields

$$\int \frac{dp}{\rho} + \frac{V^2}{2} + gz = \text{constant} \tag{6.62}$$

where for irrotational flow the constant is the same throughout the flow field. Thus, for incompressible, irrotational flow the Bernoulli equation can be written as

$$\boxed{\frac{p_1}{\gamma} + \frac{V_1^2}{2g} + z_1 = \frac{p_2}{\gamma} + \frac{V_2^2}{2g} + z_2} \tag{6.63}$$

between *any two points in the flow field*. Equation 6.63 is exactly the same form as Eq. 6.58 but is not limited to application along a streamline.

6.4.5   The Velocity Potential

For an irrotational flow the velocity gradients are related through Eqs. 6.59, 6.60, and 6.61. It follows that in this case the velocity components can be expressed in terms of a scalar function $\phi\ (x, y, z, t)$ as

$$u = \frac{\partial \phi}{\partial x} \qquad v = \frac{\partial \phi}{\partial y} \qquad w = \frac{\partial \phi}{\partial z} \tag{6.64}$$

where ϕ is called the **velocity potential.** Direct substitution of these expressions for the velocity components into Eqs. 6.59, 6.60, and 6.61 will verify that a velocity field defined by Eqs. 6.64 is indeed irrotational. In vector form, Eq. 6.64 can be written as

$$\boxed{\mathbf{V} = \nabla \phi} \tag{6.65}$$

so that for an irrotational flow the velocity is expressible as the gradient of a scalar function ϕ.

The velocity potential is a consequence of the irrotationality of the flow field, whereas the stream function (see Sect. 6.2.3) is a consequence of conservation of mass. It is to be noted, however, that the velocity potential can be defined for a general three-dimensional flow, whereas the stream function is restricted to two-dimensional flows.

For an incompressible fluid, we know from conservation of mass that

$$\nabla \cdot \mathbf{V} = 0$$

and therefore for incompressible, irrotational flow (with $\mathbf{V} = \nabla \phi$) it follows that

$$\nabla^2 \phi = 0 \tag{6.66}$$

where $\nabla^2(\ ) = \nabla \cdot \nabla(\ )$ is the *Laplacian operator.* In Cartesian coordinates

$$\frac{\partial^2 \phi}{\partial x^2} + \frac{\partial^2 \phi}{\partial y^2} + \frac{\partial^2 \phi}{\partial z^2} = 0$$

This differential equation arises in many different areas of engineering and physics and is called *Laplace's equation.* Thus, inviscid, incompressible, irrotational flow fields are governed by Laplace's equation. This type of flow is commonly called a *potential flow.* To complete the mathematical formulation of a given problem, boundary conditions have to be specified. These are usually velocities specified on the boundaries of the flow field of interest. It follows that if the potential function can be determined, then the velocity at all points in the flow field can be determined from Eq. 6.64, and the pressure at all points can be determined from the Bernoulli equation (Eq. 6.63). Although the concept of the velocity potential is applicable to both steady and unsteady flow, we will confine our attention to steady flow.

For some problems it will be convenient to use cylindrical coordinates, r, θ, and z. In this coordinate system the gradient operator is

$$\nabla(\ ) = \frac{\partial(\ )}{\partial r}\hat{\mathbf{e}}_r + \frac{1}{r}\frac{\partial(\ )}{\partial \theta}\hat{\mathbf{e}}_\theta + \frac{\partial(\ )}{\partial z}\hat{\mathbf{e}}_z \tag{6.67}$$

so that

$$\nabla\phi = \frac{\partial\phi}{\partial r}\hat{\mathbf{e}}_r + \frac{1}{r}\frac{\partial\phi}{\partial \theta}\hat{\mathbf{e}}_\theta + \frac{\partial\phi}{\partial z}\hat{\mathbf{e}}_z \tag{6.68}$$

where $\phi = \phi(r, \theta, z)$. Because

$$\mathbf{V} = v_r\hat{\mathbf{e}}_r + v_\theta\hat{\mathbf{e}}_\theta + v_z\hat{\mathbf{e}}_z \tag{6.69}$$

it follows for an irrotational flow (with $\mathbf{V} = \nabla\phi$)

$$v_r = \frac{\partial\phi}{\partial r} \qquad v_\theta = \frac{1}{r}\frac{\partial\phi}{\partial \theta} \qquad v_z = \frac{\partial\phi}{\partial z} \tag{6.70}$$

Also, Laplace's equation in cylindrical coordinates is

$$\frac{1}{r}\frac{\partial}{\partial r}\left(r\frac{\partial\phi}{\partial r}\right) + \frac{1}{r^2}\frac{\partial^2\phi}{\partial \theta^2} + \frac{\partial^2\phi}{\partial z^2} = 0 \tag{6.71}$$

EXAMPLE 6.3   Velocity Potential and Inviscid Flow Pressure

GIVEN   The two-dimensional flow of a nonviscous, incompressible fluid in the vicinity of the 90° corner of Fig. E6.3a is described by the stream function

$$\psi = 2r^2\sin 2\theta$$

where ψ has units of m^2/s when r is in meters. Assume the fluid density is 10^3 kg/m^3 and the x–y plane is horizontal— that is, there is no difference in elevation between points (1) and (2).

FIND

(a)   Determine, if possible, the corresponding velocity potential.

(b)   If the pressure at point (1) on the wall is 30 kPa, what is the pressure at point (2)?

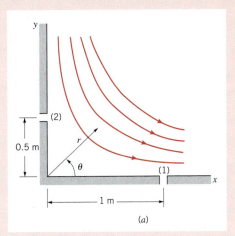

(a)

■ **FIGURE E6.3a**

SOLUTION

(a) The radial and tangential velocity components can be obtained from the stream function as (see Eq. 6.42)

$$v_r = \frac{1}{r}\frac{\partial \psi}{\partial \theta} = 4r \cos 2\theta$$

and

$$v_\theta = -\frac{\partial \psi}{\partial r} = -4r \sin 2\theta$$

Since

$$v_r = \frac{\partial \phi}{\partial r}$$

it follows that

$$\frac{\partial \phi}{\partial r} = 4r \cos 2\theta$$

and therefore by integration

$$\phi = 2r^2 \cos 2\theta + f_1(\theta) \qquad (1)$$

where $f_1(\theta)$ is an arbitrary function of θ. Similarly

$$v_\theta = \frac{1}{r}\frac{\partial \phi}{\partial \theta} = -4r \sin 2\theta$$

and integration yields

$$\phi = 2r^2 \cos 2\theta + f_2(r) \qquad (2)$$

where $f_2(r)$ is an arbitrary function of r. To satisfy both Eqs. 1 and 2, the velocity potential must have the form

$$\phi = 2r^2 \cos 2\theta + C \qquad \textbf{(Ans)}$$

where C is an arbitrary constant. As is the case for stream functions, the specific value of C is not important, and it is customary to let $C = 0$ so that the velocity potential for this corner flow is

$$\phi = 2r^2 \cos 2\theta \qquad \textbf{(Ans)}$$

COMMENT In the statement of this problem it was implied by the wording "if possible" that we might not be able to find a corresponding velocity potential. The reason for this concern is that we can always define a stream function for two-dimensional flow, but the flow must be *irrotational* if there is a corresponding velocity potential. Thus, the fact that we were able to determine a velocity potential means that the flow is irrotational. Several streamlines and lines of constant ϕ are plotted in Fig. E6.3b. These two sets of lines are *orthogonal*. The reason why streamlines and lines of constant ϕ are always orthogonal is explained in Section 6.5.

(b) Because we have an irrotational flow of a nonviscous, incompressible fluid, the Bernoulli equation can be applied

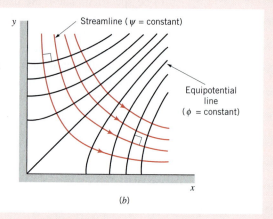

■ **FIGURE E6.3b**

between any two points. Thus, between points (1) and (2) with no elevation change

$$\frac{p_1}{\gamma} + \frac{V_1^2}{2g} = \frac{p_2}{\gamma} + \frac{V_2^2}{2g}$$

or

$$p_2 = p_1 + \frac{\rho}{2}(V_1^2 - V_2^2) \qquad (3)$$

Since

$$V^2 = v_r^2 + v_\theta^2$$

it follows that for any point within the flow field

$$\begin{aligned} V^2 &= (4r \cos 2\theta)^2 + (-4r \sin 2\theta)^2 \\ &= 16r^2(\cos^2 2\theta + \sin^2 2\theta) \\ &= 16r^2 \end{aligned}$$

This result indicates that the square of the velocity at any point depends only on the radial distance, r, to the point. Note that the constant, 16, has units of s^{-2}. Thus,

$$V_1^2 = (16\ \text{s}^{-2})(1\ \text{m})^2 = 16\ \text{m}^2/\text{s}^2$$

and

$$V_2^2 = (16\ \text{s}^{-2})(0.5\ \text{m})^2 = 4\ \text{m}^2/\text{s}^2$$

Substitution of these velocities into Eq. 3 gives

$$p_2 = 30 \times 10^3\ \text{N/m}^2 + \frac{10^3\ \text{kg/m}^3}{2}(16\ \text{m}^2/\text{s}^2 - 4\ \text{m}^2/\text{s}^2)$$

$$= 36\ \text{kPa} \qquad \textbf{(Ans)}$$

COMMENT The stream function used in this example could also be expressed in Cartesian coordinates as

$$\psi = 2r^2 \sin 2\theta = 4r^2 \sin\theta \cos\theta$$

or

$$\psi = 4xy$$

since $x = r \cos \theta$ and $y = r \sin \theta$. However, in the cylindrical polar form the results can be generalized to describe flow in the vicinity of a corner of angle α (see Fig. E6.3c) with the equations

$$\psi = Ar^{\pi/\alpha} \sin \frac{\pi\theta}{\alpha}$$

and

$$\phi = Ar^{\pi/\alpha} \cos \frac{\pi\theta}{\alpha}$$

where A is a constant.

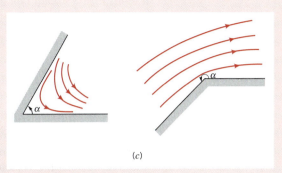

(c)

■ **FIGURE E6.3c**

6.5   Some Basic, Plane Potential Flows

A major advantage of Laplace's equation is that it is a linear partial differential equation. Because it is linear, various solutions can be added to obtain other solutions—that is, if $\phi_1(x, y, z)$ and $\phi_2(x, y, z)$ are two solutions to Laplace's equation, then $\phi_3 = \phi_1 + \phi_2$ is also a solution. The practical implication of this result is that if we have certain basic solutions we can combine them to obtain more complicated and interesting solutions. In this section several basic velocity potentials, which describe some relatively simple flows, will be determined. In the next section these basic potentials will be combined to represent more complicated flows.

For simplicity, only plane (two-dimensional) flows will be considered. In this case, as shown by the figures in the margin, for Cartesian coordinates

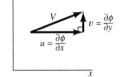

$$u = \frac{\partial\phi}{\partial x} \qquad v = \frac{\partial\phi}{\partial y} \tag{6.72}$$

or for cylindrical coordinates

$$v_r = \frac{\partial\phi}{\partial r} \qquad v_\theta = \frac{1}{r}\frac{\partial\phi}{\partial\theta} \tag{6.73}$$

Since we can define a stream function for plane flow, we can also let

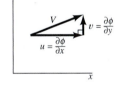

$$u = \frac{\partial\psi}{\partial y} \qquad v = -\frac{\partial\psi}{\partial x} \tag{6.74}$$

or

$$v_r = \frac{1}{r}\frac{\partial\psi}{\partial\theta} \qquad v_\theta = -\frac{\partial\psi}{\partial r} \tag{6.75}$$

where the stream function was defined previously in Eqs. 6.37 and 6.42. We know that by defining the velocities in terms of the stream function, conservation of mass is identically satisfied. If we now impose the condition of irrotationality, it follows from Eq. 6.59 that

$$\frac{\partial u}{\partial y} = \frac{\partial v}{\partial x}$$

and in terms of the stream function

$$\frac{\partial}{\partial y}\left(\frac{\partial \psi}{\partial y}\right) = \frac{\partial}{\partial x}\left(-\frac{\partial \psi}{\partial x}\right)$$

or

$$\frac{\partial^2 \psi}{\partial x^2} + \frac{\partial^2 \psi}{\partial y^2} = 0$$

Thus, for a plane, irrotational flow we can use either the velocity potential or the stream function—both must satisfy Laplace's equation in two dimensions. It is apparent from these results that the velocity potential and the stream function are somehow related. We have previously shown that lines of constant ψ are streamlines; that is,

$$\left.\frac{dy}{dx}\right|_{\text{along }\psi\,=\,\text{constant}} = \frac{v}{u} \tag{6.76}$$

The change in ϕ as we move from one point (x, y) to a nearby point $(x + dx, y + dy)$ is given by the relationship:

$$d\phi = \frac{\partial \phi}{\partial x}\,dx + \frac{\partial \phi}{\partial y}\,dy = u\,dx + v\,dy$$

Along a line of constant ϕ we have $d\phi = 0$ so that

$$\left.\frac{dy}{dx}\right|_{\text{along }\phi\,=\,\text{constant}} = -\frac{u}{v} \tag{6.77}$$

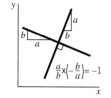

A comparison of Eqs. 6.76 and 6.77 shows that lines of constant ϕ (called *equipotential lines*) are orthogonal to lines of constant ψ (streamlines) at all points where they intersect. (Recall, as shown by the figure in the margin, that two lines are orthogonal if the product of their slopes is minus one.) For any potential flow field a "*flow net*" can be drawn that consists of a family of streamlines and equipotential lines. The flow net is useful in visualizing flow patterns and can be used to obtain graphical solutions by sketching in streamlines and equipotential lines and adjusting the lines until the lines are approximately orthogonal at all points where they intersect. An example of a flow net is shown in Fig. 6.14. Velocities

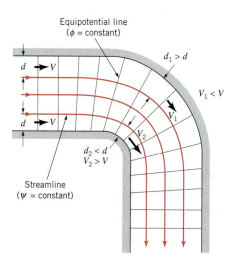

■ **FIGURE 6.14** Flow net for a 90° bend. (From Ref. 3, used by permission.)

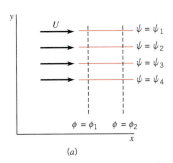 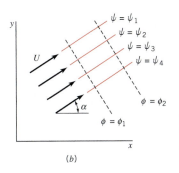

■ **FIGURE 6.15**
Uniform flow: (*a*) **in the**
x direction and (*b*) in an
arbitrary direction, α.

can be estimated from the flow net, as the velocity is inversely proportional to the stream-line spacing. Thus, for example, from Fig. 6.14 we can see that the velocity near the inside corner will be higher than the velocity along the outer part of the bend.

6.5.1   Uniform Flow

The simplest plane flow is one for which the streamlines are all straight and parallel, and the magnitude of the velocity is constant. This type of flow is called a ***uniform flow.*** For example, consider a uniform flow in the positive x direction as illustrated in Fig. 6.15*a*. In this instance, $u = U$ and $v = 0$, and in terms of the velocity potential

$$\frac{\partial \phi}{\partial x} = U \qquad \frac{\partial \phi}{\partial y} = 0$$

These two equations can be integrated to yield

$$\phi = Ux + C$$

where C is an arbitrary constant, which can be set equal to zero. Thus, for a uniform flow in the positive x direction

$$\phi = Ux \tag{6.78}$$

The corresponding stream function can be obtained in a similar manner, as

$$\frac{\partial \psi}{\partial y} = U \qquad \frac{\partial \psi}{\partial x} = 0$$

and, therefore,

$$\psi = Uy \tag{6.79}$$

These results can be generalized to provide the velocity potential and stream function for a uniform flow at an angle α with the x axis, as in Fig. 6.15*b*. For this case

$$\phi = U(x \cos \alpha + y \sin \alpha) \tag{6.80}$$

and

$$\psi = U(y \cos \alpha - x \sin \alpha) \tag{6.81}$$

6.5.2   Source and Sink

Consider a fluid flowing radially outward from a line through the origin perpendicular to the x–y plane as shown in Fig. 6.16. Let m be the volume rate of flow emanating from the

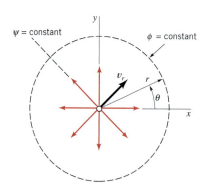

■ **FIGURE 6.16** The streamline pattern for a source.

line (per unit length), and therefore to satisfy conservation of mass

$$(2\pi r)v_r = m$$

or

$$v_r = \frac{m}{2\pi r}$$

Also, because the flow is a purely radial flow, $v_\theta = 0$, the corresponding velocity potential can be obtained by integrating the equations

$$\frac{\partial \phi}{\partial r} = \frac{m}{2\pi r} \qquad \frac{1}{r}\frac{\partial \phi}{\partial \theta} = 0$$

It follows that

$$\phi = \frac{m}{2\pi} \ln r \tag{6.82}$$

If m is positive, the flow is radially outward, and the flow is considered to be a ***source*** flow. If m is negative, the flow is toward the origin, and the flow is considered to be a ***sink*** flow. The flowrate, m, is the *strength* of the source or sink.

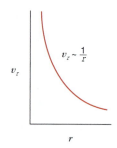

As shown by the figure in the margin, at the origin where $r = 0$ the velocity becomes infinite, which is, of course, physically impossible. Thus, sources and sinks do not really exist in real flow fields, and the line representing the source or sink is a mathematical *singularity* in the flow field. However, some real flows can be approximated at points away from the origin by using sources or sinks. Also, the velocity potential representing this hypothetical flow can be combined with other basic velocity potentials to describe approximately some real flow fields. This idea is discussed further in Section 6.6.

The stream function for the source can be obtained by integrating the relationships

$$\frac{1}{r}\frac{\partial \psi}{\partial \theta} = \frac{m}{2\pi r} \qquad \frac{\partial \psi}{\partial r} = 0$$

to yield

$$\psi = \frac{m}{2\pi} \theta \tag{6.83}$$

It is apparent from Eq. 6.83 that the streamlines (lines of ψ = constant) are radial lines, and from Eq. 6.82 the equipotential lines (lines of ϕ = constant) are concentric circles centered at the origin.

EXAMPLE 6.4   Potential Flow—Sink

GIVEN A nonviscous, incompressible fluid flows between wedge-shaped walls into a small opening as shown in Fig. E6.4. The velocity potential (in ft²/s), which approximately describes this flow, is

$$\phi = -2 \ln r$$

FIND Determine the volume rate of flow (per unit length) into the opening.

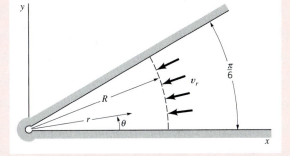

■ **FIGURE E6.4**

SOLUTION

The components of velocity are

$$v_r = \frac{\partial \phi}{\partial r} = -\frac{2}{r} \qquad v_\theta = \frac{1}{r}\frac{\partial \phi}{\partial \theta} = 0$$

which indicates we have a purely radial flow. The flowrate per unit width, q, crossing the arc of length $R\pi/6$ can thus be obtained by integrating the expression

$$q = \int_0^{\pi/6} v_r R \, d\theta = -\int_0^{\pi/6} \left(\frac{2}{R}\right) R \, d\theta = -\frac{\pi}{3} = -1.05 \text{ ft}^2/\text{s}$$

(Ans)

COMMENT Note that the radius R is arbitrary since the flowrate crossing any curve between the two walls must be the same. The negative sign indicates that the flow is toward the opening, that is, in the negative radial direction.

6.5.3   Vortex

We next consider a flow field in which the streamlines are concentric circles—that is, we interchange the velocity potential and stream function for the source. Thus, let

$$\phi = K\theta \qquad\qquad (6.84)$$

and

$$\psi = -K \ln r \qquad\qquad (6.85)$$

where K is a constant. In this case the streamlines are concentric circles as illustrated in Fig. 6.17, with $v_r = 0$ and

$$v_\theta = \frac{1}{r}\frac{\partial \phi}{\partial \theta} = -\frac{\partial \psi}{\partial r} = \frac{K}{r} \qquad\qquad (6.86)$$

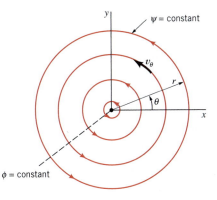

■ **FIGURE 6.17**   The streamline pattern for a vortex.

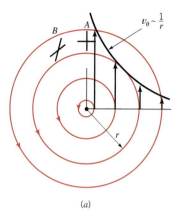
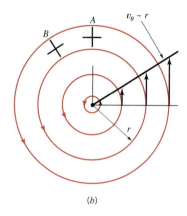

(a) (b)

■ **FIGURE 6.18**
Motion of fluid element from *A* to *B*: (*a*) for an irrotational (free) vortex and (*b*) for a rotational (forced) vortex.

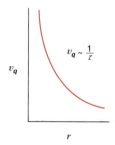

This result indicates that the tangential velocity varies inversely with the distance from the origin, as shown in the figure in the margin, with a singularity occurring at $r = 0$ (where the velocity becomes infinite).

It may seem strange that this **vortex** motion is irrotational (and it is since the flow field is described by a velocity potential). However, it must be recalled that rotation refers to the orientation of a fluid element and not the path followed by the element. Thus, for an irrotational vortex, if a pair of small sticks were placed in the flow field at location *A*, as indicated in Fig. 6.18*a*, the sticks would rotate as they move to location *B*. One of the sticks, the one that is aligned along the streamline, would follow a circular path and rotate in a counterclockwise direction. The other stick would rotate in a clockwise direction due to the nature of the flow field—that is, the part of the stick nearest the origin moves faster than the opposite end. Although both sticks are rotating, the average angular velocity of the two sticks is zero since the flow is irrotational.

If the fluid were rotating as a rigid body, such that $v_\theta = K_1 r$ where K_1 is a constant, the sticks similarly placed in the flow field would rotate as is illustrated in Fig. 6.18*b*. This type of vortex motion is *rotational* and cannot be described with a velocity potential. The rotational vortex is commonly called a *forced vortex,* whereas the irrotational vortex is usually called a *free vortex.* The swirling motion of the water as it drains from a bathtub is similar to that of a free vortex, whereas the motion of a liquid contained in a tank that is rotated about its axis with angular velocity ω corresponds to a forced vortex.

F l u i d s   i n   t h e   N e w s

Some hurricane facts One of the most interesting, yet potentially devastating, naturally occurring fluid flow phenomena is a hurricane. Broadly speaking a hurricane is a rotating mass of air circulating around a low-pressure central core. In some respects the motion is similar to that of a *free vortex.* The Caribbean and Gulf of Mexico experience the most hurricanes, with the official hurricane season being from June 1 to November 30. Hurricanes are usually 300 to 400 miles wide and are structured around a central eye in which the air is relatively calm. The eye is surrounded by an eye wall which is the region of strongest winds and precipitation. As one goes from the eye wall to the eye, the wind speeds decrease sharply and within the eye the air is relatively calm and

clear of clouds. However, in the eye the pressure is at a minimum and may be 10% less than standard atmospheric pressure. This low pressure creates strong downdrafts of dry air from above. Hurricanes are classified into five categories based on their wind speeds:

Category one—74–95 mph

Category two—96–110 mph

Category three—111–130 mph

Category four—131–155 mph

Category five—greater than 155 mph

(See Problem 6.23.) ■

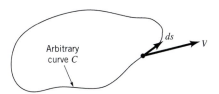

■ **FIGURE 6.19**   The notation for determining circulation around closed curve C.

A *combined vortex* is one with a forced vortex as a central core and a velocity distribution corresponding to that of a free vortex outside the core. Thus, for a combined vortex

$$v_\theta = \omega r \qquad r \leq r_0 \tag{6.87}$$

and

$$v_\theta = \frac{K}{r} \qquad r > r_0 \tag{6.88}$$

where K and ω are constants and r_0 corresponds to the radius of the central core. The pressure distribution in both the free and the forced vortex was considered previously in Example 3.3.

A mathematical concept commonly associated with vortex motion is that of *circulation.* The circulation, Γ, is defined as the line integral of the tangential component of the velocity taken around a closed curve in the flow field. In equation form, Γ can be expressed as

$$\Gamma = \oint_C \mathbf{V} \cdot d\mathbf{s} \tag{6.89}$$

where the integral sign means that the integration is taken around a closed curve, C, in the counterclockwise direction, and $d\mathbf{s}$ is a differential length along the curve as illustrated in Fig. 6.19. For an irrotational flow, $\mathbf{V} = \nabla \phi$ so that $\mathbf{V} \cdot d\mathbf{s} = \nabla \phi \cdot d\mathbf{s} = d\phi$ and, therefore,

$$\Gamma = \oint_C d\phi = 0$$

This result indicates that for an irrotational flow the circulation will generally be zero. However, if there are singularities enclosed within the curve, the circulation may not be zero. For example, for the free vortex with $v_\theta = K/r$ the circulation around the circular path of radius r shown in Fig. 6.20 is

$$\Gamma = \int_0^{2\pi} \frac{K}{r} (r \, d\theta) = 2\pi K$$

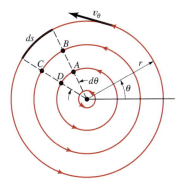

■ **FIGURE 6.20**   Circulation around various paths in a free vortex.

which shows that the circulation is nonzero and the constant $K = \Gamma/2\pi$. However, the circulation around any path that does not include the singular point at the origin will be zero. This can be easily confirmed for the closed path $ABCD$ of Fig. 6.20 by evaluating the circulation around that path.

The velocity potential and stream function for the free vortex are commonly expressed in terms of the circulation as

$$\phi = \frac{\Gamma}{2\pi}\theta \qquad (6.90)$$

and

$$\psi = -\frac{\Gamma}{2\pi}\ln r \qquad (6.91)$$

V6.2 Vortex in a beaker

The concept of circulation is often useful when evaluating the forces developed on bodies immersed in moving fluids. This application will be considered in Section 6.6.2.

EXAMPLE 6.5 ■ Potential Flow—Free Vortex

GIVEN   A liquid drains from a large tank through a small opening as illustrated in Fig. E6.5. A vortex forms whose velocity distribution away from the tank opening can be approximated as that of a free vortex having a velocity potential

$$\phi = \frac{\Gamma}{2\pi}\theta$$

FIND   Determine an expression relating the surface shape to the strength of the vortex as specified by the circulation Γ.

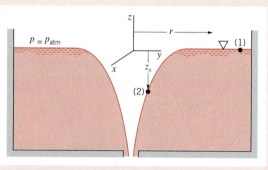

■ **FIGURE E6.5**

SOLUTION

Since the free vortex represents an irrotational flow field, the Bernoulli equation

$$\frac{p_1}{\gamma} + \frac{V_1^2}{2g} + z_1 = \frac{p_2}{\gamma} + \frac{V_2^2}{2g} + z_2$$

can be written between any two points. If the points are selected at the free surface, $p_1 = p_2 = 0$, so that

$$\frac{V_1^2}{2g} = z_s + \frac{V_2^2}{2g} \qquad (1)$$

where the free surface elevation, z_s, is measured relative to a datum passing through point (1).

The velocity is given by the equation

$$v_\theta = \frac{1}{r}\frac{\partial\phi}{\partial\theta} = \frac{\Gamma}{2\pi r}$$

We note that far from the origin at point (1) $V_1 = v_\theta \approx 0$ so that Eq. 1 becomes

$$z_s = -\frac{\Gamma^2}{8\pi^2 r^2 g} \qquad \text{(Ans)}$$

which is the desired equation for the surface profile.

COMMENT   The negative sign indicates that the surface falls as the origin is approached as shown in Fig. E6.5. This solution is not valid very near the origin since the predicted velocity becomes excessively large as the origin is approached.

6.5.4   Doublet

The final, basic potential flow to be considered is one that is formed by combining a source and sink in a special way. Consider the equal strength, source–sink pair of Fig. 6.21. The combined stream function for the pair is

$$\psi = -\frac{m}{2\pi}(\theta_1 - \theta_2)$$

which can be rewritten as

$$\tan\left(-\frac{2\pi\psi}{m}\right) = \tan(\theta_1 - \theta_2) = \frac{\tan\theta_1 - \tan\theta_2}{1 + \tan\theta_1 \tan\theta_2} \tag{6.92}$$

From Fig. 6.21 it follows that

$$\tan\theta_1 = \frac{r\sin\theta}{r\cos\theta - a}$$

and

$$\tan\theta_2 = \frac{r\sin\theta}{r\cos\theta + a}$$

These results substituted into Eq. 6.92 give

$$\tan\left(-\frac{2\pi\psi}{m}\right) = \frac{2ar\sin\theta}{r^2 - a^2}$$

so that

$$\psi = -\frac{m}{2\pi}\tan^{-1}\left(\frac{2ar\sin\theta}{r^2 - a^2}\right) \tag{6.93}$$

For small values of the distance a

$$\psi = -\frac{m}{2\pi}\frac{2ar\sin\theta}{r^2 - a^2} = -\frac{mar\sin\theta}{\pi(r^2 - a^2)} \tag{6.94}$$

since the tangent of an angle approaches the value of the angle for small angles.

    The so-called **doublet** is formed by letting the source and sink approach one another $(a \to 0)$ while increasing the strength m $(m \to \infty)$ so that the product ma/π remains constant. In this case, since $r/(r^2 - a^2) \to 1/r$, Eq. 6.94 reduces to

$$\psi = -\frac{K\sin\theta}{r} \tag{6.95}$$

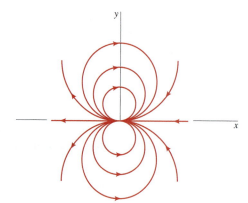

■ **FIGURE 6.22** **Streamlines for a doublet.**

where K, a constant equal to ma/π, is called the *strength* of the doublet. The corresponding velocity potential for the doublet is

$$\phi = \frac{K \cos \theta}{r} \tag{6.96}$$

Plots of lines of constant ψ reveal that streamlines for a doublet are circles through the origin tangent to the x axis as shown in Fig. 6.22. Just as sources and sinks are not physically realistic entities, neither are doublets. However, the doublet, when combined with other basic potential flows, provides a useful representation of some flow fields of practical interest. For example, we will determine in Section 6.6.2 that the combination of a uniform flow and a doublet can be used to represent the flow around a circular cylinder. Table 6.1 provides a summary of the pertinent equations for the basic, plane potential flows considered in the preceding sections.

■ **TABLE 6.1**
Summary of Basic, Plane Potential Flows

Description of Flow Field	Velocity Potential	Stream Function	Velocity Components[a]
Uniform flow at angle α with the x axis (see Fig. 6.15b)	$\phi = U(x \cos \alpha + y \sin \alpha)$	$\psi = U(y \cos \alpha - x \sin \alpha)$	$u = U \cos \alpha$ $v = U \sin \alpha$
Source or sink (see Fig. 6.16) $m > 0$ source $m < 0$ sink	$\phi = \dfrac{m}{2\pi} \ln r$	$\psi = \dfrac{m}{2\pi} \theta$	$v_r = \dfrac{m}{2\pi r}$ $v_\theta = 0$
Free vortex (see Fig. 6.17) $\Gamma > 0$ counterclockwise motion $\Gamma < 0$ clockwise motion	$\phi = \dfrac{\Gamma}{2\pi} \theta$	$\psi = -\dfrac{\Gamma}{2\pi} \ln r$	$v_r = 0$ $v_\theta = \dfrac{\Gamma}{2\pi r}$
Doublet (see Fig. 6.22)	$\phi = \dfrac{K \cos \theta}{r}$	$\psi = -\dfrac{K \sin \theta}{r}$	$v_r = -\dfrac{K \cos \theta}{r^2}$ $v_\theta = -\dfrac{K \sin \theta}{r^2}$

[a]Velocity components are related to the velocity potential and stream function through the relationships:

$$u = \frac{\partial \phi}{\partial x} = \frac{\partial \psi}{\partial y} \qquad v = \frac{\partial \phi}{\partial y} = -\frac{\partial \psi}{\partial x} \qquad v_r = \frac{\partial \phi}{\partial r} = \frac{1}{r}\frac{\partial \psi}{\partial \theta} \qquad v_\theta = \frac{1}{r}\frac{\partial \phi}{\partial \theta} = -\frac{\partial \psi}{\partial r}$$

6.6 Superposition of Basic, Plane Potential Flows

As was discussed in the previous section, potential flows are governed by Laplace's equation, which is a linear partial differential equation. It therefore follows that the various basic velocity potentials and stream functions can be combined to form new potentials and stream functions. (Why is this true?) Whether such combinations yield useful results remains to be seen. It is to be noted that *any streamline in an inviscid flow field can be considered as a solid boundary,* as the conditions along a solid boundary and a streamline are similar—that is, there is no flow through the boundary or the streamline. Thus, if we can combine some of the basic velocity potentials or stream functions to yield a streamline that corresponds to a particular body shape of interest, that combination can be used to describe in detail the flow around that body. This method of solving some interesting flow problems, commonly called the **method of superposition,** is illustrated in the following three sections.

6.6.1 Source in a Uniform Stream—Half-Body

Consider the superposition of a source and a uniform flow as shown in Fig. 6.23a. The resulting stream function is

$$\psi = \psi_{\text{uniform flow}} + \psi_{\text{source}}$$
$$= Ur \sin \theta + \frac{m}{2\pi} \theta \tag{6.97}$$

and the corresponding velocity potential is

$$\phi = Ur \cos \theta + \frac{m}{2\pi} \ln r \tag{6.98}$$

V6.3 Half-body

It is clear that at some point along the negative x axis the velocity due to the source will just cancel that due to the uniform flow and a stagnation point will be created. For the source alone

$$v_r = \frac{m}{2\pi r}$$

so that the stagnation point will occur at $x = -b$ where

$$U = \frac{m}{2\pi b}$$

or

$$b = \frac{m}{2\pi U} \tag{6.99}$$

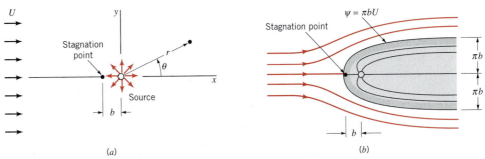

(a)   (b)

■ **FIGURE 6.23** The flow around a half-body: (*a*) superposition of a source and a uniform flow; (*b*) replacement of streamline $\psi = \pi bU$ with solid boundary to form half-body.

The value of the stream function at the stagnation point can be obtained by evaluating ψ at $r = b$ and $\theta = \pi$, which yields from Eq. 6.97

$$\psi_{\text{stagnation}} = \frac{m}{2}$$

Since $m/2 = \pi b U$ (from Eq. 6.99) it follows that the equation of the streamline passing through the stagnation point is

$$\pi b U = U r \sin \theta + b U \theta$$

or

$$r = \frac{b(\pi - \theta)}{\sin \theta} \tag{6.100}$$

where θ can vary between 0 and 2π. A plot of this streamline is shown in Fig. 6.23b. If we replace this streamline with a solid boundary, as indicated in Fig. 6.23b, then it is clear that this combination of a uniform flow and a source can be used to describe the flow around a streamlined body placed in a uniform stream. The body is open at the downstream end and thus is called a ***half-body.*** Other streamlines in the flow field can be obtained by setting $\psi =$ constant in Eq. 6.97 and plotting the resulting equation. A number of these streamlines are shown in Fig. 6.23b. Although the streamlines inside the body are shown, they are actually of no interest in this case, as we are concerned with the flow field outside the body. It should be noted that the singularity in the flow field (the source) occurs inside the body, and there are no singularities in the flow field of interest (outside the body).

The width of the half-body asymptotically approaches $2\pi b$. This follows from Eq. 6.100, which can be written as

$$y = b(\pi - \theta)$$

so that as $\theta \to 0$ or $\theta \to 2\pi$ the half-width approaches $\pm b\pi$. With the stream function (or velocity potential) known, the velocity components at any point can be obtained. For the half-body, using the stream function given by Eq. 6.97,

$$v_r = \frac{1}{r} \frac{\partial \psi}{\partial \theta} = U \cos \theta + \frac{m}{2\pi r}$$

and

$$v_\theta = -\frac{\partial \psi}{\partial r} = -U \sin \theta$$

Thus, the square of the magnitude of the velocity, V, at any point is

$$V^2 = v_r^2 + v_\theta^2 = U^2 + \frac{Um \cos \theta}{\pi r} + \left(\frac{m}{2\pi r}\right)^2$$

and since $b = m/2\pi U$

$$V^2 = U^2\left(1 + 2\frac{b}{r} \cos \theta + \frac{b^2}{r^2}\right) \tag{6.101}$$

With the velocity known, the pressure at any point can be determined from the Bernoulli equation, which can be written between any two points in the flow field since the flow is irrotational. Thus, applying the Bernoulli equation between a point far from the body, where the pressure is p_0 and the velocity is U, and some arbitrary point with pressure p and velocity V, it follows that

$$p_0 + \tfrac{1}{2}\rho U^2 = p + \tfrac{1}{2}\rho V^2 \tag{6.102}$$

where elevation changes have been neglected. Equation 6.101 can now be substituted into Eq. 6.102 to obtain the pressure at any point in terms of the reference pressure, p_0, and velocity, U.

This relatively simple potential flow provides some useful information about the flow around the front part of a streamlined body, such as a bridge pier or strut placed in a uniform stream. An important point to be noted is that the velocity tangent to the surface of the body is not zero; that is, the fluid "slips" by the boundary. This result is a consequence of neglecting viscosity, the fluid property that causes real fluids to stick to the boundary, thus creating a "no-slip" condition. All potential flows differ from the flow of real fluids in this respect and do not accurately represent the velocity very near the boundary. However, outside this very thin boundary layer the velocity distribution will generally correspond to that predicted by potential flow theory if flow separation does not occur. (See Sect. 9.2.6) Also, the pressure distribution along the surface will closely approximate that predicted from the potential flow theory, as the boundary layer is thin and there is little opportunity for the pressure to vary through the thin layer. In fact, as discussed in more detail in Chapter 9, the pressure distribution obtained from the potential flow theory is used in conjunction with the viscous flow theory to determine the nature of flow within the boundary layer.

EXAMPLE 6.6   Potential Flow—Half Body

GIVEN A 40 mi/hr wind blows toward a hill arising from a plain that can be approximated with the top section of a half-body as illustrated in Fig. E6.6a. The height of the hill approaches 200 ft as shown. Assume an air density of 0.00238 slugs/ft³.

FIND

(a) What is the magnitude of the air velocity at a point on the hill directly above the origin [point (2)]?

(b) What is the elevation of point (2) above the plain and what is the difference in pressure between point (1) on the plain far from the hill and point (2)?

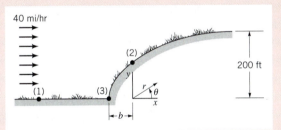

■ **FIGURE E6.6a**

SOLUTION

(a) The velocity is given by Eq. 6.101 as

$$V^2 = U^2\left(1 + 2\frac{b}{r}\cos\theta + \frac{b^2}{r^2}\right)$$

At point (2), $\theta = \pi/2$, and since this point is on the surface (Eq. 6.100)

$$r = \frac{b(\pi - \theta)}{\sin\theta} = \frac{\pi b}{2} \tag{1}$$

Thus,

$$V_2^2 = U^2\left[1 + \frac{b^2}{(\pi b/2)^2}\right]$$

$$= U^2\left(1 + \frac{4}{\pi^2}\right)$$

and the magnitude of the velocity at (2) for a 40-mi/hr approaching wind is

$$V_2 = \left(1 + \frac{4}{\pi^2}\right)^{1/2}(40 \text{ mi/hr}) = 47.4 \text{ mi/hr} \quad \text{(Ans)}$$

(b) The elevation at (2) above the plain is given by Eq. 1 as

$$y_2 = \frac{\pi b}{2}$$

Since the height of the hill approaches 200 ft and this height is equal to πb, it follows that

$$y_2 = \frac{200 \text{ ft}}{2} = 100 \text{ ft} \quad \text{(Ans)}$$

From the Bernoulli equation (with the y axis the vertical axis)

$$\frac{p_1}{\gamma} + \frac{V_1^2}{2g} + y_1 = \frac{p_2}{\gamma} + \frac{V_2^2}{2g} + y_2$$

so that

$$p_1 - p_2 = \frac{\rho}{2}(V_2^2 - V_1^2) + \gamma(y_2 - y_1)$$

and with

$$V_1 = (40 \text{ mi/hr})\left(\frac{5280 \text{ ft/mi}}{3600 \text{ s/hr}}\right) = 58.7 \text{ ft/s}$$

and

$$V_2 = (47.4 \text{ mi/hr})\left(\frac{5280 \text{ ft/mi}}{3600 \text{ s/hr}}\right) = 69.5 \text{ ft/s}$$

it follows that

$$p_1 - p_2 = \frac{(0.00238 \text{ slugs/ft}^3)}{2}[(69.5 \text{ ft/s})^2 - (58.7 \text{ ft/s})^2]$$

$$+ (0.00238 \text{ slugs/ft}^3)(32.2 \text{ ft/s}^2)(100 \text{ ft} - 0 \text{ ft})$$

$$= 9.31 \text{ lb/ft}^2 = 0.0647 \text{ psi} \qquad \textbf{(Ans)}$$

COMMENTS This result indicates that the pressure on the hill at point (2) is slightly lower than the pressure on the plain at some distance from the base of the hill with a 0.0533 psi dif-ference due to the elevation increase and a 0.0114 psi differ-ence due to the velocity increase.

By repeating the calculations for various values of the upstream wind speed, U, the results shown in Fig. E6.6b are obtained. Note that as the wind speed increases the pressure differ-ence increases from the calm conditions of $p_1 - p_2 = 0.0533$ psi.

The maximum velocity along the hill surface does not occur at point (2) but further up the hill at $\theta = 63°$. At this point $V_{surface} = 1.26U$ (Problem 6.32). The minimum velocity ($V = 0$) and maximum pressure occur at point (3), the stagnation point.

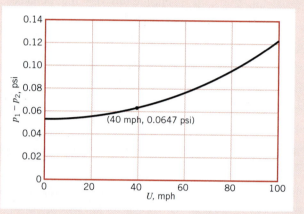

■ **FIGURE E6.6b**

The half-body is a body that is "open" at one end. To study the flow around a closed body, a source and a sink of equal strength can be combined with a uniform flow. These bodies have an oval shape and are termed *Rankine ovals*.

6.6.2   Flow around a Circular Cylinder

A uniform flow in the positive x direction combined with a doublet can be used to repre-sent flow around a circular cylinder. This combination gives for the stream function

$$\psi = Ur \sin \theta - \frac{K \sin \theta}{r} \qquad \textbf{(6.103)}$$

and for the velocity potential

$$\phi = Ur \cos \theta + \frac{K \cos \theta}{r} \qquad \textbf{(6.104)}$$

In order for the stream function to represent flow around a circular cylinder, it is necessary that $\psi = $ constant for $r = a$, where a is the radius of the cylinder. Because Eq. 6.103 can be written as

$$\psi = \left(U - \frac{K}{r^2}\right)r \sin \theta$$

it follows that $\psi = 0$ for $r = a$ if

$$U - \frac{K}{a^2} = 0$$

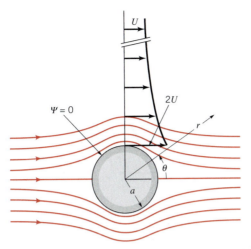

■ **FIGURE 6.24** **Flow around a circu-lar cylinder.**

which indicates that the doublet strength, K, must be equal to Ua^2. Thus, the stream function for flow around a circular cylinder can be expressed as

$$\psi = Ur\left(1 - \frac{a^2}{r^2}\right)\sin\theta \qquad (6.105)$$

and the corresponding velocity potential is

$$\phi = Ur\left(1 + \frac{a^2}{r^2}\right)\cos\theta \qquad (6.106)$$

A sketch of the streamlines for this flow field is shown in Fig. 6.24.

The velocity components can be obtained from either Eq. 6.105 or 6.106 as

$$v_r = \frac{\partial\phi}{\partial r} = \frac{1}{r}\frac{\partial\psi}{\partial\theta} = U\left(1 - \frac{a^2}{r^2}\right)\cos\theta \qquad (6.107)$$

and

$$v_\theta = \frac{1}{r}\frac{\partial\phi}{\partial\theta} = -\frac{\partial\psi}{\partial r} = -U\left(1 + \frac{a^2}{r^2}\right)\sin\theta \qquad (6.108)$$

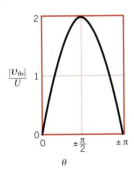

On the surface of the cylinder ($r = a$) it follows from Eqs. 6.107 and 6.108 that $v_r = 0$ and

$$v_{\theta s} = -2U\sin\theta$$

As shown by the figure in the margin, the maximum velocity occurs at the top and bottom of the cylinder ($\theta = \pm\pi/2$) and has a magnitude of twice the upstream velocity, U. As we move away from the cylinder along the ray $\theta = \pi/2$ the velocity varies, as illustrated in Fig. 6.24.

The pressure distribution on the cylinder surface is obtained from the Bernoulli equation written from a point far from the cylinder where the pressure is p_0 and the velocity is U so that

$$p_0 + \tfrac{1}{2}\rho U^2 = p_s + \tfrac{1}{2}\rho v_{\theta s}^2$$

where p_s is the surface pressure. Elevation changes are neglected. Because $v_{\theta s} = -2U\sin\theta$, the surface pressure can be expressed as

$$p_s = p_0 + \tfrac{1}{2}\rho U^2(1 - 4\sin^2\theta) \qquad (6.109)$$

A comparison of this theoretical, symmetrical pressure distribution expressed in dimensionless form with a typical measured distribution is shown in Fig. 6.25, which clearly reveals that only on the upstream part of the cylinder is there approximate agreement

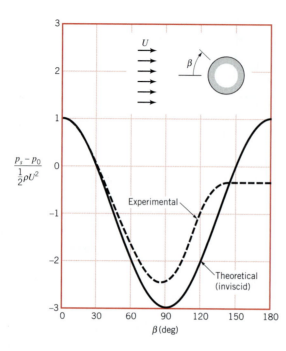

between the potential flow and the experimental results. Because of the viscous boundary layer that develops on the cylinder, the main flow separates from the surface of the cylinder, leading to the large difference between the theoretical, frictionless fluid solution and the experimental results on the downstream side of the cylinder (see Chapter 9).

The resultant force (per unit length) developed on the cylinder can be determined by integrating the pressure over the surface. From Fig. 6.26 it can be seen that

$$F_x = -\int_0^{2\pi} p_s \cos \theta \, a \, d\theta \tag{6.110}$$

and

$$F_y = -\int_0^{2\pi} p_s \sin \theta \, a \, d\theta \tag{6.111}$$

where F_x is the *drag* (force parallel to direction of the uniform flow) and F_y is the *lift* (force perpendicular to the direction of the uniform flow). Substitution for p_s from Eq. 6.109 into these two equations, and subsequent integration, reveals that $F_x = 0$ and $F_y = 0$.

These results indicate that both the drag and the lift as predicted by potential theory for a fixed cylinder in a uniform stream are zero. Since the pressure distribution is symmetrical around the cylinder, this is not really a surprising result. However, we know from experience that there is a significant drag developed on a cylinder when it is placed in a moving fluid. This discrepancy is known as d'Alembert's paradox (see Chapter 9).

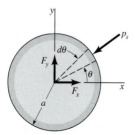

■ **FIGURE 6.26** **The notation for determining lift and drag on a circular cylinder.**

EXAMPLE 6.7   Potential Flow—Cylinder

GIVEN When a circular cylinder is placed in a uniform stream, a stagnation point is created on the cylinder as shown in Fig. E6.7a. If a small hole is located at this point, the stagnation pressure, p_{stag}, can be measured and used to determine the approach velocity, U.

FIND

(a) Show how p_{stag} and U are related.

(b) If the cylinder is misaligned by an angle α (Figure E6.7b), but the measured pressure is still interpreted as the stagnation pressure, determine an expression for the ratio of the true velocity, U, to the predicted velocity, U'. Plot this ratio as a function of α for the range $-20° \leq \alpha \leq 20°$.

SOLUTION

(a) The velocity at the stagnation point is zero so the Bernoulli equation written between a point on the stagnation streamline upstream from the cylinder and the stagnation point gives

$$\frac{p_0}{\gamma} + \frac{U^2}{2g} = \frac{p_{stag}}{\gamma}$$

Thus,

$$U = \left[\frac{2}{\rho}(p_{stag} - p_0)\right]^{1/2} \qquad \textbf{(Ans)}$$

COMMENT A measurement of the difference between the pressure at the stagnation point and the upstream pressure can be used to measure the approach velocity. This is, of course, the same result that was obtained in Section 3.5 for Pitot-static tubes.

(b) If the direction of the fluid approaching the cylinder is not known precisely, it is possible that the cylinder is misaligned by some angle, α. In this instance, the pressure actually measured, p_α, will be different from the stagnation pressure, but if the misalignment is not recognized, the predicted approach velocity, U', would still be calculated as

$$U' = \left[\frac{2}{\rho}(p_\alpha - p_0)\right]^{1/2}$$

Thus,

$$\frac{U(\text{true})}{U'(\text{predicted})} = \left(\frac{p_{stag} - p_0}{p_\alpha - p_0}\right)^{1/2} \qquad \textbf{(1)}$$

The velocity on the surface of the cylinder, v_θ, where $r = a$, is obtained from Eq. 6.108 as

$$v_\theta = -2U \sin \theta$$

If we now write the Bernoulli equation between a point upstream of the cylinder and the point on the cylinder where $r = a, \theta = \alpha$, it follows that

$$p_0 + \frac{1}{2}\rho U^2 = p_\alpha + \frac{1}{2}\rho(-2U \sin \alpha)^2$$

and, therefore,

$$p_\alpha - p_0 = \frac{1}{2}\rho U^2(1 - 4\sin^2\alpha) \qquad \textbf{(2)}$$

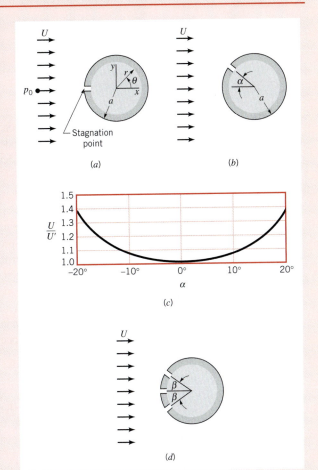

■ **FIGURE E6.7**

Since $p_{stag} - p_0 = \frac{1}{2}\rho U^2$ it follows from Eqs. 1 and 2 that

$$\frac{U(\text{true})}{U'(\text{predicted})} = (1 - 4\sin^2\alpha)^{-1/2} \qquad \textbf{(Ans)}$$

This velocity ratio is plotted as a function of the misalignment angle α in Fig. E6.7c.

COMMENT   It is clear from these results that significant errors can arise if the stagnation pressure tap is not aligned with the stagnation streamline. If two additional, symmetrically located holes are drilled on the cylinder, as illustrated in Fig. E6.7*d*, the correct orientation of the cylinder can be determined. The cylinder is rotated until the pressures in the two symmetrically placed holes are equal, thus indicating that the center hole coincides with the stagnation streamline. For $\beta = 30°$ the pressure at the two holes theoretically corresponds to the upstream pressure, p_0. With this orientation a measurement of the difference in pressure between the center hole and the side holes can be used to determine U.

An additional, interesting potential flow can be developed by adding a free vortex to the stream function or velocity potential for the flow around a cylinder. In this case

$$\psi = Ur\left(1 - \frac{a^2}{r^2}\right)\sin\theta - \frac{\Gamma}{2\pi}\ln r \tag{6.112}$$

and

$$\phi = Ur\left(1 + \frac{a^2}{r^2}\right)\cos\theta + \frac{\Gamma}{2\pi}\theta \tag{6.113}$$

where Γ is the circulation. We note that the circle $r = a$ will still be a streamline (and thus can be replaced with a solid cylinder), as the streamlines for the added free vortex are all circular. However, the tangential velocity, v_θ, on the surface of the cylinder ($r = a$) now becomes

$$v_{\theta s} = -\left.\frac{\partial\psi}{\partial r}\right|_{r=a} = -2U\sin\theta + \frac{\Gamma}{2\pi a} \tag{6.114}$$

This type of flow field could be approximately created by placing a rotating cylinder in a uniform stream. Because of the presence of viscosity in any real fluid, the fluid in contact with the rotating cylinder would rotate with the same velocity as the cylinder, and the resulting flow field would resemble that developed by the combination of a uniform flow past a cylinder and a free vortex.

A variety of streamline patterns can be developed, depending on the vortex strength, Γ. For example, from Eq. 6.114 we can determine the location of stagnation points on the surface of the cylinder. These points will occur at $\theta = \theta_{\text{stag}}$ where $v_\theta = 0$ and therefore from Eq. 6.114

$$\sin\theta_{\text{stag}} = \frac{\Gamma}{4\pi Ua} \tag{6.115}$$

If $\Gamma = 0$, then $\theta_{\text{stag}} = 0$ or π—that is, the stagnation points occur at the front and rear of the cylinder as are shown in Fig. 6.27*a*. However, for $-1 \le \Gamma/4\pi Ua \le 1$, the stagnation points will occur at some other location on the surface, as illustrated in Figs. 6.27*b,c*. If the absolute value of the parameter $\Gamma/4\pi Ua$ exceeds 1, Eq. 6.115 cannot be satisfied, and the stagnation point is located away from the cylinder, as shown in Fig. 6.27*d*.

The force per unit length developed on the cylinder can again be obtained by integrating the differential pressure forces around the circumference as in Eqs. 6.110 and 6.111. For the cylinder with circulation the surface pressure, p_s, is obtained from the Bernoulli equation (with the surface velocity given by Eq. 6.114)

$$p_0 + \frac{1}{2}\rho U^2 = p_s + \frac{1}{2}\rho\left(-2U\sin\theta + \frac{\Gamma}{2\pi a}\right)^2$$

or

$$p_s = p_0 + \frac{1}{2}\rho U^2\left(1 - 4\sin^2\theta + \frac{2\Gamma\sin\theta}{\pi aU} - \frac{\Gamma^2}{4\pi^2 a^2 U^2}\right) \tag{6.116}$$

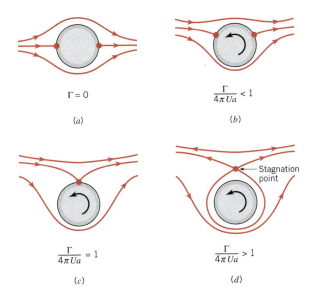

■ **FIGURE 6.27** The location of stagnation points on a circular cylinder: (*a*) without circulation and (*b, c, d*) with circulation.

Equation 6.116 substituted into Eq. 6.110 for the drag, and integrated, again yields

$$F_x = 0$$

That is, even for the rotating cylinder no force in the direction of the uniform flow is developed. However, use of Eq. 6.116 with the equation for the lift, F_y (Eq. 6.111), yields

$$F_y = -\rho U \Gamma \qquad \textbf{(6.117)}$$

Thus, for the cylinder with circulation, lift is developed equal to the product of the fluid density, the upstream velocity, and the circulation. The negative sign means that if U is positive (in the positive x direction) and Γ is positive (a free vortex with counterclockwise rotation), the direction of the F_y is downward.

Of course, if the cylinder is rotated in the clockwise direction ($\Gamma < 0$) the direction of F_y would be upward. This can be seen by studying the surface pressure distribution (Eq. 6.116), which is plotted in Fig. 6.28 for two situations. One has $\Gamma/4\pi Ua = 0$, which corresponds to no rotation of the cylinder. The other has $\Gamma/4\pi Ua = -0.25$, which corresponds to clockwise

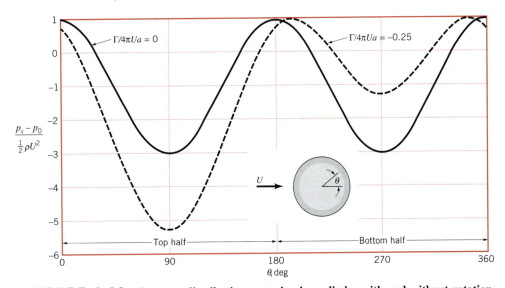

■ **FIGURE 6.28** Pressure distribution on a circular cylinder with and without rotation.

rotation of the cylinder. With no rotation the flow is symmetrical both top to bottom and front to back on the cylinder. With rotation the flow is symmetrical front to back, but not top to bottom. In this case the two stagnation points [i.e., $(p_s - p_0)/(\rho U^2/2) = 1$] are located on the bottom of the cylinder and the average pressure on the top half of the cylinder is less than that on the bottom half. The result is an upward lift force. It is this force acting in a direction perpendicular to the direction of the approach velocity that causes baseballs and golf balls to curve when they spin as they are propelled through the air. The development of this lift on rotating bodies is called the *Magnus effect*. (See Section 9.4 for further comments.)

Although Eq. 6.117 was developed for a cylinder with circulation, it gives the lift per unit length for a right cylinder of any cross-sectional shape placed in a uniform, inviscid stream. The circulation is determined around any closed curve containing the body. The generalized equation relating lift to fluid density, velocity, and circulation is called the *Kutta–Joukowski law*, and is commonly used to determine the lift on airfoils (see Section 9.4.2 and Refs. 2–6).

F l u i d s   i n   t h e   N e w s

A sailing ship without sails A sphere or cylinder spinning about its axis when placed in an airstream develops a force at right angles to the direction of the airstream. This phenomenon is commonly referred to as the *Magnus effect* and is responsible for the curved paths of baseballs and golf balls. Another lesser-known application of the Magnus effect was proposed by a German physicist and engineer, Anton Flettner, in the 1920s. Flettner's idea was to use the Magnus effect to make a ship move. To demonstrate the practicality of the "rotor-ship" he purchased a sailing schooner and replaced the ship's masts and rig-

ging with two vertical cylinders that were 50 feet high and 9 feet in diameter. The cylinders looked like smokestacks on the ship. Their spinning motion was developed by 45-hp motors. The combination of a wind and the rotating cylinders created a force (Magnus effect) to push the ship forward. The ship, named the *Baden Baden*, made a successful voyage across the Atlantic arriving in New York Harbor on May 9, 1926. Although the feasibility of the rotor-ship was clearly demonstrated, it proved to be less efficient and practical than more conventional vessels and the idea was not pursued. (See Problem 6.37.) ■

6.7 Other Aspects of Potential Flow Analysis

In the preceding section the method of superposition of basic potentials has been used to obtain detailed descriptions of irrotational flow around certain body shapes immersed in a uniform stream. It is possible to extend the idea of superposition by considering a *distribution* of sources and sinks, or doublets, which when combined with a uniform flow, can describe the flow around bodies of arbitrary shape. Techniques are available to determine the required distribution to give a prescribed body shape. Also, for plane, potential flow problems, it can be shown that the complex variable theory (the use of real and imaginary numbers) can be used effectively to obtain solutions to a great variety of important flow problems. There are, of course, numerical techniques that can be used to solve, not only plane two-dimensional problems, but the more general three-dimensional problems. Because potential flow is governed by Laplace's equation, any procedure that is available for solving this equation can be applied to the analysis of the irrotational flow of frictionless fluids. The potential flow theory is an old and well-established discipline within the general field of fluid mechanics. The interested reader can find many detailed references on this subject, including Refs. 2–6 given at the end of this chapter.

V6.4 Potential flow

An important point to remember is that regardless of the particular technique used to obtain a solution to a potential flow problem, the solution remains approximate because of the fundamental assumption of a frictionless fluid. The general differential equations that describe viscous fluid behavior and some simple solutions to these equations are considered in the remaining sections of this chapter.

6.8   Viscous Flow

To incorporate viscous effects into the differential analysis of fluid motion we must return to the previously derived general equations of motion, Eqs. 6.50. Since these equations include both stresses and velocities, there are more unknowns than equations, and therefore before proceeding it is necessary to establish a relationship between stresses and velocities.

6.8.1   Stress–Deformation Relationships

For incompressible, Newtonian fluids it is known that the stresses are linearly related to the rates of deformation and can be expressed in Cartesian coordinates as (for normal stresses)

$$\sigma_{xx} = -p + 2\mu \frac{\partial u}{\partial x} \tag{6.118a}$$

$$\sigma_{yy} = -p + 2\mu \frac{\partial v}{\partial y} \tag{6.118b}$$

$$\sigma_{zz} = -p + 2\mu \frac{\partial w}{\partial z} \tag{6.118c}$$

(for shearing stresses)

$$\tau_{xy} = \tau_{yx} = \mu \left(\frac{\partial u}{\partial y} + \frac{\partial v}{\partial x} \right) \tag{6.118d}$$

$$\tau_{yz} = \tau_{zy} = \mu \left(\frac{\partial v}{\partial z} + \frac{\partial w}{\partial y} \right) \tag{6.118e}$$

$$\tau_{zx} = \tau_{xz} = \mu \left(\frac{\partial w}{\partial x} + \frac{\partial u}{\partial z} \right) \tag{6.118f}$$

where p is the pressure, the negative of the average of the three normal stresses; that is $-p = (\frac{1}{3})(\sigma_{xx} + \sigma_{yy} + \sigma_{zz})$. For viscous fluids in motion the normal stresses are not necessarily the same in different directions—thus, the need to define the pressure as the average of the three normal stresses. For fluids at rest, or frictionless fluids, the normal stresses are equal in all directions. (We have made use of this fact in the chapter on fluid statics and in developing the equations for inviscid flow.) Detailed discussions of the development of these stress–velocity gradient relationships can be found in Refs. 3, 7, and 8. An important point to note is that whereas for elastic solids the stresses are linearly related to the deformation (or strain), for Newtonian fluids the stresses are linearly related to the rate of deformation (or rate of strain).

In cylindrical polar coordinates the stresses for Newtonian, incompressible fluids are expressed as (for normal stresses)

$$\sigma_{rr} = -p + 2\mu \frac{\partial v_r}{\partial r} \tag{6.119a}$$

$$\sigma_{\theta\theta} = -p + 2\mu \left(\frac{1}{r} \frac{\partial v_\theta}{\partial \theta} + \frac{v_r}{r} \right) \tag{6.119b}$$

$$\sigma_{zz} = -p + 2\mu \frac{\partial v_z}{\partial z} \tag{6.119c}$$

(for shearing stresses)

$$\tau_{r\theta} = \tau_{\theta r} = \mu \left[r \frac{\partial}{\partial r} \left(\frac{v_\theta}{r} \right) + \frac{1}{r} \frac{\partial v_r}{\partial \theta} \right] \qquad \textbf{(6.119d)}$$

$$\tau_{\theta z} = \tau_{z\theta} = \mu \left(\frac{\partial v_\theta}{\partial z} + \frac{1}{r} \frac{\partial v_z}{\partial \theta} \right) \qquad \textbf{(6.119e)}$$

$$\tau_{zr} = \tau_{rz} = \mu \left(\frac{\partial v_r}{\partial z} + \frac{\partial v_z}{\partial r} \right) \qquad \textbf{(6.119f)}$$

The double subscript has a meaning similar to that of stresses expressed in Cartesian coordinates—that is, the first subscript indicates the plane on which the stress acts, and the second subscript the direction. Thus, for example, σ_{rr} refers to a stress acting on a plane perpendicular to the radial direction and in the radial direction (thus a normal stress). Similarly, $\tau_{r\theta}$ refers to a stress acting on a plane perpendicular to the radial direction but in the tangential (θ direction) and is, therefore, a shearing stress.

6.8.2 The Navier–Stokes Equations

The stresses as defined in the preceding section can be substituted into the differential equations of motion (Eqs. 6.50) and simplified by using the continuity equation for incompressible flow (Eq. 6.31) to obtain (x direction)

$$\rho \left(\frac{\partial u}{\partial t} + u \frac{\partial u}{\partial x} + v \frac{\partial u}{\partial y} + w \frac{\partial u}{\partial z} \right) = -\frac{\partial p}{\partial x} + \rho g_x + \mu \left(\frac{\partial^2 u}{\partial x^2} + \frac{\partial^2 u}{\partial y^2} + \frac{\partial^2 u}{\partial z^2} \right) \qquad \textbf{(6.120a)}$$

(y direction)

$$\rho \left(\frac{\partial v}{\partial t} + u \frac{\partial v}{\partial x} + v \frac{\partial v}{\partial y} + w \frac{\partial v}{\partial z} \right) = -\frac{\partial p}{\partial y} + \rho g_y + \mu \left(\frac{\partial^2 v}{\partial x^2} + \frac{\partial^2 v}{\partial y^2} + \frac{\partial^2 v}{\partial z^2} \right) \qquad \textbf{(6.120b)}$$

(z direction)

$$\rho \left(\frac{\partial w}{\partial t} + u \frac{\partial w}{\partial x} + v \frac{\partial w}{\partial y} + w \frac{\partial w}{\partial z} \right) = -\frac{\partial p}{\partial z} + \rho g_z + \mu \left(\frac{\partial^2 w}{\partial x^2} + \frac{\partial^2 w}{\partial y^2} + \frac{\partial^2 w}{\partial z^2} \right) \qquad \textbf{(6.120c)}$$

where we have rearranged the equations so the acceleration terms are on the left side and the force terms are on the right. These equations are commonly called the *Navier–Stokes equations,* named in honor of the French mathematician L. M. H. Navier and the English mechanician Sir G. G. Stokes, who were responsible for their formulation. These three equations of motion, when combined with the conservation of mass equation (Eq. 6.31), provide a complete mathematical description of the flow of incompressible, Newtonian fluids. We have four equations and four unknowns (u, v, w, and p), and therefore the problem is "well posed" in mathematical terms. Unfortunately, because of the general complexity of the Navier–Stokes equations (they are nonlinear, second order, partial differential equations), they are not amenable to exact mathematical solutions except in a few instances. However, in those few instances in which solutions have been obtained and compared with experimental results, the results have been in close agreement. Thus, the Navier–Stokes equations are considered to be the governing differential equations of motion for incompressible, Newtonian fluids.

In terms of cylindrical polar coordinates (see Fig. 6.6) the Navier–Stokes equations can be written as (r direction)

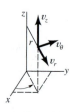

$$\rho \left(\frac{\partial v_r}{\partial t} + v_r \frac{\partial v_r}{\partial r} + \frac{v_\theta}{r} \frac{\partial v_r}{\partial \theta} - \frac{v_\theta^2}{r} + v_z \frac{\partial v_r}{\partial z} \right)$$

$$= -\frac{\partial p}{\partial r} + \rho g_r + \mu \left[\frac{1}{r} \frac{\partial}{\partial r} \left(r \frac{\partial v_r}{\partial r} \right) - \frac{v_r}{r^2} + \frac{1}{r^2} \frac{\partial^2 v_r}{\partial \theta^2} - \frac{2}{r^2} \frac{\partial v_\theta}{\partial \theta} + \frac{\partial^2 v_r}{\partial z^2} \right] \qquad \textbf{(6.121a)}$$

(θ direction)

$$\rho\left(\frac{\partial v_\theta}{\partial t} + v_r\frac{\partial v_\theta}{\partial r} + \frac{v_\theta}{r}\frac{\partial v_\theta}{\partial \theta} + \frac{v_r v_\theta}{r} + v_z\frac{\partial v_\theta}{\partial z}\right)$$
$$= -\frac{1}{r}\frac{\partial p}{\partial \theta} + \rho g_\theta + \mu\left[\frac{1}{r}\frac{\partial}{\partial r}\left(r\frac{\partial v_\theta}{\partial r}\right) - \frac{v_\theta}{r^2} + \frac{1}{r^2}\frac{\partial^2 v_\theta}{\partial \theta^2} + \frac{2}{r^2}\frac{\partial v_r}{\partial \theta} + \frac{\partial^2 v_\theta}{\partial z^2}\right] \quad \textbf{(6.121b)}$$

(z direction)

$$\rho\left(\frac{\partial v_z}{\partial t} + v_r\frac{\partial v_z}{\partial r} + \frac{v_\theta}{r}\frac{\partial v_z}{\partial \theta} + v_z\frac{\partial v_z}{\partial z}\right)$$
$$= -\frac{\partial p}{\partial z} + \rho g_z + \mu\left[\frac{1}{r}\frac{\partial}{\partial r}\left(r\frac{\partial v_z}{\partial r}\right) + \frac{1}{r^2}\frac{\partial^2 v_z}{\partial \theta^2} + \frac{\partial^2 v_z}{\partial z^2}\right] \quad \textbf{(6.121c)}$$

To provide a brief introduction to the use of the Navier–Stokes equations, a few of the simplest exact solutions are developed in the next section. Although these solutions will prove to be relatively simple, this is not the case in general. In fact, only a few other exact solutions have been obtained.

6.9   Some Simple Solutions for Viscous, Incompressible Fluids

A principal difficulty in solving the Navier–Stokes equations is because of their nonlinearity arising from the convective acceleration terms (i.e., $u\,\partial u/\partial x$, $w\,\partial v/\partial z$). There are no general analytical schemes for solving nonlinear partial differential equations (e.g., superposition of solutions cannot be used), and each problem must be considered individually. For most practical flow problems, fluid particles do have accelerated motion as they move from one location to another in the flow field. Thus, the convective acceleration terms are usually important. However, there are a few special cases for which the convective acceleration vanishes because of the nature of the geometry of the flow system. In these cases exact solutions are usually possible. The Navier–Stokes equations apply to both laminar and turbulent flow, but for turbulent flow each velocity component fluctuates randomly with respect to time, and this added complication makes an analytical solution intractable. Thus, the exact solutions referred to are for laminar flows in which the velocity is either independent of time (steady flow) or dependent on time (unsteady flow) in a well-defined manner.

6.9.1   Steady, Laminar Flow between Fixed Parallel Plates

We first consider flow between the two horizontal, infinite parallel plates of Fig. 6.29a. For this geometry the fluid particles move in the x direction parallel to the plates, and there is no velocity in the y or z direction—that is, $v = 0$ and $w = 0$. In this case it follows from the continuity equation (Eq. 6.31) that $\partial u/\partial x = 0$. Furthermore, there would be no variation of u in the z direction for infinite plates, and for steady flow $\partial u/\partial t = 0$ so that $u = u(y)$. If these conditions are used in the Navier–Stokes equations (Eqs. 6.120), they reduce to

$$0 = -\frac{\partial p}{\partial x} + \mu\left(\frac{\partial^2 u}{\partial y^2}\right) \quad \textbf{(6.122)}$$

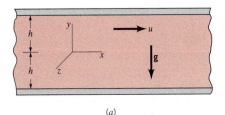
(a)

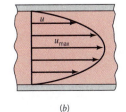
(b)

■ **FIGURE 6.29**
Viscous flow between parallel plates: (a) coordinate system and notation used in analysis and (b) parabolic velocity distribution for flow between parallel fixed plates.

$$0 = -\frac{\partial p}{\partial y} - \rho g \qquad (6.123)$$

$$0 = -\frac{\partial p}{\partial z} \qquad (6.124)$$

where we have set $g_x = 0$, $g_y = -g$, and $g_z = 0$. That is, the y axis points up. We see that for this particular problem the Navier-Stokes equations reduce to some rather simple equations. Equations 6.123 and 6.124 can be integrated to yield

$$p = -\rho g y + f_1(x) \qquad (6.125)$$

which shows that the pressure varies hydrostatically in the y direction. Equation 6.122, rewritten as

$$\frac{d^2 u}{dy^2} = \frac{1}{\mu} \frac{\partial p}{\partial x}$$

can be integrated to give

$$\frac{du}{dy} = \frac{1}{\mu} \left(\frac{\partial p}{\partial x} \right) y + c_1$$

and integrated again to yield

$$u = \frac{1}{2\mu} \left(\frac{\partial p}{\partial x} \right) y^2 + c_1 y + c_2 \qquad (6.126)$$

V6.5 No-slip
boundary condition

Note that for this simple flow the pressure gradient, $\partial p / \partial x$, is treated as constant as far as the integration is concerned, as (as shown in Eq. 6.125) it is not a function of y. The two constants, c_1 and c_2, must be determined from the boundary conditions. For example, if the two plates are fixed, then $u = 0$ for $y = \pm h$ (because of the no-slip condition for viscous fluids). To satisfy this condition, $c_1 = 0$ and

$$c_2 = -\frac{1}{2\mu} \left(\frac{\partial p}{\partial x} \right) h^2$$

Thus, the velocity distribution becomes

$$u = \frac{1}{2\mu} \left(\frac{\partial p}{\partial x} \right) (y^2 - h^2) \qquad (6.127)$$

Equation 6.127 shows that the velocity profile between the two fixed plates is parabolic as illustrated in Fig. 6.29b.

The volume rate of flow, q, passing between the plates (for a unit width in the z direction) is obtained from the relationship

$$q = \int_{-h}^{h} u \, dy = \int_{-h}^{h} \frac{1}{2\mu} \left(\frac{\partial p}{\partial x} \right) (y^2 - h^2) \, dy$$

or

$$q = -\frac{2h^3}{3\mu} \left(\frac{\partial p}{\partial x} \right) \qquad (6.128)$$

The pressure gradient $\partial p / \partial x$ is negative, as the pressure decreases in the direction of flow. If we let Δp represent the pressure *drop* between two points a distance ℓ apart, then

$$\frac{\Delta p}{\ell} = -\frac{\partial p}{\partial x}$$

and Eq. 6.128 can be expressed as

$$q = \frac{2h^3 \, \Delta p}{3\mu\ell} \qquad \text{(6.129)}$$

The flow is proportional to the pressure gradient, inversely proportional to the viscosity, and strongly dependent ($\sim h^3$) on the gap width. In terms of the mean velocity, V, where $V = q/2h$, Eq. 6.129 becomes

$$V = \frac{h^2 \, \Delta p}{3\mu\ell} \qquad \text{(6.130)}$$

Equations 6.129 and 6.130 provide convenient relationships for relating the pressure drop along a parallel-plate channel and the rate of flow or mean velocity. The maximum velocity, u_{max}, occurs midway ($y = 0$) between the two plates so that from Eq. 6.127

$$u_{max} = -\frac{h^2}{2\mu}\left(\frac{\partial p}{\partial x}\right)$$

or

$$u_{max} = \tfrac{3}{2}V \qquad \text{(6.131)}$$

Details of the steady, laminar flow between infinite parallel plates are completely predicted by this solution to the Navier–Stokes equations. For example, if the pressure gradient, viscosity, and plate spacing are specified, then from Eq. 6.127 the velocity profile can be determined, and from Eqs. 6.129 and 6.130 the corresponding flowrate and mean velocity determined. In addition, from Eq. 6.125 it follows that

$$f_1(x) = \left(\frac{\partial p}{\partial x}\right)x + p_0$$

where p_0 is a reference pressure at $x = y = 0$, and the pressure variation throughout the fluid can be obtained from

$$p = -\rho g y + \left(\frac{\partial p}{\partial x}\right)x + p_0 \qquad \text{(6.132)}$$

For a given fluid and reference pressure, p_0, the pressure at any point can be predicted. This relatively simple example of an exact solution illustrates the detailed information about the flow field that can be obtained. The flow will be laminar if the Reynolds number, $Re = \rho V(2h)/\mu$, remains below about 1400. For flow with larger Reynolds numbers the flow becomes turbulent, and the preceding analysis is not valid since the flow field is complex, three dimensional, and unsteady.

F l u i d s   i n   t h e   N e w s

10 tons on 8 psi Place a golf ball on the end of a garden hose and then slowly turn the water on a small amount until the ball just barely lifts off the end of the hose, leaving a small gap between the ball and the hose. The ball is free to rotate. This is the idea behind the new "floating ball water fountains" developed in Finland. Massive, 10-ton, 6-ft-diameter stone spheres are supported by the pressure force of the water on the curved surface within a pedestal and rotate so easily that even a small child can change their direction of rotation. The key to the fountain design is the ability to grind and polish stone to an accuracy of a few thousandts of an inch. This allows the gap between the ball and its pedestal to be very small (on the order of 5/1000 in.) and the water flowrate correspondingly small (on the order of 5 gallons per minute). Due to the small gap, the flow in the gap is essentially that of *flow between parallel plates*. Although the sphere is very heavy, the pressure under the sphere within the pedestal needs to be only about 8 psi. (See Problem 6.44.) ∎

6.9.2 Couette Flow

Another simple parallel-plate flow can be developed by fixing one plate and letting the other plate move with a constant velocity, U, as illustrated in Fig. 6.30a. The Navier–Stokes equations reduce to the same form as those in the preceding section, and the solutions for pressure and velocity distribution are still given by Eqs. 6.125 and 6.126, respectively. However, for the moving plate problem the boundary conditions for the velocity are different. For this case we locate the origin of the coordinate system at the bottom plate and designate the distance between the two plates as b (see Fig. 6.30a). The two constants, c_1 and c_2, in Eq. 6.126 can be determined from the boundary conditions, $u = 0$ at $y = 0$ and $u = U$ at $y = b$. It follows that

$$u = U\frac{y}{b} + \frac{1}{2\mu}\left(\frac{\partial p}{\partial x}\right)(y^2 - by) \tag{6.133}$$

or, in dimensionless form,

$$\frac{u}{U} = \frac{y}{b} - \frac{b^2}{2\mu U}\left(\frac{\partial p}{\partial x}\right)\left(\frac{y}{b}\right)\left(1 - \frac{y}{b}\right) \tag{6.134}$$

The actual velocity profile will depend on the dimensionless parameter

$$P = -\frac{b^2}{2\mu U}\left(\frac{\partial p}{\partial x}\right)$$

Several profiles are shown in Fig. 6.30b. This type of flow is called ***Couette flow.***

The simplest type of Couette flow is one for which the pressure gradient is zero; that is, the fluid motion is caused by the fluid being dragged along by the moving boundary. In this case, with $\partial p/\partial x = 0$, Eq. 6.133 simply reduces to

$$u = U\frac{y}{b} \tag{6.135}$$

which indicates that the velocity varies linearly between the two plates as shown in Fig. 6.30b for $P = 0$.

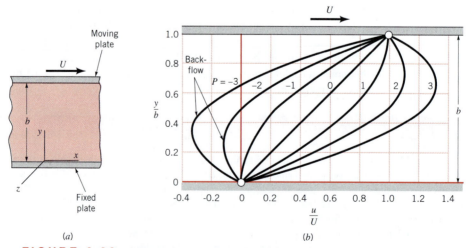

(a)　　　　　　　　　　　　　　(b)

■ **FIGURE 6.30** **Viscous flow between parallel plates with the bottom plate fixed and the upper plate moving (Couette flow): (a) coordinate system and notation used in analysis and (b) velocity distribution as a function of parameter, P, where $P = -(b^2/2\mu U)\partial p/\partial x$. (From Ref. 8, used by permission.)**

EXAMPLE 6.8   **Plane Couette Flow**

GIVEN A wide moving belt passes through a container of a viscous liquid. The belt moves vertically upward with a constant velocity, V_0, as illustrated in Fig. E6.8a. Because of viscous forces the belt picks up a film of fluid of thickness h. Gravity tends to make the fluid drain down the belt. Assume that the flow is laminar, steady, and fully developed.

FIND Use the Navier–Stokes equations to determine an expression for the average velocity of the fluid film as it is dragged up the belt.

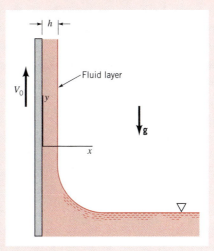

■ **FIGURE E6.8a**

SOLUTION

Since the flow is assumed to be uniform, the only velocity component is in the y direction (the v component) so that $u = w = 0$. It follows from the continuity equation that $\partial v / \partial y = 0$, and for steady flow $\partial v / \partial t = 0$, so that $v = v(x)$. Under these conditions the Navier–Stokes equations for the x direction (Eq. 6.120a) and the z direction (perpendicular to the paper) (Eq. 6.120c) simply reduce to

$$\frac{\partial p}{\partial x} = 0 \qquad \frac{\partial p}{\partial z} = 0$$

This result indicates that the pressure does not vary over a horizontal plane, and because the pressure on the surface of the film ($x = h$) is atmospheric, the pressure throughout the film must be atmospheric (or zero gage pressure). The equation of motion in the y direction (Eq. 6.120b) thus reduces to

$$0 = -\rho g + \mu \frac{d^2 v}{dx^2}$$

or

$$\frac{d^2 v}{dx^2} = \frac{\gamma}{\mu} \tag{1}$$

Integration of Eq. 1 yields

$$\frac{dv}{dx} = \frac{\gamma}{\mu} x + c_1 \tag{2}$$

On the film surface ($x = h$) we assume the shearing stress is zero—that is, the drag of the air on the film is negligible. The shearing stress at the free surface (or any interior parallel surface) is designated as τ_{xy} where from Eq. 6.118d

$$\tau_{xy} = \mu \left(\frac{dv}{dx} \right)$$

Thus, if $\tau_{xy} = 0$ at $x = h$, it follows from Eq. 2 that

$$c_1 = -\frac{\gamma h}{\mu}$$

A second integration of Eq. 2 gives the velocity distribution in the film as

$$v = \frac{\gamma}{2\mu} x^2 - \frac{\gamma h}{\mu} x + c_2$$

At the belt ($x = 0$) the fluid velocity must match the belt velocity, V_0, so that

$$c_2 = V_0$$

and the velocity distribution is therefore

$$v = \frac{\gamma}{2\mu} x^2 - \frac{\gamma h}{\mu} x + V_0 \tag{3}$$

With the velocity distribution known we can determine the flowrate per unit width, q, from the relationship

$$q = \int_0^h v \, dx = \int_0^h \left(\frac{\gamma}{2\mu} x^2 - \frac{\gamma h}{\mu} x + V_0 \right) dx$$

and thus,

$$q = V_0 h - \frac{\gamma h^3}{3\mu}$$

The average film velocity, V (where $q = Vh$), is therefore

$$V = V_0 - \frac{\gamma h^2}{3\mu} \tag{Ans}$$

COMMENT Equation (3) can be written in dimensionless form as

$$\frac{v}{V_0} = c\left(\frac{x}{h}\right)^2 - 2c\left(\frac{x}{h}\right) + 1$$

where $c = \gamma h^2 / 2\mu V_0$. This velocity profile is shown in Fig. E6.8*b*. Note that even though the belt is moving upward, for $c > 1$ (e.g., for fluids with small enough viscosity or with a small enough belt speed) there are portions of the fluid that flow downward (as indicated by $v/V_0 < 0$).

It is interesting to note from this result that there will be a net upward flow of liquid (positive V) only if $V_0 > \gamma h^2/3\mu$. It takes a relatively large belt speed to lift a small viscosity fluid.

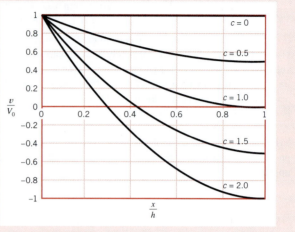

■ **FIGURE E6.8*b***

6.9.3   Steady, Laminar Flow in Circular Tubes

Probably the best known exact solution to the Navier–Stokes equations is for steady, incompressible, laminar flow through a straight circular tube of constant cross section. This type of flow is commonly called *Hagen–Poiseuille flow,* or simply *Poiseuille flow.* It is named in honor of J. L. Poiseuille, a French physician, and G. H. L. Hagen, a German hydraulic engineer. Poiseuille was interested in blood flow through capillaries and deduced experimentally the resistance laws for laminar flow through circular tubes. Hagen's investigation of flow in tubes was also experimental. It was actually after the work of Hagen and Poiseuille that the theoretical results presented in this section were determined, but their names are commonly associated with the solution of this problem.

Consider the flow through a horizontal circular tube of radius R as shown in Fig. 6.31*a*. Because of the cylindrical geometry, it is convenient to use cylindrical coordinates. We assume that the flow is parallel to the walls so that $v_r = 0$ and $v_\theta = 0$, and from the continuity equation (Eq. 6.35) $\partial v_z/\partial z = 0$. Also, for steady, axisymmetric flow, v_z is not a function of t or θ so the velocity, v_z, is only a function of the radial position within the tube—that is, $v_z = v_z(r)$. Under these conditions the Navier–Stokes equations (Eqs. 6.121) reduce to

$$0 = -\rho g \sin\theta - \frac{\partial p}{\partial r} \tag{6.136}$$

$$0 = -\rho g \cos\theta - \frac{1}{r}\frac{\partial p}{\partial \theta} \tag{6.137}$$

$$0 = -\frac{\partial p}{\partial z} + \mu \left[\frac{1}{r}\frac{\partial}{\partial r}\left(r\frac{\partial v_z}{\partial r}\right)\right] \tag{6.138}$$

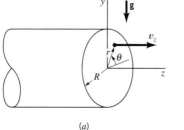

(a)

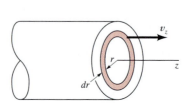

(b)

■ **FIGURE 6.31**
Viscous flow in a horizontal, circular tube: (*a*) coordinate system and notation used in analysis and (*b*) flow through a differential annular ring.

where we have used the relationships $g_r = -g \sin \theta$ and $g_\theta = -g \cos \theta$ (with θ measured from the horizontal plane).

Equations 6.136 and 6.137 can be integrated to give

$$p = -\rho g(r \sin \theta) + f_1(z)$$

or

$$p = -\rho g y + f_1(z) \tag{6.139}$$

Equation 6.139 indicates that the pressure is distributed hydrostatically at any particular cross section, and the z component of the pressure gradient, $\partial p/\partial z$, is not a function of r or θ.

The equation of motion in the z direction (Eq. 6.138) can be written in the form

$$\frac{1}{r}\frac{\partial}{\partial r}\left(r\frac{\partial v_z}{\partial r}\right) = \frac{1}{\mu}\frac{\partial p}{\partial z}$$

and integrated (using the fact that $\partial p/\partial z = $ constant) to give

$$r\frac{\partial v_z}{\partial r} = \frac{1}{2\mu}\left(\frac{\partial p}{\partial z}\right)r^2 + c_1$$

Integrating again we obtain

$$v_z = \frac{1}{4\mu}\left(\frac{\partial p}{\partial z}\right)r^2 + c_1 \ln r + c_2 \tag{6.140}$$

Since we wish v_z to be finite at the center of the tube ($r = 0$), it follows that $c_1 = 0$ [since $\ln (0) = -\infty$]. At the wall ($r = R$) the velocity must be zero so that

$$c_2 = -\frac{1}{4\mu}\left(\frac{\partial p}{\partial z}\right)R^2$$

and the velocity distribution becomes

$$v_z = \frac{1}{4\mu}\left(\frac{\partial p}{\partial z}\right)(r^2 - R^2) \tag{6.141}$$

V6.6 Laminar flow

Thus, at any cross section the velocity distribution is parabolic.

To obtain a relationship between the volume rate of flow, Q, passing through the tube and the pressure gradient, we consider the flow through the differential, washer-shaped ring of Fig. 6.31b. Since v_z is constant on this ring, the volume rate of flow through the differential area $dA = (2\pi r)\,dr$ is

$$dQ = v_z(2\pi r)\,dr$$

and therefore

$$Q = 2\pi \int_0^R v_z r\,dr \tag{6.142}$$

Equation 6.141 for v_z can be substituted into Eq. 6.142, and the resulting equation integrated to yield

$$Q = -\frac{\pi R^4}{8\mu}\left(\frac{\partial p}{\partial z}\right) \tag{6.143}$$

This relationship can be expressed in terms of the pressure *drop,* Δp, which occurs over a length, ℓ, along the tube, as

$$\frac{\Delta p}{\ell} = -\frac{\partial p}{\partial z}$$

and therefore

$$Q = \frac{\pi R^4 \Delta p}{8 \mu \ell} \tag{6.144}$$

For a given pressure drop per unit length, the volume rate of flow is inversely proportional to the viscosity and proportional to the tube radius to the fourth power. A doubling of the tube radius produces a sixteenfold increase in flow! Equation 6.144 is commonly called *Poiseuille's law.*

In terms of mean velocity, V, where $V = Q/\pi R^2$, Eq. 6.144 becomes

$$V = \frac{R^2 \, \Delta p}{8 \mu \ell} \tag{6.145}$$

The maximum velocity, v_{max}, occurs at the center of the tube, where from Eq. 6.141

$$v_{max} = -\frac{R^2}{4\mu}\left(\frac{\partial p}{\partial z}\right) = \frac{R^2 \, \Delta p}{4 \mu \ell} \tag{6.146}$$

so that

$$v_{max} = 2V$$

The velocity distribution can be written in terms of v_{max} as

$$\frac{v_z}{v_{max}} = 1 - \left(\frac{r}{R}\right)^2 \tag{6.147}$$

As was true for the similar case of flow between parallel plates (sometimes referred to as *plane Poiseuille flow*), a very detailed description of the pressure and velocity distribution in tube flow results from this solution to the Navier–Stokes equations. Numerous experiments performed to substantiate the theoretical results show that the theory and experiment are in agreement for the laminar flow of Newtonian fluids in circular tubes or pipes. In general, the flow remains laminar for Reynolds numbers, Re $= \rho V(2R)/\mu$, below 2100. Turbulent flow in tubes is considered in Chapter 8.

F l u i d s   i n   t h e   N e w s

Poiseuille's law revisited Poiseuille's law governing *laminar flow* of fluids in tubes has an unusual history. It was developed in 1842 by a French physician, J. L. M. Poiseuille, who was interested in the flow of blood in capillaries. Poiseuille, through a series of carefully conducted experiments using water flowing through very small tubes, arrived at the formula, $Q = K\Delta p \, D^4/\ell$. In this formula Q is the flowrate, K an empirical constant, Δp the pressure drop over the length ℓ, and D the tube diameter. Another formula was given for the value of K as a function of the water temperature. It was not until the concept of viscosity was introduced at a later date that Poiseuille's law was derived mathematically and the constant K found to be equal to $\pi/8\mu$, where μ is the fluid viscosity. The experiments by Poiseuille have long been admired for their accuracy and completeness considering the laboratory instrumentation available in the mid-nineteenth century. ■

6.10   Other Aspects of Differential Analysis

In this chapter the basic differential equations that govern the flow of fluids have been developed. The Navier–Stokes equations, which can be expressed compactly in vector notation as

$$\rho\left(\frac{\partial \mathbf{V}}{\partial t} + \mathbf{V} \cdot \nabla\mathbf{V}\right) = -\nabla p + \rho\mathbf{g} + \mu \nabla^2\mathbf{V} \tag{6.148}$$

along with the continuity equation

$$\nabla \cdot \mathbf{V} = 0 \tag{6.149}$$

are the general equations of motion for incompressible, Newtonian fluids. Although we have restricted our attention to incompressible fluids, these equations can be readily extended to include compressible fluids. It is well beyond the scope of this introductory text to consider in depth the variety of analytical and numerical techniques that can be used to obtain both exact and approximate solutions to the Navier–Stokes equations.

Numerical techniques using digital computers are, of course, used commonly to solve a wide variety of flow problems. Numerous laminar flow solutions based on the full Navier–Stokes equations have been obtained by using both finite element and finite difference methods. Flow characteristics of highly nonlinear problems can be obtained from numerical solutions. However, these solutions typically require the use of powerful digital computers, and nonlinearities in the equations represent a complication that challenges the ingenuity of the numerical analyst. Practical turbulent flow problems are not amenable to a completely numerical solution because of the extreme complexity of the motion. The solution of real turbulent flows usually involves the use of some type of empirical turbulence model.

The general field of computational fluid dynamics (CFD), in which computers and numerical analysis are combined to solve fluid flow problems, represents an extremely important subject area in advanced fluid mechanics. A brief introduction to CFD is given in Appendix A.

V6.7 CFD example

6.11 Chapter Summary and Study Guide

volumetric dilatation rate
vorticity
irrotational flow
continuity equation
stream function
Euler's equations of motion
ideal fluid
Bernoulli equation
velocity potential
equipotential lines

Differential analysis of fluid flow is concerned with the development of concepts and techniques that can be used to provide a detailed, point by point, description of a flow field. Concepts related to the motion and deformation of a fluid element are introduced, including the Eulerian method for describing the velocity and acceleration of fluid particles. Linear deformation and angular deformation of a fluid element are described through the use of flow characteristics such as the volumetric dilatation rate, rate of angular deformation, and vorticity. The differential form of the conservation of mass equation (continuity equation) is derived in both rectangular and cylindrical polar coordinates.

Use of the stream function for the study of steady, incompressible, plane, two-dimensional flow is introduced. The general equations of motion are developed, and for inviscid flow these equations are reduced to the simpler Euler equations of motion. The Euler equations are integrated to give the Bernoulli equation, and the concept of irrotational flow is introduced. Use of the velocity potential for describing irrotational flow is considered in detail, and several basic velocity potentials are described, including those for a uniform flow, source or sink, vortex, and doublet. The technique of using various combinations of these basic velocity potentials, by superposition, to form new potentials is described. Flows around a half-body and around a circular cylinder are obtained using this superposition technique.

Basic differential equations describing incompressible, viscous flow (the Navier–Stokes equations) are introduced. Several relatively simple solutions for steady, viscous, laminar flow between parallel plates and through circular tubes are included.

The following checklist provides a study guide for this chapter. When your study of the entire chapter and end-of-chapter exercises has been completed you should be able to

- write out meanings of the terms listed here in the margin and understand each of the related concepts. These terms are particularly important and are set in color and bold type in the text.
- determine the acceleration of a fluid particle, given the equation for the velocity field.
- determine the volumetric dilatation rate, vorticity, and rate of angular deformation for a fluid element, given the equation for the velocity field.
- show that a given velocity field satisfies the continuity equation.
- use the concept of the stream function to describe a flow field.
- use the concept of the velocity potential to describe a flow field.
- use superposition of basic velocity potentials to describe simple potential flow fields.
- use the Navier–Stokes equations to determine the detailed flow characteristics of incompressible, steady, laminar, viscous flow between parallel plates and through circular tubes.

References

1. White, F. M., *Fluid Mechanics,* Fifth Edition, McGraw-Hill, New York, 2003.
2. Street, V. L., *Fluid Dynamics,* McGraw-Hill, New York, 1948.
3. Rouse, H., *Advanced Mechanics of Fluids,* Wiley, New York, 1959.
4. Milne-Thomson, L. M., *Theoretical Hydrodynamics,* Fourth Edition, Macmillan, New York, 1960.
5. Robertson, J. M., *Hydrodynamics in Theory and Application,* Prentice-Hall, Englewood Cliffs, N.J., 1965.
6. Panton, R. L., *Incompressible Flow,* Third Edition, Wiley, New York, 2005.
7. Li, W. H., and Lam, S. H., *Principles of Fluid Mechanics,* Addison-Wesley, Reading, Mass., 1964.
8. Schlichting, H., *Boundary-Layer Theory,* Eighth Edition, McGraw-Hill, New York, 2000.

Review Problems

Go to Appendix F for a set of review problems with answers. Detailed solutions can be found in *Student Solution Manual for a Brief Introduction to Fluid Mechanics,* by Young et al. (© 2006 John Wiley and Sons, Inc.)

Problems

Note: Unless otherwise indicated, use the values of fluid properties found in the tables on the inside of the front cover. Problems designated with an (*) are intended to be solved with the aid of a programmable calculator or a computer.

Answers to the even-numbered problems are listed at the end of the book. Access to the videos that accompany problems can be obtained through the book's web site, www.wiley.com/college/young. FE problems can also be accessed on this web site.

Section 6.1 Fluid Element Kinematics

6.1 The velocity in a certain flow field is given by the equation

$$\mathbf{V} = (3x^2 + 1)\hat{\mathbf{i}} - 6xy\hat{\mathbf{j}}$$

Determine the expressions for the two rectangular components of acceleration.

6.2 Determine an expression for the vorticity of the flow field described by

$$\mathbf{V} = -4xy^3\,\hat{\mathbf{i}} + y^4\hat{\mathbf{j}}$$

Is the flow irrotational?

6.3 The three components of velocity in a flow field are given by

$$u = x^2 + y^2 + z^2$$
$$v = xy + yz + z^2$$
$$w = -3xz - z^2/2 + 4$$

(a) Determine the volumetric dilatation rate and interpret the results. **(b)** Determine an expression for the rotation vector. Is this an irrotational flow field?

6.4 An incompressible viscous fluid is placed between two large parallel plates as shown in Fig. P6.4. The bottom plate is fixed and the upper plate moves with a constant velocity, U. For these conditions the velocity distribution between the plates is linear and can be expressed as

$$u = U\frac{y}{b}$$

Determine **(a)** the volumetric dilatation rate, **(b)** the rotation vector, **(c)** the vorticity, and **(d)** the rate of angular deformation.

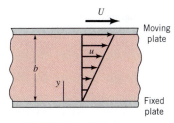

■ **FIGURE P6.4**

6.5 A viscous fluid is contained in the space between concentric cylinders. The inner wall is fixed and the outer wall rotates with an angular velocity ω. (See Fig. P6.5a and **Video V6.1**.) Assume that the velocity distribution in the gap is linear as illustrated in Fig. P6.5b. For the small rectangular element shown

in Fig. P6.5b, determine the rate of change of the right angle γ due to the fluid motion. Express your answer in terms of r_0, r_i, and ω.

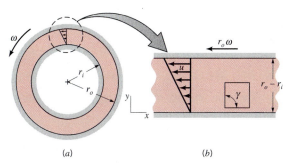

■ **FIGURE P6.5**

Section 6.2 Conservation of Mass

6.6 For a certain incompressible, two-dimensional flow field the velocity component in the y direction is given by the equation

$$v = 3xy - x^2y$$

Determine the velocity component in the x direction so that the continuity equation is satisfied.

6.7 The velocity components of an incompressible, two-dimensional velocity field are given by the equations

$$u = y^2 - x(1 + x)$$
$$v = y(2x + 1)$$

Show that the flow is irrotational and satisfies conservation of mass.

6.8 For a certain two-dimensional flow field

$$u = 0$$
$$v = V$$

(a) What are the corresponding radial and tangential velocity components? **(b)** Determine the corresponding stream function expressed in Cartesian coordinates and in cylindrical polar coordinates.

6.9 For each of the following stream functions, with units of m^2/s, determine the magnitude and the angle the velocity vector makes with the x axis at $x = 1$ m, $y = 2$ m. Locate any stagnation points in the flow field.

(a) $\psi = xy$
(b) $\psi = -2x^2 + y$

6.10 In a two-dimensional, incompressible flow field, the x component of velocity is given by the equation $u = 2x$. **(a)** Determine the corresponding equation for the y component of velocity if $v = 0$ along the x axis. **(b)** For this flow field, what is the magnitude of the average velocity of the fluid crossing the surface OA of Fig. P6.10? Assume that the velocities are in feet per second when x and y are in feet.

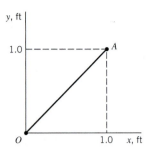

■ **FIGURE P6.10**

6.11 The radial velocity component in an incompressible, two-dimensional flow field ($v_z = 0$) is

$$v_r = 2r + 3r^2 \sin \theta$$

Determine the corresponding tangential velocity component, v_θ, required to satisfy conservation of mass.

6.12 A two-dimensional, incompressible flow is given by $u = -y$ and $v = x$. Show that the streamline passing through the point $x = 10$ and $y = 0$ is a circle centered at the origin.

6.13 The velocity components in an incompressible, two-dimensional flow field are given by the equations.

$$u = x^2$$
$$v = -2xy + x$$

Determine, if possible, the corresponding stream function.

6.14 It is proposed that a two-dimensional, incompressible flow field be described by the velocity components

$$u = Ay$$
$$v = Bx$$

where A and B are both positive constants. **(a)** Will the continuity equation be satisfied? **(b)** Is the flow irrotational? **(c)** Determine the equation for the streamlines and show a sketch of the streamline that passes through the origin. Indicate the direction of flow along this streamline.

Section 6.4 Inviscid Flow

6.15 The stream function for a given two-dimensional flow field is

$$\psi = 5x^2 y - (5/3)y^3$$

Determine the corresponding velocity potential.

6.16 It is known that the velocity distribution for two-dimensional flow of a viscous fluid between wide, fixed parallel plates (Fig. P6.16) is parabolic; that is

$$u = U_c \left[1 - \left(\frac{y}{h} \right)^2 \right]$$

with $v = 0$. Determine, if possible, the corresponding stream function and velocity potential.

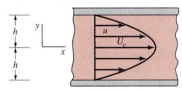

■ **FIGURE P6.16**

6.17 The velocity potential for a certain inviscid, incompressible flow field is given by the equation

$$\phi = 2x^2 y - \left(\tfrac{2}{3} \right) y^3$$

where ϕ has the units of m²/s when x and y are in meters. Determine the pressure at the point $x = 2$ m, $y = 2$ m if the pressure at $x = 1$ m, $y = 1$ m is 200 kPa. Elevation changes can be neglected and the fluid is water.

6.18 Consider the incompressible, two-dimensional flow of a nonviscous fluid between the boundaries shown in Fig. P6.18. The velocity potential for this flow field is

$$\phi = x^2 - y^2$$

(a) Determine the corresponding stream function. **(b)** What is the relationship between the discharge, q (per unit width normal to plane of paper), passing between the walls and the coordinates x_i, y_i of any point on the curved wall? Neglect body forces.

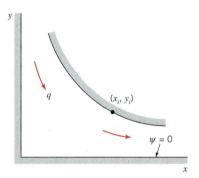

■ **FIGURE P6.18**

6.19 The velocity components in an ideal, two-dimensional velocity field are given by the equations

$$u = 3(x^2 - y^2)$$
$$v = -6xy$$

All body forces are negligible. **(a)** Does this velocity field satisfy the continuity equation? **(b)** Determine the equation for the pressure gradient in the y direction at any point in the field.

6.20 The streamlines for an incompressible, inviscid, two-dimensional flow field are all concentric circles, and the velocity varies directly with the distance from the common center of the streamlines; that is

$$v_\theta = Kr$$

where K is a constant. **(a)** For this *rotational* flow, determine, if possible, the stream function. **(b)** Can the pressure difference between the origin and any other point be determined from the Bernoulli equation? Explain.

Section 6.5 Some Basic, Plane Potential Flows

6.21 Two sources, one of strength m and the other with strength 3m, are located on the x axis as shown in Fig. P6.21. Determine the location of the stagnation point in the flow produced by these sources.

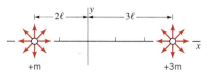

■ **FIGURE P6.21**

6.22 Water flows through a two-dimensional diffuser having a 20° expansion angle as shown in Fig. P6.22. Assume that the flow in the diffuser can be treated as a radial flow emanating from a source at the origin O. **(a)** If the velocity at the entrance is 20 m/s, determine an expression for the pressure gradient along the diffuser walls. **(b)** What is the pressure rise between the entrance and exit?

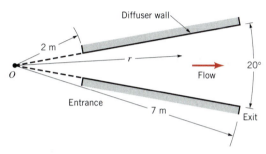

■ **FIGURE P6.22**

6.23 (See Fluids in the News article titled "Some hurricane facts," Section 6.5.3) Consider a category five hurricane that has a maximum wind speed of 160 mph at the eye wall, 10 miles from the center of the hurricane. If the flow in the hurricane outside of the hurricane's eye is approximate as a free vortex, determine the wind speeds at the location 20 mi, 30 mi, and 40 mi from the center of the storm.

6.24 The velocity distribution in a horizontal, two-dimensional bend through which an ideal fluid flows can be approximated with a free vortex as shown in Fig. P6.24. Show how the discharge (per unit width normal to plane of paper) through the channel can be expressed as

$$q = C \sqrt{\frac{\Delta p}{\rho}}$$

where $\Delta p = p_B - p_A$. Determine the value of the constant C for the bend dimensions given.

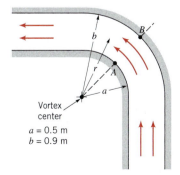

■ **FIGURE P6.24**

6.25 For a free vortex (see **Video V6.2**) determine an expression for the pressure gradient **(a)** along a streamline and **(b)** normal to a streamline. Assume that the streamline is in a horizontal plane, and express your answer in terms of the circulation.

Section 6.6 Superposition of Basic, Plane Potential Flows

6.26 Consider a uniform flow in the positive x direction combined with a free vortex located at the origin of the coordinate system. The streamline $\psi = 0$ passes through the point $x = 4$, $y = 0$. Determine the equation of this streamline.

6.27 Water flows over a flat surface at 5 ft/s as shown in Fig. P6.27. A pump draws off water through a narrow slit at a volume rate of 0.1 ft^3/s per foot length of the slit. Assume that the fluid is incompressible and inviscid and can be represented by the combination of a uniform flow and a sink. Locate the stagnation point on the wall (point A), and determine the equation for the stagnation streamline. How far above the surface, H, must the fluid be so that it does not get sucked into the slit?

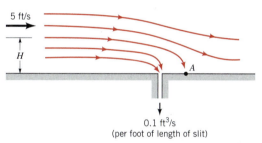

■ **FIGURE P6.27**

6.28 Potential flow against a flat plate (Fig. P6.28a) can be described with the stream function

$$\psi = Axy$$

where A is a constant. This type of flow is commonly called a "stagnation point" flow since it can be used to describe the flow in the vicinity of the stagnation point at O. By adding a source of strength m at O, stagnation point flow against a flat plate with a "bump" is obtained as illustrated in Fig. P6.28b. Determine the relationship between the bump height, h, the constant, A, and the source strength, m.

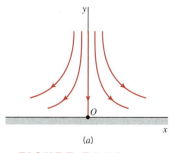

■ **FIGURE P6.28a**

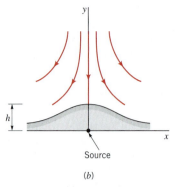

(b)

■ **FIGURE P6.28b**

6.29 The combination of a uniform flow and a source can be used to describe flow around a streamlined body called a half-body. (See **Video V6.3.**) Assume that a certain body has the shape of a half-body with a thickness of 0.5 m. If this body is placed in an air stream moving at 15 m/s, what source strength is required to simulate flow around the body?

6.30 A vehicle windshield is to be shaped as a portion of a half-body with the dimensions shown in Fig. P6.30. **(a)** Make a scale drawing of the windshield shape. **(b)** For a free stream velocity of 55 mph, determine the velocity of the air at points A and B.

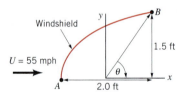

■ **FIGURE P6.30**

6.31 One end of a pond has a shoreline that resembles a half-body as shown in Fig. P6.31. A vertical porous pipe is located near the end of the pond so that water can be pumped out. When water is pumped at the rate of 0.06 m³/s through a 3-m-long pipe, what will be the velocity at point A? *Hint:* Consider the flow *inside* a half-body.

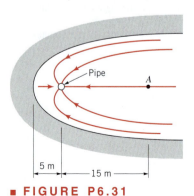

■ **FIGURE P6.31**

*6.32 For the half-body described in Section 6.6.1, show on a plot how the magnitude of the velocity on the surface, V_s, varies

as a function of the distance, s (measured along the surface), from the stagnation point. Use the dimensionless variables V_s/U and s/b where U and b are defined in Fig. 6.23.

6.33 Assume that the flow around the long circular cylinder of Fig. P6.33 is nonviscous and incompressible. Two pressures, p_1 and p_2, are measured on the surface of the cylinder, as illustrated. It is proposed that the free-stream velocity, U, can be related to the pressure difference $\Delta p = p_1 - p_2$ by the equation

$$U = C\sqrt{\frac{\Delta p}{\rho}}$$

where ρ is the fluid density. Determine the value of the constant C. Neglect body forces.

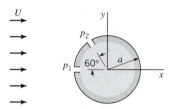

■ **FIGURE P6.33**

6.34 An ideal fluid flows past an infinitely long semicircular "hump" located along a plane boundary as shown in Fig. P6.34. Far from the hump the velocity field is uniform, and the pressure is p_0. **(a)** Determine expressions for the maximum and minimum values of the pressure along the hump, and indicate where these points are located. Express your answer in terms of ρ, U, and p_0. **(b)** If the solid surface is the $\psi = 0$ streamline, determine the equation of the streamline passing through the point $\theta = \pi/2$, $r = 2a$.

■ **FIGURE P6.34**

6.35 Water flows around a 6-ft-diameter bridge pier with a velocity of 12 ft/s. Estimate the force (per unit length) that the water exerts on the pier. Assume that the flow can be approximated as an ideal fluid flow around the front half of the cylinder, but due to flow separation (see **Video V6.4**), the average pressure on the rear half is constant and approximately equal to one-half the pressure at point A (see Fig. P6.35).

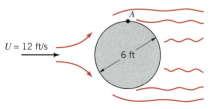

■ **FIGURE P6.35**

*6.36 Consider the steady potential flow around the circular cylinder shown in Fig. 6.24. Show on a plot the variation of the magnitude of the dimensionless fluid velocity, V/U, along the positive y axis. At what distance, y/a (along the y axis), is the velocity within 1% of the free-stream velocity?

6.37 (See Fluids in the News article titled "A sailing ship without sails," Section 6.6.2.) Determine the magnitude of the total force developed by the two rotating cylinders on the Flettner "rotor-ship" due to the Magnus effect. Assume a wind speed relative to the ship of **(a)** 10 mph and **(b)** 30 mph. Each cylinder has a diameter of 9 ft, a length of 50 ft, and rotates at 750 rev/min. Use Eq. 6.117 and calculate the circulation by assuming the air sticks to the rotating cylinders. *Note:* This calculated force is at right angles to the direction of the wind, and it is the component of this force in the direction of motion of the ship that gives the propulsive thrust. Also, due to viscous effects, the actual propulsive thrust will be smaller than that calculated from Eq. 6.117, which is based on inviscid flow theory.

6.38 Typical inviscid flow solutions for flow around bodies indicate that the fluid flows smoothly around the body, even for blunt bodies as shown in **Video V6.4**. However, experience reveals that due to the presence of viscosity, the main flow may actually separate from the body, creating a wake behind the body. As discussed in a later section (Section 9.2.6), whether separation takes place depends on the pressure gradient along the surface of the body, as calculated by inviscid flow theory. If the pressure decreases in the direction of flow (a *favorable* pressure gradient), no separation will occur. However, if the pressure increases in the direction of flow (an *adverse* pressure gradient), separation may occur. For the circular cylinder of Fig. P6.38 placed in a uniform stream with velocity, U, determine an expression for the pressure gradient in the direction flow on the surface of the cylinder. For what range of values for the angle θ will an adverse pressure gradient occur?

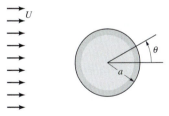

■ **FIGURE P6.38**

Section 6.8 Viscous Flow

6.39 The two-dimensional velocity field for an incompressible, Newtonian fluid is described by the relationship

$$\mathbf{V} = (12xy^2 - 6x^3)\hat{\mathbf{i}} + (18x^2y - 4y^3)\hat{\mathbf{j}}$$

where the velocity has units of meters per second when x and y are in meters. Determine the stresses σ_{xx}, σ_{yy}, and τ_{xy} at the point $x = 0.5$ m, $y = 1.0$ m if pressure at this point is 6 kPa and the fluid is glycerin at $20°$ C. Show these stresses on a sketch.

6.40 The velocity of a fluid particle moving along a horizontal streamline that coincides with the x axis in a plane, two-dimensional incompressible flow field was experimentally

found to be described by the equation $u = x^2$. Along this steamline determine an expression for the rate of change of the v-component of velocity with respect to y, **(b)** the acceleration of the particle, and **(c)** the pressure gradient in the x direction. The fluid is Newtonian.

6.41 The stream function for a certain incompressible, two-dimensional flow field is

$$\psi = 3r^3 \sin 2\theta + 2\theta$$

where ψ is in ft²/s when r is in feet and θ in radians. Determine the shearing stress, $\tau_{r\theta}$, at the point $r = 2$ ft, $\theta = \pi/3$ radians if the fluid is water.

Section 6.9.1 Steady, Laminar Flow between Fixed Parallel Plates

6.42 Two fixed, horizontal, parallel plates are spaced 0.2 in. apart. A viscous liquid ($\mu = 8 \times 10^{-3}$ lb·s/ft², $SG = 0.9$) flows between the plates with a mean velocity of 0.9 ft/s. Determine the pressure drop per unit length in the direction of flow. What is the maximum velocity in the channel?

6.43 Oil (SAE 30) at 15.6 °C flows steadily between fixed, horizontal, parallel plates. The pressure drop per unit length along the channel is 30 kPa/m, and the distance between the plates is 4 mm. The flow is laminar. Determine **(a)** the volume rate of flow (per meter of width), **(b)** the magnitude and direction of shearing stress acting on the bottom plate, and **(c)** the velocity along the centerline of the channel.

6.44 (See Fluids in the News article titled "10 tons on 8 psi," Section 6.9.1.) A massive, precisely machined, 6-ft-diameter granite sphere rests upon a 4-ft-diameter cylindrical pedestal as shown in Fig P.6.44. When the pump is turned on and the water pressure within the pedestal reaches 8 psi, the sphere rises off the pedestal, creating a 0.005-in. gap through which the water

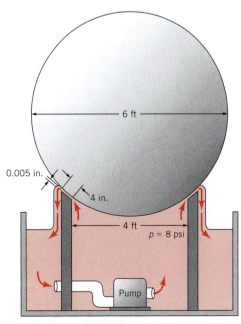

■ **FIGURE P6.44**

flows. The sphere can then be rotated about any axis with minimal friction. **(a)** Estimate the pump flowrate, Q_0, required to accomplish this. Assume the flow in the gap between the sphere and the pedestal is essentially viscous flow between fixed, parallel plates. **(b)** Describe what would happen if the pump flowrate were increased to $2Q_0$.

Section 6.9.2 Couette Flow

6.45 Two horizontal, infinite, parallel plates are spaced a distance b apart. A viscous liquid is contained between the plates. The bottom plate is fixed, and the upper plate moves parallel to the bottom plate with a velocity U. Because of the no-slip boundary condition (see **Video V6.5**), the liquid motion is caused by the liquid being dragged along by the moving boundary. There is no pressure gradient in the direction of flow. Note that this is a so-called simple *Couette flow* discussed in Section 6.9.2. **(a)** Start with the Navier–Stokes equations and determine the velocity distribution between the plates. **(b)** Determine an expression for the flowrate passing between the plates (for a unit width). Express your answer in terms of b and U.

6.46 Due to the no-slip condition, as a solid is pulled out of a viscous liquid some of the liquid is also pulled along as described in Example 6.8 and shown in **Video V6.5**. Based on the results given in Example 6.8, show on a dimensionless plot the velocity distribution in the fluid film (v/V_0 vs. x/h) when the average film velocity, V, is 10% of the belt velocity, V_0.

6.47 An incompressible, viscous fluid is placed between horizontal, infinite, parallel plates as is shown in Fig. P6.47. The two plates move in opposite directions with constant velocities, U_1 and U_2, as shown. The pressure gradient in the x direction is zero, and the only body force is due to the fluid weight. Use the Navier–Stokes equations to derive an expression for the velocity distribution between the plates. Assume laminar flow.

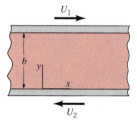

■ **FIGURE P6.47**

6.48 The viscous, incompressible flow between the parallel plates shown in Fig. P6.48 is caused by both the motion of the bottom plate and a pressure gradient, $\partial p/\partial x$. As noted in

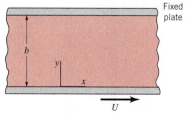

■ **FIGURE P6.48**

Section 6.9.2, an important dimensionless parameter for this type of problem is $P = -(b^2/2\,\mu U)\,(\partial p/\partial x)$ where μ is the fluid viscosity. Make a plot of the dimensionless velocity distribution (similar to that shown in Fig. 6.30b) for $P = 3$. For this case where does the maximum velocity occur?

6.49 A viscous fluid (specific weight = 80 lb/ft^3; viscosity = 0.03 lb · s/ft^2) is contained between two infinite, horizontal parallel plates as shown in Fig. P6.49. The fluid moves between the plates under the action of a pressure gradient, and the upper plate moves with a velocity U while the bottom plate is fixed. A U-tube manometer connected between two points along the bottom indicates a differential reading of 0.1 in. If the upper plate moves with a velocity of 0.02 ft/s, at what distance from the bottom plate does the maximum velocity in the gap between the two plates occur? Assume laminar flow.

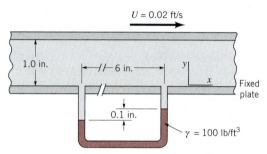

■ **FIGURE P6.49**

6.50 A viscous fluid is contained between two long concentric cylinders. The geometry of the system is such that the flow between the cylinders is approximately the same as the laminar flow between two infinite parallel plates. **(a)** Determine an expression for the torque required to rotate the outer cylinder with an angular velocity ω. The inner cylinder is fixed. Express your answer in terms of the geometry of the system, the viscosity of the fluid, and the angular velocity. **(b)** For a small, rectangular element located at the fixed wall determine an expression for the rate of angular deformation of this element. (See **Video V6.1** and Fig. P6.5.)

*6.51** Oil (SAE 30) flows between parallel plates spaced 5 mm apart. The bottom plate is fixed but the upper plate moves with a velocity of 0.2 m/s in the positive x direction. The pressure gradient is 60 kPa/m and is negative. Compute the velocity at various points across the channel and show the results on a plot. Assume laminar flow.

Section 6.9.3 Steady, Laminar Flow in Circular Tubes

6.52 It is known that the velocity distribution for steady, laminar flow in circular tubes (either horizontal or vertical) is parabolic. (See **Video V6.6**.) Consider a 10-mm-diameter horizontal tube through which ethyl alcohol is flowing with a steady mean velocity 0.15 m/s. **(a)** Would you expect the velocity distribution to be parabolic in this case? Explain. **(b)** What is the pressure drop per unit length along the tube?

6.53 A simple flow system to be used for steady flow tests consists of a constant head tank connected to a length of 4-mm-diameter tubing as shown in Fig. P6.53. The liquid has a viscosity of 0.015·s/m^2 and a density of 1200 kg/m^3, and discharges

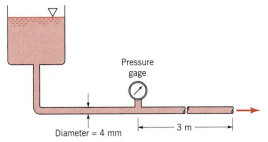

■ **FIGURE P6.53**

into the atmosphere with a mean velocity of 1 m/s. **(a)** Verify that the flow will be laminar. **(b)** The flow is fully developed in the last 3 m of the tube. What is the pressure at the pressure gage? **(c)** What is the magnitude of the wall shearing stress, τ_{rz}, in the fully developed region?

6.54 A highly viscous Newtonian liquid ($\rho = 1300$ kg/m^3; $\mu = 6.0$ N · s/m^2) is contained in a long, vertical, 150-mm-diameter tube. Initially, the liquid is at rest, but when a valve at the bottom of the tube is opened flow commences. Although the flow is slowly changing with time, at any instant the velocity distribution is parabolic; that is, the flow is quasi-steady. (See **Video V6.6.**) Some measurements show that the average velocity, V, is changing in accordance with the equation $V = 0.1$ t, with V in m/s when t is in seconds. **(a)** Show on a plot the velocity distribution (v_z vs. r) at $t = 2$ s, where v_z is the velocity and r is the radius from the center of the tube. **(b)** Verify that the flow is laminar at this instant.

6.55 A liquid (viscosity $= 0.002$ N·s/m^2; density $= 1000$ kg/m^3) is forced through the circular tube shown in Fig. P6.55. A differential manometer is connected to the tube as shown to measure the pressure drop along the tube. When the differential reading, Δh, is 9 mm, what is the mean velocity in the tube?

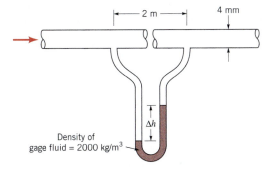

■ **FIGURE P6.55**

6.56 **(a)** Show that for Poiseuille flow in a tube of radius R the magnitude of the wall shearing stress, τ_{rz}, can be obtained from the relationship

$$|(\tau_{rz})_{\text{wall}}| = \frac{4\mu Q}{\pi R^3}$$

for a Newtonian fluid of viscosity μ. The volume rate of flow is Q. **(b)** Determine the magnitude of the wall shearing stress for a fluid having a viscosity of 0.003 N·s/m^2 flowing with an average velocity of 100 mm/s in a 2-mm-diameter tube.

6.57 As is shown by Eq. 6.143 the pressure gradient for laminar flow through a tube of constant radius is given by the expression:

$$\frac{\partial p}{\partial z} = -\frac{8\mu Q}{\pi R^4}$$

For a tube whose radius is changing very gradually, such as the one illustrated in Fig. P6.57, it is expected that this equation can be used to approximate the pressure change along the tube if the actual radius, $R(z)$, is used at each cross section. The following measurements were obtained along a particular tube.

z/ℓ	0	0.1	0.2	0.3	0.4	0.5	0.6	0.7	0.8	0.9	1.0
$R(z)/R_o$	1.00	0.73	0.67	0.65	0.67	0.80	0.80	0.71	0.73	0.77	1.00

Compare the pressure drop over the length ℓ for this nonuniform tube with one having the constant radius R_0. *Hint:* To solve this problem you will need to numerically integrate the equation for the pressure gradient given above.

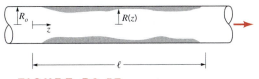

■ **FIGURE P6.57**

FE Exam Problems

Sample FE (Fundamentals of Engineering) exam questions for fluid mechanics are provided on the book's web site, www.wiley.com/college/young.

CHAPTER 7

Similitude, Dimensional Analysis, and Modeling

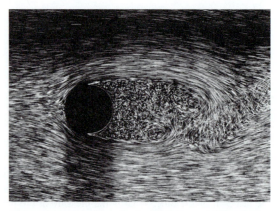

Flow past a circular cylinder with Re = 2000: *The streaklines of flow past any circular cylinder (regardless of size, velocity, or fluid) are as shown, provided that the dimensionless parameter called the Reynolds number,* Re = $\rho VD/\mu$, *is equal to 2000. For other values of* Re *the flow pattern will be different (air bubbles in water). (Photograph courtesy of ONERA, France.)*

LEARNING OBJECTIVES

After completing this chapter, you should be able to:

- apply the Buckingham pi theorem.
- develop a set of dimensionless variables for a given flow situation.
- discuss the use of dimensionless variables in data analysis.
- apply the concepts of modeling and similitude to develop prediction equations.

Although many practical engineering problems involving fluid mechanics can be solved by using the equations and analytical procedures described in the preceding chapters, a large number of problems remain that rely on experimentally obtained data for their solution. An obvious goal of any experiment is to make the results as widely applicable as possible. To achieve this end, the concept of *similitude* is often used so that measurements made on one system (for example, in the laboratory) can be used to describe the behavior of other similar systems (outside the laboratory). The laboratory systems are usually thought of as *models* and are used to study the phenomenon of interest under carefully controlled conditions. From these model studies, empirical formulations can be developed or specific predictions of one or more characteristics of some other similar system can be made. To do this, it is necessary to establish the relationship between the laboratory model and the "other" system. In the following sections, we find out how this can be accomplished in a systematic manner.

Model study of New Orleans levee breach caused by Hurricane Katrina Much of the devastation to New Orleans from Hurricane Katrina in 2005 was a result of flood waters that surged through a breach of the 17 Street Outfall Canal. To better understand why this occurred and to determine what can be done to prevent future occurrences, the U.S. Army Engineer Research and Development Center Coastal and Hydraulics Laboratory is conducting tests on a large (1:50 length scale) 15,000 square foot hydraulic *model* that replicates 0.5 mile of the canal surrounding the breach and more than a mile of the adjacent Lake Pontchartrain front. The objective of the study is to obtain information regarding the effect that waves had on the breaching of the canal and to investigate the surging water currents within the canals. The waves are generated by computer-controlled wave generators that can produce waves of varying heights, periods, and directions similar to the storm conditions that occurred during the hurricane. Data from the study will be used to calibrate and validate information that will be fed into various numerical model studies of the disaster ■

7.1   Dimensional Analysis

To illustrate a typical fluid mechanics problem in which experimentation is required, consider the steady flow of an incompressible, Newtonian fluid through a long, smooth-walled, horizontal, circular pipe. An important characteristic of this system, which would be of interest to an engineer designing a pipeline, is the pressure drop per unit length that develops along the pipe as a result of friction. Although this would appear to be a relatively simple flow problem, it cannot generally be solved analytically (even with the aid of large computers) without the use of experimental data.

The first step in the planning of an experiment to study this problem would be to decide on the factors, or variables, that will have an effect on the pressure drop per unit length, Δp_ℓ, which has dimensions of (lb/ft^2)/ft or (N/m^2)/m. We expect the list to include the pipe diameter, D, the fluid density, ρ, fluid viscosity, μ, and the mean velocity, V, at which the fluid is flowing through the pipe. Thus, we can express this relationship as

$$\Delta p_\ell = f(D, \rho, \mu, V) \tag{7.1}$$

which simply indicates mathematically that we expect the pressure drop per unit length to be some function of the factors contained within the parentheses. At this point the nature of the function is unknown, and the objective of the experiments to be performed is to determine the nature of this function.

To perform the experiments in a meaningful and systematic manner, it would be necessary to change one of the variables, such as the velocity, while holding all others constant and measure the corresponding pressure drop. This approach to determining the functional relationship between the pressure drop and the various factors that influence it, although logical in concept, is fraught with difficulties. Some of the experiments would be hard to carry out—for example, it would be necessary to vary fluid density while holding viscosity constant. How would you do this? Finally, once we obtained the various curves, how could we combine these data to obtain the desired general functional relationship among Δp_ℓ, D, ρ, μ, and V that would be valid for any similar pipe system?

Fortunately, there is a much simpler approach to this problem that will eliminate the difficulties just described. In the following sections we will show that rather than working with the original list of variables, as described in Eq. 7.1, we can collect these into two nondimensional combinations of variables (called *dimensionless products* or *dimensionless groups*) so that as shown by the figure in the margin

$$\frac{D\,\Delta p_\ell}{\rho V^2} = \phi\left(\frac{\rho V D}{\mu}\right) \tag{7.2}$$

$\dfrac{D\Delta p_\ell}{\rho V^2}$ vs $\dfrac{\rho V D}{\mu}$

Thus, instead of having to work with five variables, we now have only two. The necessary experiment would simply consist of varying the dimensionless product $\rho VD/\mu$ and determining the corresponding value of $D \, \Delta p_\ell/\rho V^2$. The results of the experiment could then be represented by a single, universal curve.

The basis for this simplification lies in a consideration of the dimensions of the variables involved. As discussed in Chapter 1, a qualitative description of physical quantities can be given in terms of *basic dimensions* such as mass, M, length, L, and time, T.[1] Alternatively, we could use force, F, L, and T as basic dimensions, since from Newton's second law

$$F \doteq MLT^{-2}$$

(Recall from Chapter 1 that the notation $\doteq$ is used to indicate dimensional equality.) The dimensions of the variables in the pipe flow example are $\Delta p_\ell \doteq (FL^{-2})/L \doteq FL^{-3}$, $D \doteq L$, $\rho \doteq FL^{-4}T^2$, $\mu \doteq FL^{-2}T$, and $V \doteq LT^{-1}$. A quick check of the dimensions of the two groups that appear in Eq. 7.2 shows that they are in fact *dimensionless* products; that is,

$$\frac{D \, \Delta p_\ell}{\rho V^2} \doteq \frac{L(F/L^3)}{(FL^{-4}T^2)(LT^{-1})^2} \doteq F^0 L^0 T^0$$

and

$$\frac{\rho VD}{\mu} \doteq \frac{(FL^{-4}T^2)(LT^{-1})(L)}{(FL^{-2}T)} \doteq F^0 L^0 T^0$$

Not only have we reduced the numbers of variables from five to two, but the new groups are dimensionless combinations of variables, which means that the results will be independent of the system of units we choose to use. This type of analysis is called *dimensional analysis,* and the basis for its application to a wide variety of problems is found in the *Buckingham pi theorem* described in the following section.

7.2 Buckingham Pi Theorem

A fundamental question we must answer is how many dimensionless products are required to replace the original list of variables. The answer to this question is supplied by the basic theorem of dimensional analysis that states the following:

> If an equation involving k variables is dimensionally homogeneous, it can be reduced to a relationship among $k - r$ independent dimensionless products, where r is the minimum number of reference dimensions required to describe the variables.

The dimensionless products are frequently referred to as **"pi terms,"** and the theorem is called the **Buckingham pi theorem.** Edgar Buckingham (1867–1940), who stimulated interest in the use of dimensional analysis, used the symbol Π to represent a dimensionless product, and this notation is commonly used. Although the pi theorem is a simple one, its proof is not so simple and we will not include it here. Many entire books have been devoted to the subject of similitude and dimensional analysis, and a number of these are listed at the end of this chapter (Refs. 1–5). Students interested in pursuing the subject in more depth (including the proof of the pi theorem) can refer to one of these books.

---

[1]As noted in Chapter 1, we will use T to represent the basic dimension of time, although T is also used for temperature in thermodynamic relationships (such as the ideal gas law).

The pi theorem is based on the idea of dimensional homogeneity, which was introduced in Chapter 1. Essentially we assume that for any physically meaningful equation involving k variables, such as

$$u_1 = f(u_2, u_3, \ldots, u_k)$$

the dimensions of the variable on the left side of the equal sign must be equal to the dimensions of any term that stands by itself on the right side of the equal sign. It then follows that we can rearrange the equation into a set of dimensionless products (pi terms) so that

$$\Pi_1 = \phi(\Pi_2, \Pi_3, \ldots, \Pi_{k-r})$$

The required number of pi terms is fewer than the number of original variables by r, where r is determined by the minimum number of reference dimensions required to describe the original list of variables. Usually the reference dimensions required to describe the variables will be the basic dimensions $M, L,$ and T or $F, L,$ and T. However, in some instances perhaps only two dimensions, such as L and T, are required, or maybe just one, such as L. Also, in a few rare cases the variables may be described by some combination of basic dimensions, such as M/T^2, and L, and in this case r would be equal to two rather than three. Although the use of the pi theorem may appear to be a little mysterious and complicated, we will actually develop a simple, systematic procedure for developing the pi terms for a given problem.

7.3   Determination of Pi Terms

Several methods can be used to form the dimensionless products, or pi terms, that arise in a dimensional analysis. Essentially we are looking for a method that will allow us to systematically form the pi terms so that we are sure that they are dimensionless and independent, and that we have the right number. The method we will describe in detail in this section is called the *method of repeating variables.*

It will be helpful to break the repeating variable method down into a series of distinct steps that can be followed for any given problem. With a little practice you will be able to readily complete a dimensional analysis for your problem.

Step 1.   **List all the variables that are involved in the problem.** This step is the most difficult one, and it is, of course, vitally important that all pertinent variables be included. Otherwise the dimensional analysis will not be correct! We are using the term "variable" to include any quantity, including dimensional and nondimensional constants, that plays a role in the phenomenon under investigation. All such quantities should be included in the list of "variables" to be considered for the dimensional analysis. The determination of the variables must be accomplished by the experimenter's knowledge of the problem and the physical laws that govern the phenomenon. Typically the variables will include those that are necessary to describe the *geometry* of the system (such as a pipe diameter), to define any *fluid properties* (such as a fluid viscosity), and to indicate *external effects* that influence the system (such as a driving pressure). These general classes of variables are intended as broad categories that should be helpful in identifying variables. It is likely, however, that there will be variables that do not fit easily into one of these categories, and each problem needs to be analyzed carefully.

Since we wish to keep the number of variables to a minimum, so that we can minimize the amount of laboratory work, it is important that all variables be independent. For example, if in a certain problem the cross-sectional area of a pipe is an important variable, either the area or the pipe diameter could be used, but not both, since they are obviously not independent. Similarly, if both fluid density, ρ, and specific weight, γ, are important variables, we could list ρ and γ, or ρ and g (acceleration of gravity), or γ and g. However, it would be incorrect to use all three since $\gamma = \rho g$; that is, ρ, γ, and g are not independent. Note that although g would normally be constant in a given experiment, that fact is irrelevant as far as a dimensional analysis is concerned.

Step 2. **Express each of the variables in terms of basic dimensions.** For the typical fluid mechanics problem the basic dimensions will be either M, L, and T or F, L, and T. Dimensionally these two sets are related through Newton's second law ($\mathbf{F} = m\mathbf{a}$) so that $F \doteq MLT^{-2}$. For example, $\rho \doteq ML^{-3}$ or $\rho \doteq FL^{-4}T^2$. Thus, either set can be used. The basic dimensions for typical variables found in fluid mechanics problems are listed in Table 1.1 in Chapter 1.

Step 3. **Determine the required number of pi terms.** This can be accomplished by means of the Buckingham pi theorem, which indicates that the number of pi terms is equal to $k - r$, where k is the number of variables in the problem (which is determined from Step 1) and r is the number of reference dimensions required to describe these variables (which is determined from Step 2). The reference dimensions usually correspond to the basic dimensions and can be determined by an inspection of the dimensions of the variables obtained in Step 2. As previously noted, there may be occasions (usually rare) in which the basic dimensions appear in combinations so that the number of reference dimensions is less than the number of basic dimensions.

Step 4. **Select a number of repeating variables. The number required is equal to the number of reference dimensions.** Essentially what we are doing here is selecting from the original list of variables, several of which can be combined with each of the remaining variables to form a pi term. All of the required reference dimensions must be included within the group of repeating variables, and each repeating variable must be dimensionally independent of the others (i.e., the dimensions of one repeating variable cannot be reproduced by some combination of products of powers of the remaining repeating variables). This means that the repeating variables cannot themselves be combined to form a dimensionless product.

For any given problem we usually are interested in determining how one particular variable is influenced by the other variables. We would consider this variable to be the dependent variable, and we would want this to appear in only one pi term. Thus, do *not* choose the dependent variable as one of the repeating variables, since the repeating variables will generally appear in more than one pi term.

Step 5. **Form a pi term by multiplying one of the nonrepeating variables by the product of the repeating variables, each raised to an exponent that will make the combination dimensionless.** Essentially, each pi term will be of the form $u_i u_1^{a_i} u_2^{b_i} u_3^{c_i}$ where u_i is one of the nonrepeating variables; u_1, u_2, and u_3 are the repeating variables; and the exponents a_i, b_i, and c_i are determined so that the combination is dimensionless.

Step 6. **Repeat Step 5 for each of the remaining nonrepeating variables.** The resulting set of pi terms will correspond to the required number obtained from Step 3. If not, check your work—you have made a mistake!

Step 7. **Check all the resulting pi terms to make sure they are dimensionless.** It is easy to make a mistake in forming the pi terms. However, this can be checked by simply substituting the dimensions of the variables into the pi terms to confirm that they are dimensionless. One good way to do this is to express the variables in terms of M, L, and T if the basic dimensions F, L, and T were used initially, or vice versa, and then check to make sure the pi terms are dimensionless.

Step 8. **Express the final form as a relationship among the pi terms and think about what it means.** Typically the final form can be written as

$$\Pi_1 = \phi(\Pi_2, \Pi_3, \ldots, \Pi_{k-r})$$

where Π_1 would contain the dependent variable in the numerator. It should be emphasized that if you started out with the correct list of variables (and the other steps were completed correctly), then the relationship in terms of the pi terms can be used to describe the problem. You need only work with the pi terms—not with the individual variables. However, it should be clearly noted that this is as far as we can go with the dimensional analysis; that is, the actual functional relationship among the pi terms must be determined by experiment.

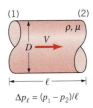

(1)   (2)

ρ, μ

D V

ℓ

$\Delta p_\ell = (p_1 - p_2)/\ell$

To illustrate these various steps we will again consider the problem discussed earlier in this chapter that was concerned with the steady flow of an incompressible, Newtonian fluid through a long, smooth-walled horizontal, circular pipe. We are interested in the pressure drop per unit length, Δp_ℓ, along the pipe. First (Step 1), we must list all of the pertinent variables that are involved based on the experimenter's knowledge of the problem. As shown by the figure in the margin, we assume that

$$\Delta p_\ell = f(D, \rho, \mu, V)$$

where D is the pipe diameter, ρ and μ are the fluid density and viscosity, respectively, and V is the mean velocity.

Next (Step 2) we express all the variables in terms of basic dimensions. Using F, L, and T as basic dimensions it follows that

$$\Delta p_\ell \doteq (FL^{-2})/L = FL^{-3}$$
$$D \doteq L$$
$$\rho \doteq FL^{-4}T^2$$
$$\mu \doteq FL^{-2}T$$
$$V \doteq LT^{-1}$$

We could also use M, L, and T as basic dimensions if desired—the final result will be the same! Note that for density, which is a mass per unit volume (ML^{-3}), we have used the relationship $F \doteq MLT^{-2}$ to express the density in terms of F, L, and T. Do not mix the basic dimensions; that is, use either F, L, and T or M, L, and T.

We can now apply the pi theorem to determine the required number of pi terms (Step 3). An inspection of the dimensions of the variables from Step 2 reveals that all three basic dimensions are required to describe the variables. Since there are five ($k = 5$) variables (do not forget to count the dependent variable, Δp_ℓ) and three required reference dimensions ($r = 3$), then according to the pi theorem there will be ($5 - 3$), or two pi terms required.

The repeating variables to be used to form the pi terms (Step 4) need to be selected from the list D, ρ, μ, and V. Remember, we do not want to use the dependent variable as one of the repeating variables. Because three reference dimensions are required, we will need to select three repeating variables. Generally, we would try to select for repeating variables those that are the simplest, dimensionally. For example, if one of the variables has the dimension of a length, choose it as one of the repeating variables. In this example we will use D, V, and ρ as repeating variables. Note that these are dimensionally independent, since D is a length, V involves both length and time, and ρ involves force, length, and time. This means that we cannot form a dimensionless product from this set.

We are now ready to form the two pi terms (Step 5). Typically, we would start with the dependent variable and combine it with the repeating variables to form the first pi term; that is,

$$\Pi_1 = \Delta p_\ell D^a V^b \rho^c$$

Since this combination is to be dimensionless, it follows that

$$(FL^{-3})(L)^a(LT^{-1})^b(FL^{-4}T^2)^c \doteq F^0 L^0 T^0$$

The exponents a, b, and c must be determined such that the resulting exponent for each of the basic dimensions—F, L, and T—must be zero (so that the resulting combination is dimensionless). Thus, we can write

$$1 + c = 0 \qquad \text{(for } F)$$

$$-3 + a + b - 4c = 0 \qquad \text{(for } L)$$

$$-b + 2c = 0 \qquad \text{(for } T)$$

The solution of this system of algebraic equations gives the desired values for a, b, and c. It follows that $a = 1$, $b = -2$, $c = -1$, and, therefore,

$$\Pi_1 = \frac{\Delta p_\ell D}{\rho V^2}$$

The process is now repeated for the remaining nonrepeating variables (Step 6). In this example there is only one additional variable (μ) so that

$$\Pi_2 = \mu D^a V^b \rho^c$$

or

$$(FL^{-2}T)(L)^a(LT^{-1})^b(FL^{-4}T^2)^c \doteq F^0 L^0 T^0$$

and, therefore,

$$1 + c = 0 \qquad \text{(for } F)$$

$$-2 + a + b - 4c = 0 \qquad \text{(for } L)$$

$$1 - b + 2c = 0 \qquad \text{(for } T)$$

Solving these equations simultaneously it follows that $a = -1$, $b = -1$, $c = -1$ so that

$$\Pi_2 = \frac{\mu}{DV\rho}$$

Note that we end up with the correct number of pi terms as determined from Step 3.

At this point stop and check to make sure the pi terms are actually dimensionless (Step 7). Finally (Step 8), we can express the result of the dimensional analysis as

$$\frac{\Delta p_\ell D}{\rho V^2} = \tilde{\phi}\left(\frac{\mu}{DV\rho}\right)$$

This result indicates that this problem can be studied in terms of these two pi terms rather than the original five variables we started with. However, dimensional analysis will *not* provide the form of the function $\tilde{\phi}$. This can only be obtained from a suitable set of experiments. If desired, the pi terms can be rearranged; that is, the reciprocal of $\mu/DV\rho$ could be used, and of course the order in which we write the variables can be changed. Thus, for example, Π_2 could be expressed as

$$\Pi_2 = \frac{\rho V D}{\mu}$$

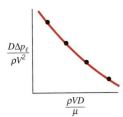

and the relationship between Π_1 and Π_2 as

$$\frac{D\,\Delta p_\ell}{\rho V^2} = \phi\left(\frac{\rho V D}{\mu}\right)$$

This form, shown by the figure in the margin, was used previously in our initial discussion of this problem (Eq. 7.2). The dimensionless product $\rho V D/\mu$ is a very famous one in fluid mechanics—the Reynolds number. This number has been alluded to briefly in Chapters 1 and 6 and will be discussed further in Section 7.6.

EXAMPLE 7.1 | Method of Repeating Variables

GIVEN A thin rectangular plate having a width w and a height h is located so that it is normal to a moving stream of fluid. Assume the drag, $\mathscr{D}$, that the fluid exerts on the plate is a function of w and h, the fluid viscosity and density, μ and ρ, respectively, and the velocity V of the fluid approaching the plate.

FIND Determine a suitable set of pi terms to study this problem experimentally.

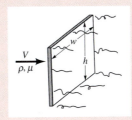

■ **FIGURE E7.1**

SOLUTION

From the statement of the problem we can write

$$\mathscr{D} = f(w, h, \mu, \rho, V)$$

where this equation expresses the general functional relationship between the drag and the several variables that will affect it. The dimensions of the variables (using the *MLT* system) are

$$\mathscr{D} \doteq MLT^{-2}$$
$$w \doteq L$$
$$h \doteq L$$
$$\mu \doteq ML^{-1}T^{-1}$$
$$\rho \doteq ML^{-3}$$
$$V \doteq LT^{-1}$$

We see that all three basic dimensions are required to define the six variables so that the Buckingham pi theorem tells us that three pi terms will be needed (six variables minus three reference dimensions, $k - r = 6 - 3$).

We will next select three repeating variables such as w, V, and ρ. A quick inspection of these three reveals that they are dimensionally independent, since each one contains a basic dimension not included in the others. Note that it would be incorrect to use both w and h as repeating variables since they have the same dimensions.

Starting with the dependent variable, $\mathscr{D}$, the first pi term can be formed by combining $\mathscr{D}$ with the repeating variables such that

$$\Pi_1 = \mathscr{D}w^a V^b \rho^c$$

and in terms of dimensions

$$(MLT^{-2})(L)^a(LT^{-1})^b(ML^{-3})^c \doteq M^0L^0T^0$$

Thus, for Π_1 to be dimensionless it follows that

$$
\begin{aligned}
1 + c &= 0 & \text{(for } M\text{)}\\
1 + a + b - 3c &= 0 & \text{(for } L\text{)}\\
-2 - b &= 0 & \text{(for } T\text{)}
\end{aligned}
$$

and, therefore, $a = -2$, $b = -2$, and $c = -1$. The pi term then becomes

$$\Pi_1 = \frac{\mathscr{D}}{w^2 V^2 \rho}$$

Next the procedure is repeated with the second nonrepeating variable, h, so that

$$\Pi_2 = hw^a V^b \rho^c$$

It follows that

$$(L)(L)^a(LT^{-1})^b(ML^{-3})^c \doteq M^0L^0T^0$$

and

$$c = 0 \quad \text{(for } M\text{)}$$
$$1 + a + b - 3c = 0 \quad \text{(for } L\text{)}$$
$$b = 0 \quad \text{(for } T\text{)}$$

so that $a = -1$, $b = 0$, $c = 0$, and therefore

$$\Pi_2 = \frac{h}{w}$$

The remaining nonrepeating variable is μ so that

$$\Pi_3 = \mu w^a V^b \rho^c$$

with

$$(ML^{-1}T^{-1})(L)^a(LT^{-1})^b(ML^{-3})^c \doteq M^0L^0T^0$$

and, therefore,

$$1 + c = 0 \quad \text{(for } M\text{)}$$
$$-1 + a + b - 3c = 0 \quad \text{(for } L\text{)}$$
$$-1 - b = 0 \quad \text{(for } T\text{)}$$

Solving for the exponents we obtain $a = -1$, $b = -1$, $c = -1$ so that

$$\Pi_3 = \frac{\mu}{wV\rho}$$

Now that we have the three required pi terms we should check to make sure they are dimensionless. To make this check we use F, L, and T, which will also verify the correctness of the original dimensions used for the variables. Thus,

$$\Pi_1 = \frac{\mathcal{D}}{w^2V^2\rho} \doteq \frac{(F)}{(L)^2(LT^{-1})^2(FL^{-4}T^2)} \doteq F^0L^0T^0$$

$$\Pi_2 = \frac{h}{w} \doteq \frac{(L)}{(L)} \doteq F^0L^0T^0$$

$$\Pi_3 = \frac{\mu}{wV\rho} \doteq \frac{(FL^{-2}T)}{(L)(LT^{-1})(FL^{-4}T^2)} \doteq F^0L^0T^0$$

If these do not check, go back to the original list of variables and make sure you have the correct dimensions for each of the variables and then check the algebra you used to obtain the exponents a, b, and c.

Finally, we can express the results of the dimensional analysis in the form

$$\frac{\mathcal{D}}{w^2V^2\rho} = \tilde{\phi}\left(\frac{h}{w}, \frac{\mu}{wV\rho}\right) \tag{Ans}$$

Since at this stage in the analysis the nature of the function $\tilde{\phi}$ is unknown, we could rearrange the pi terms if we so desired. For example, we could express the final result in the form

$$\frac{\mathcal{D}}{w^2\rho V^2} = \phi\left(\frac{w}{h}, \frac{\rho Vw}{\mu}\right) \tag{Ans}$$

which would be more conventional, since the ratio of the plate width to height, w/h, is called the *aspect ratio* and $\rho Vw/\mu$ is the Reynolds number.

COMMENT To proceed, it would be necessary to perform a set of experiments to determine the nature of the function ϕ, as discussed in Section 7.7.

7.4 Some Additional Comments about Dimensional Analysis

The preceding section provides a systematic procedure for performing a dimensional analysis. Other methods could be used, although we think the method of repeating variables is the easiest for the beginning student to use. Pi terms can also be formed by inspection, as is discussed in Section 7.5. Regardless of the specific method used for the dimensional analysis, there are certain aspects of this important engineering tool that must seem a little baffling and mysterious to the student (and sometimes to the experienced investigator as well). In this section we will attempt to elaborate on some of the more subtle points that, based on our experience, can prove to be puzzling to students.

7.4.1 Selection of Variables

One of the most important, and difficult, steps in applying dimensional analysis to any given problem is the selection of the variables that are involved. As noted previously, for convenience we will use the term *variable* to indicate any quantity involved, including dimensional and nondimensional constants. There is no simple procedure whereby the variables can be easily identified. Generally, one must rely on a good understanding of the phenomenon involved and the governing physical laws.

For most engineering problems (including areas outside of fluid mechanics), pertinent variables can be classified into three general groups—geometry, material properties, and external effects.

Geometry. The geometric characteristics can usually be prescribed by a series of lengths and angles. In most problems the geometry of the system plays an important role, and a sufficient number of geometric variables must be included to describe the system. These variables can usually be readily identified.

Material Properties. Since the response of a system to applied external effects such as forces, pressures, and changes in temperature is dependent on the nature of the materials involved in the system, the material properties that relate the external effects and the responses must be included as variables. For example, for Newtonian fluids the viscosity of the fluid is the property that relates the applied forces to the rates of deformation of the fluid.

External Effects. This terminology is used to denote any variable that produces, or tends to produce, a change in the system. For example, in structural mechanics, forces (either concentrated or distributed) applied to a system tend to change its geometry, and such forces would need to be considered as pertinent variables. For fluid mechanics, variables in this class would be related to pressures, velocities, or gravity.

7.4.2   Determination of Reference Dimensions

For any given problem it is obviously desirable to reduce the number of pi terms to a minimum and, therefore, we wish to reduce the number of variables to a minimum; that is, we certainly do not want to include extraneous variables. It is also important to know how many reference dimensions are required to describe the variables. As we have seen in the preceding examples, F, L, and T appear to be a convenient set of basic dimensions for characterizing fluid-mechanical quantities. There is, however, really nothing "fundamental" about this set, and as previously noted M, L, and T would also be suitable. Of course, in some problems only one or two of these is required.

7.4.3   Uniqueness of Pi Terms

A little reflection on the process used to determine pi terms by the method of repeating variables reveals that the specific pi terms obtained depend on the somewhat arbitrary selection of repeating variables. For example, in the problem of studying the pressure drop in a pipe, we selected D, V, and ρ as repeating variables. This led to the formulation of the problem in terms of pi terms as

$$\frac{\Delta p_\ell D}{\rho V^2} = \phi\left(\frac{\rho V D}{\mu}\right) \tag{7.3}$$

What if we had selected D, V, and μ as repeating variables? A quick check will reveal that the pi term involving Δp_ℓ becomes

$$\frac{\Delta p_\ell D^2}{V \mu}$$

and the second pi term remains the same. Thus, we can express the final result as

$$\frac{\Delta p_\ell D^2}{V\mu} = \phi_1\left(\frac{\rho V D}{\mu}\right) \tag{7.4}$$

Both results are correct, and both would lead to the same final equation for Δp_ℓ. Note, however, that the functions ϕ and ϕ_1 in Eqs. 7.3 and 7.4 will be different because the dependent pi terms are different for the two relationships.

We can conclude from this illustration that there is *not* a unique set of pi terms that arises from a dimensional analysis. However, the required *number* of pi terms is fixed.

7.5 Determination of Pi Terms by Inspection

One method for forming pi terms, the method of repeating variables, has been presented in Section 7.3. This method provides a step-by-step procedure that, if executed properly, will provide a correct and complete set of pi terms. Although this method is simple and straightforward, it is rather tedious, particularly for problems in which large numbers of variables are involved. Since the only restrictions placed on the pi terms are that they be (1) correct in number, (2) dimensionless, and (3) independent, it is possible to simply form the pi terms by inspection, without resorting to the more formal procedure.

To illustrate this approach, we again consider the pressure drop per unit length along a smooth pipe. Regardless of the technique to be used, the starting point remains the same—determine the variables, which in this case are

$$\Delta p_\ell = f(D, \rho, \mu, V)$$

Next, the dimensions of the variables are listed:

$$\Delta p_\ell \doteq FL^{-3}$$

$$D \doteq L$$

$$\rho \doteq FL^{-4}T^2$$

$$\mu \doteq FL^{-2}T$$

$$V \doteq LT^{-1}$$

and subsequently the number of reference dimensions determined. Application of the pi theorem then tells us how many pi terms are required. In this problem, since there are five variables and three reference dimensions, two pi terms are needed. Thus, the required number of pi terms can be easily determined, and the determination of this number should always be done at the beginning of the analysis.

Once the number of pi terms is known, we can form each pi term by inspection, simply making use of the fact that each pi term must be dimensionless. We will always let Π_1 contain the dependent variable, which in this example is Δp_ℓ. Since this variable has the dimensions FL^{-3}, we need to combine it with other variables so that a nondimensional product will result. One possibility is

$$\Pi_1 = \frac{\Delta p_\ell D}{\rho V^2}$$

Next, we will form the second pi term by selecting the variable that was not used in Π_1, which in this case is μ. We simply combine μ with the other variables to make the combination dimensionless (but do not use Δp_ℓ in Π_2, since we want the dependent variable to appear only in Π_1). For example, divide μ by ρ (to eliminate F), then by V (to eliminate T), and finally by D (to eliminate L). Thus,

$$\Pi_2 = \frac{\mu}{\rho V D}$$

and, therefore,

$$\frac{\Delta p_\ell D}{\rho V^2} = \phi\left(\frac{\mu}{\rho V D}\right)$$

which is, of course, the same result we obtained by using the method of repeating variables.

Although forming pi terms by inspection is essentially equivalent to the repeating variable method, it is less structured. With a little practice the pi terms can be readily formed by inspection, and this method offers an alternative to the more formal procedure.

7.6   Common Dimensionless Groups in Fluid Mechanics

At the top of Table 7.1 is a list of variables that commonly arise in fluid mechanics problems. The list is obviously not exhaustive, but does indicate a broad range of variables likely to be found in a typical problem. Fortunately, not all of these variables would be encountered

■ **TABLE 7.1**

Some Common Variables and Dimensionless Groups in Fluid Mechanics

Variables: Acceleration of gravity, g; Bulk modulus, E_v; Characteristic length, ℓ; Density, ρ; Frequency of oscillating flow, ω; Pressure, p (or Δp); Speed of sound, c; Surface tension, σ; Velocity, V; Viscosity, μ

Dimensionless Groups	Name	Interpretation (Index of Force Ratio Indicated)	Types of Applications
$\dfrac{\rho V \ell}{\mu}$	Reynolds number, Re	$\dfrac{\text{inertia force}}{\text{viscous force}}$	Generally of importance in all types of fluid dynamics problems
$\dfrac{V}{\sqrt{g\ell}}$	Froude number, Fr	$\dfrac{\text{inertia force}}{\text{gravitational force}}$	Flow with a free surface
$\dfrac{p}{\rho V^2}$	Euler number, Eu	$\dfrac{\text{pressure force}}{\text{inertia force}}$	Problems in which pressure or pressure differences are of interest
$\dfrac{\rho V^2}{E_v}$	Cauchy number,[a] Ca	$\dfrac{\text{inertia force}}{\text{compressibility force}}$	Flows in which the compressibility of the fluid is important
$\dfrac{V}{c}$	Mach number,[a] Ma	$\dfrac{\text{inertia force}}{\text{compressibility force}}$	Flows in which the compressibility of the fluid is important
$\dfrac{\omega \ell}{V}$	Strouhal number, St	$\dfrac{\text{inertia (local) force}}{\text{inertia (convective) force}}$	Unsteady flow with a characteristic frequency of oscillation
$\dfrac{\rho V^2 \ell}{\sigma}$	Weber number, We	$\dfrac{\text{inertia force}}{\text{surface tension force}}$	Problems in which surface tension is important

[a]The Cauchy number and the Mach number are related, and either can be used as an index of the relative effects of inertia and compressibility.

in each problem. However, when combinations of these variables are present, it is standard practice to combine them into some of the common dimensionless groups (pi terms) given in Table 7.1. These combinations appear so frequently that special names are associated with them as indicated in Table 7.1.

It is also often possible to provide a physical interpretation to the dimensionless groups, which can be helpful in assessing their influence in a particular application. For example, the Froude number is an index of the ratio of the force due to the acceleration of a fluid particle (inertial force) to the force due to gravity (weight). A similar interpretation in terms of indices of force ratios can be given to the other dimensionless groups, as indicated in Table 7.1. The Reynolds number is undoubtedly the most famous dimensionless parameter in fluid mechanics. It is named in honor of Osborne Reynolds, a British engineer, who first demonstrated that this combination of variables could be used as a criterion to distinguish between laminar and turbulent flow. In most fluid flow problems there will be a characteristic length, ℓ, and a velocity, V, as well as the fluid properties of density, ρ, and viscosity, μ, which are relevant variables in the problem. Thus, with these variables the Reynolds number

V7.1 Reynolds number

$$Re = \frac{\rho V \ell}{\mu}$$

arises naturally from the dimensional analysis. The Reynolds number is a measure of the ratio of the inertia force on an element of fluid to the viscous force on an element. When these two types of forces are important in a given problem, the Reynolds number will play an important role.

7.7 Correlation of Experimental Data

One of the most important uses of dimensional analysis is as an aid in the efficient handling, interpretation, and correlation of experimental data. Since the field of fluid mechanics relies heavily on empirical data, it is not surprising that dimensional analysis is such an important tool in this field. As noted previously, a dimensional analysis cannot provide a complete answer to any given problem, since the analysis only provides the dimensionless groups describing the phenomenon and not the specific relationship among the groups. To determine this relationship, suitable experimental data must be obtained. The degree of difficulty involved in this process depends on the number of pi terms and the nature of the experiments. (How hard is it to obtain the measurements?) The simplest problems are obviously those involving the fewest pi terms, and the following sections indicate how the complexity of the analysis increases with the increasing number of pi terms.

7.7.1 Problems with One Pi Term

Application of the pi theorem indicates that if the number of variables minus the number of reference dimensions is equal to unity, then only *one* pi term is required to describe the phenomenon. The functional relationship that must exist for one pi term is

$$\Pi_1 = C$$

where C is a constant. This is one situation in which a dimensional analysis reveals the specific form of the relationship and, as is illustrated by the following example, shows how the individual variables are related. The value of the constant, however, must still be determined by experiment.

EXAMPLE 7.2   Flow with Only One Pi Term

GIVEN As shown in Fig. E7.2, assume that the drag, $\mathcal{D}$, acting on a spherical particle that falls very slowly through a viscous fluid is a function of the particle diameter, D, the particle velocity, V, and the fluid viscosity, μ.

FIND Determine, with the aid of dimensional analysis, how the drag depends on the particle velocity.

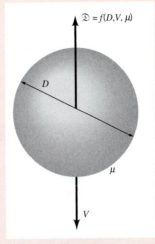

■ **FIGURE E7.2**

SOLUTION

From the information given, it follows that

$$\mathcal{D} = f(D, V, \mu)$$

and the dimensions of the variables are

$$\mathcal{D} \doteq F$$
$$D \doteq L$$
$$V \doteq LT^{-1}$$
$$\mu \doteq FL^{-2}T$$

We see that there are four variables and three reference dimensions (F, L, and T) required to describe the variables. Thus, according to the pi theorem, one pi term is required. This pi term can be easily formed by inspection and can be expressed as

$$\Pi_1 = \frac{\mathcal{D}}{\mu V D}$$

Because there is only one pi term, it follows that

$$\frac{\mathcal{D}}{\mu V D} = C$$

or

$$\mathcal{D} = C\mu V D$$

Thus, for a given particle and fluid, the drag varies directly with the velocity so that

$$\mathcal{D} \propto V \qquad \textbf{(Ans)}$$

COMMENTS Actually, the dimensional analysis reveals that the drag not only varies directly with the velocity, but it also varies directly with the particle diameter and the fluid

viscosity. We could not, however, predict the value of the drag, since the constant, C, is unknown. An experiment would have to be performed in which the drag and the corresponding velocity are measured for a given particle and fluid. Although in principle we would only have to run a single test, we would certainly want to repeat it several times to obtain a reliable value for C. It should be emphasized that once the value of C is determined it is not necessary to run similar tests by using different spherical particles and fluids; that is, C is a universal constant so long as the drag is a function only of particle diameter, velocity, and fluid viscosity.

An approximate solution to this problem can also be obtained theoretically, from which it is found that $C = 3\pi$ so that

$$\mathcal{D} = 3\pi\mu V D$$

This equation is commonly called *Stokes law* and is used in the study of the settling of particles. Our experiments would reveal that this result is only valid for small Reynolds numbers ($\rho V D/\mu \ll 1$). This follows, since in the original list of variables, we have neglected inertial effects (fluid density is not included as a variable). The inclusion of an additional variable would lead to another pi term so that there would be two pi terms rather than one.

7.7.2   Problems with Two or More Pi Terms

If a given phenomenon can be described with two pi terms such that

$$\Pi_1 = \phi(\Pi_2)$$

the functional relationship among the variables can then be determined by varying Π_2 and measuring the corresponding values of Π_1. For this case the results can be conveniently

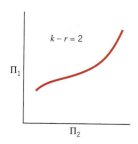

$k - r = 2$

Π_1

Π_2

presented in graphical form by plotting Π_1 versus Π_2 as shown by the figure in the margin. It should be emphasized that the resulting curve would be a "universal" one for the particular phenomenon studied. This means that if the variables and the resulting dimensional analysis are correct, then there is only a single relationship between Π_1 and Π_2.

In addition to presenting data graphically, it may be possible (and desirable) to obtain an empirical equation relating Π_1 and Π_2 by using a standard curve-fitting technique.

EXAMPLE 7.3 Dimensionless Correlation of Experimental Data

GIVEN The relationship between the pressure drop per unit length along a smooth-walled, horizontal pipe and the variables that affect the pressure drop is to be determined experimentally. In the laboratory the pressure drop was measured over a 5-ft length of smooth-walled pipe having an inside diameter of 0.496 in. The fluid used was water at 60° F ($\mu = 2.34 \times 10^{-5}$ lb·s/ft^2, $\rho = 1.94$ slugs/ft^3). Tests were run in which the velocity was varied and the corresponding pressure drop measured. The results of these tests are shown here:

Velocity (ft/s)	Pressure drop for 5-ft length (lb/ft^2)
1.17	6.26
1.95	15.6
2.91	30.9
5.84	106
11.13	329
16.92	681
23.34	1200
28.73	1730

FIND Make use of these data to obtain a general relationship between the pressure drop per unit length and the other variables.

To determine the form of the relationship, we need to vary the Reynolds number, Re $= \rho VD/\mu$, and to measure the corresponding values of $D\,\Delta p_\ell/\rho V^2$. The Reynolds number could be varied by changing any one of the variables, ρ, V, D, or μ, or any combination of them. However, the simplest way to do this is to vary the velocity, since this will allow us to use the same fluid and pipe. Based on the data given, values for the two pi terms can be computed with the result:

$D\Delta p_\ell/\rho V^2$	$\rho VD/\mu$
0.0195	4.01×10^3
0.0175	6.68×10^3
0.0155	9.97×10^3
0.0132	2.00×10^4
0.0113	3.81×10^4
0.0101	5.80×10^4
0.00939	8.00×10^4
0.00893	9.85×10^4

SOLUTION

The first step is to perform a dimensional analysis during the planning stage *before* the experiments are actually run. As discussed in Section 7.3, we will assume that the pressure drop per unit length, Δp_ℓ, is a function of the pipe diameter, D, fluid density, ρ, fluid viscosity, μ, and the velocity, V. Thus,

$$\Delta p_\ell = (D, \rho, \mu, V)$$

and application of the pi theorem yields two pi terms

$$\Pi_1 = \frac{D\,\Delta p_\ell}{\rho V^2} \quad \text{and} \quad \Pi_2 = \frac{\rho VD}{\mu}$$

Hence,

$$\frac{D\,\Delta p_\ell}{\rho V^2} = \phi\left(\frac{\rho VD}{\mu}\right)$$

These are dimensionless groups so that their values are independent of the system of units used so long as a consistent system is used. For example, if the velocity is in ft/s, then the diameter should be in feet, not inches or meters.

A plot of these two pi terms can now be made with the results shown in Fig. E7.3a. The correlation appears to be quite good, and if it was not, this would suggest that either we had large experimental measurement errors or we had perhaps omitted an important variable. The curve shown in Fig. E7.3a represents the general relationship between the pressure drop and the other factors in the Reynolds number range $4.10 \times 10^3 \leq$ Re $\leq 9.85 \times 10^4$. Thus, for this range of Reynolds numbers it is *not* necessary to repeat the tests for other pipe sizes or other fluids provided the assumed independent variables (D, ρ, μ, V) are the only important ones.

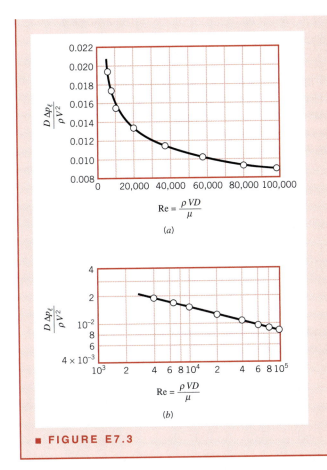

(a)

(b)

■ **FIGURE E7.3**

Because the relationship between Π_1 and Π_2 is nonlinear, it is not immediately obvious what form of empirical equation might be used to describe the relationship. If, however, the same data are plotted on logarithmic graph paper, as shown in Fig. E7.3b, the data form a straight line, suggesting that a suitable equation is of the form $\Pi_1 = A\Pi_2^n$ where A and n are empirical constants to be determined from the data by using a suitable curve-fitting technique, such as a nonlinear regression program. For the data given in this example, a good fit of the data is obtained with the equation

$$\Pi_1 = 0.150\,\Pi_2^{-0.25} \qquad \textbf{(Ans)}$$

COMMENT In 1911, H. Blasius, a German fluid mechanician, established a similar empirical equation that is used widely for predicting the pressure drop in smooth pipes in the range $4 \times 10^3 < \text{Re} < 10^5$. This equation can be expressed in the form

$$\frac{D\,\Delta p_\ell}{\rho V^2} = 0.1582\left(\frac{\rho VD}{\mu}\right)^{-1/4}$$

The so-called Blasius formula is based on numerous experimental results of the type used in this example. Flow in pipes is discussed in more detail in the next chapter, where it is shown how pipe roughness (which introduces another variable) may affect the results given in this example (which is for smooth-walled pipes).

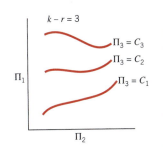

As the number of required pi terms increases, it becomes more difficult to display the results in a convenient graphical form and to determine a specific empirical equation that describes the phenomenon. For problems involving three pi terms

$$\Pi_1 = \phi(\Pi_2, \Pi_3)$$

it is still possible to show data correlations on simple graphs by plotting families of curves as shown by the figure in the margin. This is an informative and useful way of representing the data in a general way. It may also be possible to determine a suitable empirical equation relating the three pi terms. However, as the number of pi terms continues to increase, corresponding to an increase in the general complexity of the problem of interest, both the graphical presentation and the determination of a suitable empirical equation become intractable. For these more complicated problems, it is often more feasible to use models to predict specific characteristics of the system rather than to try to develop general correlations.

7.8 Modeling and Similitude

Models are widely used in fluid mechanics. Major engineering projects involving structures, aircraft, ships, rivers, harbors, dams, air and water pollution, and so on frequently involve the use of models. Although the term "model" is used in many different contexts, the "engineering model" generally conforms to the following definition. *A **model** is a representation* *of a physical system that may be used to predict the behavior of the system in some desired respect.* The physical system for which the predictions are to be made is called the ***prototype.*** Although *mathematical* or *computer* models may also conform to this definition, our interest will be in physical models, that is, models that resemble the prototype but are generally of a different size, may involve different fluids, and often operate under different conditions (pressures, velocities, etc.). Usually a model is smaller than the prototype. Therefore, it is handled more easily in the laboratory and is less expensive to construct and operate than a large prototype. With the successful development of a valid model, it is possible to predict the behavior of the prototype under a certain set of conditions.

In the following sections we will develop the procedures for designing models so that the model and prototype will behave in a similar fashion.

7.8.1 Theory of Models

The theory of models can be developed readily by using the principles of dimensional analysis. It has been shown that any given problem can be described in terms of a set of pi terms as

$$\Pi_1 = \phi(\Pi_2, \Pi_3, \ldots, \Pi_n) \tag{7.5}$$

In formulating this relationship, only a knowledge of the general nature of the physical phenomenon, and the variables involved, is required. Specific values for variables (size of components, fluid properties, and so on) are not needed to perform the dimensional analysis. Thus, Eq. 7.5 applies to any system that is governed by the same variables. If Eq. 7.5 describes the behavior of a particular prototype, a similar relationship can be written for a model of this prototype; that is,

$$\Pi_{1m} = \phi(\Pi_{2m}, \Pi_{3m}, \ldots, \Pi_{nm}) \tag{7.6}$$

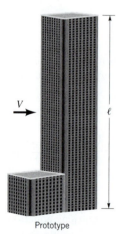

Prototype

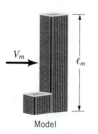

Model

where the form of the function will be the same as long as the same phenomenon is involved in both the prototype and the model. As shown by the figure in the margin, variables or pi terms without a subscript will refer to the prototype, whereas the subscript m will be used to designate the model variables or pi terms.

The pi terms can be developed so that Π_1 contains the variable that is to be predicted from observations made on the model. Therefore, if the model is designed and operated under the following conditions

$$\Pi_{2m} = \Pi_2$$
$$\Pi_{3m} = \Pi_3 \tag{7.7}$$
$$\vdots$$
$$\Pi_{nm} = \Pi_n$$

then with the presumption that the form of ϕ is the same for model and prototype, it follows that

$$\Pi_1 = \Pi_{1m} \tag{7.8}$$

Equation 7.8 is the desired ***prediction equation*** and indicates that the measured value of Π_{1m} obtained with the model will be equal to the corresponding Π_1 for the prototype as long as the other pi terms are equal. The conditions specified by Eqs. 7.7 provide the ***model design conditions,*** also called ***similarity requirements*** or ***modeling laws.***

As an example of the procedure, consider the problem of determining the drag, $\mathcal{D}$, on a thin rectangular plate ($w \times h$ in size) placed normal to a fluid with velocity, V. The

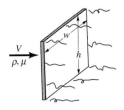

dimensional analysis of this problem was performed in Example 7.1, where it was assumed that

$$\mathscr{D} = f(w, h, \mu, \rho, V)$$

Application of the pi theorem yielded

$$\frac{\mathscr{D}}{w^2 \rho V^2} = \phi\left(\frac{w}{h}, \frac{\rho V w}{\mu}\right) \tag{7.9}$$

We are now concerned with designing a model that could be used to predict the drag on a certain prototype (which presumably has a different size than the model). Since the relationship expressed by Eq. 7.9 applies to both prototype and model, Eq. 7.9 is assumed to govern the prototype, with a similar relationship

$$\frac{\mathscr{D}_m}{w_m^2 \rho_m V_m^2} = \phi\left(\frac{w_m}{h_m}, \frac{\rho_m V_m w_m}{\mu_m}\right) \tag{7.10}$$

for the model. The model design conditions, or similarity requirements, are therefore

$$\frac{w_m}{h_m} = \frac{w}{h} \qquad \frac{\rho_m V_m w_m}{\mu_m} = \frac{\rho V w}{\mu}$$

The size of the model is obtained from the first requirement, which indicates that

$$w_m = \frac{h_m}{h} w \tag{7.11}$$

We are free to establish the height ratio h_m/h, but then the model plate width, w_m, is fixed in accordance with Eq. 7.11.

The second similarity requirement indicates that the model and prototype must be operated at the same Reynolds number. Thus, the required velocity for the model is obtained from the relationship

$$V_m = \frac{\mu_m}{\mu} \frac{\rho}{\rho_m} \frac{w}{w_m} V \tag{7.12}$$

Note that this model design requires not only geometric scaling, as specified by Eq. 7.11, but also the correct scaling of the velocity in accordance with Eq. 7.12. This result is typical of most model designs—there is more to the design than simply scaling the geometry!

With the foregoing similarity requirements satisfied, the prediction equation for the drag is

$$\frac{\mathscr{D}}{w^2 \rho V^2} = \frac{\mathscr{D}_m}{w_m^2 \rho_m V_m^2}$$

or

$$\mathscr{D} = \left(\frac{w}{w_m}\right)^2 \left(\frac{\rho}{\rho_m}\right) \left(\frac{V}{V_m}\right)^2 \mathscr{D}_m$$

Thus, a measured drag on the model, $\mathscr{D}_m$, must be multiplied by the ratio of the square of the plate widths, the ratio of the fluid densities, and the ratio of the square of the velocities to obtain the predicted value of the prototype drag, $\mathscr{D}$.

Generally, as is illustrated in this example, to achieve similarity between model and prototype behavior, *all the corresponding pi terms must be equated between model and prototype.* Usually, one or more of these pi terms will involve ratios of important lengths (such as w/h in the foregoing example); that is, they are purely geometrical. Thus, when we equate

V7.2 Environmental models

the pi terms involving length ratios we are requiring that there be complete *geometric similarity* between the model and the prototype. This means that the model must be a scaled version of the prototype. Geometric scaling may extend to the finest features of the system, such as surface roughness or small protuberances on a structure, since these kinds of geometric features may significantly influence the flow.

Another group of typical pi terms (such as the Reynolds number in the foregoing example) involves force ratios as noted in Table 7.1. The equality of these pi terms requires the ratio of like forces in model and prototype to be the same. Thus, for flows in which the Reynolds numbers are equal, the ratio of viscous forces in model and prototype is equal to the ratio of inertia forces. If other pi terms are involved, such as the Froude number or Weber number, a similar conclusion can be drawn; that is, the equality of these pi terms requires the ratio of like forces in model and prototype to be the same. Thus, when these types of pi terms are equal in model and prototype, we have *dynamic similarity* between model and prototype. It follows that with both geometric and dynamic similarity the streamline patterns will be the same and corresponding velocity ratios (V_m/V) and acceleration ratios (a_m/a) are constant throughout the flow field. Thus, *kinematic similarity* exists between model and prototype. To have complete similarity between model and prototype, we must maintain geometric, kinematic, and dynamic similarity between the two systems. This will automatically follow if all the important variables are included in the dimensional analysis and if all the similarity requirements based on the resulting pi terms are satisfied.

F l u i d s   i n   t h e   N e w s

Modeling parachutes in a water tunnel The first use of a parachute with a free-fall jump from an aircraft occurred in 1914, although parachute jumps from hot air balloons had occurred since the late 1700s. In more modern times parachutes are commonly used by the military, and for safety and sport. It is not surprising that there remains interest in the design and characteristics of parachutes, and researchers at the Worcester Polytechnic Institute have been studying various aspects of the aerodynamics associated with parachutes. An unusual part of their study is that they are using small-scale

parachutes tested in a *water tunnel*. The *model parachutes* are reduced in size by a factor of 30 to 60 times. Various types of tests can be performed ranging from the study of the velocity fields in the wake of the canopy with a steady free-stream velocity to the study of conditions during rapid deployment of the canopy. According to the researchers, the advantage of using water as the working fluid, rather than air, is that the velocities and deployment dynamics are slower than in the atmosphere, thus providing more time to collect detailed experimental data. (See Problem 7.26.) ■

EXAMPLE 7.4 ▮ Prediction of Prototype Performance from Model Data

GIVEN A long structural component of a bridge has the cross section shown in Fig. E7.4. It is known that when a steady wind blows past this type of bluff body, vortices may develop on the downwind side that are shed in a regular fashion at some definite frequency. Because these vortices can create harmful periodic forces acting on the structure, it is important to determine the shedding frequency. For the specific structure of interest, $D = 0.1$ m, $H = 0.3$ m, and a representative wind velocity is $V = 50$ km/hr. Standard air can be assumed. The shedding frequency is to be determined through the use of a small-scale model that is to be tested in a water tunnel. For the model, $D_m = 20$ mm and the water temperature is 20° C.

FIND Determine the model dimension, H_m, and the velocity at which the test should be performed. If the shedding frequency for the model is found to be 49.9 Hz, what is the corresponding frequency for the prototype?

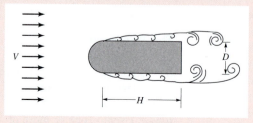

■ **FIGURE E7.4**

SOLUTION

We expect the shedding frequency, ω, to depend on the lengths D and H, the approach velocity, V, and the fluid density, ρ, and viscosity, μ. Thus,

$$\omega = f(D, H, V, \rho, \mu)$$

where

$$\omega \doteq T^{-1}$$
$$D \doteq L$$
$$H \doteq L$$
$$V \doteq LT^{-1}$$
$$\rho \doteq ML^{-3}$$
$$\mu \doteq ML^{-1}T^{-1}$$

Because there are six variables and three reference dimensions (MLT), three pi terms are required. Application of the pi theorem yields

$$\frac{\omega D}{V} = \phi\left(\frac{D}{H}, \frac{\rho VD}{\mu}\right)$$

We recognize the pi term on the left as the Strouhal number (see Table 7.1), and the dimensional analysis indicates that the Strouhal number is a function of the geometric parameter, D/H, and the Reynolds number. Thus, to maintain similarity between model and prototype

$$\frac{D_m}{H_m} = \frac{D}{H}$$

and

$$\frac{\rho_m V_m D_m}{\mu_m} = \frac{\rho VD}{\mu}$$

From the first similarity requirement

$$H_m = \frac{D_m}{D} H$$

$$= \frac{(20 \times 10^{-3}\text{ m})}{(0.1\text{ m})}(0.3\text{ m})$$

$$H_m = 60 \times 10^{-3}\text{ m} = 60\text{ mm} \qquad \text{(Ans)}$$

The second similarity requirement indicates that the Reynolds number must be the same for model and prototype so that the model velocity must satisfy the condition

$$V_m = \frac{\mu_m}{\mu} \frac{\rho}{\rho_m} \frac{D}{D_m} V \qquad \qquad \text{(1)}$$

For air at standard conditions, $\mu = 1.79 \times 10^{-5}$ kg/m·s, $\rho = 1.23$ kg/m³, and for water at 20° C, $\mu = 1.00 \times 10^{-3}$ kg/m·s, $\rho = 998$ kg/m³. The fluid velocity for the prototype is

$$V = \frac{(50 \times 10^3\text{ m/hr})}{(3600\text{ s/hr})} = 13.9\text{ m/s}$$

The required velocity can now be calculated from Eq. 1 as

$$V_m = \frac{[1.00 \times 10^{-3}\text{ kg/(m·s)}](1.23\text{ kg/m}^3)}{[1.79 \times 10^{-5}\text{ kg/(m·s)}](998\text{ kg/m}^3)}$$

$$\times \frac{(0.1\text{ m})}{(20 \times 10^{-3}\text{ m})}(13.9\text{ m/s})$$

$$V_m = 4.79\text{ m/s} \qquad \text{(Ans)}$$

This is a reasonable velocity that could be readily achieved in a water tunnel.

With the two similarity requirements satisfied, it follows that the Strouhal numbers for prototype and model will be the same so that

$$\frac{\omega D}{V} = \frac{\omega_m D_m}{V_m}$$

and the predicted prototype vortex shedding frequency is

$$\omega = \frac{V}{V_m} \frac{D_m}{D} \omega_m$$

$$= \frac{(13.9\text{ m/s})}{(4.79\text{ m/s})} \frac{(20 \times 10^{-3}\text{ m})}{(0.1\text{ m})}(49.9\text{ Hz})$$

$$\omega = 29.0\text{ Hz} \qquad \text{(Ans)}$$

COMMENT This same model could also be used to predict the drag per unit length, $\mathcal{D}_\ell$, on the prototype, since the drag would depend on the same variables as those used for the frequency. Thus, the similarity requirements would be the same, and with these requirements satisfied it follows that the drag per unit length expressed in dimensionless form, such as $\mathcal{D}_\ell/D\rho V^2$, would be equal in model and prototype. The measured drag per unit length on the model could then be related to the corresponding drag on the prototype through the relationship

$$\mathcal{D}_\ell = \left(\frac{D}{D_m}\right)\left(\frac{\rho}{\rho_m}\right)\left(\frac{V}{V_m}\right)^2 \mathcal{D}_{\ell m}$$

7.8.2 Model Scales

It is clear from the preceding section that the ratio of like quantities for the model and prototype naturally arises from the similarity requirements. For example, if in a given problem there are two length variables ℓ_1 and ℓ_2, the resulting similarity requirement based on a pi term obtained from these variables is

$$\frac{\ell_1}{\ell_2} = \frac{\ell_{1m}}{\ell_{2m}}$$

so that

$$\frac{\ell_{1m}}{\ell_1} = \frac{\ell_{2m}}{\ell_2}$$

We define the ratio ℓ_{1m}/ℓ_1 or ℓ_{2m}/ℓ_2 as the ***length scale.*** For true models there will be only one length scale, and all lengths are fixed in accordance with this scale. There are, however, other scales, such as the velocity scale, V_m/V, density scale, ρ_m/ρ, viscosity scale, μ_m/μ, and so on. In fact, we can define a scale for each of the variables in the problem. Thus, it is actually meaningless to talk about a "scale" of a model without specifying which scale.

We will designate the length scale as λ_ℓ, and other scales as λ_V, λ_ρ, λ_μ, and so on, where the subscript indicates the particular scale. Also, we will take the ratio of the model value to the prototype value as the scale (rather than the inverse). Length scales are often specified, for example, as 1:10 or as a $\frac{1}{10}$-scale model. The meaning of this specification is that the model is one-tenth the size of the prototype, and the tacit assumption is that all relevant lengths are scaled accordingly so the model is geometrically similar to the prototype.

F l u i d s   i n   t h e   N e w s

"Galloping Gertie" One of the most dramatic bridge collapses occurred in 1940 when the Tacoma Narrows bridge, located near Tacoma, Washington, failed due to aerodynamic instability. The bridge had been nicknamed "Galloping Gertie" due to its tendency to sway and move in high winds. On the fateful day of the collapse the wind speed was 65 km/hr. This particular combination of a high wind and the aeroelastic properties of the bridge created large oscillations leading to its failure. The bridge was replaced in 1950, and plans are underway to add a second bridge parallel to the existing structure. To determine possible wind interference effects due to two bridges in close proximity, wind tunnel tests were run in a 9 m × 9 m wind tunnel operated by the National Research Council of Canada. *Models* of the two side-by-side bridges, each having a length scale of 1 : 211, were tested under various wind conditions. Since the failure of the original Tacoma Narrows bridge, it is now common practice to use wind tunnel model studies during the design process to evaluate any bridge that is to be subjected to wind-induced vibrations. (See Problem 7.39.) ■

7.8.3   Distorted Models

Although the general idea behind establishing similarity requirements for models is straightforward (we simply equate pi terms), it is not always possible to satisfy all the known requirements. If one or more of the similarity requirements are not met, for example, if $\Pi_{2m} \neq \Pi_2$, then it follows that the prediction equation $\Pi_1 = \Pi_{1m}$ is not true; that is, $\Pi_1 \neq \Pi_{1m}$. Models for which one or more of the similarity requirements are not satisfied are called ***distorted models.***

Distorted models are rather commonplace and can arise for a variety of reasons. For example, perhaps a suitable fluid cannot be found for the model. The classic example of a distorted model occurs in the study of open channel or free-surface flows. Typically in these problems both the Reynolds number, $\rho V \ell / \mu$, and the Froude number, $V/\sqrt{g\ell}$, are involved.

Froude number similarity requires

$$\frac{V_m}{\sqrt{g_m \ell_m}} = \frac{V}{\sqrt{g\ell}}$$

If the model and prototype are operated in the same gravitational field, then the required velocity scale is

$$\frac{V_m}{V} = \sqrt{\frac{\ell_m}{\ell}} = \sqrt{\lambda_\ell}$$

Reynolds number similarity requires

$$\frac{\rho_m V_m \ell_m}{\mu_m} = \frac{\rho V \ell}{\mu}$$

and the velocity scale is

$$\frac{V_m}{V} = \frac{\mu_m}{\mu} \frac{\rho}{\rho_m} \frac{\ell}{\ell_m}$$

V7.3 Model of fish hatchery pond

Since the velocity scale must be equal to the square root of the length scale, it follows that

$$\frac{\mu_m/\rho_m}{\mu/\rho} = \frac{\nu_m}{\nu} = (\lambda_\ell)^{3/2} \tag{7.13}$$

where the ratio μ/ρ is the kinematic viscosity, ν. Although it may be possible to satisfy this design condition in principle, it may be quite difficult, if not impossible, to find a suitable model fluid, particularly for small-length scales. For problems involving rivers, spillways, and harbors, for which the prototype fluid is water, the models are also relatively large so that the only practical model fluid is water. However, in this case (with the kinematic viscosity scale equal to unity) Eq. 7.13 will not be satisfied, and a distorted model will result. Generally, hydraulic models of this type are distorted and are designed on the basis of the Froude number, with the Reynolds number different in model and prototype.

Distorted models can be used successfully, but the interpretation of results obtained with this type of model is obviously more difficult than the interpretation of results obtained with *true models* for which all similarity requirements are met.

F l u i d s   i n   t h e   N e w s

Old Man River in (large) miniature One of the world's largest scale models, a Mississippi River model, resides near Jackson, Mississippi. It is a detailed, complex model that covers many acres and replicates the 1,250,000-acre Mississippi River basin. Built by the Army Corps of Engineers and used from 1943 to 1973, today it has mostly gone to ruin. As with many hydraulic models, this is a *distorted model*, with a horizontal scale of 1 to 2000 and a vertical scale of 1 to 100. One step along the model river corresponds to one mile along the river. All essential river basin elements such as geological features, levees, and railroad embankments were sculpted by hand to match the actual contours. The main purpose of the model was to predict floods. This was done by supplying specific amounts of water at prescribed locations along the model and then measuring the water depths up and down the model river. Because of the length scale, there is a difference in the time taken by the corresponding model and prototype events. Although it takes days for the actual floodwaters to travel from Sioux City, Iowa, to Omaha, Nebraska, it would take only minutes for the simulated flow in the model. ■

7.9 Some Typical Model Studies

Models are used to investigate many different types of fluid mechanics problems, and it is difficult to characterize in a general way all necessary similarity requirements, as each problem is unique. We can, however, broadly classify many of the problems on the basis of the general nature of the flow and subsequently develop some general characteristics of model designs in each of these classifications. The following sections consider models for the study of (1) flow through closed circuits, (2) flow around immersed bodies, and (3) flow with a free surface. Turbomachine models are considered in Chapter 11.

7.9.1 Flow through Closed Conduits

Common examples of this type of flow include pipe flow and flow through valves, fittings, and metering devices. Although the conduits are often circular, they could have other shapes as well and may contain expansions or contractions. Since there are no fluid interfaces or free surfaces, the dominant forces are inertial and viscous so that the Reynolds number is an important similarity parameter. For low Mach numbers (Ma < 0.3), compressibility effects are negligible for the flow of both liquids or gases. For this class of problems, geometric similarity between model and prototype must be maintained. Generally the geometric characteristics can be described by a series of length terms, $\ell_1, \ell_2, \ell_3, \dots, \ell_i$, and ℓ, where ℓ is some particular length dimension for the system. Such a series of length terms leads to a set of pi terms of the form

$$\Pi_i = \frac{\ell_i}{\ell}$$

where $i = 1, 2, \dots$, and so on. In addition to the basic geometry of the system, the roughness of the internal surfaces in contact with the fluid may be important. If the average height of surface roughness elements is defined as ε, then the pi term representing roughness would be ε/ℓ. This parameter indicates that for complete geometric similarity, surface roughness would also have to be scaled.

It follows from this discussion that for flow in closed conduits at low Mach numbers, any dependent pi term (the one that contains the particular variable of interest, such as pressure drop) can be expressed as

$$\text{Dependent pi term} = \phi\left(\frac{\ell_i}{\ell}, \frac{\varepsilon}{\ell}, \frac{\rho V \ell}{\mu}\right) \tag{7.14}$$

This is a general formulation for this type of problem.

With the similarity requirements satisfied, it follows that the dependent pi term will be equal in model and prototype. For example, if the dependent variable of interest is the pressure differential, Δp, between two points along a closed conduit, then the dependent pi term could be expressed as

$$\Pi_1 = \frac{\Delta p}{\rho V^2}$$

The prototype pressure drop would then be obtained from the relationship

$$\Delta p = \frac{\rho}{\rho_m}\left(\frac{V}{V_m}\right)^2 \Delta p_m$$

so that from a measured pressure differential in the model, Δp_m, the corresponding pressure differential for the prototype could be predicted. Note that in general $\Delta p \neq \Delta p_m$.

EXAMPLE 7.5   Reynolds Number Similarity

GIVEN Model tests are to be performed to study the flow through a large check valve having a 2-ft-diameter inlet and carrying water at a flowrate of 30 cfs as shown in Fig E7.5a. The working fluid in the model is water at the same temperature as that in the prototype. Complete geometric similarity exists between model and prototype, and the model inlet diameter is 3 in.

FIND Determine the required flowrate in the model.

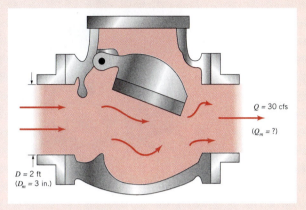

FIGURE E7.5a

SOLUTION

To ensure dynamic similarity, the model tests should be run so that

$$Re_m = Re$$

or

$$\frac{V_m D_m}{\nu_m} = \frac{VD}{\nu}$$

where V and D correspond to the inlet velocity and diameter, respectively. Since the same fluid is to be used in model and prototype, $\nu = \nu_m$, and therefore

$$\frac{V_m}{V} = \frac{D}{D_m}$$

The discharge, Q, is equal to VA, where A is the inlet area, so

$$\frac{Q_m}{Q} = \frac{V_m A_m}{VA} = \left(\frac{D}{D_m}\right)\frac{[(\pi/4)D_m^2]}{[(\pi/4)D^2]}$$

$$= \frac{D_m}{D}$$

and for the data given

$$Q_m = \frac{(3/12 \text{ ft})}{(2 \text{ ft})}(30 \text{ ft}^3/\text{s})$$

$$Q_m = 3.75 \text{ cfs} \qquad \textbf{(Ans)}$$

COMMENT As indicated by the above analysis, to maintain Reynolds number similarity using the same fluid in model and prototype, the required velocity scale is inversely proportional to the length scale, that is, $V_m/V = (D_m/D)^{-1}$. This strong influence of the length scale on the velocity scale is shown in Fig. E7.5b. For this particular example, $D_m/D = 0.125$, and the corresponding velocity scale is 8 (see

Fig. E7.5b). Thus, with the prototype velocity equal to $V = (30 \text{ ft}^3/\text{s})/(\pi/4)(2 \text{ ft})^2 = 9.50 \text{ ft/s}$, the required model velocity is $V_m = 76.4 \text{ ft/s}$. Although this is a relatively large velocity, it could be attained in a laboratory facility. It is to be noted that if we tried to use a smaller model, say one with $D = 1 \text{ in.}$, the required model velocity is 229 ft/s, a very high velocity that would be difficult to achieve. These results are indicative of one of the difficulties encountered in maintaining Reynolds number similarity—the required model velocities may be impractical to obtain.

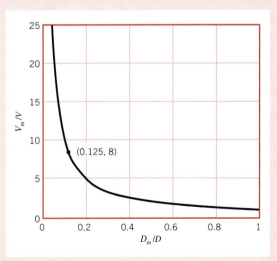

FIGURE E7.5b

■ FIGURE 7.1 Model of the National Bank of Commerce, San Antonio, Texas, for measurement of peak, rms, and mean pressure distributions. The model is located in a long-test-section, meteorological wind tunnel. (Photograph courtesy of Cermak Peterka Petersen, Inc.)

7.9.2 Flow around Immersed Bodies

Models have been widely used to study the flow characteristics associated with bodies that are completely immersed in a moving fluid. Examples include flow around aircraft, automobiles, golf balls, and buildings. (These types of models are usually tested in wind tunnels as is illustrated in Fig. 7.1.) Modeling laws for these problems are similar to those described in the preceding section; that is, geometric and Reynolds number similarity is required. Since there are no fluid interfaces, surface tension (and therefore the Weber number, see Table 7.1) is not important. Also, gravity will not affect the flow patterns, so the Froude number need not be considered. The Mach number will be important for high-speed flows in which compressibility becomes an important factor, but for incompressible fluids (such as liquids or for gases at relatively low speeds) the Mach number can be omitted as a similarity requirement. In this case, a general formulation for these problems is

$$\text{Dependent pi term} = \phi\left(\frac{\ell_i}{\ell}, \frac{\varepsilon}{\ell}, \frac{\rho V \ell}{\mu}\right) \tag{7.15}$$

V7.4 Wind engineering models

where ℓ is some characteristic length of the system and ℓ_i represents other pertinent lengths, ε/ℓ is the relative roughness of the surface (or surfaces), and $\rho V \ell/\mu$ is the Reynolds number.

Frequently, the dependent variable of interest for this type of problem is the drag, $\mathscr{D}$, developed on the body, and in this situation the dependent pi term would usually be expressed in the form of a *drag coefficient, C_D,* where

$$C_D = \frac{\mathscr{D}}{\frac{1}{2}\rho V^2 \ell^2}$$

The numerical factor, $\frac{1}{2}$, is arbitrary but commonly included, and ℓ^2 is usually taken as some representative area of the object. Thus, drag studies can be undertaken with the formulation

$$\frac{\mathscr{D}}{\frac{1}{2}\rho V^2 \ell^2} = C_D = \phi\left(\frac{\ell_i}{\ell}, \frac{\epsilon}{\ell}, \frac{\rho V \ell}{\mu}\right) \tag{7.16}$$

EXAMPLE 7.6 Model Design Conditions and Predicted Prototype Performance

GIVEN The drag on the airplane shown in Fig. E7.6 cruising at 240 mph in standard air is to be determined from tests on a 1:10 scale model placed in a pressurized wind tunnel. To minimize compressibility effects, the air speed in the wind tunnel is also to be 240 mph.

FIND Determine

(a) the required air pressure in the tunnel (assuming the same air temperature for model and prototype) and

(b) the drag on the prototype corresponding to a measured force of 1 lb on the model.

■ **FIGURE E7.6**

SOLUTION

(a) From Eq. 7.16 it follows that drag can be predicted from a geometrically similar model if the Reynolds numbers in model and prototype are the same. Thus,

$$\frac{\rho_m V_m \ell_m}{\mu_m} = \frac{\rho V \ell}{\mu}$$

For this example, $V_m = V$ and $\ell_m/\ell = \frac{1}{10}$ so that

$$\frac{\rho_m}{\rho} = \frac{\mu_m}{\mu} \frac{V}{V_m} \frac{\ell}{\ell_m}$$

$$= \frac{\mu_m}{\mu} (1)(10)$$

and therefore,

$$\frac{\rho_m}{\rho} = 10 \frac{\mu_m}{\mu}$$

This result shows that the same fluid with $\rho_m = \rho$ and $\mu_m = \mu$ cannot be used if Reynolds number similarity is to be maintained. One possibility is to pressurize the wind tunnel to increase the density of the air. We assume that an increase in pressure does not significantly change the viscosity so that the required increase in density is given by the relationship

$$\frac{\rho_m}{\rho} = 10$$

For an ideal gas, $p = \rho RT$ so that

$$\frac{p_m}{p} = \frac{\rho_m}{\rho}$$

for constant temperature ($T = T_m$). Therefore, the wind tunnel would need to be pressurized so that

$$\frac{p_m}{p} = 10$$

Since the prototype operates at standard atmospheric pressure, the required pressure in the wind tunnel is 10 atmospheres or

$$p_m = 10 \, (14.7 \text{ psia})$$
$$= 147 \text{ psia} \qquad \textbf{(Ans)}$$

COMMENT Thus, we see that a high pressure would be required and this could not be achieved easily or inexpensively. However, under these conditions, Reynolds similarity would be attained.

(b) The drag could be obtained from Eq. 7.16 so that

$$\frac{\mathscr{D}}{\frac{1}{2} \rho V^2 \ell^2} = \frac{\mathscr{D}_m}{\frac{1}{2} \rho_m V_m^2 \ell_m^2}$$

or

$$\mathscr{D} = \frac{\rho}{\rho_m} \left(\frac{V}{V_m}\right)^2 \left(\frac{\ell}{\ell_m}\right)^2 \mathscr{D}_m$$

$$= \left(\frac{1}{10}\right) (1)^2 (10)^2 \mathscr{D}_m$$

$$= 10 \mathscr{D}_m$$

Thus, for a drag of 1 lb on the model the corresponding drag on the prototype is

$$\mathscr{D} = 10 \text{ lb} \qquad \textbf{(Ans)}$$

For problems involving high velocities in which the Mach number is greater than about 0.3, the influence of compressibility, and therefore the Mach number (or Cauchy number), becomes significant. In this case complete similarity requires not only geometric and Reynolds number similarity, but also Mach number similarity so that

$$\frac{V_m}{c_m} = \frac{V}{c} \tag{7.17}$$

This similarity requirement, when combined with that for Reynolds number similarity, yields

$$\frac{c}{c_m} = \frac{\nu}{\nu_m}\frac{\ell_m}{\ell} \tag{7.18}$$

V7.5 Wind tunnel train model

Clearly the same fluid with $c = c_m$ and $\nu = \nu_m$ cannot be used in the model and prototype unless the length scale is unity (which means that we are running tests on the prototype). In high-speed aerodynamics the prototype fluid is usually air and it is difficult to satisfy Eq. 7.18 for reasonable length scales. Thus, models involving high-speed flows are often distorted with respect to Reynolds number similarity, but Mach number similarity is maintained.

7.9.3   Flow with a Free Surface

Flows in canals, rivers, spillways, and stilling basins, as well as flow around ships, are all examples of flow phenomena involving a free surface. For this class of problems, both gravitational and inertial forces are important and, therefore, the Froude number becomes an important similarity parameter. Also, because there is a free surface with a liquid–air interface, forces due to surface tension may be significant, and the Weber number becomes another similarity parameter that needs to be considered along with the Reynolds number. Geometric variables will obviously still be important. Thus a general formulation for problems involving flow with a free surface can be expressed as

V7.6 River flow model

$$\text{Dependent pi term} = \phi\left(\frac{\ell_i}{\ell}, \frac{\varepsilon}{\ell}, \frac{\rho V \ell}{\mu}, \frac{V}{\sqrt{g\ell}}, \frac{\rho V^2 \ell}{\sigma}\right) \tag{7.19}$$

As discussed previously, ℓ is some characteristic length of the system, ℓ_i represents other pertinent lengths, and ε/ℓ is the relative roughness of the various surfaces. Because gravity is the driving force in these problems, Froude number similarity is definitely required so that

$$\frac{V_m}{\sqrt{g_m \ell_m}} = \frac{V}{\sqrt{g\ell}}$$

The model and prototype are expected to operate in the same gravitational field ($g_m = g$), and therefore it follows that

$$\frac{V_m}{V} = \sqrt{\frac{\ell_m}{\ell}} = \sqrt{\lambda_\ell} \tag{7.20}$$

Thus, when models are designed on the basis of Froude number similarity, the velocity scale is determined by the square root of the length scale. As is discussed in Section 7.8.3, to

■ **FIGURE 7.2**   **A scale hydraulic model (1:197) of the Guri Dam in Venezuela, which is used to simulate the characteristics of the flow over and below the spillway and the erosion below the spillway. (Photograph courtesy of St. Anthony Falls Laboratory.)**

simultaneously have Reynolds and Froude number similarity it is necessary that the kinematic viscosity scale be related to the length scale as

$$\frac{\nu_m}{\nu} = (\lambda_\ell)^{3/2} \tag{7.21}$$

V7.7 Boat model

The working fluid for the prototype is normally either freshwater or seawater, and the length scale is small. Under these circumstances it is virtually impossible to satisfy Eq. 7.21, so models involving free-surface flows are usually distorted. The problem is further complicated if an attempt is made to model surface tension effects, as this requires equality, of Weber numbers. Fortunately, in many problems involving free-surface flows, both surface tension and viscous effects are small, and consequently strict adherence to Weber and Reynolds number similarity is not required.

For large hydraulic structures, such as dam spillways, the Reynolds numbers are large so that viscous forces are small in comparison to the forces due to gravity and inertia. In this case Reynolds number similarity is not maintained and models are designed on the basis of Froude number similarity. Care must be taken to ensure that the model Reynolds numbers are also large, but they are not required to be equal to those of the prototype. This type of hydraulic model is usually made as large as possible so that the Reynolds number will be large. A spillway model is shown in Fig. 7.2.

EXAMPLE 7.7   °Froude Number Similarity

GIVEN   A certain spillway for a dam is 20 m wide and is designed to carry 125 m^3/s at flood stage. A 1:15 model is constructed to study the flow characteristics through the spillway. The effects of surface tension and viscosity are to be neglected.

FIND

(a)   Determine the required model width and flowrate.

(b)   What operating time for the model corresponds to a 24-hr period in the prototype?

SOLUTION

The width, w_m, of the model spillway is obtained from the length scale, λ_ℓ, so that

$$\frac{w_m}{w} = \lambda_\ell$$

$$= \frac{1}{15}$$

and

$$w_m = \frac{20 \text{ m}}{15} = 1.33 \text{ m} \qquad \text{(Ans)}$$

Of course, all other geometric features (including surface roughness) of the spillway must be scaled in accordance with the same length scale.

With the neglect of surface tension and viscosity, Eq. 7.19 indicates that dynamic similarity will be achieved if the Froude numbers are equal between model and prototype. Thus,

$$\frac{V_m}{\sqrt{g_m \ell_m}} = \frac{V}{\sqrt{g\ell}}$$

and for $g_m = g$

$$\frac{V_m}{V} = \sqrt{\frac{\ell_m}{\ell}}$$

Since the flowrate is given by $Q = VA$, where A is an appropriate cross-sectional area, it follows that

$$\frac{Q_m}{Q} = \frac{V_m A_m}{VA} = \sqrt{\frac{\ell_m}{\ell}} \left(\frac{\ell_m}{\ell}\right)^2$$

$$= (\lambda_\ell)^{5/2}$$

where we have made use of the relationship $A_m/A = (\ell_m/\ell)^2$. For $\lambda_\ell = \frac{1}{15}$ and $Q = 125$ m^3/s

$$Q_m = \left(\tfrac{1}{15}\right)^{5/2} (125 \text{ m}^3/\text{s}) = 0.143 \text{ m}^3/\text{s} \qquad \text{(Ans)}$$

The time scale can be obtained from the velocity scale, since the velocity is distance divided by time ($V = \ell/t$), and therefore

$$\frac{V}{V_m} = \frac{\ell}{t} \frac{t_m}{\ell_m}$$

or

$$\frac{t_m}{t} = \frac{V}{V_m} \frac{\ell_m}{\ell} = \sqrt{\frac{\ell_m}{\ell}} = \sqrt{\lambda_\ell}$$

This result indicates that time intervals in the model will be smaller than the corresponding intervals in the prototype if $\lambda_\ell < 1$. For $\lambda_\ell = \frac{1}{15}$ and a prototype time interval of 24 hr

$$t_m = \sqrt{\tfrac{1}{15}} \,(24 \text{ hr}) = 6.20 \text{ hr} \qquad \text{(Ans)}$$

COMMENT   As indicated by the above analysis, the time scale varies directly as the square root of the length scale. Thus, as shown in Fig. E7.7, the model time interval, t_m, corresponding to a 24-hr prototype time interval can be varied by changing the length scale, λ_ℓ. The ability to scale times may be very useful, since it is possible to "speed up" events in the model which may occur over a relatively long time in the prototype. There is of course a practical limit to how small the length scale (and the corresponding time scale) can become. For example, if the length scale is too small, then surface tension effects may become important in the model whereas they are not in the prototype. In such a case the present model design, based simply on Froude number similarity, would not be adequate.

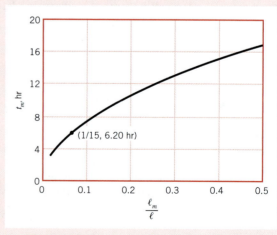

■ **FIGURE E7.7**

F l u i d s   i n   t h e   N e w s

Jurassic Tank Geologists use *models* involving water flowing across sand- or soil-filled tanks to investigate how river beds and river valleys are formed. The only variables in these studies are the model flowrate, the bottom slope, and the type of sand or soil material used. Now researchers at the University of Minnesota have developed the "Jurassic Tank," the first apparatus to use a "sinking floor." The bottom of the 40-ft-long, 20-ft-wide, 5-ft-deep tank contains 432 honeycomb funnels on top of which rests a rubber membrane floor. The floor can be programmed to sink in any

uneven fashion via computer control by removing the supporting gravel within the honeycombs. By running sediment-loaded water into the tank and studying the patterns of sediment deposition as the basin floor is lowered, it is possible to determine how sinking of the earth's crust interacts with sediment buildup to produce the sediment layers that fill ocean sedimentary basins. The name Jurassic Tank comes from its ability to model conditions during the Jurassic era at the beginning of the formation of the Atlantic Ocean (about 160 million years ago). ∎

7.10   Chapter Summary and Study Guide

similitude
dimensionless
   product
basic dimensions
pi term
Buckingham
   pi theorem
method of
   repeating
   variables
model
prototype
prediction equation
model design
   conditions
similarity
   requirements
modeling laws
length scale
distorted models
true models

Many practical engineering problems involving fluid mechanics require experimental data for their solution. Thus, laboratory studies and experimentation play a significant role in this field. It is important to develop good procedures for the design of experiments so they can be efficiently completed with as broad applicability as possible. To achieve this end the concept of similitude is often used in which measurements made in the laboratory can be utilized for predicting the behavior of other similar systems. In this chapter, dimensional analysis is used for designing such experiments, as an aid for correlating experimental data, and as the basis for the design of physical models. As the name implies, dimensional analysis is based on a consideration of the dimensions required to describe the variables in a given problem. A discussion of the use of dimensions and the concept of dimensional homogeneity (which forms the basis for dimensional analysis) was included in Chapter 1.

Essentially, dimensional analysis simplifies a given problem described by a certain set of variables by reducing the number of variables that need to be considered. In addition to being fewer in number, the new variables are dimensionless products of the original variables. Typically these new dimensionless variables are much simpler to work with in performing the desired experiments. The Buckingham pi theorem, which forms the theoretical basis for dimensional analysis, is described. This theorem establishes the framework for reducing a given problem described in terms of a set of variables to a new set of dimensionless variables. A simple method, called the repeating variable method, is described for actually forming the dimensionless variables (often called pi terms). Forming dimensionless variables by inspection is also considered. It is shown how the use of dimensionless variables can be of assistance in planning experiments and as an aid in correlating experimental data.

For problems in which there are a large number of variables, the use of physical models is described. Models are used to make specific predictions from laboratory tests rather than formulating a general relationship for the phenomenon of interest. The correct design of a model is obviously imperative for the accurate predictions of other similar, but usually larger, systems. It is shown how dimensional analysis can be used to establish a valid model design.

The following checklist provides a study guide for this chapter. When your study of the entire chapter and end-of-chapter exercises has been completed you should be able to

- write out meanings of the terms listed here in the margin and understand each of the related concepts. These terms are particularly important and are set in color and bold type in the text.

- use the Buckingham pi theorem to determine the number of dimensionless variables needed for a given flow problem.
- form a set of dimensionless variables using the method of repeating variables.
- form a set of dimensionless variables by inspection.
- use dimensionless variables as an aid in interpreting and correlating experimental data.
- establish a set of similarity requirements (and prediction equation) for a model to be used to predict the behavior of another similar system (the prototype).

References

1. Bridgman, P. W., *Dimensional Analysis,* Yale University Press, New Haven, Conn., 1922.
2. Murphy, G., *Similitude in Engineering,* Ronald Press, New York, 1950.
3. Langhaar, H. L., *Dimensional Analysis and Theory of Models,* Wiley, New York, 1951.
4. Ipsen, D. C., *Units, Dimensions, and Dimensionless Numbers,* McGraw-Hill, New York, 1960.
5. Isaacson, E. de St. Q., and Isaacson, M. de St. Q., *Dimensional Methods in Engineering and Physics,* Wiley, New York, 1975.

Review Problems

Go to Appendix F for a set of review problems with answers. Detailed solutions can be found in *Student Solution Manual for a Brief Introduction to Fluid Mechanics,* by Young et al. (© 2007 John Wiley and Sons, Inc.).

Problems

Note: Unless otherwise indicated use the values of fluid properties found in the tables on the inside of the front cover. Problems designated with an (*) are intended to be solved with the aid of a programmable calculator or a computer. Problems designated with a (†) are "open-ended" problems and require critical thinking in that to work them one must make various assumptions and provide the necessary data. There is not a unique answer to these problems.

Answers to the even-numbered problems are listed at the end of the book. Access to the videos that accompany problems can be obtained through the book's web site, www.wiley.com/college/young. The lab-type problems, FE problems, and FlowLab problems can also be accessed on this web site.

Section 7.1 Dimensional Analysis

7.1   The Reynolds number, $\rho VD/\mu$, is a very important parameter in fluid mechanics. Verify that the Reynolds number is dimensionless, using both the *FLT* system and the *MLT* system for basic dimensions, and determine its value for ethyl alcohol flowing at a velocity of 3 m/s through a 2-in.-diameter pipe.

7.2   What are the dimensions of acceleration of gravity, density, dynamic viscosity, kinematic viscosity, specific weight, and speed of sound in **(a)** the *FLT* system and **(b)** the *MLT*

system? Compare your results with those given in Table 1.1 in Chapter 1.

7.3   For the flow of a thin film of a liquid with a depth h and a free surface, two important dimensionless parameters are the Froude number, $V/\sqrt{gh}$, and the Weber number, $\rho V^2 h/\sigma$. Determine the value of these two parameters for glycerin (at 20° C) flowing with a velocity of 0.5 m/s at a depth of 2 mm.

Section 7.3 Determination of Pi Terms

7.4   Water sloshes back and forth in a tank as shown in Fig. P7.4. The frequency of sloshing, ω, is assumed to be a function of the acceleration of gravity, g, the average depth of the water, h, and the length of the tank, ℓ. Develop a suitable set of dimensionless parameters for this problem using g and ℓ as repeating variables.

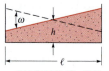

■ **FIGURE P7.4**

7.5 Water flows over a dam as illustrated in Fig. P7.5. Assume the flowrate, q, per unit length along the dam depends on the head, H, width, b, acceleration of gravity, g, fluid density, ρ, and fluid viscosity, μ. Develop a suitable set of dimensionless parameters for this problem using b, g, and ρ as repeating variables.

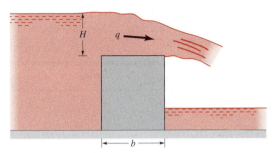

■ **FIGURE P7.5**

7.6 When a small pebble is dropped into a liquid, small waves travel outward as shown in Fig. P7.6. The speed of these waves, c, is assumed to be a function of the liquid density, ρ, the wavelength, λ, the wave height, h, and the surface tension of the liquid, σ. Use h, ρ, and σ as repeating variables to determine a suitable set of pi terms that could be used to describe this problem.

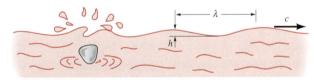

■ **FIGURE P7.6**

7.7 The pressure rise, Δp, across a pump can be expressed as

$$\Delta p = f(D, \rho, \omega, Q)$$

where D is the impeller diameter, ρ the fluid density, ω the rotational speed, and Q the flowrate. Determine a suitable set of dimensionless parameters.

7.8 The drag, $\mathcal{D}$, on a washer-shaped plate placed normal to a stream of fluid can be expressed as

$$\mathcal{D} = f(d_1, d_2, V, \mu, \rho)$$

where d_1 is the outer diameter, d_2 the inner diameter, V the fluid velocity, μ the fluid viscosity, and ρ the fluid density. Some experiments are to be performed in a wind tunnel to determine the drag. What dimensionless parameters would you use to organize these data?

7.9 Under certain conditions, wind blowing past a rectangular speed limit sign can cause the sign to oscillate with a frequency ω. (See Fig. P7.9 and **Video V9.6**.) Assume that ω is a function of the sign width, b, sign height, h, wind velocity, V, air density, ρ, and an elastic constant, k, for the supporting pole.

The constant, k, has dimensions of FL. Develop a suitable set of pi terms for this problem.

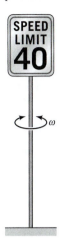

■ **FIGURE P7.9**

Section 7.5 Determination of Pi Terms by Inspection

7.10 The velocity, c, at which pressure pulses travel through arteries (pulse-wave velocity) is a function of the artery diameter, D, and wall thickness, h, the density of blood, ρ, and the modulus of elasticity, E, of the arterial wall. Determine a set of nondimensional parameters that can be used to study experimentally the relationship between the pulse-wave velocity and the variables listed. Form the nondimensional parameters by inspection.

7.11 Assume that the flowrate, Q, of a gas from a smokestack (see **Video V 5.3**) is a function of the density of the ambient air, ρ_a, the density of the gas, ρ_g, within the stack, the acceleration of gravity, g, and the height and diameter of the stack, h and d, respectively. Develop a set of pi terms that could be used to describe this problem. Form the pi terms by inspection.

7.12 As shown in Fig. P7.12 and **Video V5.4**, a jet of liquid directed against a block can tip over the block. Assume that the velocity, V, needed to tip over the block is a function of the fluid density, ρ, the diameter of the jet, D, the weight of the block, $\mathcal{W}$, the width of the block, b, and the distance, d, between the jet and the bottom of the block. (**a**) Determine a set of dimensionless parameters for this problem. Form the dimensionless parameters by inspection. (**b**) Use the momentum equation to determine an equation for V in terms of the other variables. (**c**) Compare the results of parts (**a**) and (**b**).

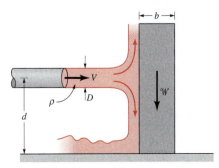

■ **FIGURE P7.12**

Section 7.7 Correlation of Experimental Data (also see Lab Problems 7.43, 7.44, 7.45, and 7.46)

7.13 The buoyant force, F_B, acting on a body submerged in a fluid is a function of the specific weight, γ, of the fluid and the volume, V, of the body. Show, by dimensional analysis, that the buoyant force must be directly proportional to the specific weight.

7.14 When a sphere of diameter d falls slowly in a highly viscous fluid, the settling velocity, V, is known to be a function of d, the fluid viscosity, μ, and the difference, $\Delta\gamma$, between the specific weight of the sphere and the specific weight of the fluid. Due to a tight budget situation, only one experiment can be performed, and the following data were obtained: $V = 0.42$ ft/s for $d = 0.1$ in., $\mu = 0.03$ lb·s/ft^2, and $\Delta\gamma = 10$ lb/ft^3. If possible, based on this limited amount of data, determine the general equation for the settling velocity. If you do not think it is possible, indicate what additional data would be required.

*7.15 The pressure drop across a short hollowed plug placed in a circular tube through which a liquid is flowing (see Fig. P7.15) can be expressed as

$$\Delta p = f(\rho, V, D, d)$$

where ρ is the fluid density and V is the mean velocity in the tube. Some experimental data obtained with $D = 0.2$ ft, $\rho = 2.0$ slugs/ft^3, and $V = 2$ ft/s are given in the following table:

d (ft)	0.06	0.08	0.10	0.15
Δp (lb/ft^2)	493.8	156.2	64.0	12.6

Plot the results of these tests, using suitable dimensionless parameters, on log–log graph paper. Use a standard curve-fitting technique to determine a general equation for Δp. What are the limits of applicability of the equation?

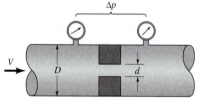

■ **FIGURE P7.15**

7.16 A liquid flows with a velocity V through a hole in the side of a large tank. Assume that

$$V = f(h, g, \rho, \sigma)$$

where h is the depth of fluid above the hole, g is the acceleration of gravity, ρ the fluid density, and σ the surface tension. The following data were obtained by changing h and measuring V, with a fluid having a density $= 10^3$ kg/m^3 and surface tension $= 0.074$ N/m.

V (m/s)	3.13	4.43	5.42	6.25	7.00
h (m)	0.50	1.00	1.50	2.00	2.50

Plot these data by using appropriate dimensionless variables. Could any of the original variables have been omitted?

*7.17 As shown in Fig. 2.16, Fig. P7.17, and **Video V2.7,** a rectangular barge floats in a stable configuration provided the distance between the center of gravity, CG, of the object (boat and load) and the center of buoyancy, C, is less than a certain amount, H. If this distance is greater than H, the boat will tip over. Assume H is a function of the boat's width, b, length, ℓ, and draft, h. **(a)** Put this relationship into dimensionless form. **(b)** The results of a set of experiments with a model barge with a width of 1.0 m are shown in the table. Plot these data in dimensionless form and determine a power-law equation relating the dimensionless parameters.

ℓ, m	h, m	H, m
2.0	0.10	0.833
4.0	0.10	0.833
2.0	0.20	0.417
4.0	0.20	0.417
2.0	0.35	0.238
4.0	0.35	0.238

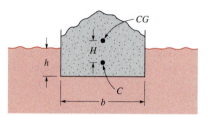

■ **FIGURE P7.17**

7.18 A fluid flows through the horizontal curved pipe of Fig. P7.18 with a velocity V. The pressure drop, Δp, between the entrance and the exit to the bend is thought to be a function of the velocity, bend radius, R, pipe diameter, D, and fluid density, ρ. The data shown in the following table were obtained in the laboratory. For these tests $\rho = 2.0$ slugs/ft^3, $R = 0.5$ ft, and $D = 0.1$ ft. Perform a dimensional analysis and based on the data given, determine if the variables used for this problem appear to be correct. Explain how you arrive at your answer.

V (ft/s)	2.1	3.0	3.9	5.1
Δp (lb/ft^2)	1.2	1.8	6.0	6.6

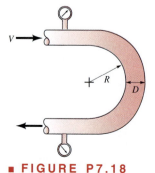

■ **FIGURE P7.18**

7.19 The time, t, it takes to pour a certain volume of liquid from a cylindrical container depends on several factors, including the viscosity of the liquid. (See **Video V1.1.**) Assume that for very viscous liquids the time it takes to pour out two-thirds of the initial volume depends on the initial liquid depth, ℓ, the cylinder diameter, D, the liquid viscosity, μ, and the liquid specific weight, γ. Data shown in the following table were obtained in the laboratory. For these tests $\ell = 45$ mm, $D = 67$ mm, and $\gamma = 9.60$ kN/m^3. **(a)** Perform a dimensional analysis and, based on the data given, determine if variables used for this problem appear to be correct. Explain how you arrived at your answer. **(b)** If possible, determine an equation relating the pouring time and viscosity for the cylinder and liquids used in these tests. If it is not possible, indicate what additional information is needed.

μ (N·s/m^2)	11	17	39	61	107
t(s)	15	23	53	83	145

Section 7.8 Modeling and Similitude

7.20 SAE 30 oil at 60° F is pumped through a 3-ft-diameter pipeline at a rate of 5700 gal/min. A model of this pipeline is to be designed using a 2-in.-diameter pipe and water at 60° F as the working fluid. To maintain Reynolds number similarity between these two systems, what fluid velocity will be required in the model?

7.21 A liquid contained in a steadily rotating cylinder with a vertical axis moves as a rigid body. In this case, as shown in Fig. P7.21 and **Video V7.1,** the fluid velocity, V, varies directly with the radius r so that $V = r\omega$ where ω is the angular velocity of the rotating cylinder. Assume that the characteristic Reynolds number for this system is based on the radius and velocity at the wall of the cylinder. **(a)** For a 12-in.-diameter cylinder rotating with an angular velocity $\omega = 0.4$, rad/s, calculate the Reynolds number if the liquid has a kinematic viscosity of (i) 0.33 ft^2/s or (ii) 0.33×10^{-2} ft^2/s. **(b)** If the cylinder were suddenly stopped, would you expect the motion of the liquids to be similar for the two liquids of part **(a)**? Explain. Do the results shown in **Video V7.1** support your conclusions?

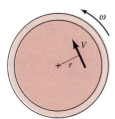

■ **FIGURE P7.21**

7.22 The design of a river model is to be based on Froude number similarity, and a river depth of 3 m is to correspond to a model depth of 100 mm. Under these conditions what is the prototype velocity corresponding to a model velocity of 2 m/s?

7.23 The drag characteristics of a torpedo are to be studied in a water tunnel using a 1:5 scale model. The tunnel operates with freshwater at 20 °C, whereas the prototype torpedo is to be used in seawater at 15.6 °C. To correctly simulate the behavior of the prototype moving with a velocity of 30 m/s, what velocity is required in the water tunnel? Assume Reynolds number similarity.

7.24 Carbon tetrachloride flows with a velocity of 0.30 m/s through a 30-mm-diameter tube. A model of this system is to be developed using standard air as the model fluid. The air velocity is to be 2 m/s. What tube diameter is required for the model if dynamic similarity is to be maintained between model and prototype?

7.25 Water flowing under the obstacle shown in Fig. P7.25 puts a vertical force, F_v, on the obstacle. This force is assumed to be a function of the flowrate, Q, the density of the water, ρ, the acceleration of gravity, g, and a length, ℓ, that characterizes the size of the obstacle. A 1/20 scale model is to be used to predict the vertical force on the prototype. **(a)** Perform a dimensional analysis for this problem. **(b)** If the prototype flowrate is 1000 ft^3/s, determine the water flowrate for the model if the flows are to be similar. **(c)** If the model force is measured as $(F_v)_m = 20$ lb, predict the corresponding force on the prototype.

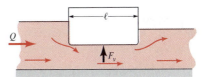

■ **FIGURE P7.25**

7.26 (See Fluids in the News article titled "Modeling parachutes in a water tunnel," Section 7.8.1.) Flow characteristics for a 30-ft-diameter prototype parachute are to be determined by tests of a 1-ft-diameter model parachute in a water tunnel. Some data collected with the model parachute indicate a drag of 17 lb when the water velocity is 4 ft/s. Use the model data to predict the drag on the prototype parachute falling through air at 10 ft/s. Assume the drag to be a function of the velocity, V, the fluid density, ρ, and the parachute diameter, D.

7.27 The drag on a sphere moving in a fluid is known to be a function of the sphere diameter, the velocity, and the fluid viscosity and density. Laboratory tests on a 4-in.-diameter sphere were performed in a water tunnel, and some model data are plotted in Fig. P7.27. For these tests the viscosity of the

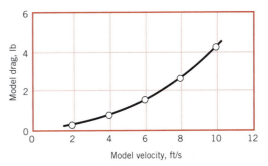

■ **FIGURE P7.27**

water was 2.3×10^{-5} lb·s/ft² and the water density was 1.94 slugs/ft³. Estimate the drag on an 8-ft-diameter balloon moving in air at a velocity of 3 ft/s. Assume the air to have a viscosity of 3.7×10^{-7} lb·s/ft² and a density of 2.38×10^{-3} slugs/ft³.

7.28   As shown in Fig. P7.28, a thin, flat plate containing a series of holes is to be placed in a pipe to filter out any particles in the liquid flowing through the pipe. There is some concern about the large pressure drop that may develop across the plate, and it is proposed to study this problem with a geometrically similar model. The following data apply.

Prototype	Model
d—hole diameter = 1.0 mm	$d = ?$
D—pipe diameter = 50 mm	$D = 10$ mm
μ—viscosity = 0.002 N·s/m²	$\mu = 0.002$ N·s/m²
ρ—density = 1000 kg/m³	$\rho = 1000$ kg/m³
V—velocity = 0.1 m/s to 2 m/s	$V = ?$

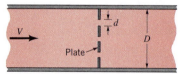

■ **FIGURE P7.28**

(a) Assuming that the pressure drop, Δp, depends on the variables listed, use dimensional analysis to develop a suitable set of dimensionless parameters for this problem. **(b)** Determine values for the model indicated in the list with a question mark. What will be the pressure drop scale, $\Delta p_m / \Delta p$?

7.29   A large, rigid, rectangular billboard is supported by an elastic column as shown in Fig. P7.29. There is concern

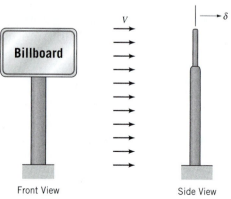

Front View   Side View

■ **FIGURE P7.29**

about the deflection, δ, of the top of the structure during a high wind of velocity V. A wind tunnel test is to be conducted with a 1:15 scale model. Assume the pertinent column variables are its length and cross-sectional dimensions, and the modulus of elasticity of the material used for the column. The only important "wind" variables are the air density and velocity. **(a)** Determine the model design conditions and the prediction equation for the deflection. **(b)** If the same structural materials are used for the model and prototype, and the wind tunnel operates under standard atmospheric conditions, what is the required wind tunnel velocity to match an 80 km/hr wind?

7.30   Assume that the wall shear stress, τ_w, created when a fluid flows through a pipe (see Fig. P7.30a) depends on the pipe diameter, D, the flowrate, Q, the fluid density, ρ, and the kinematic viscosity, ν. Some model tests run in a laboratory using water in a 0.2-ft-diameter pipe yield the τ_w vs. Q data shown in Fig. P7.30b. Perform a dimensional analysis and use model data to predict the wall shear stress in a 0.3-ft-diameter pipe through which water flows at the rate of 1.5 ft³/s.

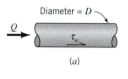

(a)

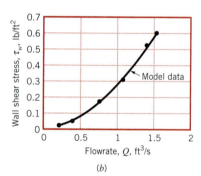

(b)

■ **FIGURE P7.30**

†7.31   If a large oil spill occurs from a tanker operating near a coastline, the time it would take for the oil to reach shore is of great concern. Design a model system that can be used to investigate this type of problem in the laboratory. Indicate all assumptions made in developing the design and discuss any difficulty that may arise in satisfying the similarity requirements arising from your model design.

7.32   The pressure rise, Δp, across a centrifugal pump of a given shape (see Fig. P7.32a) can be expressed as

$$\Delta p = f(D, \omega, \rho, Q)$$

where D is the impeller diameter, ω the angular velocity of the impeller, ρ the fluid density, and Q the volume rate of flow through the pump. A model pump having a diameter of 8 in. is tested in the laboratory using water. When operated at an angular velocity of 40π rad/s, the model pressure rises as a function of Q as shown in Fig. P7.32b. Use this curve to predict the pressure rise across a geometrically similar pump (prototype) for a prototype flowrate of 6 ft³/s. The prototype has a diameter of 12 in. and operates at an angular velocity of 60π rad/s. The prototype fluid is also water.

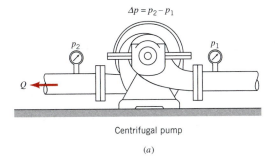

$\Delta p = p_2 - p_1$

p_2 p_1

Q

Centrifugal pump

(a)

(b)

Model data
$\omega_m = 40\pi$ rad/s
$D_m = 8$ in.

Δp_m (psi)

Q_m (ft³/s)

■ **FIGURE P7.32**

Section 7.9 Some Typical Model Studies

7.33 At a large fish hatchery the fish are reared in open, water-filled tanks. Each tank is approximately square in shape with curved corners, and the walls are smooth. To create motion in the tanks, water is supplied through a pipe at the edge of the tank. The water is drained from the tank through an opening at the center. (See **Video V7.3.**) A model with a length scale of 1:13 is to be used to determine the velocity, V, at various locations within the tank. Assume that $V = f(\ell, \ell_i, \rho, \mu, g, Q)$ where ℓ is some characteristic length such as the tank width, ℓ_i represents a series of other pertinent lengths, such as inlet pipe diameter, fluid depth, etc., ρ is the fluid density, μ is the fluid viscosity, g is the acceleration of gravity, and Q is the discharge through the tank. **(a)** Determine a suitable set of dimensionless parameters for this problem and the prediction equation for the velocity. If water is to be used for the model, can all of the similarity requirements be satisfied? Explain and support your answer with the necessary calculations. **(b)** If the flowrate into the full-sized tank is 250 gpm, determine the required value for the model discharge assuming Froude number similarity. What model depth will correspond to a depth of 32 in. in the full-sized tank?

7.34 During a storm, a snow drift is formed behind some bushes as shown in Fig. P7.34 and **Video V9.4.** Assume that the height of the drift, h, is a function of the number of inches of snow deposited by the storm, d, the height of the bush, H, the width of the bush, b, the wind speed, V, the acceleration of gravity, g, the air density, ρ, the specific weight of the snow, γ_s, and the porosity of the bush, η. Note that porosity is defined as the percentage of open area of the bush. **(a)** Determine a suitable set of dimensionless variables for this problem. **(b)** A storm with 30 mph winds deposits 16 in. of snow having a specific weight of 5.0 lb/ft³. A half-sized scale model bush is to be used to investigate the drifting behind the bush. If the air density is the same for the model and the storm, determine the required specific weight of the model snow, the required wind speed for the model, and the number of inches of model snow to be deposited.

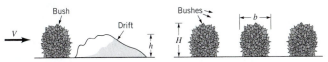

Bush   Drift   Bushes b

V h H

■ **FIGURE P7.34**

7.35 (See Fluids in the News article titled "Ice engineering," Section 7.9.3.) A model study is to be developed to determine the force exerted on bridge piers due to floating chunks of ice in a river. The piers of interest have square cross sections. Assume that the force, R, is a function of the pier width, b, the depth of the ice, d, the velocity of the ice, V, the acceleration of gravity, g, the density of the ice, ρ_i, and a measure of the strength of the ice, E_i, where E_i has the dimensions FL^{-2}. **(a)** Based on these variables determine a suitable set of dimensionless variables for this problem. **(b)** The prototype conditions of interest include an ice thickness of 12 in. and an ice velocity of 6 ft/s. What model ice thickness and velocity would be required if the length scale is to be 1/10? **(c)** If the model and prototype ice have the same density can the model ice have the same strength properties as that of the prototype ice? Explain.

7.36 Flow patterns that develop as winds blow past a vehicle, such as a train, are often studied in low-speed environmental (meteorological) wind tunnels. (See **Video V7.5.**) Typically, the air velocities in these tunnels are in the range of 0.1 m/s to 30 m/s. Consider a cross wind blowing past a train locomotive. Assume that the local wind velocity, V, is a function of the approaching wind velocity (at some distance from the locomotive), U, the locomotive length, ℓ, height, h, and width, b, the air density, ρ, and the air viscosity, μ. **(a)** Establish the similarity requirements and prediction equation for a model to be used in the wind tunnel to study the air velocity, V, around the locomotive. **(b)** If the model is to be used for cross winds gusting to $U = 25$ m/s, explain why it is not practical to maintain Reynolds number similarity for a typical length scale 1:50.

7.37 River models are used to study many different types of flow situations. (See, for example, **Video V7.6.**) A certain small river has an average width and depth of 60 and 4 ft, respectively, and carries water at a flowrate of 700 ft³/s. A model is to be designed based on Froude number similarity so that the discharge scale is 1/250. At what depth and flowrate would the model operate?

7.38 As illustrated in **Video V7.2,** models are commonly used to study the dispersion of a gaseous pollutant from an exhaust stack located near a building complex. Similarity requirements for the pollutant source involve the following independent variables: the stack gas speed, V, the wind speed, U, the density of the atmospheric air, ρ, the difference in densities between the air and the stack gas, $\rho - \rho_s$, the acceleration of gravity, g, the kinematic viscosity of the stack gas, ν_s, and the stack diameter, D. **(a)** Based on these variables, determine a suitable set of similarity requirements for modeling the pollutant source. **(b)** For this type of model a typical length scale might be 1:200. If the same fluids were used in model and prototype, would the similarity requirements be satisfied? Explain and support your answer with the necessary calculations.

7.39   (See Fluids in the News article titled "Galloping Gertie," Section 7.8.2.) The Tacoma Narrows bridge failure is a dramatic example of the possible serious effects of wind-induced vibrations. As a fluid flows around a body, vortices may be created which are shed periodically creating an oscillating force on the body. If the frequency of the shedding vortices coincides with the natural frequency of the body, large displacements of the body can be induced as was the case with the Tacoma Narrows bridge. To illustrate this type of phenomenon, consider fluid flow past a circular cylinder. Assume the frequency, n, of the shedding vortices behind the cylinder is a function of the cylinder diameter, D, the fluid velocity, V, and the fluid kinematic viscosity, ν. **(a)** Determine a suitable set of dimensionless variables for this problem. One of the dimensionless variables should be the Strouhal number, nD/V. **(b)** Some experiments, using small models (cylinders), were performed in which the shedding frequency of the vortices (in Hz) was measured. Results for a *particular* cylinder in a Newtonian, incompressible fluid are shown in Fig. P7.39. Is this a "universal curve" that can be used to predict the shedding frequency for any cylinder placed in any fluid? Explain. **(c)** A certain structural component in the form of a 1-in.-diameter, 12-ft-long rod acts as a cantilever beam with a natural frequency of 19 Hz. Based on the data in Fig. P7.39, estimate the wind speed that may cause the rod to oscillate at its natural frequency. *Hint:* Use a trial and error solution.

7.40   A 1/50 scale model is to be used in a towing tank to study the water motion near the bottom of a shallow channel as a large barge passes over. (See **Video V7.7.**) Assume that the model is operated in accordance with the Froude number criteria for dynamic similitude. The prototype barge moves at a typical speed of 15 knots. **(a)** At what speed (in ft/s) should the model be towed? **(b)** Near the bottom of the model channel a small particle is found to move 0.15 ft in 1 s so that the fluid velocity at that point is approximately 0.15 ft/s. Determine the velocity at the corresponding point in the prototype channel.

7.41   As winds blow past buildings, complex flow patterns can develop due to various factors, such as flow separation and interactions between adjacent buildings. (See **Video V7.4.**) Assume that the local gage pressure, p, at a particular location on a building is a function of the air density, ρ, the wind speed, V, some characteristic length, ℓ, and all other pertinent lengths, ℓ_i, needed to characterize the geometry of the building or building complex. **(a)** Determine a suitable set of dimensionless parameters that can be used to study the pressure distribution. **(b)** An eight-story building that is 100 ft tall is to be modeled in a wind tunnel. If a length scale of 1:300 is to be used, how tall should the model building be? **(c)** How will a measured pressure in the model be related to the corresponding prototype pressure? Assume the same air density in the model and prototype. Based on the assumed variables, does the model wind speed have to be equal to the prototype wind speed? Explain.

7.42   A square parking lot of width w is bounded on all sides by a curb of height d with only one opening of width b as shown in Fig. P7.42. During a heavy rain the lot fills with water and it is of interest to determine the time, t, it takes for the water to completely drain from the lot after the rain stops. A scale model is to be used to study this problem, and it is assumed that

$$t = f(w, b, d, g, \mu, \rho)$$

where g is the acceleration of gravity, μ is the fluid viscosity, and ρ is the fluid density. **(a)** A dimensional analysis indicates that two important dimensionless parameters are b/w and d/w. What additional dimensionless parameters are required? **(b)** For a geometrically similar model having a length scale of 1/10, what is the relationship between the drain time for the model and the corresponding drain time for the actual parking lot? Assume all similarity requirements are satisfied. Can water be used as the model fluid? Explain and justify your answer.

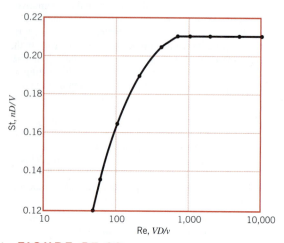

■ **FIGURE P7.39**

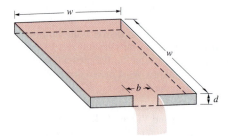

■ **FIGURE P7.42**

Lab Problems

7.43   This problem involves the time that it takes water to drain from two geometrically similar tanks. To proceed with this problem, go to the book's web site, www.wiley.com/college/young.

7.44   This problem involves determining the frequency of vortex shedding from a circular cylinder as water flows past it. To proceed with this problem, go to the book's web site, www.wiley.com/college/young.

7.45   This problem involves the determination of the head loss for flow through a valve. To proceed with this problem, go to the book's web site, www.wiley.com/college/young.

7.46   This problem involves the calibration of a rotameter. To proceed with this problem, go to the book's web site, www.wiley.com/college/young.

FlowLab Problem

*7.47   This FlowLab problem involves investigation of the Reynolds number significance in fluid dynamics through the simulation of flow past a cylinder. To proceed with this problem, go to the book's web site, www.wiley.com/college/young.

FE Exam Problems

Sample FE (Fundamentals of Engineering) exam questions for fluid mechanics are provided on the book's web site, www.wiley.com/college/young.

CHAPTER 8

Viscous Flow in Pipes

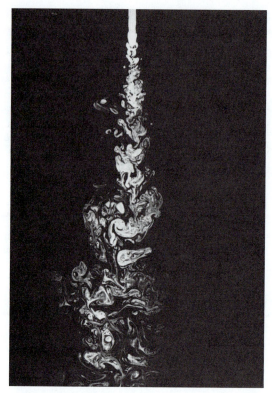

Turbulent jet: The jet of water from the pipe is turbulent. The complex, irregular, unsteady structure typical of turbulent flows is apparent. (Laser-induced fluorescence of dye in water.) (Photography by P. E. Dimotakis, R. C. Lye, and D. Z. Papantoniou.)

LEARNING OBJECTIVES

After completing this chapter, you should be able to:

- identify and understand various characteristics of the flow in pipes.
- discuss the main properties of laminar and turbulent pipe flow and appreciate their differences.
- calculate losses in straight portions of pipes as well as those in various pipe system components.
- apply appropriate equations and principles to analyze a variety of pipe flow situations.
- predict the flowrate in a pipe by use of common flowmeters.

In this chapter we apply the basic principles concerning mass, momentum, and energy to a specific, important topic—the flow of viscous incompressible fluids in pipes and ducts. Some of the basic components of a typical *pipe system* are shown in Fig. 8.1. They include the pipes themselves (perhaps of more than one diameter), the various fittings used to connect the individual pipes to form the desired system, the flowrate control devices (valves), and the pumps or turbines that add energy to or remove energy from the fluid.

8.1   General Characteristics of Pipe Flow

Before we apply the various governing equations to pipe flow examples, we will discuss some of the basic concepts of pipe flow. Unless otherwise specified, we will assume that the conduit is round, although we will show how to account for other shapes. For all flows involved in this chapter, we assume that the pipe is completely filled with the fluid being transported.

8.1.1   Laminar or Turbulent Flow

V8.1 Laminar/
turbulent pipe flow

The flow of a fluid in a pipe may be laminar flow or it may be turbulent flow. Osborne Reynolds, a British scientist and mathematician, was the first to distinguish the difference between these two classifications of flow by using a simple apparatus as shown in Fig. 8.2a. For "small enough flowrates" the dye streak (a streakline) will remain as a well-defined line as it flows along, with only slight blurring due to molecular diffusion of the dye into the surrounding water. For a somewhat larger "intermediate flowrate" the dye streak fluctuates in time and space, and intermittent bursts of irregular behavior appear along the streak. However, for "large enough flowrates" the dye streak almost immediately becomes blurred and spreads across the entire pipe in a random fashion. These three characteristics, denoted as ***laminar, transitional,*** and ***turbulent flow,*** respectively, are illustrated in Fig. 8.2b.

In the previous paragraph the term *flowrate* should be replaced by Reynolds number, $Re = \rho V D / \mu$, where V is the average velocity in the pipe. That is, the flow in a pipe is laminar, transitional, or turbulent provided the Reynolds number is "small enough," "intermediate," or "large enough." It is not only the fluid velocity that determines the character of the flow—its density, viscosity, and the pipe size are of equal importance. These parameters combine to produce the Reynolds number. For general engineering purposes (i.e., without undue precautions to eliminate disturbances), the following values are appropriate: The flow in a round pipe is laminar if the Reynolds number is less than approximately 2100. The flow in a round pipe is turbulent if the Reynolds number is greater than approximately 4000. For Reynolds numbers between these two limits, the flow may switch between laminar and turbulent conditions in an apparently random fashion (transitional flow).

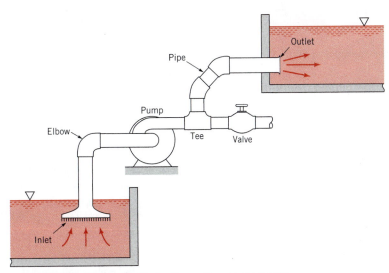

■ **FIGURE 8.1**   **Typical pipe system components.**

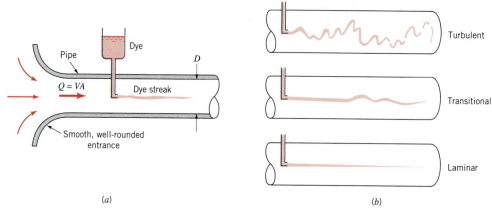

(a) *(b)*

■ **FIGURE 8.2** *(a)* **Experiment to illustrate type of flow.** *(b)* **Typical dye streaks.**

F l u i d s   i n   t h e   N e w s

Nanoscale flows The term *nanoscale* generally refers to objects with characteristic lengths from atomic dimensions up to a few hundred nanometers (nm). (Recall that 1 nm = 10^{-9} m.) Nanoscale fluid mechanics research has recently uncovered many surprising and useful phenomena. No doubt many more remain to be discovered. For example, in the future researchers envision using nanoscale tubes to push tiny amounts of water-soluble drugs to exactly where they are needed in the human body. Because of the tiny diameters involved, the *Reynolds numbers* for such flows are extremely small and the flow is definitely laminar. In addition, some standard properties of everyday flows (for example, the fact that a fluid sticks to a solid boundary) may not be valid for nanoscale flows. Also, ultratiny mechanical pumps and valves are difficult to manufacture and may become clogged by tiny particles such as biological molecules. As a possible solution to such problems, researchers have investigated the possibility of using a system that does not rely on mechanical parts. It involves using light-sensitive molecules attached to the surface of the tubes. By shining light onto the molecules, the light-responsive molecules attract water and cause motion of water through the tube. (See Problem 8.2.) ■

EXAMPLE 8.1   Laminar or Turbulent Flow

GIVEN Water at a temperature of 50° F flows through a pipe of diameter $D = 0.73$ in.

SOLUTION

(a) If the flow in the pipe is to remain laminar, the minimum time to fill the glass will occur if the Reynolds number is the maximum allowed for laminar flow, typically Re = $\rho VD/\mu$ = 2100. Thus, $V = 2100\,\mu/\rho D$, where from Table B.1, $\rho = 1.94$ slugs/ft^3 and $\mu = 2.73 \times 10^{-5}$ lb·s/ft^2 at 50° F. Therefore, the maximum average velocity for laminar flow in the pipe is

$$V = \frac{2100\mu}{\rho D} = \frac{2100(2.73 \times 10^{-5}\ \text{lb·s/ft}^2)}{(1.94\ \text{slugs/ft}^3)(0.73/12\ \text{ft})} = 0.486\ \text{lb·s/slug}$$

$$= 0.486\ \text{ft/s}$$

FIND Determine

(a) the minimum time taken to fill a 12-oz glass (volume = 0.0125 ft^3) with water if the flow in the pipe is to be laminar.

(b) the maximum time taken to fill the glass if the flow is to be turbulent.

With $\mathcal{V}$ = volume of glass and $\mathcal{V} = Qt$ we obtain

$$t = \frac{\mathcal{V}}{Q} = \frac{\mathcal{V}}{(\pi/4)D^2 V} = \frac{4(0.0125\ \text{ft}^3)}{(\pi[0.73/12]^2\text{ft}^2)(0.486\ \text{ft/s})}$$

$$= 8.85\ \text{s} \qquad \textbf{(Ans)}$$

(b) If the flow in the pipe is to be turbulent, the maximum time to fill the glass will occur if the Reynolds number is the minimum allowed for turbulent flow, Re = 4000. Thus, with the given values of ρ and μ, $V = 4000 \, \mu/\rho D = 0.925$ ft/s and

$$t = \frac{\forall}{Q} = \frac{\forall}{(\pi/4)D^2 V} = \frac{4(0.0125 \text{ ft}^3)}{(\pi[0.73/12]^2 \text{ft}^2)(0.925 \text{ ft/s})}$$

$$= 4.65 \text{ s} \qquad \textbf{(Ans)}$$

COMMENTS Note that because water is "not very viscous," the velocity must be "fairly small" to maintain laminar flow. In general, turbulent flows are encountered more often than laminar flows because of the relatively small viscosity of most common fluids (water, gasoline, air). By repeating the calculations at various water temperatures, T (i.e., with different densities and viscosities), the results shown in Fig. E8.1 are obtained. As the water temperature increases, the kinematic viscosity, $\nu = \mu/\rho$, decreases and the corresponding times to fill the glass increase as indicated.

If the flowing fluid had been honey with a kinematic viscosity ($\nu = \mu/\rho$) 3000 times greater than that of water, the velocities given earlier would be increased by a factor of 3000 and the

times reduced by the same factor. As shown in the following sections, the pressure needed to force a very viscous fluid through a pipe at such a high velocity may be unreasonably large.

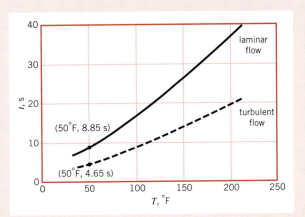

■ **FIGURE E8.1**

8.1.2   Entrance Region and Fully Developed Flow

Any fluid flowing in a pipe had to enter the pipe at some location. The region of flow near where the fluid enters the pipe is termed the *entrance region* and is illustrated in Fig. 8.3. As shown, the fluid typically enters the pipe with a nearly uniform velocity profile at section (1). As the fluid moves through the pipe, viscous effects cause it to stick to the pipe wall (the no-slip boundary condition). This is true whether the fluid is relatively inviscid air or a very viscous oil. Thus, a *boundary layer* in which viscous effects are important is produced along the pipe wall such that the initial velocity profile changes with distance along the pipe, x, until the fluid reaches the end of the entrance length, section (2), beyond which the velocity profile does not vary with x. The boundary layer has grown in thickness to completely fill the pipe.

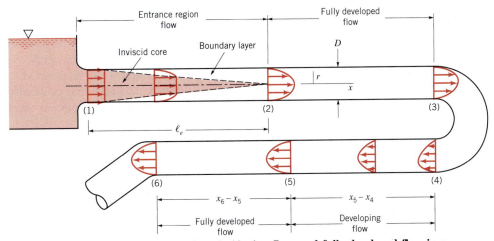

■ **FIGURE 8.3**   **Entrance region, developing flow, and fully developed flow in a pipe system.**

The shape of the velocity profile in the pipe depends on whether the flow is laminar or turbulent, as does the *entrance length,* ℓ_e. Typical entrance lengths are given by

$$\frac{\ell_e}{D} = 0.06 \, \text{Re for laminar flow} \tag{8.1}$$

and

$$\frac{\ell_e}{D} = 4.4 \, (\text{Re})^{1/6} \text{ for turbulent flow} \tag{8.2}$$

Once the fluid reaches the end of the entrance region, section (2) of Fig. 8.3, the flow is simpler to describe because the velocity is a function of only the distance from the pipe centerline, r, and independent of x. This is true until the character of the pipe changes in some way, such as a change in diameter or the fluid flows through a bend, valve, or some other component at section (3). The flow between (2) and (3) is termed *fully developed flow.* Beyond the interruption of the fully developed flow [at section (4)], the flow gradually begins its return to its fully developed character [section (5)] and continues with this profile until the next pipe system component is reached [section (6)].

8.2   Fully Developed Laminar Flow

Knowledge of the velocity profile can lead directly to other useful information such as pressure drop, head loss, flowrate, and the like. Thus, we begin by developing the equation for the velocity profile in fully developed laminar flow. If the flow is not fully developed, a theoretical analysis becomes much more complex and is outside the scope of this text. If the flow is turbulent, a rigorous theoretical analysis is as yet not possible.

8.2.1   From F = *m*a Applied Directly to a Fluid Element

We consider the fluid element at time t as is shown in Fig. 8.4. It is a circular cylinder of fluid of length ℓ and radius r centered on the axis of a horizontal pipe of diameter D. Because the velocity is not uniform across the pipe, the initially flat ends of the cylinder of fluid at time t become distorted at time $t + \delta t$ when the fluid element has moved to its new location along the pipe as shown in Fig. 8.4. If the flow is fully developed and steady, the distortion on each end of the fluid element is the same, and no part of the fluid experiences any acceleration as it flows. Every part of the fluid merely flows along its pathline parallel to the pipe walls with constant velocity, although neighboring particles have slightly different velocities. The velocity varies from one pathline to the next. This velocity variation, combined with the fluid viscosity, produces the shear stress.

If gravitational effects are neglected, the pressure is constant across any vertical cross section of the pipe, although it varies along the pipe from one section to the next. Thus, if

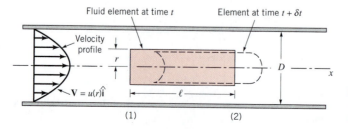

■ **FIGURE 8.4**
Motion of a cylindrical fluid element within a pipe.

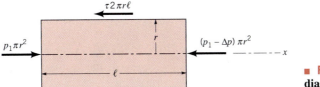

the pressure is $p = p_1$ at section (1), it is $p_2 = p_1 - \Delta p$ at section (2), where Δp is the pressure drop between sections (1) and (2). We anticipate the fact that the pressure decreases in the direction of flow so that $\Delta p > 0$. A shear stress, τ, acts on the surface of the cylinder of fluid. This viscous stress is a function of the radius of the cylinder, $\tau = \tau(r)$.

As was done in fluid statics analysis (Chapter 2), we isolate the cylinder of fluid as is shown in Fig. 8.5 and apply Newton's second law, $F_x = ma_x$. In this case even though the fluid is moving, it is not accelerating, so that $a_x = 0$. Thus, fully developed horizontal pipe flow is merely a balance between pressure and viscous forces. This can be written as

$$(p_1)\pi r^2 - (p_1 - \Delta p)\pi r^2 - (\tau)2\pi r\ell = 0$$

and simplified to give

$$\frac{\Delta p}{\ell} = \frac{2\tau}{r} \tag{8.3}$$

Since neither Δp nor ℓ is a function of the radial coordinate, r, it follows that $2\tau/r$ must also be independent of r. That is, $\tau = Cr$, where C is a constant. At $r = 0$ (the centerline of the pipe) there is no shear stress ($\tau = 0$). At $r = D/2$ (the pipe wall) the shear stress is a maximum, denoted τ_w, the *wall shear stress.* Hence, $C = 2\tau_w/D$ and the shear stress distribution throughout the pipe is a linear function of the radial coordinate

$$\tau = \frac{2\tau_w r}{D} \tag{8.4}$$

as is indicated in Fig. 8.6. As seen from Eqs. 8.3 and 8.4, the pressure drop and wall shear stress are related by

$$\Delta p = \frac{4\ell\tau_w}{D} \tag{8.5}$$

To carry the analysis further we must prescribe how the shear stress is related to the velocity. This is the critical step that separates the analysis of laminar from that of turbulent flow—from being able to solve for the laminar flow properties and not being able to solve for the turbulent flow properties without additional ad hoc assumptions. As discussed in Section 8.3, the shear stress dependence for turbulent flow is very complex. However, for laminar flow of a Newtonian fluid, the shear stress is simply proportional to the velocity

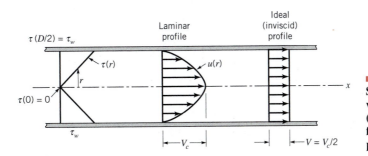

gradient, "$\tau = \mu \, du/dy$" (see Section 1.6). In the notation associated with our pipe flow, this becomes

$$\tau = -\mu \frac{du}{dr} \tag{8.6}$$

The negative sign is indicated to give $\tau > 0$ with $du/dr < 0$ (the velocity decreases from the pipe centerline to the pipe wall).

By combining Eq. 8.3 (Newton's second law of motion) and 8.6 (the definition of a Newtonian fluid) we obtain

$$\frac{du}{dr} = -\left(\frac{\Delta p}{2\mu\ell}\right) r$$

which can be integrated to give the velocity profile as follows:

$$\int du = -\frac{\Delta p}{2\mu\ell} \int r \, dr$$

or

$$u = -\left(\frac{\Delta p}{4\mu\ell}\right) r^2 + C_1$$

where C_1 is a constant. Because the fluid is viscous, it sticks to the pipe wall so that $u = 0$ at $r = D/2$. Hence, $C_1 = (\Delta p/16\mu\ell)D^2$ and the velocity profile can be written as

$$u(r) = \left(\frac{\Delta p D^2}{16\mu\ell}\right)\left[1 - \left(\frac{2r}{D}\right)^2\right] = V_c\left[1 - \left(\frac{2r}{D}\right)^2\right] \tag{8.7}$$

where $V_c = \Delta p D^2/(16\mu\ell)$ is the centerline velocity.

This velocity profile, plotted in Fig. 8.6, is parabolic in the radial coordinate, r, has a maximum velocity, V_c, at the pipe centerline, and a minimum velocity (zero) at the pipe wall. The volume flowrate through the pipe can be obtained by integrating the velocity profile across the pipe. Since the flow is axisymmetric about the centerline, the velocity is constant on small area elements consisting of rings of radius r and thickness dr which have area $dA = 2\pi r \, dr$ as shown by the figure in the margin. Thus,

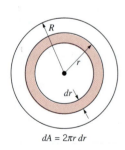

$dA = 2\pi r \, dr$

$$Q = \int u \, dA = \int_{r=0}^{r=R} u(r)2\pi r \, dr = 2\pi V_c \int_0^R \left[1 - \left(\frac{r}{R}\right)^2\right] r \, dr$$

or

$$Q = \frac{\pi R^2 V_c}{2}$$

By definition, the average velocity is the flowrate divided by the cross-sectional area, $V = Q/A = Q/\pi R^2$, so that for this flow

$$V = \frac{\pi R^2 V_c}{2\pi R^2} = \frac{V_c}{2} = \frac{\Delta p D^2}{32\mu\ell} \tag{8.8}$$

and

$$\boxed{Q = \frac{\pi D^4 \, \Delta p}{128\mu\ell}} \tag{8.9}$$

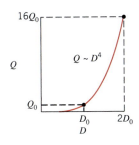

■ **FIGURE 8.7**   Free-body diagram of a fluid cylinder for flow in a nonhorizontal pipe.

This flow is termed *Hagen–Poiseuille flow.* Equation 8.9 is commonly referred to as Poiseuille's law. Recall that all of these results are restricted to laminar flow (those with Reynolds numbers less than approximately 2100) in a horizontal pipe. For a given pressure drop per unit length, the volume rate of flow is inversely proportional to the viscosity and proportional to the tube diameter to the fourth power. As indicated in the margin figure, a doubling of the tube diameter produces a sixteenfold increase in flow!

The adjustment necessary to account for nonhorizontal pipes, as shown in Fig. 8.7, can be easily included by replacing the pressure drop, Δp, by the combined effect of pressure and gravity, $\Delta p - \gamma\ell\sin\theta$, where θ is the angle between the pipe and the horizontal. This can be shown from a force balance in the x direction (along the pipe axis) on the cylinder of fluid. The method is exactly analogous to that used to obtain the Bernoulli equation (Eq. 3.6) when the streamline is not horizontal. Thus, all of the results for the horizontal pipe are valid provided the pressure gradient is adjusted for the elevation term. That is, Δp is replaced by $\Delta p - \gamma\ell\sin\theta$ so that

$$V = \frac{(\Delta p - \gamma\ell\sin\theta)D^2}{32\mu\ell} \tag{8.10}$$

and

$$Q = \frac{\pi(\Delta p - \gamma\ell\sin\theta)D^4}{128\mu\ell} \tag{8.11}$$

It is seen that the driving force for pipe flow can be either a pressure drop in the flow direction, Δp, or the component of weight in the flow direction, $-\gamma\ell\sin\theta$.

EXAMPLE 8.2   Laminar Pipe Flow

GIVEN   An oil with a viscosity of $\mu = 0.40$ N·s/m^2 and density $\rho = 900$ kg/m^3 flows in a pipe of diameter $D = 0.020$ m.

FIND   **(a)**   What pressure drop, $p_1 - p_2$, is needed to produce a flowrate of $Q = 2.0 \times 10^{-5}$ m^3/s if the pipe is horizontal with $x_1 = 0$ and $x_2 = 10$ m?

(b)   How steep a hill, θ, must the pipe be on if the oil is to flow through the pipe at the same rate as in part **(a)**, but with $p_1 = p_2$?

(c)   For the conditions of part **(b)**, if $p_1 = 200$ kPa, what is the pressure at section $x_3 = 5$ m, where x is measured along the pipe?

SOLUTION

(a)   If the Reynolds number is less than 2100 the flow is laminar and the equations derived in this section are valid. Since

the average velocity is $V = Q/A = 2.0 \times 10^{-5}$ m^3/s/ $[\pi(0.020)^2$m$^2/4] = 0.0637$ m/s, the Reynolds number is

Re = $\rho VD/\mu$ = 2.87 < 2100. Hence, the flow is laminar and from Eq. 8.9 with $\ell = x_2 - x_1 = 10$ m, the pressure drop is

$$\Delta p = p_1 - p_2 = \frac{128\mu\ell Q}{\pi D^4}$$

$$= \frac{128(0.40 \text{ N·s/m}^2)(10.0 \text{ m})(2.0 \times 10^{-5} \text{ m}^3/\text{s})}{\pi(0.020 \text{ m})^4}$$

or

$$\Delta p = 20{,}400 \text{ N/m}^2 = 20.4 \text{ kPa} \qquad \text{(Ans)}$$

(b)   If the pipe is on a hill of angle θ such that $\Delta p = p_1 - p_2 = 0$, Eq. 8.11 gives

$$\sin \theta = -\frac{128\mu Q}{\pi \rho g D^4} \qquad (1)$$

or

$$\sin \theta = \frac{-128(0.40 \text{ N·s/m}^2)(2.0 \times 10^{-5} \text{ m}^3/\text{s})}{\pi(900 \text{ kg/m}^3)(9.81 \text{ m/s}^2)(0.020 \text{ m})^4}$$

Thus, $\theta = -13.34°$. (Ans)

COMMENT   This checks with the previous horizontal result as is seen from the fact that a change in elevation of $\Delta z = \ell \sin \theta = (10 \text{ m}) \sin(-13.34°) = -2.31$ m is equivalent to a pressure change of $\Delta p = \rho g \, \Delta z = (900 \text{ kg/m}^3)(9.81 \text{ m/s}^2)(2.31 \text{ m}) = 20{,}400 \text{ N/m}^2$, which is equivalent to that needed for the horizontal pipe. For the horizontal pipe it is the work done by the pressure forces that overcomes the viscous dissipation. For the zero pressure drop pipe on the hill, it is the change in potential energy of the fluid "falling" down the hill that is converted to the energy lost by viscous dissipation. Note that if it is desired to increase the flowrate to $Q = 1.0 \times 10^{-4} \text{ m}^3/\text{s}$ with $p_1 = p_2$, the value of θ given by Eq. 1 is $\sin \theta = -1.15$. Since the sine of an angle cannot be greater than 1, this flow would not be possible. The weight of the fluid would not be large enough to offset the viscous force generated for the flowrate desired. A larger diameter pipe would be needed.

(c)   With $p_1 = p_2$ the length of the pipe, ℓ, does not appear in the flowrate equation (Eq. 1). This is a statement of the fact that for such cases the pressure is constant all along the pipe (provided the pipe lies on a hill of constant slope). This can be seen by substituting the values of Q and θ from case **(b)** into Eq. 8.11 and noting that $\Delta p = 0$ for any ℓ. For example, $\Delta p = p_1 - p_3 = 0$ if $\ell = x_3 - x_1 = 5$ m. Thus, $p_1 = p_2 = p_3$ so that

$$p_3 = 200 \text{ kPa} \qquad \text{(Ans)}$$

COMMENT   Note that if the fluid were gasoline ($\mu = 3.1 \times 10^{-4} \text{ N·s/m}^2$ and $\rho = 680 \text{ kg/m}^3$), the Reynolds number would be Re = 2790, the flow would probably not be laminar, and use of Eqs. 8.9 and 8.11 would give incorrect results. Also note from Eq. 1 that the kinematic viscosity, $v = \mu/\rho$, is the important viscous parameter. This is a statement of the fact that with constant pressure along the pipe, it is the ratio of the viscous force ($\sim\mu$) to the weight force ($\sim\gamma = \rho g$) that determines the value of θ.

8.2.2   From the Navier–Stokes Equations

In the previous section we obtained results for fully developed laminar pipe flow by applying Newton's second law and the assumption of a Newtonian fluid to a specific portion of the fluid—a cylinder of fluid centered on the axis of a long round pipe. When this governing law and assumptions are applied to a general fluid flow (not restricted to pipe flow), the result is the Navier–Stokes equations as discussed in Chapter 6. In Section 6.9.3 these equations were solved for the specific geometry of fully developed laminar flow in a round pipe. The results are the same as those given in Eq. 8.7.

8.3   Fully Developed Turbulent Flow

In the previous section various properties of fully developed laminar pipe flow were discussed. Since turbulent pipe flow is actually more likely to occur than laminar flow in practical situations, it is necessary to obtain similar information for turbulent pipe flow. However, turbulent flow is a very complex process. Numerous persons have devoted considerable effort in attempting to understand the variety of baffling aspects of turbulence. Although a considerable amount of knowledge about the topic has been developed, the field of turbulent flow still remains the least understood area of fluid mechanics.

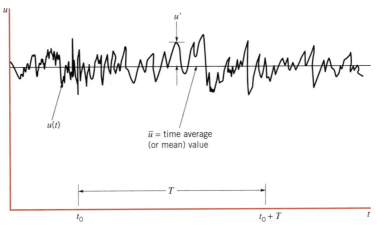

■ **FIGURE 8.8** The time averaged, $\bar{u}$, and fluctuating, u', description of a parameter for turbulent flow.

8.3.1 Transition from Laminar to Turbulent Flow

Flows are classified as laminar or turbulent. For any flow geometry, there is one (or more) dimensionless parameter such that with this parameter value below a particular value the flow is laminar, whereas with the parameter value larger than a certain value it is turbulent. For pipe flow this parameter is the Reynolds number. The value of the Reynolds number must be less than approximately 2100 for laminar flow and greater than approximately 4000 for turbulent flow.

A typical trace of the axial component of velocity, $u = u(t)$, measured at a given location in turbulent pipe flow is shown in Fig. 8.8. Its irregular, random nature is the distinguishing feature of turbulent flows. The character of many of the important properties of the flow (pressure drop, heat transfer, etc.) depends strongly on the existence and nature of the turbulent fluctuations or randomness indicated.

For example, mixing processes and heat and mass transfer processes are considerably enhanced in turbulent flow compared to laminar flow. This is due to the macroscopic scale of the randomness in turbulent flow. We are all familiar with the "rolling," vigorous eddy-type motion of the water in a pan being heated on the stove (even if it is not heated to boiling). Such finite-sized random mixing is very effective in transporting energy and mass throughout the flow field, thereby increasing the various rate processes involved. Laminar flow, however, can be thought of as very small but finite-sized fluid particles flowing smoothly in layers, one over another. The only randomness and mixing take place on the molecular scale and result in relatively small heat, mass, and momentum transfer rates.

V8.2 Turbulence
in a bowl

8.3.2 Turbulent Shear Stress

The fundamental difference between laminar and turbulent flow lies in the chaotic, random behavior of the various fluid parameters. As indicated in Fig. 8.8, such flows can be described in terms of their mean values (denoted with an overbar) on which are superimposed the fluctuations (denoted with a prime). Thus, if $u = u(x, y, z, t)$ is the x component of instantaneous velocity, then its time mean (or *time average*) value, $\bar{u}$, is

$$\bar{u} = \frac{1}{T} \int_{t_0}^{t_0 + T} u(x, y, z, t) \, dt \qquad (8.12)$$

where the time interval, T, is considerably longer than the period of the longest fluctuations, but considerably shorter than any unsteadiness of the average velocity. This is illustrated in Fig. 8.8.

It is tempting to extend the concept of viscous shear stress for laminar flow ($\tau = \mu \, du/dy$) to that of turbulent flow by replacing u, the instantaneous velocity, by $\bar{u}$, the time-average velocity. However, numerous experimental and theoretical studies have shown that such an approach leads to completely incorrect results.

That is, the shear stress in turbulent flow is not merely proportional to the gradient of the time-average velocity: $\tau \neq \mu \, d\bar{u}/dy$. It also contains a contribution due to the random fluctuations of the components of velocity. One could express the shear stress for turbulent flow in terms of a new parameter called the *eddy viscosity*, η, where

$$\tau = \eta \frac{d\bar{u}}{dy} \qquad (8.13)$$

Although the concept of an eddy viscosity is intriguing, in practice it is not an easy parameter to use. Unlike the absolute viscosity, μ, which is a known value for a given fluid, the eddy viscosity is a function of both the fluid and the flow conditions. That is, the eddy viscosity of water cannot be looked up in handbooks—its value changes from one turbulent flow condition to another and from one point in a turbulent flow to another.

Several semiempirical theories have been proposed (Ref. 1) to determine approximate values of η. For example, the turbulent process could be viewed as the random transport of bundles of fluid particles over a certain distance, ℓ_m, the *mixing length*, from a region of one velocity to another region of a different velocity. By the use of some ad hoc assumptions and physical reasoning, the eddy viscosity is then given by

$$\eta = \rho \ell_m^2 \left| \frac{d\bar{u}}{dy} \right|$$

Thus, the turbulent shear stress is

$$\tau_{\text{turb}} = \rho \ell_m^2 \left(\frac{d\bar{u}}{dy} \right)^2 \qquad (8.14)$$

The problem is thus shifted to that of determining the mixing length, ℓ_m. Further considerations indicate that ℓ_m is not a constant throughout the flow field. Near a solid surface the turbulence is dependent on the distance from the surface. Thus, additional assumptions are made regarding how the mixing length varies throughout the flow.

The net result is that as yet there is no general, all-encompassing, useful model that can accurately predict the shear stress throughout a general incompressible, viscous turbulent flow. Without such information it is impossible to integrate the force balance equation to obtain the turbulent velocity profile and other useful information, as was done for laminar flow.

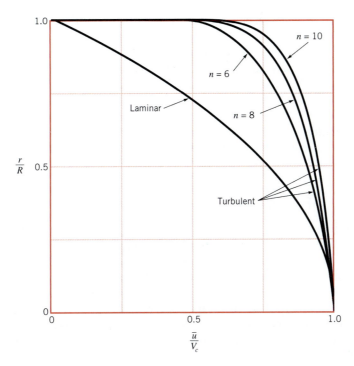

■ **FIGURE 8.9**
Typical laminar flow and turbulent flow velocity profiles.

8.3.3 Turbulent Velocity Profile

Although considerable information concerning turbulent velocity profiles has been obtained through the use of dimensional analysis, experimentation, and semiempirical theoretical efforts, there is still no general accurate expression for turbulent velocity profiles.

An often-used (and relatively easy to use) correlation is the empirical *power-law velocity profile*

$$\frac{\bar{u}}{V_c} = \left(1 - \frac{r}{R}\right)^{1/n} \tag{8.15}$$

In this representation, the value of n is a function of the Reynolds number, with typical values between $n = 6$ and $n = 10$. Typical turbulent velocity profiles based on this power-law representation are shown in Fig. 8.9. Note that the turbulent profiles are much "flatter" than the laminar profile.

A closer examination of Eq. 8.15 shows that the power-law profile cannot be valid near the wall, since according to this equation the velocity gradient is infinite there. In addition, Eq. 8.15 cannot be precisely valid near the centerline because it does not give $d\bar{u}/dr = 0$ at $r = 0$. However, it does provide a reasonable approximation to the measured velocity profiles across most of the pipe.

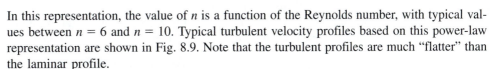

V8.3 Laminar/ turbulent velocity profiles

8.4 Dimensional Analysis of Pipe Flow

As noted previously, turbulent flow can be a very complex, difficult topic—one that as yet has defied a rigorous theoretical treatment. Thus, most turbulent pipe flow analyses are based on experimental data and semiempirical formulas. These data are expressed conveniently in dimensionless form.

It is often necessary to determine the head loss, h_L, that occurs in a pipe flow so that the energy equation, Eq. 5.57, can be used in the analysis of pipe flow problems. As shown in Fig. 8.1, a typical pipe system usually consists of various lengths of straight pipe interspersed with various types of components (valves, elbows, etc.) The overall head loss for the pipe system consists of the head loss due to viscous effects in the straight pipes, termed the *major loss* and denoted $h_{L\,major}$, and the head loss in the various pipe components, termed the *minor loss* and denoted $h_{L\,minor}$. That is,

$$h_L = h_{L\,major} + h_{L\,minor}$$

The head loss designations of "major" and "minor" do not necessarily reflect the relative importance of each type of loss. For a pipe system that contains many components and a relatively short length of pipe, the minor loss may actually be larger than the major loss.

8.4.1 Major Losses

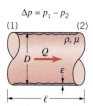

$\Delta p = p_1 - p_2$

A dimensional analysis treatment of pipe flow provides the most convenient base from which to consider turbulent, fully developed pipe flow. The pressure drop and head loss in a pipe are dependent on the wall shear stress, τ_w, between the fluid and the pipe surface. A fundamental difference between laminar and turbulent flow is that the shear stress for turbulent flow is a function of the density of the fluid, ρ. For laminar flow, the shear stress is independent of the density, leaving the viscosity, μ, as the only important fluid property.

Thus, as shown by the figure in the margin, the pressure drop, Δp, for steady, incompressible turbulent flow in a horizontal round pipe of diameter D can be written in functional form as

$$\Delta p = F(V, D, \ell, \varepsilon, \mu, \rho) \tag{8.16}$$

where V is the average velocity, ℓ is the pipe length, and ε is a measure of the roughness of the pipe wall. It is clear that Δp should be a function of V, D, and ℓ. The dependence of Δp on the fluid properties μ and ρ is expected because of the dependence of τ on these parameters.

Although the pressure drop for laminar pipe flow is found to be independent of the roughness of the pipe, it is necessary to include this parameter when considering turbulent flow. This can be shown to be due to the random velocity components that account for a momentum transfer (which is a function of the density) and, hence, a shear force.

Since there are seven variables ($k = 7$) that can be written in terms of the three reference dimensions MLT ($r = 3$), Eq. 8.16 can be written in dimensionless form in terms of $k - r = 4$ dimensionless groups. One such representation is

$$\frac{\Delta p}{\frac{1}{2}\rho V^2} = \tilde{\phi}\left(\frac{\rho VD}{\mu}, \frac{\ell}{D}, \frac{\varepsilon}{D}\right)$$

This result differs from that used for laminar flow in two ways. First, we have chosen to make the pressure dimensionless by dividing by the dynamic pressure, $\rho V^2/2$, rather than a characteristic viscous shear stress, $\mu V/D$. Second, we have introduced two additional dimensionless parameters, the Reynolds number, $Re = \rho VD/\mu$, and the *relative roughness, ε/D*, which are not present in the laminar formulation because the two parameters ρ and ε are not important in fully developed laminar pipe flow.

As was done for laminar flow, the functional representation can be simplified by imposing the reasonable assumption that the pressure drop should be proportional to the pipe length. (Such a step is not within the realm of dimensional analysis. It is merely a logical assumption supported by experiments.) The only way that this can be true is if the ℓ/D dependence is factored out as

$$\frac{\Delta p}{\frac{1}{2}\rho V^2} = \frac{\ell}{D}\phi\left(Re, \frac{\varepsilon}{D}\right)$$

The quantity $\Delta p D/(\ell \rho V^2/2)$ is termed the ***friction factor, f.*** Thus, for a horizontal pipe

$$\Delta p = f \frac{\ell}{D} \frac{\rho V^2}{2}$$ (8.17)

where

$$f = \phi\left(\text{Re}, \frac{\varepsilon}{D}\right)$$

The friction factor, f, defined by Eq. 8.17 and termed the Darcy friction factor, is widely used in engineering calculations. Infrequently, the Fanning friction factor, f_F, defined by $f_F = f/4$ is used instead. Only f is used in this book.

From Eq. 5.57 the energy equation for steady incompressible flow is

$$\frac{p_1}{\gamma} + \frac{V_1^2}{2g} + z_1 = \frac{p_2}{\gamma} + \frac{V_2^2}{2g} + z_2 + h_L$$

where h_L is the head loss between sections (1) and (2). With the assumption of a constant diameter ($D_1 = D_2$ so that $V_1 = V_2$), horizontal ($z_1 = z_2$) pipe, this becomes $\Delta p = p_1 - p_2 = \gamma h_L$, which can be combined with Eq. 8.17 to give

$$h_{L\,\text{major}} = f \frac{\ell}{D} \frac{V^2}{2g}$$ (8.18)

Equation 8.18, called the *Darcy–Weisbach equation,* is valid for any fully developed, steady, incompressible pipe flow—whether the pipe is horizontal or on a hill. However, Eq. 8.17 is valid only for horizontal pipes. In general, with $V_1 = V_2$ the energy equation gives

$$p_1 - p_2 = \gamma(z_2 - z_1) + \gamma h_L = \gamma(z_2 - z_1) + f \frac{\ell}{D} \frac{\rho V^2}{2}$$

Part of the pressure change is due to the elevation change and part is due to the head loss associated with frictional effects, which are given in terms of the friction factor, f.

It is not easy to determine the functional dependence of the friction factor on the Reynolds number and relative roughness. Much of this information is a result of experiments. Figure 8.10 shows the functional dependence of f on Re and ε/D and is called the ***Moody chart.*** Typical roughness values for various new, clean pipe surfaces are given in Table 8.1.

The following characteristics are observed from the data of Fig. 8.10. For laminar flow, $f = 64/\text{Re}$, which is independent of relative roughness. For very large Reynolds numbers, $f = \phi(\varepsilon/D)$, which is independent of the Reynolds number. Such flows are commonly termed *completely turbulent flow* (or *wholly turbulent flow*). For flows with moderate values of Re, the friction factor is indeed dependent on both the Reynolds number and relative roughness—$f = \phi(\text{Re}, \varepsilon/D)$. Note that even for *hydraulically smooth* pipes ($\varepsilon = 0$) the friction factor is not zero. That is, there is a head loss in any pipe, no matter how smooth the surface is made.

The following equation is valid for the entire nonlaminar range of the Moody chart

$$\frac{1}{\sqrt{f}} = -2.0 \log\left(\frac{\varepsilon/D}{3.7} + \frac{2.51}{\text{Re}\sqrt{f}}\right)$$ (8.19)

In fact, the Moody chart is a graphical representation of this equation, which is an empirical fit of the pipe flow pressure drop data. Equation 8.19 is called the ***Colebrook formula.*** A difficulty with its use is that it is implicit in the dependence of f. That is, for given conditions (Re and ε/D), it is not possible to solve for f without some sort of iterative scheme. With the use of modern computers and calculators, such calculations are not difficult.

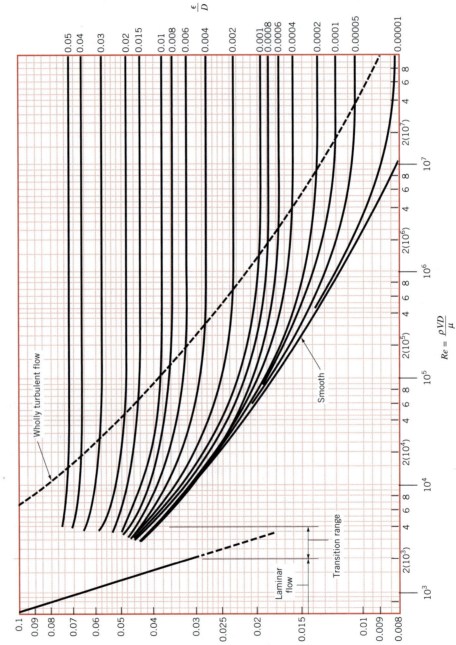

■ **FIGURE 8.10** Friction factor as a function of Reynolds number and relative roughness for round pipes—the Moody chart (data from Ref. 2 with permission).

■ **TABLE 8.1**
Equivalent Roughness for New Pipes [from Moody (Ref. 2) and Colebrook (Ref. 3)]

Pipe	Equivalent Roughness, ε	
	Feet	Millimeters
Riveted steel	0.003–0.03	0.9–9.0
Concrete	0.001–0.01	0.3–3.0
Wood stave	0.0006–0.003	0.18–0.9
Cast iron	0.00085	0.26
Galvanized iron	0.0005	0.15
Commercial steel or wrought iron	0.00015	0.045
Drawn tubing	0.000005	0.0015
Plastic, glass	0.0 (smooth)	0.0 (smooth)

EXAMPLE 8.3   Comparison of Laminar or Turbulent Pressure Drop

GIVEN Air under standard conditions flows through a 4.0-mm-diameter drawn tubing with an average velocity of $V = 50$ m/s. For such conditions the flow would normally be turbulent. However, if precautions are taken to eliminate disturbances to the flow (the entrance to the tube is very smooth, the air is dust-free, the tube does not vibrate, etc.), it may be possible to maintain laminar flow.

FIND (a) Determine the pressure drop in a 0.1-m section of the tube if the flow is laminar.

(b) Repeat the calculations if the flow is turbulent.

SOLUTION

Under standard temperature and pressure conditions the density and viscosity are $\rho = 1.23$ kg/m^3 and $\mu = 1.79 \times 10^{-5}$ N·s/m^2. Thus, the Reynolds number is

$$\mathrm{Re} = \frac{\rho V D}{\mu} = \frac{(1.23 \text{ kg/m}^3)(50 \text{ m/s})(0.004 \text{ m})}{1.79 \times 10^{-5} \text{ N·s/m}^2}$$

$$= 13{,}700$$

which would normally indicate turbulent flow.

(a) If the flow were laminar, then $f = 64/\mathrm{Re} = 64/13{,}700 = 0.00467$ and the pressure drop in a 0.1-m-long horizontal section of the pipe would be

$$\Delta p = f \frac{\ell}{D} \frac{1}{2} \rho V^2$$

$$= (0.00467) \frac{(0.1 \text{ m})}{(0.004 \text{ m})} \frac{1}{2} (1.23 \text{ kg/m}^3)(50 \text{ m/s})^2$$

or

$$\Delta p = 0.179 \text{ kPa} \qquad \textbf{(Ans)}$$

COMMENT Note that the same result is obtained from Eq. 8.8.

$$\Delta p = \frac{32 \mu \ell}{D^2} V = \frac{32(1.79 \times 10^{-5} \text{ N·s/m}^2)(0.1 \text{ m})(50 \text{ m/s})}{(0.004 \text{ m})^2}$$

$$= 179 \text{ N/m}^2$$

(b) If the flow were turbulent, then $f = \phi(\mathrm{Re}, \varepsilon/D)$, where from Table 8.1, $\varepsilon = 0.0015$ mm so that $\varepsilon/D = 0.0015$ mm/4.0 mm $= 0.000375$. From the Moody chart with $\mathrm{Re} = 1.37 \times 10^4$ and $\varepsilon/D = 0.000375$ we obtain $f = 0.028$. Thus, the pressure drop in this case would be approximately

$$\Delta p = f \frac{\ell}{D} \frac{1}{2} \rho V^2$$

$$= (0.028) \frac{(0.1 \text{ m})}{(0.004 \text{ m})} \frac{1}{2} (1.23 \text{ kg/m}^3)(50 \text{ m/s})^2$$

or

$$\Delta p = 1.076 \text{ kPa} \qquad \textbf{(Ans)}$$

COMMENTS A considerable savings in effort to force the fluid through the pipe could be realized (0.179 kPa rather than 1.076 kPa) if the flow could be maintained as laminar flow at this Reynolds number. In general this is very difficult to do, although laminar flow in pipes has been maintained up to Re ≈ 100,000 in rare instances.

An alternate method to determine the friction factor for the turbulent flow would be to use the Colebrook formula, Eq. 8.19. Thus,

$$\frac{1}{\sqrt{f}} = -2.0 \log\left(\frac{\varepsilon/D}{3.7} + \frac{2.51}{\mathrm{Re}\sqrt{f}}\right)$$

$$= -2.0 \log\left(\frac{0.000375}{3.7} + \frac{2.51}{1.37 \times 10^4 \sqrt{f}}\right)$$

or

$$\frac{1}{\sqrt{f}} = -2.0 \log\left(1.01 \times 10^{-4} + \frac{1.83 \times 10^{-4}}{\sqrt{f}}\right) \quad (1)$$

By using a root-finding technique on a computer or calculator, the solution to Eq. 1 is determined to be $f = 0.0291$, in agreement (within the accuracy of reading the graph) with the Moody chart method of $f = 0.028$.

Numerous other empirical formulas can be found in the literature (Ref. 4) for portions of the Moody chart. For example, an often-used equation, commonly referred to as the Blasius formula, for turbulent flow in smooth pipes ($\varepsilon/D = 0$) with Re $< 10^5$ is

$$f = \frac{0.316}{\mathrm{Re}^{1/4}}$$

For our case this gives

$$f = 0.316(13,700)^{-0.25} = 0.0292$$

which is in agreement with the previous results. Note that the value of f is relatively insensitive to ε/D for this particular situation. Whether the tube was smooth glass ($\varepsilon/D = 0$) or the drawn tubing ($\varepsilon/D = 0.000375$) would not make much difference in the pressure drop. For this flow, an increase in relative roughness by a factor of 30 to $\varepsilon/D = 0.0113$ (equivalent to a commercial steel surface; see Table 8.1) would give $f = 0.043$. This would represent an increase in pressure drop and head loss by a factor of $0.043/0.0291 = 1.48$ compared with that for the original drawn tubing.

The pressure drop of 1.076 kPa in a length of 0.1 m of pipe corresponds to a change in absolute pressure [assuming $p = 101$ kPa (abs) at $x = 0$] of approximately $1.076/101 = 0.0107$, or about 1%. Thus, the incompressible flow assumption on which the aforementioned calculations (and all of the formulas in this chapter) are based is reasonable. However, if the pipe were 2 m long the pressure drop would be 21.5 kPa, approximately 20% of the original pressure. In this case the density would not be approximately constant along the pipe, and a compressible flow analysis would be needed.

8.4.2 Minor Losses

Most pipe systems consist of more than straight pipes. These additional components (valves, bends, tees, and the like) add to the overall head loss of the system. Such losses are generally termed *minor losses,* with the corresponding head loss denoted $h_{L\,\text{minor}}$.

Head loss information for essentially all components is given in dimensionless form and is based on experimental data. The most common method used to determine these head losses or pressure drops is to specify the ***loss coefficient,*** K_L, which is defined as

$$K_L = \frac{h_{L\,\text{minor}}}{(V^2/2g)} = \frac{\Delta p}{\tfrac{1}{2}\rho V^2}$$

so that

$$\Delta p = K_L \tfrac{1}{2}\rho V^2$$

or

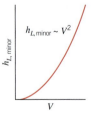

$$\boxed{h_{L\,\text{minor}} = K_L \frac{V^2}{2g}} \quad (8.20)$$

The pressure drop across a component that has a loss coefficient of $K_L = 1$ is equal to the dynamic pressure, $\rho V^2/2$. As shown by Eq. 8.20 and the figure in the margin, for a given value of K_L the head loss is proportional to the square of the velocity.

Many pipe systems contain various transition sections in which the pipe diameter changes from one size to another. Any change in flow area contributes losses that are not

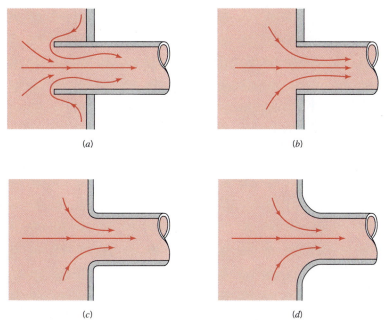

(a)   *(b)*

(c)   *(d)*

■ **FIGURE 8.11**   **Entrance flow conditions and loss coefficient
(Refs. 12, 13).** *(a)* **Reentrant,** $K_L = 0.8$, *(b)* **sharp edged,** $K_L = 0.5$, *(c)*
slightly rounded, $K_L = 0.2$ **(see Fig. 8.12), and** *(d)* **well rounded,** $K_L = 0.04$
(see Fig. 8.12).

V8.4 Entrance/
exit flows

accounted for in the fully developed head loss calculation (the friction factor). The extreme
cases involve flow into a pipe from a reservoir (an entrance) or out of a pipe into a reservoir (an exit).

A fluid may flow from a reservoir into a pipe through any number of different shaped
entrance regions as are sketched in Fig. 8.11. Each geometry has an associated loss coefficient. An obvious way to reduce the entrance loss is to round the entrance region as is shown
in Fig. 8.11c. Typical values for the loss coefficient for entrances with various amounts of
rounding of the lip are shown in Fig. 8.12. A significant reduction in K_L can be obtained with
only slight rounding.

A head loss (the exit loss) is also produced when a fluid flows from a pipe into a
tank as is shown in Fig. 8.13. In these cases the entire kinetic energy of the exiting fluid
(velocity V_1) is dissipated through viscous effects as the stream of fluid mixes with the fluid

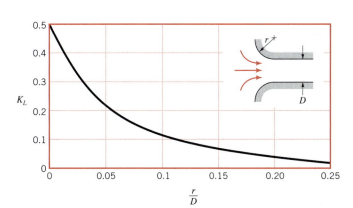

■ **FIGURE 8.12**
**Entrance loss coefficient as
a function of rounding of
the inlet edge (Ref. 5).**

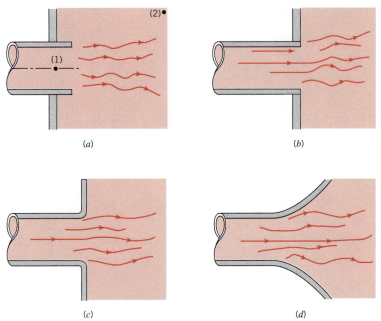

■ **FIGURE 8.13** Exit flow conditions and loss coefficient. (a) Reentrant, $K_L = 1.0$, (b) sharp edged, $K_L = 1.0$, (c) slightly rounded, $K_L = 1.0$, and (d) well rounded, $K_L = 1.0$.

in the tank and eventually comes to rest ($V_2 = 0$). The exit loss from points (1) and (2) is therefore equivalent to one velocity head, or $K_L = 1$.

Losses also occur because of a change in pipe diameter. The loss coefficient for a sudden contraction, $K_L = h_L/(V_2^2/2g)$, is a function of the area ratio, A_2/A_1, as is shown in Fig. 8.14. The value of K_L changes gradually from one extreme of a sharp-edged entrance ($A_2/A_1 = 0$ with $K_L = 0.50$) to the other extreme of no area change ($A_2/A_1 = 1$ with $K_L = 0$). The loss coefficient for a sudden expansion is shown in Fig. 8.15.

Bends in pipes produce a greater head loss than if the pipe were straight. The losses are due to the separated region of flow near the inside of the bend (especially if the bend is sharp) and the swirling secondary flow that occurs because of the imbalance of centripetal forces as a result of the curvature of the pipe centerline (see Fig. 8.16). These effects and

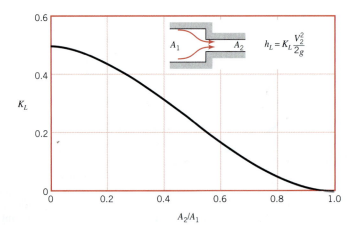

■ **FIGURE 8.14**
Loss coefficient for a sudden contraction (Ref. 6).

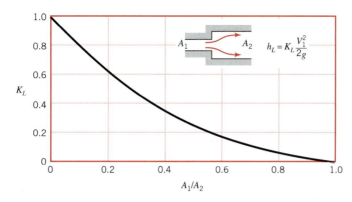

■ **FIGURE 8.15**
Loss coefficient for a sudden expansion (Ref. 6).

the associated values of K_L for large Reynolds number flows through a 90° bend are shown in Fig. 8.16. The friction loss due to the axial length of the pipe bend must be calculated and added to that given by the loss coefficient of Fig. 8.16.

For situations in which space is limited, a flow direction change is often accomplished by the use of miter bends, as is shown in Fig. 8.17, rather than smooth bends. The considerable losses in such bends can be reduced by the use of carefully designed guide vanes that help direct the flow with less unwanted swirl and disturbances.

Another important category of pipe system components is that of commercially available pipe fittings such as elbows, tees, reducers, valves, and filters. The values of K_L for such components depend strongly on the shape of the component and only very weakly on the Reynolds number for typical large Re flows. Thus, the loss coefficient for a 90° elbow depends on whether the pipe joints are threaded or flanged, but is, within the accuracy of the data, fairly independent of the pipe diameter, flow rate, or fluid properties (the Reynolds number effect). Typical values of K_L for such components are given in Table 8.2.

V8.5 Car exhaust system

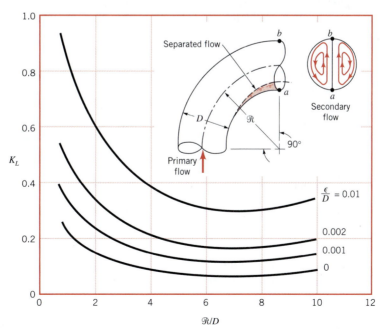

■ **FIGURE 8.16** **Character of the flow in a 90° bend and the associated loss coefficient (Ref. 4).**

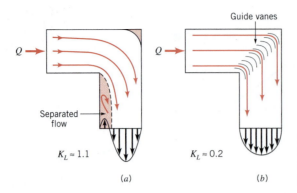

$Q \rightarrow$

Separated flow

$K_L \approx 1.1$

(a)

Guide vanes

$Q \rightarrow$

$K_L \approx 0.2$

(b)

■ **FIGURE 8.17** **Character of the flow in a 90° miter bend and the associated loss coefficient: (a) without guide vanes and (b) with guide vanes.**

■ **TABLE 8.2**

Loss Coefficients for Pipe Components $\left(h_L = K_L \dfrac{V^2}{2g} \right)$ (Data from Refs. 4, 6, 11)

Component	K_L
a. Elbows	
Regular 90°, flanged	0.3
Regular 90°, threaded	1.5
Long radius 90°, flanged	0.2
Long radius 90°, threaded	0.7
Long radius 45°, flanged	0.2
Regular 45°, threaded	0.4
b. 180° return bends	
180° return bend, flanged	0.2
180° return bend, threaded	1.5
c. Tees	
Line flow, flanged	0.2
Line flow, threaded	0.9
Branch flow, flanged	1.0
Branch flow, threaded	2.0
d. Union, threaded	0.08
***e. Valves**	
Globe, fully open	10
Angle, fully open	2
Gate, fully open	0.15
Gate, $\frac{1}{4}$ closed	0.26
Gate, $\frac{1}{2}$ closed	2.1
Gate, $\frac{3}{4}$ closed	17
Swing check, forward flow	2
Swing check, backward flow	∞
Ball valve, fully open	0.05
Ball valve, $\frac{1}{3}$ closed	5.5
Ball valve, $\frac{2}{3}$ closed	210

*See Fig. 8.18 for typical valve geometry.

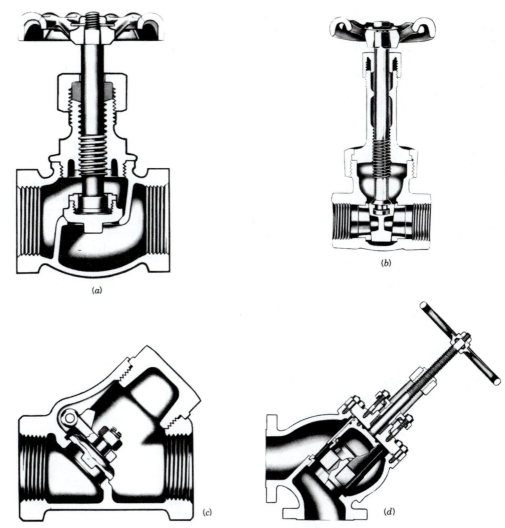

■ **FIGURE 8.18** **Internal structure of various valves: (a) globe valve, (b) gate valve, (c) swing check valve, and (d) stop check valve (courtesy of Crane Co., Valve Division).**

Valves control the flowrate by providing a means to adjust the overall system loss coefficient to the desired value. When the valve is closed, the value of K_L is infinite and no fluid flows. Opening of the valve reduces K_L, producing the desired flowrate. Typical cross sections of various types of valves are shown in Fig. 8.18. Loss coefficients for typical valves are given in Table 8.2.

EXAMPLE 8.4 | Minor Losses

GIVEN Air at standard conditions is to flow through the test section [between sections (5) and (6)] of the closed-circuit wind tunnel shown in Fig. E8.4a with a velocity of 200 ft/s. The flow is driven by a fan that essentially increases the static pressure by the amount $p_1 - p_9$ that is needed to over-come the head losses experienced by the fluid as it flows around the circuit.

FIND Estimate the value of $p_1 - p_9$ and the horsepower supplied to the fluid by the fan.

SOLUTION

The maximum velocity within the wind tunnel occurs in the test section (smallest area). Thus, the maximum Mach number of the flow is $Ma_5 = V_5/c_5$, where $V_5 = 200$ ft/s and from Eq. 1.15 the speed of sound is $c_5 = (kRT)_5^{1/2} = \{1.4(1716$ ft·lb/slug·° R) $[(460 + 59)°$ R]$\}^{1/2} = 1117$ ft/s. Thus, $Ma_5 = 200/1117 = 0.179$. As indicated in Chapter 3, most flows can be considered as incompressible if the Mach number is less than about 0.3. Hence, we can use the incompressible formulas for this problem.

The purpose of the fan in the wind tunnel is to provide the necessary energy to overcome the net head loss experienced by the air as it flows around the circuit. This can be found from the energy equation between points (1) and (9) as

$$\frac{p_1}{\gamma} + \frac{V_1^2}{2g} + z_1 = \frac{p_9}{\gamma} + \frac{V_9^2}{2g} + z_9 + h_{L_{1-9}}$$

where $h_{L_{1-9}}$ is the total head loss from (1) to (9). With $z_1 = z_9$ and $V_1 = V_9$ this gives

$$\frac{p_1}{\gamma} - \frac{p_9}{\gamma} = h_{L_{1-9}} \qquad (1)$$

Similarly, by writing the energy equation (Eq. 5.57) across the fan, from (9) to (1), we obtain

$$\frac{p_9}{\gamma} + \frac{V_9^2}{2g} + z_9 + h_p = \frac{p_1}{\gamma} + \frac{V_1^2}{2g} + z_9$$

where h_p is the actual head rise supplied by the pump (fan) to the air. Again since $z_9 = z_1$ and $V_9 = V_1$ this, when combined with Eq. 1, becomes

$$h_p = \frac{(p_1 - p_9)}{\gamma} = h_{L_{1-9}}$$

The actual power supplied to the air (horsepower, $\mathcal{P}_a$) is obtained from the fan head by

$$\mathcal{P}_a = \gamma Q h_p = \gamma A_5 V_5 h_p = \gamma A_5 V_5 h_{L_{1-9}} \qquad (2)$$

Thus, the power that the fan must supply to the air depends on the head loss associated with the flow through the wind tunnel. To obtain a reasonable, approximate answer we make the following assumptions. We treat each of the four turning corners as a mitered bend with guide vanes so that from Fig. 8.17 $K_{L_{corner}} = 0.2$. Thus, for each corner

$$h_{L_{corner}} = K_L \frac{V^2}{2g} = 0.2 \frac{V^2}{2g}$$

where, because the flow is assumed incompressible, $V = V_5 A_5/A$. The values of A and the corresponding velocities throughout the tunnel are given in Table E8.4.

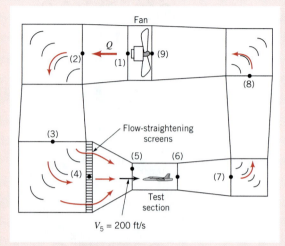

■ **FIGURE E8.4a**

■ **TABLE E8.4**

Location	Area (ft²)	Velocity (ft/s)
1	22.0	36.4
2	28.0	28.6
3	35.0	22.9
4	35.0	22.9
5	4.0	200.0
6	4.0	200.0
7	10.0	80.0
8	18.0	44.4
9	22.0	36.4

We also treat the enlarging sections from the end of the test section (6) to the beginning of the nozzle (4) as a conical diffuser with a loss coefficient of $K_{L_{dif}} = 0.6$. This value is larger than that of a well-designed diffuser (see Ref. 4, for example). Since the wind tunnel diffuser is interrupted by the four turning corners and the fan, it may not be possible to obtain a smaller value of $K_{L_{dif}}$ for this situation. Thus,

$$h_{L_{dif}} = K_{L_{dif}} \frac{V_6^2}{2g} = 0.6 \frac{V_6^2}{2g}$$

The loss coefficients for the conical nozzle between section (4) and (5) and the flow-straightening screens are assumed to be $K_{L_{noz}} = 0.2$ and $K_{L_{scr}} = 4.0$ (Ref. 14), respectively. We neglect the head loss in the relatively short test section.

Thus, the total head loss is

$$h_{L_{1-9}} = h_{L_{corner7}} + h_{L_{corner8}} + h_{L_{corner2}} + h_{L_{corner3}}$$
$$+ h_{L_{dif}} + h_{L_{noz}} + h_{L_{scr}}$$

or

$$h_{L_{1-9}} = [0.2(V_7^2 + V_8^2 + V_2^2 + V_3^2)$$
$$+ 0.6V_6^2 + 0.2V_5^2 + 4.0V_4^2]/2g$$
$$= [0.2(80.0^2 + 44.4^2 + 28.6^2 + 22.9^2) + 0.6(200)^2$$
$$+ 0.2(200)^2 + 4.0(22.9)^2] \text{ ft}^2/\text{s}^2/[2(32.2 \text{ ft/s}^2)]$$

or

$$h_{L_{1-9}} = 560 \text{ ft}$$

Hence, from Eq. 1 we obtain the pressure rise across the fan as

$$p_1 - p_9 = \gamma h_{L_{1-9}} = (0.0765 \text{ lb/ft}^3)(560 \text{ ft})$$
$$= 42.8 \text{ lb/ft}^2 = 0.298 \text{ psi} \qquad \textbf{(Ans)}$$

From Eq. 2 we obtain the power added to the fluid as

$$\mathcal{P}_a = (0.0765 \text{ lb/ft}^3)(4.0 \text{ ft}^2)(200 \text{ ft/s})(560 \text{ ft})$$
$$= 34,300 \text{ ft·lb/s}$$

or

$$\mathcal{P}_a = \frac{34,300 \text{ ft·lb/s}}{550 \text{ (ft·lb/s)/hp}} = 62.3 \text{ hp} \qquad \textbf{(Ans)}$$

COMMENTS   By repeating the calculations with various test section velocities, V_5, the results shown in Fig. E8.4b are obtained. Since the head loss varies as V_5^2 and the power varies as head loss times V_5, it follows that the power varies as the cube of the velocity. Thus, doubling the wind tunnel speed requires an eightfold increase in power.

With a closed-return wind tunnel of this type, all of the power required to maintain the flow is dissipated through viscous effects, with the energy remaining within the closed tunnel. If heat transfer across the tunnel walls is negligible, the air temperature within the tunnel will increase in time. For steady-state operations of such tunnels, it is often necessary to provide some means of cooling to maintain the temperature at acceptable levels.

It should be noted that the actual size of the motor that powers the fan must be greater than the calculated 62.3 hp

because the fan is not 100% efficient. The power calculated earlier is that needed by the fluid to overcome losses in the tunnel, excluding those in the fan. If the fan were 60% efficient, it would require a shaft power of $\mathcal{P} = 62.3 \text{ hp}/(0.60) = 104 \text{ hp}$ to run the fan. Determination of fan (or pump) efficiencies can be a complex problem that depends on the specific geometry of the fan. Introductory material about fan performance can be found in various references (Refs. 7, 8, 9, for example).

It should also be noted that the aforementioned results are only approximate. Clever, careful design of the various components (corners, diffuser, etc.) may lead to improved (i.e., lower) values of the various loss coefficients, and hence lower power requirements. Since h_L is proportional to V^2, components with the larger V tend to have the larger head loss. Thus, even though $K_L = 0.2$ for each of the four corners, the head loss for corner (7) is $(V_7/V_3)^2 = (80/22.9)^2 = 12.2$ times greater than it is for corner (3).

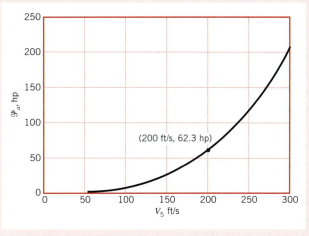

■ **FIGURE E8.4b**

8.4.3   Noncircular Conduits

Many of the conduits that are used for conveying fluids are not circular in cross section. Although the details of the flows in such conduits depend on the exact cross-sectional shape, many round pipe results can be carried over, with slight modification, to flow in conduits of other shapes.

Practical, easy-to-use results can be obtained by introducing the ***hydraulic diameter,*** $D_h = 4A/P$, defined as four times the ratio of the cross-sectional flow area divided by the wetted perimeter, P, of the pipe as is illustrated in Fig. 8.19. The hydraulic diameter is used in the definition of the friction factor, $h_L = f(\ell/D_h)V^2/2g$, the Reynolds number, $\text{Re}_h = \rho V D_h/\mu$, and the relative roughness, ε/D_h.

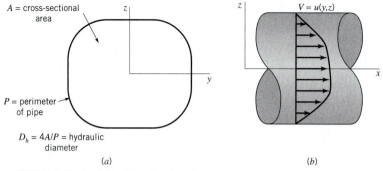

■ **FIGURE 8.19** **Noncircular duct.**

Calculations for fully developed turbulent flow in ducts of noncircular cross section are usually carried out by using the Moody chart data for round pipes with the diameter replaced by the hydraulic diameter as indicated earlier. For turbulent flow such calculations are usually accurate to within about 15%. If greater accuracy is needed, a more detailed analysis based on the specific geometry of interest is needed.

EXAMPLE 8.5   Noncircular Conduit

GIVEN Air at a temperature of 120° F and standard pressure flows from a furnace through an 8-in.-diameter pipe with an average velocity of 10 ft/s. It then passes through a transition section and into a square duct whose side is of length a.

The pipe and duct surfaces are smooth ($\varepsilon = 0$). The head loss per foot is to be the same for the pipe and the duct.

FIND Determine the duct size, a.

SOLUTION

We first determine the head loss per foot for the pipe, $h_L/\ell = (f/D)V^2/2g$, and then size the square duct to give the same value. For the given pressure and temperature we obtain (from Table B.3) $\nu = 1.89 \times 10^{-4}$ ft^2/s so that

$$\text{Re} = \frac{VD}{\nu} = \frac{(10 \text{ ft/s})(\tfrac{8}{12} \text{ ft})}{1.89 \times 10^{-4} \text{ ft}^2/\text{s}} = 35{,}300$$

With this Reynolds number and with $\varepsilon/D = 0$ we obtain the friction factor from Fig. 8.10 as $f = 0.022$ so that

$$\frac{h_L}{\ell} = 0.022 \frac{(10 \text{ ft/s})^2}{(\tfrac{8}{12} \text{ ft}) \, 2(32.2 \text{ ft/s}^2)} = 0.0512$$

Thus, for the square duct we must have

$$\frac{h_L}{\ell} = \frac{f}{D_h} \frac{V_s^2}{2g} = 0.0512 \tag{1}$$

where

$$D_h = 4A/P = 4a^2/4a = a \quad \text{and}$$

$$V_s = \frac{Q}{A} = \frac{\dfrac{\pi}{4}\left(\dfrac{8}{12}\text{ ft}\right)^2 (10 \text{ ft/s})}{a^2} = \frac{3.49}{a^2} \tag{2}$$

is the velocity in the duct.

By combining Eqs. 1 and 2 we obtain

$$0.0512 = \frac{f}{a} \frac{(3.49/a^2)^2}{2(32.2)}$$

or

$$a = 1.30 \, f^{1/5} \tag{3}$$

where a is in feet. Similarly, the Reynolds number based on the hydraulic diameter is

$$\text{Re}_h = \frac{V_s D_h}{\nu} = \frac{(3.49/a^2)a}{1.89 \times 10^{-4}} = \frac{1.85 \times 10^4}{a} \tag{4}$$

We have three unknowns (a, f, and Re$_h$) and three equations (Eqs. 3, 4, and the third equation in graphical form, Fig. 8.10, the Moody chart). Thus, a trial-and-error solution is required.

As an initial attempt, assume the friction factor for the duct is the same as for the pipe. That is, assume $f = 0.022$. From Eq. 3 we obtain $a = 0.606$ ft, while from Eq. 4 we have Re$_h = 3.05 \times 10^4$. From Fig. 8.10, with this Reynolds number and the given smooth duct we obtain $f = 0.023$, which does not quite agree with the assumed value of f. Hence, we do not have the solution. We try again, using the latest calculated value of $f = 0.023$ as our guess. The calculations are repeated until the guessed value of f agrees with the value obtained from Fig. 8.10. The final result (after only two iterations) is $f = 0.023$, $Re_h = 3.03 \times 10^4$, and

$$a = 0.611 \text{ ft} = 7.34 \text{ in.} \qquad \textbf{(Ans)}$$

COMMENTS Alternatively, we can use the Colebrook equation (rather than the Moody chart) to obtain the solution as follows. For a smooth pipe ($\varepsilon/D_h = 0$) the Colebrook equation, Eq. 8.19, becomes

$$\frac{1}{\sqrt{f}} = -2.0 \log\left(\frac{\varepsilon/D_h}{3.7} + \frac{2.51}{Re_h \sqrt{f}}\right)$$

$$= -2.0 \log\left(\frac{2.51}{Re_h \sqrt{f}}\right) \qquad \textbf{(5)}$$

where from Eq. 3,

$$f = 0.269 \, a^5 \qquad \textbf{(6)}$$

If we combine Eqs. 4, 5, and 6 and simplify, Eq. 7 is obtained for a.

$$1.928 \, a^{-5/2} = -2 \log(2.62 \times 10^{-4} \, a^{-3/2}) \qquad \textbf{(7)}$$

By using a root-finding technique on a computer or calculator, the solution to Eq. 7 is determined to be $a = 0.614$ ft, in agreement (given the accuracy of reading the Moody chart) with that obtained by the trial-and-error method given above.

Note that the length of the side of the equivalent square duct is $a/D = 7.34/8 = 0.918$, or approximately 92% of the diameter of the equivalent duct. It can be shown that this value, 92%, is a very good approximation for any pipe flow—laminar or turbulent. The cross-sectional area of the duct ($A = a^2 = 53.9$ in.^{2}) is greater than that of the round pipe ($A = \pi D^2/4 = 50.3$ in.^{2}). Also, it takes less material to form the round pipe (perimeter $= \pi D = 25.1$ in.) than the square duct (perimeter $= 4a = 29.4$ in.). Circles are very efficient shapes.

8.5   Pipe Flow Examples

The previous sections of this chapter discussed concepts concerning flow in pipes and ducts. The purpose of this section is to apply these ideas to the solutions of various practical problems.

F l u i d s   i n   t h e   N e w s

New hi-tech fountains Ancient Egyptians used fountains in their palaces for decorative and cooling purposes. Current use of fountains continues, but with a hi-tech flare. Although the basic fountain still consists of a typical *pipe system* (i.e., pump, pipe, regulating valve, nozzle, filter, and basin), recent use of computer-controlled devices has led to the design of innovative fountains with special effects. For example, by using several rows of multiple nozzles, it is possible to program and activate control valves to produce water jets that resemble symbols, letters, or the time of day. Other fountains use specially designed nozzles to produce coherent, laminar streams of water that look like glass rods flying through the air. By using fast-acting control valves in a synchronized manner it is possible to produce mesmerizing three-dimensional patterns of water droplets. The possibilities are nearly limitless. With the initial artistic design of the fountain established, the initial engineering design (i.e., the capacity and pressure requirements of the nozzles and the size of the pipes and pumps) can be carried out. It is often necessary to modify the artistic and/or engineering aspects of the design in order to obtain a functional, pleasing fountain. (See Problem 8.27.) ■

8.5.1   Single Pipes

The nature of the solution process for pipe flow problems can depend strongly on which of the various parameters are independent parameters (the "given") and which is the dependent parameter (the "determine"). The three most common types of problems are discussed here.

In a Type I problem we specify the desired flowrate or average velocity and determine the necessary pressure difference or head loss. For example, if a flowrate of 2.0 gal/min is required for a dishwasher that is connected to the water heater by a given pipe system as shown by the figure in the margin, what pressure is needed in the water heater?

In a Type II problem we specify the applied driving pressure (or, alternatively, the head loss) and determine the flowrate. For example, how many gallons per minute of hot water are supplied to the dishwasher if the pressure within the water heater is 60 psi and the pipe system details (length, diameter, roughness of the pipe; number of elbows; etc.) are specified?

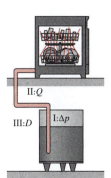

II:Q

III:D I:Δp

In a Type III problem we specify the pressure drop and the flowrate and determine the diameter of the pipe needed. For example, what diameter of pipe is needed between the water heater and dishwasher if the pressure in the water heater is 60 psi (determined by the city water system) and the flowrate is to be not less than 2.0 gal/min (determined by the manufacturer)?

EXAMPLE 8.6 (Type I, Determine Pressure Drop)

GIVEN Water at 60° F flows from the basement to the second floor through the 0.75-in. (0.0625-ft-)-diameter copper pipe (a drawn tubing) at a rate of $Q = 12.0$ gal/min $= 0.0267$ ft³/s and exits through a faucet of diameter 0.50 in. as shown in Fig. E8.6a.

FIND Determine the pressure at point (1) if

(a) all losses are neglected,

(b) the only losses included are major losses, or

(c) all losses are included.

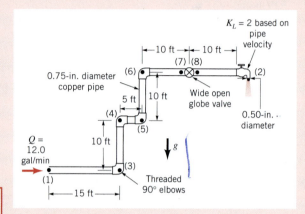

■ **FIGURE E8.6a**

SOLUTION

Since the fluid velocity in the pipe is given by $V_1 = Q/A_1 = Q/(\pi D^2/4) = (0.0267$ ft³/s)/$[\pi(0.0625$ ft)²/4$] = 8.70$ ft/s, and the fluid properties are $\rho = 1.94$ slugs/ft³ and $\mu = 2.34 \times 10^{-5}$ lb·s/ft² (see Table B.1), it follows that Re $= \rho V D/\mu = (1.94$ slugs/ft³)(8.70 ft/s)(0.0625 ft)/(2.34 × 10⁻⁵ lb·s/ft²) = 45,000. Thus, the flow is turbulent. The governing equation for case **(a)**, **(b)**, or **(c)** is the energy equation as given by Eq. 5.59,

$$\frac{p_1}{\gamma} + \alpha_1 \frac{V_1^2}{2g} + z_1 = \frac{p_2}{\gamma} + \alpha_2 \frac{V_2^2}{2g} + z_2 + h_L$$

where $z_1 = 0$, $z_2 = 20$ ft, $p_2 = 0$ (free jet), $\gamma = \rho g = 62.4$ lb/ft³, and the outlet velocity is $V_2 = Q/A_2 = (0.0267$ ft³/s)/$[\pi(0.50/12)^2$ft²/4$] = 19.6$ ft/s. We assume that the kinetic energy coefficients α_1 and α_2 are unity. This is reasonable because turbulent velocity profiles are nearly uniform across the pipe (see Section 5.3.4). Thus,

$$p_1 = \gamma z_2 + \tfrac{1}{2}\rho(V_2^2 - V_1^2) + \gamma h_L \qquad (1)$$

where the head loss is different for each of the three cases.

(a) If all losses are neglected ($h_L = 0$), Eq. 1 gives

$$p_1 = (62.4 \text{ lb/ft}^3)(20 \text{ ft})$$
$$+ \frac{1.94 \text{ slugs/ft}^3}{2}\left[\left(19.6 \frac{\text{ft}}{\text{s}}\right)^2 - \left(8.70 \frac{\text{ft}}{\text{s}}\right)^2\right]$$
$$= (1248 + 299) \text{ lb/ft}^2 = 1547 \text{ lb/ft}^2$$

or

$$p_1 = 10.7 \text{ psi} \qquad \textbf{(Ans)}$$

COMMENT Note that for this pressure drop, the amount due to elevation change (the hydrostatic effect) is $\gamma(z_2 - z_1) = 8.67$ psi and the amount due to the increase in kinetic energy is $\rho(V_2^2 - V_1^2)/2 = 2.07$ psi.

(b) If the only losses included are the major losses, the head loss is

$$h_L = f\frac{\ell}{D}\frac{V_1^2}{2g}$$

From Table 8.1 the roughness for a 0.75-in.-diameter copper pipe (drawn tubing) is $\varepsilon = 0.000005$ ft so that $\varepsilon/D = 8 \times 10^{-5}$. With this ε/D and the calculated Reynolds number (Re = 45,000), the value of f is obtained from the Moody chart as $f = 0.0215$. Note that the Colebrook equation (Eq. 8.19) would give the same value of f. Hence, with the total length of the pipe as $\ell = (15 + 5 + 10 + 10 + 20)$ ft $= 60$ ft and the elevation and kinetic energy portions the same as for part **(a)**, Eq. 1 gives

$$p_1 = \gamma z_2 + \frac{1}{2}\rho(V_2^2 - V_1^2) + \rho f\frac{\ell}{D}\frac{V_1^2}{2}$$

$$= (1248 + 299) \text{ lb/ft}^2$$

$$+ (1.94 \text{ slugs/ft}^3)(0.0215)\left(\frac{60 \text{ ft}}{0.0625 \text{ ft}}\right)\frac{(8.70 \text{ ft/s})^2}{2}$$

$$= (1248 + 299 + 1515) \text{ lb/ft}^2 = 3062 \text{ lb/ft}^2$$

or

$$p_1 = 21.3 \text{ psi} \qquad \textbf{(Ans)}$$

COMMENT Of this pressure drop, the amount due to pipe friction is approximately $(21.3 - 10.7)$ psi = 10.6 psi.

(c) If major and minor losses are included, Eq. 1 becomes

$$p_1 = \gamma z_2 + \frac{1}{2}\rho(V_2^2 - V_1^2) + f\gamma\frac{\ell}{D}\frac{V_1^2}{2g} + \sum \rho K_L\frac{V^2}{2}$$

or

$$p_1 = 21.3 \text{ psi} + \sum \rho K_L \frac{V^2}{2} \qquad (2)$$

where the 21.3-psi contribution is due to elevation change, kinetic energy change, and major losses [part **(b)**] and the last term represents the sum of all of the minor losses. The loss coefficients of the components ($K_L = 1.5$ for each elbow and $K_L = 10$ for the wide-open globe valve) are given in Table 8.2 (except for the loss coefficient of the faucet, which is given in Fig. E8.6a as $K_L = 2$). Thus,

$$\sum \rho K_L \frac{V^2}{2} = (1.94 \text{ slugs/ft}^3) \frac{(8.70 \text{ ft/s})^2}{2} [10 + 4(1.5) + 2]$$

$$= 1321 \text{ lb/ft}^2$$

or

$$\sum \rho K_L \frac{V^2}{2} = 9.17 \text{ psi} \qquad (3)$$

COMMENTS Note that we did not include an entrance or exit loss because points (1) and (2) are located within the fluid streams, not within an attaching reservoir where the kinetic energy is zero. Thus, by combining Eqs. 2 and 3 we obtain the entire pressure drop as

$$p_1 = (21.3 + 9.17) \text{ psi} = 30.5 \text{ psi} \qquad \textbf{(Ans)}$$

This pressure drop calculated by including all losses should be the most realistic answer of the three cases considered.

More detailed calculations will show that the pressure distribution along the pipe is as illustrated in Fig. E8.6b for cases **(a)** and **(c)**—neglecting all losses or including all losses. Note that not all of the pressure drop, $p_1 - p_2$, is a "pressure loss." The pressure change due to the elevation and velocity changes is completely reversible. The portion due to the major and minor losses is irreversible.

This flow can be illustrated in terms of the energy line and hydraulic grade line concepts introduced in Section 3.7. As is shown in Fig. E8.6c, for case **(a)** there are no losses and the energy line (EL) is horizontal, one velocity head ($V^2/2g$) above the hydraulic grade line (HGL), which is one pressure head (γz) above the pipe itself. For case **(c)** the energy line is not horizontal. Each bit of friction in the pipe or loss in a component reduces the available energy, thereby lowering the energy line. Thus, for case **(a)** the total head remains constant throughout the flow with a value of

$$H = \frac{p_1}{\gamma} + \frac{V_1^2}{2g} + z_1 = \frac{(1547 \text{ lb/ft}^2)}{(62.4 \text{ lb/ft}^3)} + \frac{(8.70 \text{ ft/s})^2}{2(32.2 \text{ ft/s}^2)} + 0$$

$$= 26.0 \text{ ft}$$

$$= \frac{p_2}{\gamma} + \frac{V_2^2}{2g} + z_2 = \frac{p_3}{\gamma} + \frac{V_3^2}{2g} + z_3 = \cdots$$

For case **(c)** the energy line starts at

$$\begin{aligned} H_1 &= \frac{p_1}{\gamma} + \frac{V_1^2}{2g} + z_1 \\ &= \frac{(30.5 \times 144)\text{lb/ft}^2}{(62.4 \text{ lb/ft}^3)} + \frac{(8.70 \text{ ft/s})^2}{2(32.2 \text{ ft/s}^2)} + 0 \\ &= 71.6 \text{ ft} \end{aligned}$$

and falls to a final value of

$$\begin{aligned} H_2 &= \frac{p_2}{\gamma} + \frac{V_2^2}{2g} + z_2 = 0 + \frac{(19.6 \text{ ft/s})^2}{2(32.2 \text{ ft/s}^2)} + 20 \text{ ft} \\ &= 26.0 \text{ ft} \end{aligned}$$

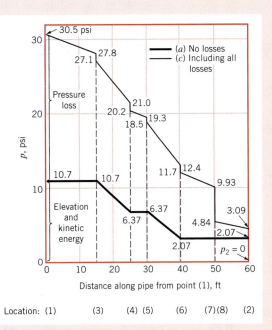

■ FIGURE E8.6b

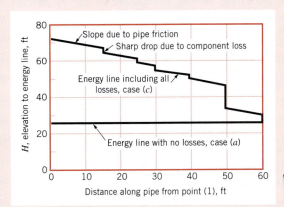

■ FIGURE E8.6c

The elevation of the energy line can be calculated at any point along the pipe. For example, at point (7), 50 ft from point (1),

$$H_7 = \frac{p_7}{\gamma} + \frac{V_7^2}{2g} + z_7$$

$$= \frac{(9.93 \times 144) \text{ lb/ft}^2}{(62.4 \text{ lb/ft}^3)} + \frac{(8.70 \text{ ft/s})^2}{2(32.2 \text{ ft/s}^2)} + 20 \text{ ft}$$

$$= 44.1 \text{ ft}$$

The head loss per foot of pipe is the same all along the pipe. That is,

$$\frac{h_L}{\ell} = f\frac{V^2}{2gD} = \frac{0.0215(8.70 \text{ ft/s})^2}{2(32.2 \text{ ft/s}^2)(0.0625 \text{ ft})} = 0.404 \text{ ft/ft}$$

Thus, the energy line is a set of straight-line segments of the same slope separated by steps whose height equals the head loss of the minor component at that location. As is seen from Fig. E8.6c, the globe valve produces the largest of all the minor losses.

Although the governing pipe flow equations are quite simple, they can provide very reasonable results for a variety of applications, as shown in the next example.

EXAMPLE 8.7   (Type I, Determine Head Loss)

GIVEN As shown in Fig. E8.7a, crude oil at 140°F with $\gamma = 53.7$ lb/ft³ and $\mu = 8 \times 10^{-5}$ lb·s/ft² (about four times the viscosity of water) is pumped across Alaska through the Alaskan pipeline, a 799-mile-long, 4-ft-diameter steel pipe, at a maximum rate of $Q = 2.4$ million barrels/day = 117 ft³/s.

FIND Determine the horsepower needed for the pumps that drive this large system.

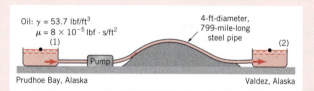

■ **FIGURE E8.7a**

SOLUTION

From the energy equation (Eq. 5.57) we obtain

$$\frac{p_1}{\gamma} + \frac{V_1^2}{2g} + z_1 + h_p = \frac{p_2}{\gamma} + \frac{V_2^2}{2g} + z_2 + h_L$$

where points (1) and (2) represent locations within the large holding tanks at either end of the line and h_p is the head provided to the oil by the pumps. We assume that $z_1 = z_2$ (pumped from sea level to sea level), $p_1 = p_2 = V_1 = V_2 = 0$ (large, open tanks) and $h_L = (f\ell/D)(V^2/2g)$. Minor losses are negligible because of the large length-to-diameter ratio of the relatively straight, uninterrupted pipe; $\ell/D = (799 \text{ mi})(5280 \text{ ft/mi})/(4 \text{ ft}) = 1.05 \times 10^6$. Thus,

$$h_p = h_L = f\frac{\ell}{D}\frac{V^2}{2g}$$

where $V = Q/A = (117 \text{ ft}^3/\text{s})/[\pi (4 \text{ ft})^2/4] = 9.31$ ft/s. From Fig. 8.10, $f = 0.0125$ since $\varepsilon/D = (0.00015 \text{ ft})/(4 \text{ ft}) = 0.0000375$ (see Table 8.1) and Re $= \rho VD/\mu = [(53.7/32.2) \text{ slugs/ft}^3](9.31 \text{ ft/s})(4.0 \text{ ft})/(8 \times 10^{-5} \text{ lb·s/ft}^2) = 7.76 \times 10^5$. Thus,

$$h_p = 0.0125(1.05 \times 10^6)\frac{(9.31 \text{ ft/s})^2}{2(32.2 \text{ ft/s}^2)} = 17,700 \text{ ft}$$

and the actual power supplied to the fluid, $\mathcal{P}_a$, is

$$\mathcal{P}_a = \gamma Q h_p = (53.7 \text{ lb/ft}^3)(117 \text{ ft}^3/\text{s})(17,700 \text{ ft})$$

$$= 1.11 \times 10^8 \text{ ft·lb/s}\left(\frac{1 \text{ hp}}{550 \text{ ft·lb/s}}\right)$$

$$= 202,000 \text{ hp} \qquad \textbf{(Ans)}$$

COMMENTS There are many reasons why it is not practical to drive this flow with a single pump of this size. First, there are no pumps this large! Second, the pressure at the pump outlet would need to be $p = \gamma h_L = (53.7 \text{ lb/ft}^3)(17,700 \text{ ft})$ (1 ft²/144 in.²) = 6600 psi. No practical 4-ft-diameter pipe would withstand this pressure. An equally unfeasible alternative would be to place the holding tank at the beginning of the pipe on top of a hill at height $h_L = 17,700$ ft and let gravity force the oil through the 799-mi pipe! How much power would it take to lift the oil to the top of the hill?

To produce the desired flow, the actual system contains 12 pumping stations positioned at strategic locations along the pipeline. Each station contains four pumps, three of which operate at any one time (the fourth is in reserve in case of emergency). Each pump is driven by a 13,500-hp motor, thereby producing a total horsepower of $\mathcal{P} = 12$ stations (3 pumps/station) (13,500 hp/pump) = 486,000 hp. If we assume that the pump/motor combination is approximately 60% efficient, there is a total of 0.60 (486,000) hp = 292,000 hp available to drive the fluid. This number compares favorably with the 202,000-hp answer calculated earlier.

The assumption of a 140 °F oil temperature may not seem reasonable for flow across Alaska. Note, however, that the oil is warm when it is pumped from the ground and that the 202,000 hp needed to pump the oil is dissipated as a head loss (and therefore a temperature rise) along the pipe. However, if the oil temperature were 70 °F rather than 140 °F, the viscosity would be approximately 16×10^{-5} lb·s/ft² (twice as large), but the friction factor would only increase from $f = 0.0125$ at 140 °F (Re = 7.76×10^5) to $f = 0.0140$ at 70 °F (Re = 3.88×10^5). This doubling of viscosity would result in only an 11% increase in power (from 202,000 to 226,000 hp). Because of the large Reynolds numbers involved, the shear stress is due mostly to the turbulent nature of the flow. That is, the value of Re for this flow is large enough (on the relatively flat part of the Moody chart) so that f is nearly independent of Re (or viscosity).

By repeating the calculations for various values of the pipe diameter, D, the results shown in Fig. E8.7b are obtained. Clearly the required pump power, $\mathcal{P}_a$, is a strong function of the pipe diameter, with $\mathcal{P}_a \sim D^{-4}$ if the friction factor is constant. The actual 4-ft-diameter pipe used represents a compromise between using smaller diameter pipes which are less expensive to make but require considerably more pump power and larger diameter pipes which require less pump power but are very expensive to make and maintain.

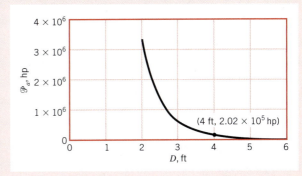

■ **FIGURE E8.7b**

Pipe flow problems in which it is desired to determine the flowrate for a given set of conditions (Type II problems) often require trial-and-error solution techniques. This is because it is necessary to know the value of the friction factor to carry out the calculations, but the friction factor is a function of the unknown velocity (flowrate) in terms of the Reynolds number. The solution procedure is indicated in Example 8.8.

EXAMPLE 8.8   (Type II, Determine Flowrate)

GIVEN   Air at a temperature of 100 °F and standard pressure flows from a clothes dryer. According to the appliance manufacturer, the 4-in.-diameter galvanized iron vent on the clothes dryer is not to contain more than 20 ft of pipe and four 90° elbows.

FIND   Under these conditions determine the air flowrate if the pressure within the dryer is 0.20 in. of water.

SOLUTION

Application of the energy equation (Eq. 5.57) between the inside of the dryer, point (1), and the exit of the vent pipe, point (2), gives

$$\frac{p_1}{\gamma} + \frac{V_1^2}{2g} + z_1 = \frac{p_2}{\gamma} + \frac{V_2^2}{2g} + z_2 + f\frac{\ell}{D}\frac{V^2}{2g} + \sum K_L \frac{V^2}{2g} \quad (1)$$

where K_L for the entrance is assumed to be 0.5 and that for each elbow is assumed to be 1.5. In addition we assume that $V_1 = 0$ and $z_1 = z_2$. (The change in elevation is often negligible for gas flows.) Also, $p_2 = 0$, and $p_1/\gamma_{H_2O} = 0.2$ in., or

$$p_1 = (0.2 \text{ in.})\left(\frac{1 \text{ ft}}{12 \text{ in.}}\right)(62.4 \text{ lb/ft}^3) = 1.04 \text{ lb/ft}^2$$

Thus, with $\gamma = 0.0709$ lb/ft³ (see Table B.3) and $V_2 = V$ (the air velocity in the pipe), Eq. 1 becomes

$$\frac{(1.04 \text{ lb/ft}^2)}{(0.0709 \text{ lb/ft}^3)} = \left[1 + f\frac{(20 \text{ ft})}{(\frac{4}{12} \text{ ft})} + 0.5\right. $$
$$\left. + 4(1.5)\right]\frac{V^2}{2(32.2 \text{ ft/s}^2)}$$

or

$$945 = (7.5 + 60f)V^2 \quad (2)$$

where V is in feet per second.

The value of f is dependent on Re, which is dependent on V, an unknown. However, from Table B.3, $\nu = 1.79 \times 10^{-4}$ ft²/s and we obtain

$$\text{Re} = \frac{VD}{\nu} = \frac{(\frac{4}{12}\text{ ft})V}{1.79 \times 10^{-4}\text{ ft}^2/\text{s}}$$

or

$$\text{Re} = 1860\,V \qquad (3)$$

where again V is in feet per second.

Also, since $\varepsilon/D = (0.0005\text{ ft})/(4/12\text{ ft}) = 0.0015$ (see Table 8.1 for the value of ε), we know which particular curve of the Moody chart is pertinent to this flow. Thus, we have three relationships (Eqs. 2, 3, and the $\varepsilon/D = 0.0015$ curve of Fig. 8.10) from which we can solve for the three unknowns f, Re, and V. This is done easily by an iterative scheme as follows.

It is usually simplest to assume a value of f, calculate V from Eq. 2, calculate Re from Eq. 3, and look up the appropriate value of f in the Moody chart for this value of Re. If the assumed f and the new f do not agree, the assumed answer is not correct—we do not have the solution to the three equations. Although values of f, V, or Re could be assumed as starting values, it is usually simplest to assume a value of f because the correct value often lies on the relatively flat portion of the Moody chart for which f is quite insensitive to Re.

Thus, we assume $f = 0.022$, approximately the large Re limit for the given relative roughness. From Eq. 2 we obtain

$$V = \left[\frac{945}{7.5 + 60(0.022)}\right]^{1/2} = 10.4\text{ ft/s}$$

and from Eq. 3

$$\text{Re} = 1860(10.4) = 19{,}300$$

With this Re and ε/D, Fig. 8.10 gives $f = 0.029$, which is not equal to the assumed solution $f = 0.022$ (although it is close!). We try again, this time with the newly obtained value of $f = 0.029$, which gives $V = 10.1$ ft/s and Re $= 18{,}800$. With these values, Fig. 8.10 gives $f = 0.029$, which agrees with the assumed value. Thus, the solution is $V = 10.1$ ft/s, or

$$Q = AV = \frac{\pi}{4}(\tfrac{4}{12}\text{ ft})^2(10.1\text{ ft/s}) = 0.881\text{ ft}^3/\text{s} \quad \textbf{(Ans)}$$

COMMENTS Note that the need for the iteration scheme is because one of the equations, $f = \phi(\text{Re}, \varepsilon/D)$, is in graphical form (the Moody chart, Fig. 8.10). If the dependence of f on Re and ε/D is known in equation form, this graphical dependency is eliminated, and the solution technique may be easier. Such is the case if the flow is laminar so that the friction factor is simply $f = 64/\text{Re}$. For turbulent flow, we can use the Colebrook equation rather than the Moody chart, although this will normally require an iterative scheme also because of the complexity of the equation. As is shown here, such a formulation is ideally suited for an iterative computer solution.

We keep Eqs. 2 and 3 and use the Colebrook equation (Eq. 8.19, rather than the Moody chart) with $\varepsilon/D = 0.0015$ to give

$$\frac{1}{\sqrt{f}} = -2.0 \log\left(\frac{\varepsilon/D}{3.7} + \frac{2.51}{\text{Re}\sqrt{f}}\right)$$
$$= -2.0 \log\left(4.05 \times 10^{-4} + \frac{2.51}{\text{Re}\sqrt{f}}\right) \qquad (4)$$

From Eq. 2 we have $V = [945/(7.5 + 60f)]^{1/2}$, which can be combined with Eq. 3 to give

$$\text{Re} = \frac{57{,}200}{\sqrt{7.5 + 60f}} \qquad (5)$$

The combination of Eqs. 4 and 5 provides a single equation for the determination of f

$$\frac{1}{\sqrt{f}} = -2.0 \log\Bigg(4.05 \times 10^{-4} + 4.39 \times 10^{-5}\sqrt{60 + \frac{7.5}{f}}\,\Bigg) \qquad (6)$$

By using a root-finding technique on a computer or calculator, the solution to this equation is determined to be $f = 0.029$, in agreement with the above solution which used the Moody chart.

Note that unlike the Alaskan pipeline example (Example 8.7) in which we assumed minor losses are negligible, minor losses are of importance in this example because of the relatively small length-to-diameter ratio: $\ell/D = 20/(4/12) = 60$. The ratio of minor to major losses in this case is $K_L/(f\ell/D) = 6.5/[0.029(60)] = 3.74$. The elbows and entrance produce considerably more loss than the pipe itself.

In pipe flow problems for which the diameter is the unknown (Type III), an iterative technique is required since neither the Reynolds number, Re $= \rho VD/\mu = 4\rho Q/\pi\mu D$, nor the relative roughness, ε/D, is known unless D is known. Example 8.9 illustrates this.

EXAMPLE 8.9 (Type III Without Minor Losses, Determine Diameter)

GIVEN Air at standard temperature and pressure flows through a horizontal, galvanized iron pipe ($\varepsilon = 0.0005$ ft) at a rate of 2.0 ft^3/s. The pressure drop is to be no more than 0.50 psi per 100 ft of pipe.

FIND Determine the minimum pipe diameter.

SOLUTION

We assume the flow to be incompressible with $\rho = 0.00238$ slugs/ft^3 and $\mu = 3.74 \times 10^{-7}$ lb·s/ft^2. Note that if the pipe were too long, the pressure drop from one end to the other, $p_1 - p_2$, would not be small relative to the pressure at the beginning, and compressible flow considerations would be required. For example, a pipe length of 200 ft gives $(p_1 - p_2)/p_1 = [(0.50$ psi)/(100 ft)](200 ft)/14.7 psi = 0.068 = 6.8%, which is probably small enough to justify the incompressible assumption.

With $z_1 = z_2$ and $V_1 = V_2$ the energy equation (Eq. 5.57) becomes

$$p_1 = p_2 + f \frac{\ell}{D} \frac{\rho V^2}{2} \tag{1}$$

where $V = Q/A = 4Q/(\pi D^2) = 4(2.0 \text{ ft}^3/\text{s})/\pi D^2$, or

$$V = \frac{2.55}{D^2}$$

where D is in feet. Thus, with $p_1 - p_2 = (0.5 \text{ lb/in.}^2)(144$ in.^{2}/ft^2) and $\ell = 100$ ft, Eq. 1 becomes

$$p_1 - p_2 = (0.5)(144) \text{ lb/ft}^2$$
$$= f \frac{(100 \text{ ft})}{D} (0.00238 \text{ slugs/ft}^3) \frac{1}{2} \left(\frac{2.55}{D^2} \frac{\text{ft}}{\text{s}} \right)^2$$

or

$$D = 0.404 f^{1/5} \tag{2}$$

where D is in feet. Also Re $= \rho VD/\mu = (0.00238 \text{ slugs/ft}^3)$ [$(2.55/D^2)$ ft/s]$D/(3.74 \times 10^{-7}$ lb·s/ft^2), or

$$\text{Re} = \frac{1.62 \times 10^4}{D} \tag{3}$$

and

$$\frac{\varepsilon}{D} = \frac{0.0005}{D} \tag{4}$$

Thus, we have four equations [Eqs. 2, 3, 4, and either the Moody chart (Fig. 8.10) or the Colebrook equation (Eq. 8.19)] and four unknowns (f, D, ε/D, and Re) from which the solution can be obtained by trial-and-error methods.

If we use the Moody chart, it is probably easiest to assume a value of f, use Eqs. 2, 3, and 4 to calculate D, Re, and ε/D, and then compare the assumed f with that from the Moody chart. If they do not agree, try again. Thus, we assume $f = 0.02$, a typi-

cal value, and obtain $D = 0.404(0.02)^{1/5} = 0.185$ ft, which gives $\varepsilon/D = 0.0005/0.185 = 0.0027$ and Re $= 1.62 \times 10^4/0.185 = 8.76 \times 10^4$. From the Moody chart we obtain $f = 0.027$ for these values of ε/D and Re. Since this is not the same as our assumed value of f, we try again. With $f = 0.027$, we obtain $D = 0.196$ ft, $\varepsilon/D = 0.0026$, and Re $= 8.27 \times 10^4$, which in turn give $f = 0.027$, in agreement with the assumed value. Thus, the diameter of the pipe should be

$$D = 0.196 \text{ ft} \tag{Ans}$$

COMMENT If we use the Colebrook equation (Eq. 8.19) with $\varepsilon/D = 0.0005/0.404 f^{1/5} = 0.00124/f^{1/5}$ and Re $= 1.62 \times 10^4/0.404 f^{1/5} = 4.01 \times 10^4/f^{1/5}$, we obtain

$$\frac{1}{\sqrt{f}} = -2.0 \log \left(\frac{\varepsilon/D}{3.7} + \frac{2.51}{\text{Re}\sqrt{f}} \right)$$

or

$$\frac{1}{\sqrt{f}} = -2.0 \log \left(\frac{3.35 \times 10^{-4}}{f^{1/5}} + \frac{6.26 \times 10^{-5}}{f^{3/10}} \right)$$

By using a root-finding technique on a computer or calculator, the solution to this equation is determined to be $f = 0.027$, and hence $D = 0.196$ ft, in agreement with the Moody chart method.

By repeating the calculations for various values of the flowrate, Q, the results shown in Fig. E8.9 are obtained. Although an increase in flowrate requires a larger diameter pipe (for the given pressure drop), the increase in diameter is minimal. For example, if the flowrate is doubled from 1 ft^3/s to 2 ft^3/s, the diameter increases from 0.151 ft to 0.196 ft.

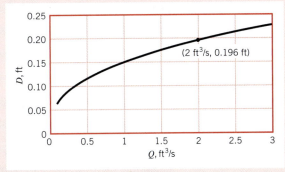

■ **FIGURE E8.9**

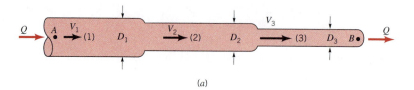

(a)

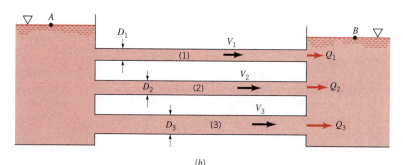

(b)

■ **FIGURE 8.20** Series (*a*) and parallel (*b*) pipe systems.

8.5.2 Multiple Pipe Systems

In many pipe systems there is more than one pipe involved. The governing mechanisms for the flow in ***multiple pipe systems*** are the same as for single pipes.

One of the simplest multiple pipe systems is that containing pipes in *series,* as is shown in Fig. 8.20*a*. Every fluid particle that passes through the system passes through each of the pipes. Thus, the flowrate (but not the velocity) is the same in each pipe, and the head loss from point *A* to point *B* is the sum of the head losses in each of the pipes. The governing equations can be written as follows

$$Q_1 = Q_2 = Q_3$$

and

$$h_{L_{A-B}} = h_{L_1} + h_{L_2} + h_{L_3}$$

where the subscripts refer to each of the pipes.

Another common multiple pipe system contains pipes in *parallel,* as is shown in Fig. 8.20*b*. In this system a fluid particle traveling from *A* to *B* may take any of the paths available, with the total flowrate equal to the sum of the flowrates in each pipe. However, by writing the energy equation between points *A* and *B* it is found that the head loss experienced by any fluid particle traveling between these locations is the same, independent of the path taken. Thus, the governing equations for parallel pipes are

$$Q = Q_1 + Q_2 + Q_3$$

and

$$h_{L_1} = h_{L_2} = h_{L_3}$$

The flow in a relatively simple-looking multiple-pipe system may be more complex than it appears initially. The branching system termed the *three-reservoir problem* shown in Fig. 8.21 is such a system. Three reservoirs at known elevations are connected together with three pipes of known properties (lengths, diameters, and roughnesses). The problem is to determine the flowrates into or out of the reservoirs. In general, the flow direction (whether the fluid flows into or out of reservoir B) is not obvious, and the solution process must include the determination of this direction.

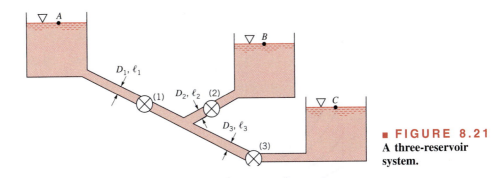

■ **FIGURE 8.21**
A three-reservoir system.

8.6 Pipe Flowrate Measurement

Three of the most common devices used to measure the instantaneous flowrate in pipes are the orifice meter, the nozzle meter, and the Venturi meter as shown in Figs. 8.22, 8.24, and 8.26. As discussed in Section 3.6.3, each of these meters operates on the principle that a decrease in flow area in a pipe causes an increase in velocity that is accompanied by a decrease in pressure. Correlation of the pressure difference with the velocity provides a means of measuring the flowrate.

Based on the results of the previous sections of this chapter, we anticipate that there is a head loss between sections (1) and (2) (see Fig. 8.22) so that the governing equations become

$$Q = A_1 V_1 = A_2 V_2$$

and

$$\frac{p_1}{\gamma} + \frac{V_1^2}{2g} = \frac{p_2}{\gamma} + \frac{V_2^2}{2g} + h_L$$

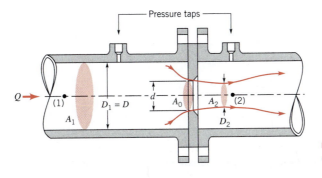

■ **FIGURE 8.22** **Typical orifice meter construction.**

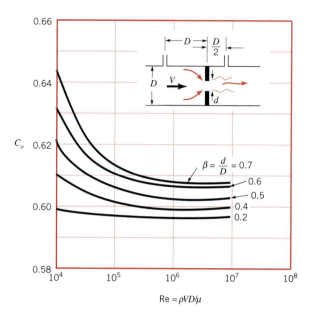

■ **FIGURE 8.23** Orifice meter discharge coefficient (Ref. 10).

The ideal situation has $h_L = 0$ and results in

$$Q_{ideal} = A_2 V_2 = A_2 \sqrt{\frac{2(p_1 - p_2)}{\rho(1 - \beta^4)}} \tag{8.21}$$

where $\beta = D_2/D_1$ (see Section 3.6.3). The difficulty in including the head loss is that there is no accurate expression for it. The net result is that empirical coefficients are used in the flowrate equations to account for the complex "real-world" effect brought on by the nonzero viscosity. The coefficients are discussed in this section.

A typical *orifice meter* is constructed by inserting a flat plate with a hole between two flanges of a pipe, as shown in Fig. 8.22. An *orifice discharge coefficient, C_o,* is used to take nonideal effects into account. That is,

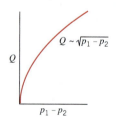

$$Q = C_o Q_{ideal} = C_o A_o \sqrt{\frac{2(p_1 - p_2)}{\rho(1 - \beta^4)}} \tag{8.22}$$

where $A_0 = \pi d^2/4$ is the area of the hole in the orifice plate. The value of C_o is a function of $\beta = d/D$ and the Reynolds number Re $= \rho VD/\mu$, where $V = Q/A_1$. Typical values of C_o are given in Fig. 8.23. As shown by Eq. 8.22 and the figure in the margin, for a given value of C_o, the flowrate is proportional to the square root of the pressure difference.

Another type of pipe flowmeter that is based on the same principles used in the orifice meter is the *nozzle meter,* three variations of which are shown in Fig. 8.24. The

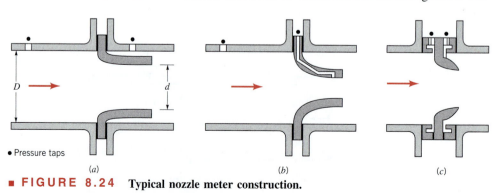

■ **FIGURE 8.24** Typical nozzle meter construction.

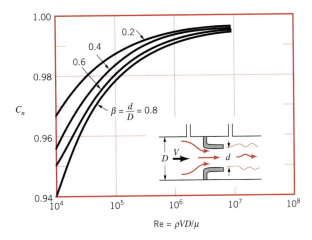

■ **FIGURE 8.25** Nozzle meter discharge coefficient (Ref. 10).

flow pattern for the nozzle meter is closer to ideal than the orifice meter flow, but there still are viscous effects. These are accounted for by use of the *nozzle discharge coefficient, C_n,* where

$$Q = C_n Q_{ideal} = C_n A_n \sqrt{\frac{2(p_1 - p_2)}{\rho(1 - \beta^4)}}$$

(8.23)

with $A_n = \pi d^2/4$. As with the orifice meter, the value of C_n is a function of the diameter ratio, $\beta = d/D$, and the Reynolds number, Re $= \rho VD/\mu$. Typical values obtained from experiments are shown in Fig. 8.25. Note that $C_n > C_o$; the nozzle meter is more efficient (less energy dissipated) than the orifice meter.

The most precise and most expensive of the three obstruction-type flowmeters is the ***Venturi meter*** shown in Fig. 8.26. It is designed to reduce head losses to a minimum. Most of the head loss that occurs in a well-designed Venturi meter is due to friction losses along the walls rather than losses associated with separated flows and the inefficient mixing motion that accompanies such flow.

The flowrate through a Venturi meter is given by

$$Q = C_v Q_{ideal} = C_v A_T \sqrt{\frac{2(p_1 - p_2)}{\rho(1 - \beta^4)}}$$

(8.24)

where $A_T = \pi d^2/4$ is the throat area. The range of values of C_v, the *Venturi discharge coefficient,* is given in Fig. 8.27. The throat-to-pipe diameter ratio ($\beta = d/D$), the Reynolds number, and the shape of the converging and diverging sections of the meter are among the parameters that affect the value of C_v.

Again, the precise values of C_n, C_o, and C_v depend on the specific geometry of the devices used. Considerable information concerning the design, use, and installation of standard flowmeters can be found in various books (Refs. 10, 15, 16, 17).

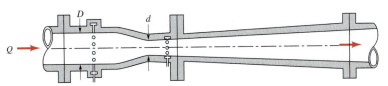

■ **FIGURE 8.26** **Typical Venturi meter construction.**

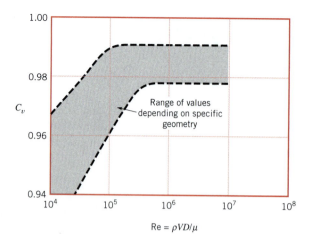

■ **FIGURE 8.27** Venturi meter discharge coefficient (Ref. 15).

$$\text{Re} = \rho V D / \mu$$

EXAMPLE 8.10   Nozzle Flow Meter

GIVEN Ethyl alcohol flows through a pipe of diameter $D = 60$ mm in a refinery. The pressure drop across the nozzle meter used to measure the flowrate is to be $\Delta p = 4.0$ kPa when the flowrate is $Q = 0.003$ m³/s.

FIND Determine the diameter, d, of the nozzle.

SOLUTION

From Table 1.5 the properties of ethyl alcohol are $\rho = 789$ kg/m³ and $\mu = 1.19 \times 10^{-3}$ N·s/m². Thus,

$$\text{Re} = \frac{\rho V D}{\mu} = \frac{4\rho Q}{\pi D \mu}$$

$$= \frac{4(789 \text{ kg/m}^3)(0.003 \text{ m}^3/\text{s})}{\pi(0.06 \text{ m})(1.19 \times 10^{-3} \text{ N·s/m}^2)} = 42{,}200$$

From Eq. 8.23 the flowrate through the nozzle is

$$Q = 0.003 \text{ m}^3/\text{s} = C_n \frac{\pi}{4} d^2 \sqrt{\frac{2(4 \times 10^3 \text{ N/m}^2)}{789 \text{ kg/m}^3(1 - \beta^4)}}$$

or

$$1.20 \times 10^{-3} = \frac{C_n d^2}{\sqrt{1 - \beta^4}} \tag{1}$$

where d is in meters. Note that $\beta = d/D = d/0.06$. Equation 1 and Fig. 8.25 represent two equations for the two unknowns d and C_n that must be solved by trial and error.

As a first approximation we assume that the flow is ideal, or $C_n = 1.0$, so that Eq. 1 becomes

$$d = (1.20 \times 10^{-3} \sqrt{1 - \beta^4})^{1/2} \tag{2}$$

In addition, for many cases $1 - \beta^4 \approx 1$, so that an approximate value of d can be obtained from Eq. 2 as

$$d = (1.20 \times 10^{-3})^{1/2} = 0.0346 \text{ m}$$

Hence, with an initial guess of $d = 0.0346$ m or $\beta = d/D = 0.0346/0.06 = 0.577$, we obtain from Fig. 8.25 (using Re = 42,200) a value of $C_n = 0.972$. Clearly this does not agree with our initial assumption of $C_n = 1.0$. Thus, we do not have the solution to Eq. 1 and Fig. 8.25. Next we assume $\beta = 0.577$ and $C_n = 0.972$ and solve for d from Eq. 1 to obtain

$$d = \left(\frac{1.20 \times 10^{-3}}{0.972} \sqrt{1 - 0.577^4}\right)^{1/2}$$

or $d = 0.0341$ m. With the new value of $\beta = 0.0341/0.060 = 0.568$ and Re = 42,200, we obtain (from Fig. 8.25) $C_n \approx 0.972$ in agreement with the assumed value. Thus,

$$d = 34.1 \text{ mm} \tag{Ans}$$

COMMENTS If numerous cases are to be investigated, it may be much easier to replace the discharge coefficient data of Fig. 8.25 by the equivalent equation, $C_n = \phi(\beta, \text{Re})$, and use a computer to iterate for the answer. Such equations are available in the literature (Ref. 10). This would be similar to using the

Colebrook equation rather than the Moody chart for pipe friction problems.

By repeating the calculations, the nozzle diameters, d, needed for the same flowrate and pressure drop but with different fluids are shown in Fig. E8.10. The diameter is a function of the fluid viscosity because the nozzle coefficient, C_n, is a function of the Reynolds number (see Fig. 8.25). In addition, the diameter is a function of the density because of this Reynolds number effect and, perhaps more importantly, because the density is involved directly in the flowrate equation, Eq. 8.23. These factors all combine to produce the results shown in the figure.

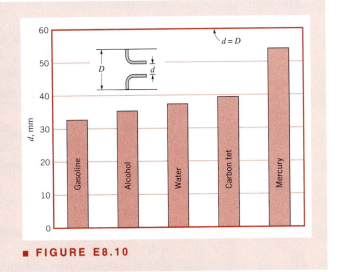

■ **FIGURE E8.10**

V8.6 Rotameter

V8.7 Water meter

Numerous other devices are used to measure the flowrate in pipes. Many of these devices use principles other than the high-speed/low-pressure concept of the orifice, nozzle, and Venturi meters (see Ref. 16). In some instances, it is necessary to know the amount (volume or mass) of fluid that has passed through a pipe during a given time period, rather than the instantaneous flowrate. There are several quantity-measuring devices that provide such information. These include the nutating disk meter used to determine the amount of water used in your house or the amount of gasoline pumped into your car's fuel tank and the bellows meter used to determine the amount of natural gas delivered to the furnace in your house.

8.7 Chapter Summary and Study Guide

This chapter discussed the flow of a viscous fluid in a pipe. General characteristics of laminar, turbulent, fully developed, and entrance flows are considered. Poiseuille's equation is obtained to describe the relationship among the various parameters for fully developed laminar flow.

Various characteristics of turbulent pipe flow are introduced and contrasted to laminar flow. It is shown that the head loss for laminar or turbulent pipe flow can be written in terms of the friction factor (for major losses) and the loss coefficients (for minor losses). In general, the friction factor is obtained from the Moody chart or the Colebrook formula and is a function of the Reynolds number and the relative roughness. The minor loss coefficients are a function of the flow geometry for each system component.

laminar flow
transitional flow
turbulent flow
entrance length
fully developed
* flow*
wall shear stress
Hagen–Poiseuille
* flow*
major loss

Analysis of noncircular conduits is carried out by use of the hydraulic diameter concept. Various examples involving flow in single pipe systems and flow in multiple pipe systems are presented. The inclusion of viscous effects and losses in the analysis of orifice, nozzle, and Venturi flow meters is discussed.

The following checklist provides a study guide for this chapter. When your study of the entire chapter and end-of-chapter exercises has been completed you should be able to

■ write out meanings of the terms listed here in the margin and understand each of the related concepts. These terms are particularly important and are set in color and bold type in the text.

minor loss
relative roughness
friction factor
Moody chart
Colebrook formula
loss coefficient
hydraulic diameter
multiple pipe
  systems
orifice meter
nozzle meter
Venturi meter

- determine which of the following types of flow will occur: entrance flow, or fully developed flow; laminar flow, or turbulent flow.
- use the Poiseuille equation in appropriate situations and understand its limitations.
- explain the main properties of turbulent pipe flow and how they are different from or similar to laminar pipe flow.
- use the Moody chart and the Colebrook equation to determine major losses in pipe systems.
- use minor loss coefficients to determine minor losses in pipe systems.
- determine the head loss in noncircular conduits.
- incorporate major and minor losses into the energy equation to solve a variety of pipe flow problems, including Type I problems (determine the pressure drop or head loss), Type II problems (determine the flow rate), and Type III problems (determine the pipe diameter).
- solve problems involving multiple pipe systems.
- determine the flowrate through orifice, nozzle, and Venturi flowmeters as a function of the pressure drop across the meter.

References

1. Schlichting, H., *Boundary Layer Theory,* Eighth Edition, McGraw-Hill, New York, 2000.
2. Moody, L. F., "Friction Factors for Pipe Flow," *Transactions of the ASME,* Vol. 66, 1944.
3. Colebrook, C. F., "Turbulent Flow in Pipes with Particular Reference to the Transition Between the Smooth and Rough Pipe Laws," *Journal of the Institute of Civil Engineers London,* Vol. 11, 1939.
4. White, F. M., *Viscous Fluid Flow,* Third Edition, McGraw-Hill, New York, 2006.
5. *ASHRAE Handbook of Fundamentals,* ASHRAE, Atlanta, 1981.
6. Streeter, V. L., ed., *Handbook of Fluid Dynamics,* McGraw-Hill, New York, 1961.
7. Balje, O. E., *Turbomachines: A Guide to Design, Selection and Theory,* Wiley, New York, 1981.
8. Wallis, R. A., *Axial Flow Fans and Ducts,* Wiley, New York, 1983.
9. Karassick, I. J., et al., *Pump Handbook,* Second Edition, McGraw-Hill, New York, 1985.
10. "Measurement of Fluid Flow by Means of Orifice Plates, Nozzles, and Venturi Tubes Inserted in Circular Cross Section Conduits Running Full," Int. Organ. Stand. Rep. DIS-5167, Geneva, 1976.
11. Hydraulic Institute, *Engineering Data Book,* First Edition, Cleveland Hydraulic Institute, 1979.
12. Harris, C. W., *University of Washington Engineering Experimental Station Bulletin,* 48, 1928.
13. Hamilton, J. B., *University of Washington Engineering Experimental Station Bulletin,* 51, 1929.
14. Laws, E. M., and Livesey, J. L., "Flow Through Screens," *Annual Review of Fluid Mechanics,* Vol. 10, Annual Reviews, Inc., Palo Alto, CA, 1978.
15. Bean, H. S., ed., *Fluid Meters: Their Theory and Application,* Sixth Edition, American Society of Mechanical Engineers, New York, 1971.

16. Goldstein, R. J., ed., *Fluid Mechanics Measurements,* Second Edition, Taylor and Francis, Philadelphia, 1996.

17. Spitzer, D. W., ed., *Flow Measurement: Practical Guides for Measurement and Control,* Instrument Society of America, Research Triangle Park, North Carolina, 1991.

Review Problems

Go to Appendix F for a set of review problems with answers. Detailed solutions can be found in *Student Solution Manual for a Brief Introduction to Fluid Mechanics,* by Young et al. (© 2006 John Wiley and Sons, Inc.).

Problems

Note: Unless otherwise indicated use the values of fluid properties found in the tables on the inside of the front cover. Problems designated with an (*) are intended to be solved with the aid of a programmable calculator or a computer. Problems designated with a (†) are "open-ended" problems and require critical thinking in that to work them one must make various assumptions and provide the necessary data. There is not a unique answer to these problems.

Answers to the even-numbered problems are listed at the end of the book. Access to the videos that accompany problems can be obtained through the book's web site, www.wiley.com/college/young. The lab-type problems, FE problems, and FlowLab problems can also be accessed on this web site.

Section 8.1 General Characteristics of Pipe Flow (also see Lab Problem 8.77)

8.1 Rainwater runoff from a parking lot flows through a 3-ft-diameter pipe, completely filling it. Whether flow in a pipe is laminar or turbulent depends on the value of the Reynolds number. (See **Video V8.1.**) Would you expect the flow to be laminar or turbulent? Support your answer with appropriate calculations.

8.2 (See Fluids in the News article titled "Nanoscale flows," Section 8.1.1.) (a) Water flows in a tube that has a diameter of $D = 0.1$ m. Determine the Reynolds number if the average velocity is 10 diameters per second. (b) Repeat the calculations if the tube is a nanoscale tube with a diameter of $D = 100$ nm.

8.3 Air at 200 °F flows at standard atmospheric pressure in a pipe at a rate of 0.08 lb/s. Determine the minimum diameter allowed if the flow is to be laminar.

8.4 Blue and yellow streams of paint at 60 °F (each with a density of 1.6 slugs/ft^3 and a viscosity 1000 times greater than water) enter a pipe with an average velocity of 4 ft/s as shown in

Fig. P8.4. Would you expect the paint to exit the pipe as green paint or separate streams of blue and yellow paint? Explain. Repeat the problem if the paint were "thinned" so that it is only 10 times more viscous than water. Assume the density remains the same.

8.5 A soft drink with the properties of 10 °C water is sucked through a 4-mm-diameter, 0.25-m-long straw at a rate of 4 cm^3/s. Is the flow at the outlet of the straw laminar? Is it fully developed? Explain.

8.6 To cool a given room it is necessary to supply 4 ft^3/s of air through an 8-in.-diameter pipe. Approximately how long is the entrance length in this pipe?

8.7 The pressure distribution measured along a straight, horizontal portion of a 50-mm-diameter pipe attached to a tank is shown below. Approximately how long is the entrance length? In the fully developed portion of the flow, what is the value of the wall shear stress?

x (m) (± 0.01 m)	p (mm H$_2$O) (± 5 mm)
0 (tank exit)	520
0.5	427
1.0	351
1.5	288
2.0	236
2.5	188
3.0	145
3.5	109
4.0	73
4.5	36
5.0 (pipe exit)	0

Section 8.2 Fully Developed Laminar Flow

8.8 The pressure drop needed to force water through a horizontal 1-in.-diameter pipe is 0.60 psi for every 12-ft length of pipe. Determine the shear stress on the pipe wall. Determine the shear stress at distances 0.3 and 0.5 in. away from the pipe wall.

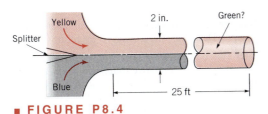

■ **FIGURE P8.4**

8.9 Water flows in a constant diameter pipe with the following conditions measured: At section (a) p_a = 32.4 psi and z_a = 56.8 ft; at section (b) p_b = 29.7 psi and z_b = 68.2 ft. Is the flow from (a) to (b) or from (b) to (a)? Explain.

8.10 A fluid of specific gravity 0.96 flows steadily in a long, vertical 1-in.-diameter pipe with an average velocity of 0.50 ft/s. If the pressure is constant throughout the fluid, what is the viscosity of the fluid? Determine the shear stress on the pipe wall.

8.11 A fluid flows through a horizontal 0.1-in.-diameter pipe. When the Reynolds number is 1500, the head loss over a 20-ft length of the pipe is 6.4 ft. Determine the fluid velocity.

8.12 Glycerin at 20° C flows upward in a vertical 75-mm-diameter pipe with a centerline velocity of 1.0 m/s. Determine the head loss and pressure drop in a 10-m length of the pipe.

8.13 Laminar flow in a horizontal pipe of diameter D gives a flowrate of Q if the pressure gradient is $\partial p/\partial x = -K$. The fluid is cooled so that the density increases by a factor of 1.04 and the viscosity increases by a factor of 3.8. Determine the new pressure gradient required (in terms of K) if the flowrate remains the same.

8.14 A large artery in a person's body can be approximated by a tube of diameter 9 mm and length 0.35 m. Also assume that blood has a viscosity of approximately 4×10^{-3} N·s/m², a specific gravity of 1.0, and that the pressure at the beggining of the artery is equivalent to 120 mm Hg. If the flow were steady (it is not) with V = 0.2 m/s, determine the pressure at the end of the artery if it is oriented (a) vertically up (flow up) or (b) horizontal.

8.15 At time t = 0 the level of water in tank A shown in Fig P8.15 is 2 ft above that in tank B. Plot the elevation of the water in tank A as a function of time until the free surfaces in both tanks are at the same elevation. Assume quasisteady conditions—that is, the steady pipe flow equations are assumed valid at any time, even though the flowrate does change (slowly) in time. Neglect minor losses. *Note:* Verify and use the fact that the flow is laminar

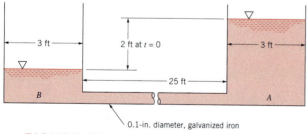

■ **FIGURE P8.15**

8.16 Water flows downhill through a 3-in.-diameter steel pipe. The slope of the hill is such that for each mile (5280 ft) of horizontal distance, the change in elevation is Δz ft. Determine the maximum value of Δz if the flow is to remain laminar and the pressure all along the pipe is constant.

Section 8.3 Fully Developed Turbulent Flow

8.17 As shown in **Video V8.3** and Fig. P8.17, the velocity profile for laminar flow in a pipe is quite different from that for turbulent flow. With laminar flow the velocity profile is parabolic; with turbulent flow at Re = 10,000 the velocity profile can be approximated by the power-law profile shown in Fig. P8.17. **(a)** For laminar flow, determine at what radial location you would place a Pitot tube if it is to measure the average velocity in the pipe. **(b)** Repeat part **(a)** for turbulent flow with Re = 10,000.

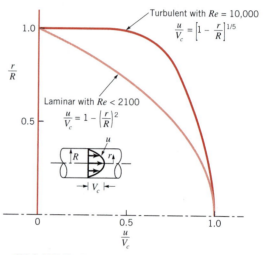

■ **FIGURE P8.17**

8.18 When soup is stirred in a bowl, there is considerable turbulence in the resulting motion (see **Video V8.2**). From a very simplistic standpoint, this turbulence consists of numerous intertwined swirls, each involving a characteristic diameter and velocity. As time goes by, the smaller swirls (the fine scale structure) die out relatively quickly, leaving the large swirls that continue for quite some time. Explain why this is to be expected.

Section 8.4.1 Major Losses (also see Lab Problem 8.73)

8.19 During a heavy rainstorm, water from a parking lot completely fills an 18-in.-diameter, smooth, concrete storm sewer. If the flowrate is 10 ft³/s, determine the pressure drop in a 100-ft horizontal section of the pipe. Repeat the problem if there is a 2-ft change in elevation of the pipe per 100 ft of its length.

8.20 Carbon dioxide at a temperature of 0° C and a pressure of 600 kPa (abs) flows through a horizontal 40-mm-diameter pipe with an average velocity of 2 m/s. Determine the friction factor if the pressure drop is 235 N/m² per 10-m length of pipe.

8.21 Water flows in a cast-iron pipe of 200-mm diameter at a rate of 0.10 m³/s. Determine the friction factor for this flow.

8.22 Water flows at a rate of 2.0 ft³/s in an old, rusty 6-in.-diameter pipe that has a relative roughness of 0.010. It is proposed that by inserting a smooth plastic liner with an inside diameter of 5 in. into the old pipe as shown in Fig. P8.22 the pressure drop per mile can be reduced. Is it true that the lined pipe can carry the required 2.0 ft³/s at a lower pressure drop than in the old pipe? Support your answer with appropriate calculations.

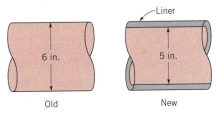

■ **FIGURE P8.22**

8.23 Repeat Problem 8.15 if the pipe diameter is changed to 0.1 ft rather than 0.1 in. *Note:* The flow may not be laminar for this case.

8.24 A 3-ft-diameter duct is used to carry ventilating air into a vehicular tunnel at a rate of 9000 ft³/min. Tests show that the pressure drop is 1.5 in. of water per 1500 ft of duct. What is the value of the friction factor for this duct and the approximate size of the equivalent roughness of the surface of the duct?

Section 8.4.2 Minor Losses (also see Lab Problem 8.78)

8.25 Air at standard temperature and pressure flows through a 1-in.-diameter galvanized iron pipe with an average velocity of 10 ft/s. What length of pipe produces a head loss equivalent to **(a)** a flanged 90° elbow, **(b)** a wide-open angle valve, or **(c)** a sharp-edged entrance?

8.26 To conserve water and energy, a "flow reducer" is installed in the shower head as shown in Fig. P8.26. If the pressure at point (1) remains constant and all losses except for that in the "flow reducer" are neglected, determine the value of the loss coefficient (based on the velocity in the pipe) of the "flow reducer" if its presence is to reduce the flowrate by a factor of 2. Neglect gravity.

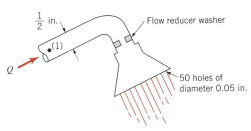

■ **FIGURE P8.26**

8.27 (See Fluids in the News article titled "New hi-tech fountains," Section 8.5.) The fountain shown in Fig. P8.27 is designed to provide a stream of water that rises $h = 10$ ft to $h = 20$ ft above the nozzle exit in a periodic fashion. To do this

the water from the pool enters a pump, passes through a pressure regulator that maintains a constant pressure ahead of the flow control valve. The valve is electronically adjusted to provide the desired water height. With $h = 10$ ft the loss coefficient for the valve is $K_L = 50$. Determine the valve loss coefficient needed for $h = 20$ ft. All losses except for the flow control valve are negligible. The area of the pipe is 5 times the area of the exit nozzle.

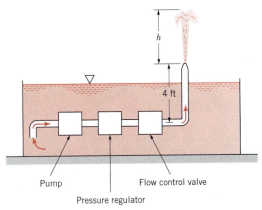

■ **FIGURE P8.27**

8.28 As shown in Fig. P8.28, water flows from one tank to another through a short pipe whose length is n times the pipe diameter. Head losses occur in the pipe and at the entrance and exit. (See **Video V8.4.**) Determine the maximum value of n if the major loss is to be no more than 10% of the minor loss and the friction factor is 0.02.

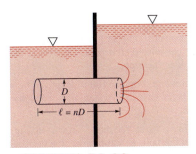

■ **FIGURE P8.28**

8.29 Air at 80 °F and standard atmospheric pressure flows through a furnace filter with an average velocity of 2.4 ft/s. If the pressure drop across the filter is 0.11 in. of water, what is the loss coefficient for the filter?

8.30 Water flows steadily through the 0.75-in.-diameter galvanized iron pipe system shown in **Video V8.6** and Fig. P8.30 at a rate of 0.020 cfs. Your boss suggests that friction losses in the straight pipe sections are negligible compared to losses in the threaded elbows and fittings of the system. Do you agree or disagree with your boss? Support your answer with appropriate calculations.

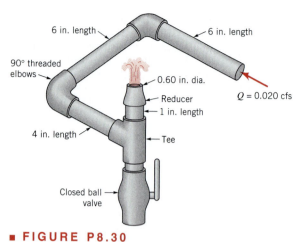

■ **FIGURE P8.30**

Section 8.4.3 Noncircular Conduits

8.31 Air at standard temperature and pressure flows at a rate of 7.0 cfs through a horizontal, galvanized iron duct that has a rectangular cross-sectional shape of 12 in. by 6 in. Estimate the pressure drop per 200 ft of duct.

8.32 Air at standard conditions flows through a horizontal 1 ft by 1.5 ft rectangular wooden duct at a rate of 5000 ft^3/min. Determine the head loss, pressure drop, and power supplied by the fan to overcome the flow resistance in 500 ft of the duct.

Section 8.5.1 Single Pipes—Determine Pressure Drop

8.33 A 70-ft-long, 0.5-in.-diameter hose with a roughness of $\varepsilon = 0.0009$ ft is fastened to a water faucet where the pressure is p_1. Determine p_1 if there is no nozzle attached and the average velocity in the hose is 6 ft/s. Neglect minor losses and elevation changes.

8.34 Water flows at a rate of 10 gal per minute in a new horizontal 0.75-in.-diameter galvanized iron pipe. Determine the pressure gradient, $\Delta p/\ell$, along the pipe.

8.35 Determine the pressure drop per 100-m length of horizontal new 0.20-m-diameter cast iron water pipe when the average velocity is 1.7 m/s.

8.36 When water flows from the tank shown in Fig. P8.36, the water depth in the tank as a function of time is as indicated. Determine the cross-sectional area of the tank. The total length of the 0.60-in.-diameter pipe is 20 ft, and the friction factor is 0.03. The loss coefficients are 0.50 for the entrance, 1.5 for each elbow, and 10 for the valve.

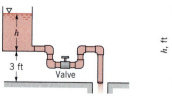

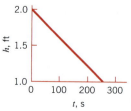

■ **FIGURE P8.36**

8.37 Natural gas ($\rho = 0.0044$ slugs/ft^3 and $\nu = 5.2 \times 10^{-5}$ ft^2/s) is pumped through a horizontal 6-in.-diameter cast-iron pipe at a rate of 800 lb/hr. If the pressure at section (1) is 50 psi (abs), determine the pressure at section (2) 8 mi downstream if the flow is assumed incompressible. Is the incompressible assumption reasonable? Explain.

8.38 Gasoline flows in a smooth pipe of 40 mm diameter at a rate of 0.001 m^3/s. If it were possible to prevent turbulence from occurring, what would be the ratio of the head loss for the actual turbulent flow compared to that if it were laminar flow?

8.39 Water flows from the container shown in Fig. P8.39. Determine the loss coefficient needed in the valve if the water is to "bubble up" 3 in. above the outlet pipe.

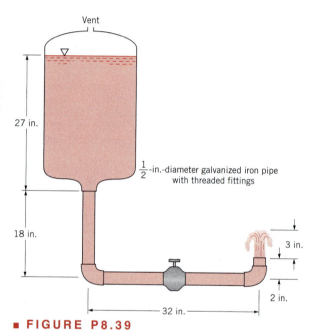

■ **FIGURE P8.39**

8.40 Water flows from a lake as is shown in Fig. P8.40 at a rate of 4.0 cfs. Is the device inside the building a pump or a turbine? Explain and determine the horsepower of the device. Neglect all minor losses and assume the friction factor is 0.025.

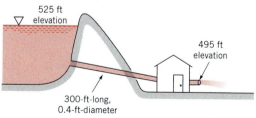

■ **FIGURE P8.40**

8.41 At a ski resort water at 40° F is pumped through a 3-in.-diameter, 2000-ft-long steel pipe from a pond at an elevation of 4286 ft to a snow-making machine at an elevation of 4623 ft at a rate of 0.26 ft^3/s. If it is necessary to maintain a pressure of 180 psi at the snow-making machine, determine the horsepower added to the water by the pump. Neglect minor losses.

8.42 The exhaust from your car's engine flows through a complex pipe system as shown in Fig. P8.42 and **Video V8.5.** Assume that the pressure drop through this system is Δp_1 when the engine is idling at 1000 rpm at a stop sign. Estimate the pressure drop (in terms of Δp_1) with the engine at 3000 rpm when you are driving on the highway. List all the assumptions that you made to arrive at your answer.

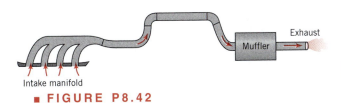

Exhaust

Muffler

Intake manifold

■ **FIGURE P8.42**

8.43 Water at 40° F flows through the coils of the heat exchanger as shown in Fig. P8.43 at a rate of 0.9 gal/min. Determine the pressure drop between the inlet and outlet of the horizontal device.

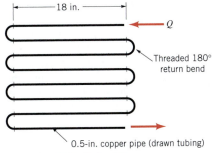

18 in.

Q

Threaded 180°
return bend

0.5-in. copper pipe (drawn tubing)

■ **FIGURE P8.43**

8.44 The $\frac{1}{2}$-in.-diameter hose shown in Fig. P8.44 can withstand a maximum pressure of 200 psi without rupturing. Determine the maximum length, ℓ, allowed if the friction factor is 0.022 and the flowrate is 0.010 cfs. Neglect minor losses. The fluid is water.

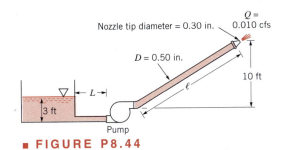

Nozzle tip diameter = 0.30 in.

$Q = 0.010$ cfs

$D = 0.50$ in.

10 ft

L

ℓ

3 ft

Pump

■ **FIGURE P8.44**

8.45 The hose shown in Fig. P8.44 will collapse if the pressure within it is lower than 10 psi below atmospheric pressure.

Determine the maximum length, L, allowed if the friction factor is 0.015 and the flowrate is 0.010 cfs. Neglect minor losses.

8.46 As shown in **Video V8.6** and Fig. P8.46, water "bubbles up" 3 in. above the exit of the vertical pipe attached to three horizontal pipe segments. The total length of the 0.75-in.-diameter galvanized iron pipe between point (1) and the exit is 21 inches. Determine the pressure needed at point (1) to produce this flow.

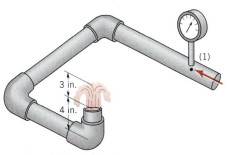

(1)

3 in.

4 in.

■ **FIGURE P8.46**

8.47 Estimate the pressure drop associated with the air flow from the furnace to the hot air register in your room (see Figure P8.47). List all assumptions and show all calculations.

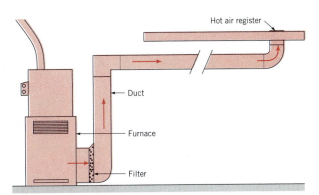

Hot air register

Duct

Furnace

Filter

■ **FIGURE P8.47**

8.48 As shown in Fig. P8.48, a standard household water meter is incorporated into a lawn irrigation system to measure the volume of water applied to the lawn. Note that these meters measure volume, not volume flowrate. (See **Video V8.7.**) With an upstream pressure of $p_1 = 50$ psi the meter registered that 120 ft³ of water was delivered to the lawn during an "on" cycle. Estimate the upstream pressure, p_1, needed if it is desired to have 150 ft³ delivered during an "on" cycle. List any assumptions needed to arrive at your answer.

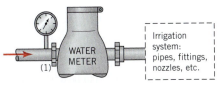

WATER METER

(1)

Irrigation system:
pipes, fittings,
nozzles, etc.

■ **FIGURE P8.48**

8.49 A fan is to produce a constant air speed of 40 m/s throughout the pipe loop shown in Fig. P8.49. The 3-m-diameter pipes are smooth, and each of the four 90° elbows has a loss coefficient of 0.30. Determine the power that the fan adds to the air.

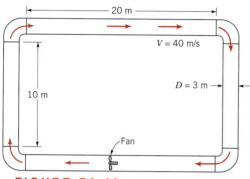

■ **FIGURE P8.49**

Section 8.5.1 Single Pipes—Determine Flowrate (also see Lab Problems 8.75 and 8.76)

8.50 Water at 40° F is pumped from a lake as shown in Fig. P8.50. What is the maximum flowrate possible without cavitation occurring in the pipe?

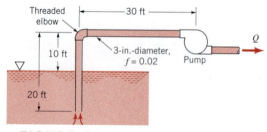

■ **FIGURE P8.50**

8.51 The pump shown in Fig. P8.51 delivers a head of 250 ft to the water. Determine the power that the pump adds to the water. The difference in elevation of the two ponds is 200 ft.

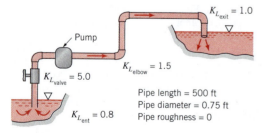

■ **FIGURE P8.51**

8.52 Water flows through two sections of the vertical pipe shown in Fig. P8.52. The bellows connection cannot support any force in the vertical direction. The 0.4-ft-diameter pipe weighs 0.2 lb/ft, and the friction factor is assumed to be 0.02. At what velocity will the force, F, required to hold the pipe be zero?

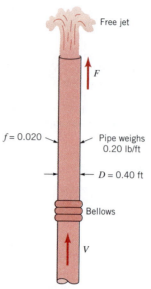

■ **FIGURE P8.52**

*8.53 Water flows from a large open tank through a sharp-edged entrance and into a horizontal 0.5-in.-diameter copper pipe (drawn tubing) of length ℓ. The water exits the pipe as a free jet a distance of 4.0 ft below the free surface of the tank. Plot a log–log graph of the flow rate through the pipe, Q, as a function of ℓ for $10 \le \ell \le 10,000$ ft. Comment on your results.

8.54 The pump shown in Fig. P8.54 adds 25 kW to the water and causes a flowrate of 0.04 m³/s. Determine the flowrate expected if the pump is removed from the system. Assume $f = 0.016$ for either case and neglect minor losses.

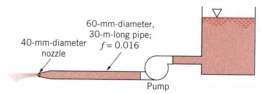

■ **FIGURE P8.54**

8.55 Air, assumed incompressible, flows through the two pipes shown in Fig. P8.55. Determine the flowrate if minor losses are neglected and the friction factor in each pipe is 0.020. Determine the flowrate if the 0.5-in.-diameter pipe were replaced by a 1-in.-diameter pipe. Comment on the assumption of incompressibility.

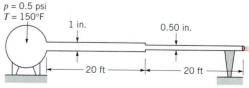

■ **FIGURE P8.55**

Section 8.5.1 Single Pipes—Determine Diameter

8.56   According to fire regulations in a town, the pressure drop in a commercial steel, horizontal pipe must not exceed 1.0 psi per 150-ft pipe for flowrates up to 500 gal/min. If the water temperature is never below 50° F, what diameter pipe is needed?

8.57   Determine the diameter of a steel pipe that is to carry 2000 gal/min of gasoline with a pressure drop of 5 psi per 100 ft of horizontal pipe.

8.58   Water is to be moved from a large, closed tank in which the air pressure is 20 psi into a large, open tank through 2000 ft of smooth pipe at the rate of 3 ft³/s. The fluid level in the open tank is 150 ft below that in the closed tank. Determine the required diameter of the pipe. Neglect minor losses.

8.59   Repeat Problem 8.58 if there are numerous components (valves, elbows, etc.) along the pipe so that the minor loss is equal to 40 velocity heads.

Section 8.5.2 Multiple Pipe Systems

8.60   The flowrate between tank A and tank B shown in Fig. P8.60 is to be increased by 30% (i.e., from Q to $1.30Q$) by the addition of a second pipe (indicated by the dotted lines) running from node C to tank B. If the elevation of the free surface in tank A is 25 ft above that in tank B, determine the diameter, D, of this new pipe. Neglect minor losses and assume that the friction factor for each pipe is 0.02.

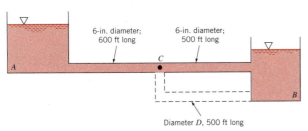

■ FIGURE P8.60

8.61   (See Fluids in the News article titled "Deepwater pipeline," Section 8.5.2.) Five oil fields, each producing an output of Q barrels per day, are connected to the 28-in.-diameter "main line pipe" ($A-B-C$) by 16-in.-diameter "lateral pipes" as shown in Fig. P8.61. The friction factor is the same for each of the pipes and elevation effects are negligible. (**a**) For section $A-B$ determine the ratio of the pressure drop per mile in the main line pipe to that in the lateral pipes. (**b**) Repeat the calculations for section $B-C$.

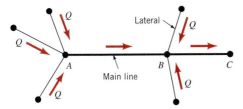

■ FIGURE P8.61

8.62   With the valve closed, water flows from tank A to tank B as shown in Fig. P8.62. What is the flowrate into tank B when the valve is opened to allow water to flow into tank C also? Neglect all minor losses and assume that the friction factor is 0.02 for all pipes.

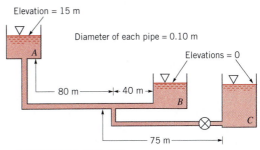

■ FIGURE P8.62

8.63   The three water-filled tanks shown in Fig. P8.63 are connected by pipes as indicated. If minor losses are neglected, determine the flowrate in each pipe.

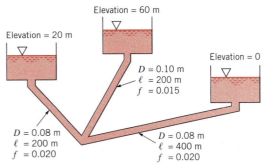

■ FIGURE P8.63

†8.64   As shown in Fig. P8.64, cold water ($T = 50°$ F) flows from the water meter to either the shower or the hot water heater. In the hot water heater it is heated to a temperature of

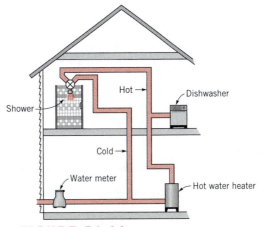

■ FIGURE P8.64

150° F. Thus, with equal amounts of hot and cold water, the shower is at a comfortable 100° F. However, when the dishwasher is turned on, the shower water becomes too cold. Indicate how you would predict this new shower temperature (assume the shower faucet is not adjusted). State any assumptions needed in your analysis.

Section 8.6 Pipe Flowrate Measurement (also see Lab Problem 8.74)

8.65   Gasoline flows through a 35-mm-diameter pipe at a rate of 0.0032 m³/s. Determine the pressure drop across a flow nozzle placed in the line if the nozzle diameter is 20 mm.

8.66   Air to ventilate an underground mine flows through a large 2-m-diameter pipe. A crude flowrate meter is constructed by placing a sheet metal "washer" between two sections of the pipe. Estimate the flowrate if the hole in the sheet metal has a diameter of 1.6 m and the pressure difference across the sheet metal is 8.0 mm of water.

8.67   A 2.5-in.-diameter nozzle meter is installed in a 3.8-in.-diameter pipe that carries water at 160° F. If the inverted air-water U-tube manometer used to measure the pressure difference across the meter indicates a reading of 3.1 ft, determine the flowrate.

8.68   Water flows through the Venturi meter shown in Fig. P8.68. The specific gravity of the manometer fluid is 1.52. Determine the flowrate.

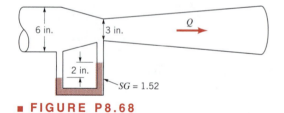

■ **FIGURE P8.68**

8.69   A 2-in.-diameter orifice plate is inserted in a 3-in.-diameter pipe. If the water flowrate through the pipe is 0.70 cfs, determine the pressure difference indicated by a manometer attached to the flow meter.

8.70   Water flows through the orifice meter shown in Fig. P8.70 at a rate of 0.10 cfs. If $d = 0.1$ ft, determine the value of h.

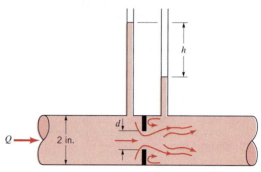

■ **FIGURE P8.70**

8.71   Water flows through the orifice meter shown in Fig. P8.70 at a rate of 0.10 cfs. If $h = 3.8$ ft, determine the value of d.

8.72   Water flows through the orifice meter shown in Fig. P8.70 such that $h = 1.6$ ft with $d = 1.5$ in. Determine the flowrate.

Lab Problems

8.73   This problem involves the determination of the friction factor in a pipe for laminar and transitional flow conditions. To proceed with this problem, go to the book's web site, www.wiley.com/college/young.

8.74   This problem involves the calibration of an orifice meter and a Venturi meter. To proceed with this problem, go to the book's web site, www.wiley.com/college/young.

8.75   This problem involves the flow of water from a tank and through a pipe system. To proceed with this problem, go to the book's web site, www.wiley.com/college/young.

8.76   This problem involves the flow of water pumped from a tank and through a pipe system. To proceed with this problem, go to the book's web site, www.wiley.com/college/young.

8.77   This problem involves the pressure distribution in the entrance region of a pipe. To proceed with this problem, go to the book's web site, www.wiley.com/college/young.

8.78   This problem involves the power loss due to friction in a coiled pipe. To proceed with this problem, go to the book's web site, www.wiley.com/college/young.

FlowLab Problems

*8.79   This FlowLab problem involves simulating the flow in the entrance region of a pipe and looking at basic concepts involved with the flow regime. To proceed with this problem, go to the book's web site, www.wiley.com/college/young.

*8.80   This FlowLab problem involves investigation of the centerline pressure distribution along a pipe. To proceed with this problem, go to the book's web site, www.wiley.com/college/young.

*8.81   This FlowLab problem involves conducting a parametric study to see how Reynolds number affects the entrance length of a pipe. To proceed with this problem, go to the book's web site, www.wiley.com/college/young.

*8.82   This FlowLab problem involves investigation of pressure drop in the entrance region of a pipe as a function of Reynolds number as well as comparing simulation results to analytic values. To proceed with this problem, go to the book's web site, www.wiley.com/college/young.

*8.83   This FlowLab problem involves the simulation of fully developed pipe flow and how the Reynolds number affects the wall friction. To proceed with this problem, go to the book's web site, www.wiley.com/college/young.

*8.84   This FlowLab problem involves conducting a parametric study on the effects of a sudden pipe expansion on the overall pressure drop in a pipe. To proceed with this problem, go to the book's web site, www.wiley.com/college/young.

*8.85   This FlowLab problem involves investigation of effects of the pipe expansion ratio on flow separation. To proceed with this problem, go to the book's web site, www.wiley.com/college/young.

FE Exam Problems

Sample FE (Fundamentals of Engineering) exam questions for fluid mechanics are provided on the book's web site, www.wiley.com/college/young.

CHAPTER 9

Flow over Immersed Bodies

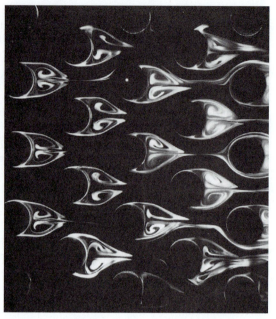

Impulsive start of flow past an array of cylinders: The complex structure of laminar flow past a relatively simple geometric structure illustrates why it is often difficult to obtain exact analytical results for external flows. (Dye in water.) (Photograph courtesy of ONERA, France.)

LEARNING OBJECTIVES

After completing this chapter, you should be able to:

- identify and discuss the features of external flow.
- explain the fundamental characteristics of a boundary layer, including laminar, transitional, and turbulent regimes.
- calculate boundary layer parameters for flow past a flat plate.
- provide a description of boundary layer separation.
- calculate the lift and drag for various objects.

In this chapter we consider various aspects of the flow over bodies that are immersed in a fluid. Examples include the flow of air around airplanes, automobiles, and falling snowflakes, or the flow of water around submarines and fish. In these situations the object is completely surrounded by the fluid, and the flows are termed *external flows*. Theoretical (i.e., analytical and numerical) techniques can provide much of the needed information about such flows. However, because of the complexities of the governing equations and the complexities of the geometry of the objects involved, the amount of information obtained from purely theoretical methods is limited. Thus, much of the information about external flows comes from experiments carried out, for the most part, on scale models of the actual objects.

9.1   General External Flow Characteristics

A body immersed in a moving fluid experiences a resultant force due to the interaction between the body and the fluid surrounding it. We can fix the coordinate system in the body and treat the situation as fluid flowing past a stationary body with velocity U, the *upstream velocity*.

Three general categories of bodies are shown in Fig. 9.1. They include (a) two-dimensional objects (infinitely long and of constant cross-sectional size and shape), (b) axisymmetric bodies (formed by rotating their cross-sectional shape about the axis of symmetry), and (c) three-dimensional bodies that may possess a line or plane of symmetry.

Another classification of body shape can be made depending on whether the body is streamlined or blunt. The flow characteristics depend strongly on the amount of streamlining present. In general, *streamlined bodies* (i.e., airfoils, racing cars) have little effect on the

V9.1 Space shuttle landing

F l u i d s   i n   t h e   N e w s

Armstrong's aerodynamic bike and suit Lance Armstrong rode to his seventh straight Tour de France victory in 2005 using specially designed lightweight, *streamlined* bikes and suits. The bicycle manufacturer, Trek, used computational fluid dynamics analysis and wind tunnel testing to find ways to make the frame more streamlined to reduce drag. The two specially designed bikes were the lightest and most stream-lined the company has made in its 27-year history. When racing in a pack, the more efficient aerodynamics are not very important. However, in a 200-km stage race where the riders are "on their own," not drafting, the new frame theoretically saved Armstrong 10 watts. In addition, Nike designed special skinsuits that have "zoned" fabrics to make the flow past the arms, thighs, and torso more streamlined. They have directional seams that follow airflow lines (no seams crossing the flow) and use materials selected to avoid wrinkles when the rider is in the racing position. Although it is hard to quantify, the results of the new aerodynamic bikes and suits could have made the difference between winning or losing the Tour for Armstrong. (See Problem 9.30.) ■

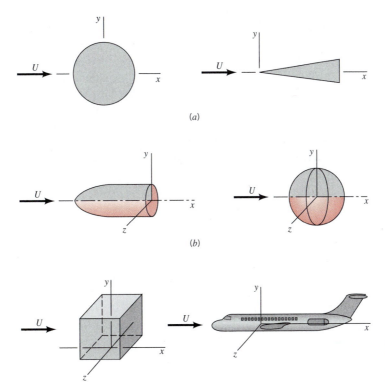

■ **FIGURE 9.1**
Flow classification:
(*a*) two-dimensional,
(*b*) axisymmetric, and
(*c*) three-dimensional.

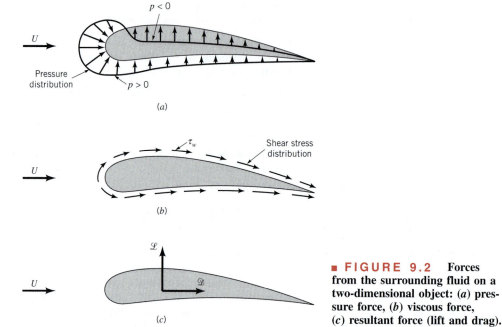

(a)

(b)

(c)

■ **FIGURE 9.2** Forces from the surrounding fluid on a two-dimensional object: (a) pressure force, (b) viscous force, (c) resultant force (lift and drag).

surrounding fluid compared with the effect that *blunt bodies* (i.e., parachutes, buildings) have on the fluid.

9.1.1   Lift and Drag Concepts

When any body moves through a fluid, an interaction between the body and the fluid occurs; this effect can be described in terms of the forces at the fluid–body interface. This can be described in terms of the stresses—wall shear stresses, τ_w, due to viscous effects and normal stresses due to the pressure, p. Typical shear stress and pressure distributions are shown in Figs. 9.2a and 9.2b. Both τ_w and p vary in magnitude and direction along the surface.

The resultant force in the direction of the upstream velocity is termed the *drag,* $\mathcal{D}$, and the resultant force normal to the upstream velocity is termed the *lift,* $\mathcal{L}$, as is indicated in Fig. 9.2c. The resultant of the shear stress and pressure distributions can be obtained by integrating the effect of these two quantities on the body surface as is indicated in Fig. 9.3. The net x and y components of the force on the object are

$$\mathcal{D} = \int dF_x = \int p \cos \theta \, dA + \int \tau_w \sin \theta \, dA \tag{9.1}$$

and

$$\mathcal{L} = \int dF_y = -\int p \sin \theta \, dA + \int \tau_w \cos \theta \, dA \tag{9.2}$$

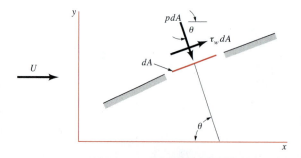

■ **FIGURE 9.3** Pressure and shear forces on a small element of the surface of a body.

Of course, to carry out the integrations and determine the lift and drag, we must know the body shape (i.e., θ as a function of location along the body) and the distribution of τ_w and p along the surface. These distributions are often extremely difficult to obtain, either experimentally or theoretically.

F l u i d s   i n   t h e   N e w s

Pressure-sensitive paint For many years, the conventional method for measuring *surface pressure* has been to use static pressure taps consisting of small holes on the surface connected by hoses from the holes to a pressure-measuring device. Pressure-sensitive paint (PSP) is now gaining acceptance as an alternative to the static surface pressure ports. The PSP material is typically a luminescent compound that is sensitive to the pressure on it and can be excited by an appropriate light which is captured by special video imaging equipment. Thus, it provides a quantitative measure of the surface

pressure. One of the biggest advantages of PSP is that it is a global measurement technique, measuring pressure over the entire surface, as opposed to discrete points. PSP also has the advantage of being nonintrusive to the flow field. Although static pressure port holes are small, they do alter the surface and can slightly alter the flow, thus affecting downstream ports. In addition, the use of PSP eliminates the need for a large number of pressure taps and connecting tubes. This allows pressure measurements to be made in less time and at a lower cost. ■

EXAMPLE 9.1 — Drag from Pressure and Shear Stress Distributions

GIVEN Air at standard conditions flows past a flat plate as is indicated in Fig. E9.1. In case **(a)** the plate is parallel to the upstream flow, and in case **(b)** it is perpendicular to the upstream flow. The pressure and shear stress distributions on the surface are as indicated (obtained either by experiment or theory).

FIND Determine the lift and drag on the plate.

SOLUTION

For either orientation of the plate, the lift and drag are obtained from Eqs. 9.1 and 9.2. With the plate parallel to the upstream flow we have $\theta = 90°$ on the top surface and $\theta = 270°$ on the bottom surface so that the lift and drag are given by

$$\mathcal{L} = -\int_{\text{top}} p \, dA + \int_{\text{bottom}} p \, dA = 0$$

and

$$\mathcal{D} = \int_{\text{top}} \tau_w \, dA + \int_{\text{bottom}} \tau_w \, dA = 2\int_{\text{top}} \tau_w \, dA \qquad (1)$$

where we have used the fact that because of symmetry the shear stress distribution is the same on the top and the bottom surfaces, as is the pressure also [whether we use gage ($p = 0$) or absolute ($p = p_{\text{atm}}$) pressure]. There is no lift generated— the plate does not know up from down. With the given shear stress distribution, Eq. 1 gives

$$\mathcal{D} = 2 \int_{x=0}^{4\text{ ft}} \left(\frac{1.24 \times 10^{-3}}{x^{1/2}} \text{ lb/ft}^2\right)(10 \text{ ft}) \, dx$$

or

$$\mathcal{D} = 0.0992 \text{ lb} \qquad \text{(Ans)}$$

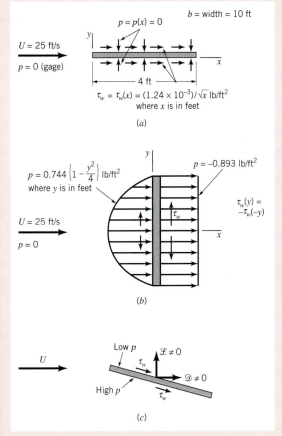

■ **FIGURE E9.1**

With the plate perpendicular to the upstream flow, we have $\theta = 0°$ on the front and $\theta = 180°$ on the back. Thus, from Eqs. 9.1 and 9.2

$$\mathcal{L} = \int_{\text{front}} \tau_w \, dA - \int_{\text{back}} \tau_w \, dA = 0$$

and

$$\mathcal{D} = \int_{\text{front}} p \, dA - \int_{\text{back}} p \, dA$$

Again there is no lift because the pressure forces act parallel to the upstream flow (in the direction of $\mathcal{D}$ not $\mathcal{L}$) and the shear stress is symmetrical about the center of the plate. With the given relatively large pressure on the front of the plate (the center of the plate is a stagnation point) and the negative pressure (less than the upstream pressure) on the back of the plate, we obtain the following drag

$$\mathcal{D} = \int_{y=-2}^{2\,\text{ft}} \left[0.744\left(1 - \frac{y^2}{4}\right) \text{lb/ft}^2 \right.$$
$$\left. - (-0.893)\text{lb/ft}^2 \right](10\,\text{ft})\, dy$$

or

$$\mathcal{D} = 55.6 \text{ lb} \qquad \text{(Ans)}$$

COMMENT Clearly two mechanisms are responsible for the drag. On the ultimately streamlined body (a zero thickness flat plate parallel to the flow) the drag is entirely due to the shear stress at the surface and, in this example, is relatively small. For the ultimately blunted body (a flat plate normal to the upstream flow) the drag is entirely due to the pressure difference between the front and the back portions of the object and, in this example, is relatively large.

If the flat plate were oriented at an arbitrary angle relative to the upstream flow as indicated in Fig. E9.1c, there would be both a lift and a drag, each of which would be dependent on both the shear stress and the pressure. Both the pressure and the shear stress distributions would be different for the top and bottom surfaces.

Without detailed information concerning the shear stress and pressure distributions on a body, Eqs. 9.1 and 9.2 cannot be used. The widely used alternative is to define dimensionless lift and drag coefficients and determine their approximate values by means of a simplified analysis, some numerical technique, or an appropriate experiment. The *lift coefficient, C_L*, and *drag coefficient, C_D*, are defined as

$$C_L = \frac{\mathcal{L}}{\frac{1}{2}\rho U^2 A}$$

and

$$C_D = \frac{\mathcal{D}}{\frac{1}{2}\rho U^2 A}$$

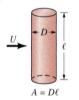

$A = D\ell$

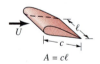

$A = c\ell$

where A is a characteristic area of the object (see Chapter 7). Typically, as shown in the figure in the margin, A is taken to be *frontal area*—the projected area seen by a person looking toward the object from a direction parallel to the upstream velocity, U. In other situations A is taken to be the *planform area*—the projected area seen by an observer looking toward the object from a direction normal to the upstream velocity (i.e., from "above" it).

9.1.2 Characteristics of Flow Past an Object

External flows past objects encompass an extremely wide variety of fluid mechanics phenomena. For a given-shaped object, the characteristics of the flow depend very strongly on various parameters, such as size, orientation, speed, and fluid properties. As discussed in Chapter 7, according to dimensional analysis arguments, the character of the flow should depend on the various dimensionless parameters involved. For typical external flows the most important of these parameters are the Reynolds number, the Mach number, and the Froude number.

For the present, we consider how the external flow and its associated lift and drag vary as a function of Reynolds number. Recall that the Reynolds number represents the ratio of inertial effects to viscous effects. The nature of the flow past a body depends strongly on whether Re $\gg$ 1 or Re $\ll$ 1.

Flows past three flat plates of length ℓ with $\text{Re} = \rho U \ell / \mu = 0.1$, 10, and 10^7 are shown in Fig. 9.4. If the Reynolds number is small, the viscous effects are relatively strong and the plate affects the uniform upstream flow far ahead, above, below, and behind the plate. To reach that portion of the flow field where the velocity has been altered by less than 1% of its undisturbed value (i.e., $U - u < 0.01U$) we must travel relatively far from the plate. In low Reynolds number flows the viscous effects are felt far from the object in all directions.

As the Reynolds number is increased (by increasing U, for example), the region in which viscous effects are important becomes smaller in all directions except downstream, as shown in Fig. 9.4b. One does not need to travel very far ahead, above, or below the plate to reach areas in which the viscous effects of the plate are not felt. The streamlines

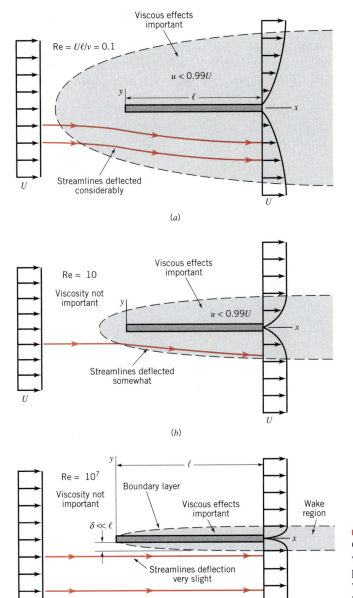

(a)

(b)

(c)

■ **FIGURE 9.4**
Character of the steady, viscous flow past a flat plate parallel to the upstream velocity: (a) low Reynolds number flow, (b) moderate Reynolds number flow, and (c) large Reynolds number flow.

are displaced from their original uniform upstream conditions, but the displacement is not as great as for the Re = 0.1 situation shown in Fig. 9.4a.

As suggested by Ludwig Prandtl in 1904, if the Reynolds number is large (but not infinite), the flow is dominated by inertial effects and the viscous effects are negligible everywhere except in a region very close to the plate and in the relatively thin *wake region* behind the plate, as shown in Fig. 9.4c. Since the fluid viscosity is not zero (Re < ∞), it follows that the fluid must stick to the solid surface (the no-slip boundary condition). There is a thin *boundary layer* region of thickness $\delta = \delta(x) \ll \ell$ (i.e., thin relative to the length of the plate) next to the plate in which the fluid velocity changes from the upstream value of $u = U$ to zero velocity on the plate. The existence of the plate has very little effect on the streamlines outside of the boundary layer—ahead, above, or below the plate. However, the wake region is due entirely to the viscous interaction between the fluid and the plate.

As with the flow past the flat plate described earlier, the flow past a blunt object (such as a circular cylinder) also varies with Reynolds number. In general, the larger the Reynolds number, the smaller the region of the flow field in which viscous effects are important. For objects that are not sufficiently streamlined, however, an additional characteristic of the flow is observed. This is termed *flow separation* and is illustrated in Fig. 9.5.

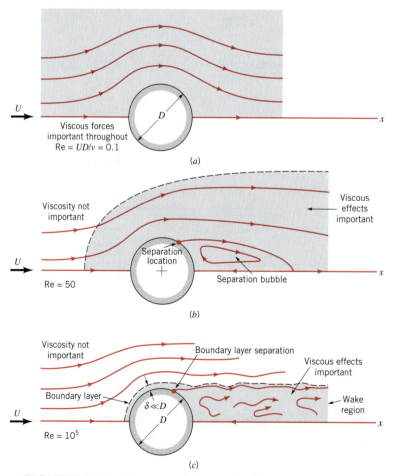

■ **FIGURE 9.5** **Character of the steady, viscous flow past a circular cylinder: (a) low Reynolds number flow, (b) moderate Reynolds number flow, and (c) large Reynolds number flow.**

Low Reynolds number flow (Re = $UD/\nu < 1$) past a circular cylinder is characterized by the fact that the presence of the cylinder and the accompanying viscous effects are felt throughout a relatively large portion of the flow field. As is indicated in Fig. 9.5a, for Re = $UD/\nu = 0.1$, the viscous effects are important several diameters in any direction from the cylinder.

V9.2 Streamlined and blunt bodies

As the Reynolds number is increased, the region ahead of the cylinder in which viscous effects are important becomes smaller, with the viscous region extending only a short distance ahead of the cylinder. The viscous effects are convected downstream and the flow loses its symmetry. Another characteristic of external flows becomes important—the flow separates from the body at the *separation location* as indicated in Fig. 9.5b.

At still larger Reynolds numbers, the area affected by the viscous forces is forced farther downstream until it involves only a thin ($\delta \ll D$) boundary layer on the front portion of the cylinder and an irregular, unsteady (perhaps turbulent) wake region that extends far downstream of the cylinder. The fluid in the region outside of the boundary layer and wake region flows as if it were inviscid.

9.2 Boundary Layer Characteristics

As was discussed in the previous section, it is often possible to treat flow past an object as a combination of viscous flow in the boundary layer and inviscid flow elsewhere. If the Reynolds number is large enough, viscous effects are important only in the boundary layer regions near the object (and in the wake region behind the object). The boundary layer is needed to allow for the no-slip boundary condition that requires the fluid to cling to any solid surface that it flows past. Outside of the boundary layer the velocity gradients normal to the flow are relatively small, and the fluid acts as if it were inviscid, even though the viscosity is not zero. A necessary condition for this structure of the flow is that the Reynolds number be large.

9.2.1 Boundary Layer Structure and Thickness on a Flat Plate

This section considers the situation in which the boundary layer is formed on a long flat plate along which flows a viscous, incompressible fluid as is shown in Fig. 9.6. For a finite length plate, it is clear that the plate length, ℓ, can be used as the characteristic length, with the Reynolds number as Re = $U\ell/\nu$. For the infinitely long flat plate extending from $x = 0$ to $x = \infty$, it is not obvious how to define the Reynolds number because there is no characteristic length. The plate has no thickness and is not of finite length!

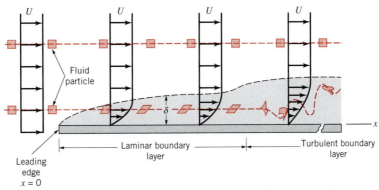

■ **FIGURE 9.6** Distortion of a fluid particle as it flows within the boundary layer.

For an infinitely long plate we use x, the coordinate distance along the plate from the leading edge, as the characteristic length and define the Reynolds number as $\text{Re}_x = Ux/\nu$. Thus, for any fluid or upstream velocity the Reynolds number will be sufficiently large for boundary layer type flow (i.e., Fig. 9.4c) if the plate is long enough. Physically, this means that the flow situations illustrated in Fig. 9.4 could be thought of as occurring on the same plate, but should be viewed by looking at longer portions of the plate as we step away from the plate to see the flows in Fig. 9.4a, 9.4b, and 9.4c, respectively. If the plate is sufficiently long, the Reynolds number $\text{Re} = U\ell/\nu$ is sufficiently large so that the flow takes on its boundary layer character (except very near the leading edge).

An appreciation of the structure of the boundary layer flow can be obtained by considering what happens to a fluid particle that flows into the boundary layer. As is indicated in Fig. 9.6, a small rectangular particle retains its original shape as it flows in the uniform flow outside of the boundary layer. Once it enters the boundary layer, the particle begins to distort because of the velocity gradient within the boundary layer—the top of the particle has a larger speed than its bottom. Near the leading edge of the plate there is *laminar boundary layer* flow.

V9.3 Laminar/ turbulent transition

At some distance downstream from the leading edge, the boundary layer flow becomes turbulent and the fluid particles become greatly distorted because of the random, irregular nature of the turbulence. The *transition* from laminar to *turbulent boundary layer* flow occurs at a critical value of the Reynolds number, Re_{xcr}, on the order of 2×10^5 to 3×10^6, depending on the roughness of the surface and the amount of turbulence in the upstream flow, as discussed in Section 9.2.4.

The purpose of the boundary layer is to allow the fluid to change its velocity from the upstream value of U to zero on the surface. Thus, $\mathbf{V} = 0$ at $y = 0$ and $\mathbf{V} \approx U\hat{\mathbf{i}}$ at the edge of the boundary layer, with the velocity profile, $u = u(x, y)$ bridging the boundary layer thickness. This boundary layer characteristic occurs in a variety of flow situations, not just on flat plates. For example, boundary layers form on the surfaces of cars, in the water running down the gutter of the street, and in the atmosphere as the wind blows across the surface of the earth (land or water).

F l u i d s   i n   t h e   N e w s

The Albatross: Nature's Aerodynamic Solution for Long Flights The albatross is a phenomenal seabird that soars just above ocean waves, taking advantage of the local *boundary layer* to travel incredible distances with little to no wing flapping. This limited physical exertion results in minimal energy consumption and, combined with aerodynamic optimization, allows the albatross to easily travel 1000 km (620 miles) per day, with some tracking data showing almost double that amount. The albatross has high aspect ratio wings (up to 11 ft in wingspan) and a lift-to-drag ratio ($\mathcal{L}/\mathcal{D}$) of approximately 27, both similar to high-performance sailplanes. With this aerodynamic configuration, the albatross then makes use of a technique called "dynamic soaring" to take advantage of the wind profile over the ocean surface. Based on the boundary layer profile, the albatross uses the rule of dynamic soaring, which is to climb when pointed upwind and dive when pointed downwind, thus constantly exchanging kinetic and potential energy. Though the albatross loses energy to drag, it can periodically regain energy due to vertical and directional motions within the boundary layer by changing local airspeed and direction. This is not a direct line of travel, but it does provide the most fuel-efficient method of long-distance flight. ■

We define the *boundary layer thickness,* δ, as that distance from the plate at which the fluid velocity is within some arbitrary value of the upstream velocity. Typically, as indicated in Fig. 9.7a,

$$\delta = y \quad \text{where} \quad u = 0.99U$$

To remove this arbitrariness (i.e., what is so special about 99%; why not 98%?), the following definitions are introduced. Figure 9.7b shows two velocity profiles for flow past a flat

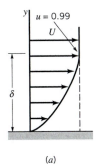
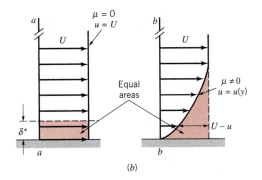

■ **FIGURE 9.7**
Boundary layer thickness: (*a*) **standard boundary layer thickness and** (*b*) **boundary layer displacement thickness.**

plate—one if there were no viscosity (a uniform profile) and the other if there is viscosity and zero slip at the wall (the boundary layer profile). Because of the velocity deficit, $U - u$, within the boundary layer, the flowrate across section b–b is less than that across section a–a. However, if we displace the plate at section a–a by an appropriate amount δ^*, the *boundary layer displacement thickness,* the flowrate across each section will be identical. This is true if

$$\delta^* b U = \int_0^\infty (U - u) b \, dy$$

where b is the plate width. Thus,

$$\delta^* = \int_0^\infty \left(1 - \frac{u}{U}\right) dy \tag{9.3}$$

The displacement thickness represents the amount that the thickness of the body must be increased so that the fictitious uniform inviscid flow has the same mass flowrate properties as the actual viscous flow. It represents the outward displacement of the streamlines caused by the viscous effects on the plate.

Another boundary layer thickness definition, the *boundary layer momentum thickness,* Θ, is often used when determining the drag on an object. Again because of the velocity deficit, $U - u$, in the boundary layer, the momentum flux across section b–b in Fig. 9.7 is less than that across section a–a. This deficit in momentum flux for the actual boundary layer flow is given by

$$\int \rho u (U - u) \, dA = \rho b \int_0^\infty u (U - u) \, dy$$

which by definition is the momentum flux in a layer of uniform speed U and thickness Θ. That is,

$$\rho b U^2 \Theta = \rho b \int_0^\infty u (U - u) \, dy$$

or

$$\Theta = \int_0^\infty \frac{u}{U} \left(1 - \frac{u}{U}\right) dy \tag{9.4}$$

All three boundary layer thickness definitions, δ, δ^*, and Θ, are of use in boundary layer analyses.

9.2.2   Prandtl/Blasius Boundary Layer Solution

In theory, the details of viscous incompressible flow past any object can be obtained by solving the governing Navier–Stokes equations discussed in Section 6.8.2. For steady, two-dimensional laminar flows with negligible gravitational effects, these equations (Eqs. 6.120*a, b,* and *c*) reduce to the following

$$u \frac{\partial u}{\partial x} + v \frac{\partial u}{\partial y} = -\frac{1}{\rho} \frac{\partial p}{\partial x} + \nu \left(\frac{\partial^2 u}{\partial x^2} + \frac{\partial^2 u}{\partial y^2} \right)$$

(9.5)

$$u \frac{\partial v}{\partial x} + v \frac{\partial v}{\partial y} = -\frac{1}{\rho} \frac{\partial p}{\partial y} + \nu \left(\frac{\partial^2 v}{\partial x^2} + \frac{\partial^2 v}{\partial y^2} \right)$$

(9.6)

which express Newton's second law. In addition, the conservation of mass equation, Eq. 6.31, for incompressible flow is

$$\frac{\partial u}{\partial x} + \frac{\partial v}{\partial y} = 0$$

(9.7)

The appropriate boundary conditions are that the fluid velocity far from the body is the upstream velocity and that the fluid sticks to the solid body surfaces. Although the mathematical problem is well-posed, no one has obtained an analytical solution to these equations for flow past any shaped body!

By using boundary layer concepts introduced in the previous sections, Prandtl was able to impose certain approximations (valid for large Reynolds number flows), and thereby to simplify the governing equations. In 1908, H. Blasius, one of Prandtl's students, was able to solve these simplified equations for the boundary layer flow past a flat plate parallel to the flow. Details may be found in the literature (Refs. 1, 2, 3).

From the Blasius solution it is found that the boundary layer thickness is

$$\delta = 5 \sqrt{\frac{\nu x}{U}}$$

(9.8)

or

$$\frac{\delta}{x} = \frac{5}{\sqrt{\text{Re}_x}}$$

where $\text{Re}_x = Ux/\nu$. It can also be shown that the displacement and momentum thicknesses are given by

$$\frac{\delta^*}{x} = \frac{1.721}{\sqrt{\text{Re}_x}}$$

(9.9)

and

$$\frac{\Theta}{x} = \frac{0.664}{\sqrt{\text{Re}_x}}$$

(9.10)

As postulated, the boundary layer is thin provided that Re_x is large (i.e., $\delta/x \to 0$ as $\text{Re}_x \to \infty$).

With the velocity profile known it is an easy matter to determine the wall shear stress, $\tau_w = \mu(\partial u/\partial y)_{y=0}$, where the velocity gradient is evaluated at the plate. The value of $\partial u/\partial y$ at $y = 0$ can be obtained from the Blasius solution to give

$$\tau_w = 0.332 U^{3/2} \sqrt{\frac{\rho \mu}{x}}$$

(9.11)

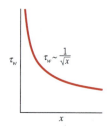

As indicated by Eq. 9.11 and illustrated in the figure in the margin, the shear stress decreases with increasing *x* because of the increasing thickness of the boundary layer—the velocity gradient at the wall decreases with increasing *x*. Also, τ_w varies as $U^{3/2}$, not as U as it does for fully developed laminar pipe flow.

For a flat plate of length ℓ and width b, the net friction drag, $\mathcal{D}_f$, can be expressed in terms of the *friction drag coefficient, C_{Df},* as

$$C_{Df} = \frac{\mathcal{D}_f}{\frac{1}{2}\rho U^2 b\ell} = \frac{b\int_0^\ell \tau_w\,dx}{\frac{1}{2}\rho U^2 b\ell} \qquad (9.12)$$

For the Blasius solution Eq. 9.12 gives

$$C_{Df} = \frac{1.328}{\sqrt{Re_\ell}}$$

where $Re_\ell = U\ell/\nu$ is the Reynolds number based on the plate length.

9.2.3 Momentum Integral Boundary Layer Equation for a Flat Plate

One of the important aspects of boundary layer theory is determination of the drag caused by shear forces on a body. As discussed in the previous section, such results can be obtained from the governing differential equations for laminar boundary layer flow. Since these solutions are extremely difficult to obtain, it is of interest to have an alternative approximate method. The momentum integral method described in this section provides such an alternative.

We consider the uniform flow past a flat plate and the fixed control volume as shown in Fig. 9.8. In agreement with advanced theory and experiment, we assume that the pressure is constant throughout the flow field. The flow entering the control volume at the leading edge of the plate [section (1)] is uniform, whereas the velocity of the flow exiting the control volume [section (2)] varies from the upstream velocity at the edge of the boundary layer to zero velocity on the plate.

Fluid adjacent to the plate makes up the lower portion of the control surface. The upper surface coincides with the streamline just outside the edge of the boundary layer at section (2). It need not (in fact, does not) coincide with the edge of the boundary layer except at section (2). If we apply the x component of the momentum equation (Eq. 5.17) to the steady flow of fluid within this control volume we obtain

$$\sum F_x = \sum u_2 \rho A_2 u_2 - \sum u_1 \rho A_1 u_1$$

Because the flow is uniform over the inlet, section (1), it follows that

$$\sum u_1 \rho A_1 u_1 = \rho U^2 bh$$

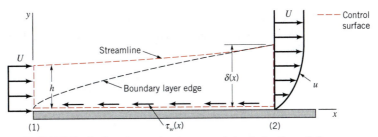

■ **FIGURE 9.8** Control volume used in derivation of the momentum integral equation for boundary layer flow.

where b is the plate width. Because of the variable velocity over the outlet, section (2), we can write the outlet, momentum flowrate as

$$\sum u_2 \rho A_2 u_2 = \rho \int_{(2)} u^2 \, dA = \rho b \int_0^\delta u^2 \, dy$$

Hence, the x component of the momentum equation can be written as

$$\sum F_x = \rho b \int_0^\delta u^2 \, dy - \rho U^2 bh \qquad (9.13)$$

In addition the net force that the plate exerts on the fluid is the drag, $\mathcal{D}$, where

$$\sum F_x = -\mathcal{D} = -\int_{\text{plate}} \tau_w \, dA = -b \int_{\text{plate}} \tau_w \, dx \qquad (9.14)$$

Thus, by combining Eqs. 9.13 and 9.14,

$$\mathcal{D} = \rho U^2 bh - \rho b \int_0^\delta u^2 \, dy \qquad (9.15)$$

Although the height h is not known, it is known that for conservation of mass the flowrate through section (1) must equal that through section (2), or

$$Uh = \int_0^\delta u \, dy$$

which can be written as

$$\rho U^2 bh = \rho b \int_0^\delta Uu \, dy \qquad (9.16)$$

Thus, by combining Eqs. 9.15 and 9.16 we obtain the drag in terms of the deficit of momentum flux across the outlet of the control volume as

$$\mathcal{D} = \rho b \int_0^\delta u(U - u) \, dy \qquad (9.17)$$

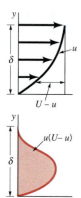

A typical boundary layer profile and the corresponding integrand of the integral in Eq. 9.17 are shown in the figure in the margin.

If the flow were inviscid, the drag would be zero, since we would have $u \equiv U$ and the right-hand side of Eq. 9.17 would be zero. (This is consistent with the fact that $\tau_w = 0$ if $\mu = 0$.) Equation 9.17 points out the important fact that boundary layer flow on a flat plate is governed by a balance between shear drag (the left-hand side of Eq. 9.17) and a decrease in the momentum of the fluid (the right-hand side of Eq. 9.17). By comparing Eqs. 9.17 and 9.4 we see that the drag can be written in terms of the momentum thickness, Θ, as

$$\mathcal{D} = \rho b U^2 \Theta \qquad (9.18)$$

Note that this equation is valid for laminar or turbulent flows.

The shear stress distribution can be obtained from Eq. 9.18 by differentiating both sides with respect to x to obtain

$$\frac{d\mathcal{D}}{dx} = \rho b U^2 \frac{d\Theta}{dx} \qquad (9.19)$$

Since $d\mathcal{D} = \tau_w \, b \, dx$ (see Eq. 9.14) it follows that

$$\frac{d\mathcal{D}}{dx} = b\tau_w \qquad (9.20)$$

Hence, by combining Eqs. 9.19 and 9.20 we obtain the *momentum integral equation* for the boundary layer flow on a flat plate

$$\tau_w = \rho U^2 \frac{d\Theta}{dx} \tag{9.21}$$

The usefulness of this relationship lies in the ability to obtain approximate boundary layer results easily using rather crude assumptions. This method is illustrated in Example 9.2.

EXAMPLE 9.2   Momentum Integral Boundary Layer Equation

GIVEN Consider the laminar flow of an incompressible fluid past a flat plate at $y = 0$. The boundary layer velocity profile is approximated as $u = Uy/\delta$ for $0 \le y \le \delta$ and $u = U$ for $y > \delta$, as is shown in Fig. E9.2.

FIND Determine the shear stress by using the momentum integral equation. Compare these results with the Blasius results given by Eq. 9.11.

SOLUTION

From Eq. 9.21 the shear stress is given by

$$\tau_w = \rho U^2 \frac{d\Theta}{dx} \tag{1}$$

while for laminar flow we know that $\tau_w = \mu(\partial u/\partial y)_{y=0}$. For the assumed profile we have

$$\tau_w = \mu \frac{U}{\delta} \tag{2}$$

and from Eq. 9.4

$$\Theta = \int_0^\infty \frac{u}{U}\left(1 - \frac{u}{U}\right) dy = \int_0^\delta \frac{u}{U}\left(1 - \frac{u}{U}\right) dy$$

$$= \int_0^\delta \left(\frac{y}{\delta}\right)\left(1 - \frac{y}{\delta}\right) dy$$

or

$$\Theta = \frac{\delta}{6} \tag{3}$$

Note that as yet we do not know the value of δ (but suspect that it should be a function of x).

By combining Eqs. 1, 2, and 3 we obtain the following differential equation for δ:

$$\frac{\mu U}{\delta} = \frac{\rho U^2}{6} \frac{d\delta}{dx}$$

or

$$\delta \, d\delta = \frac{6\mu}{\rho U} dx$$

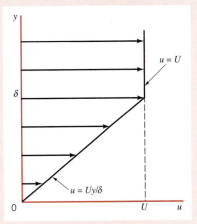

■ **FIGURE E9.2**

This can be integrated from the leading edge of the plate, $x = 0$, where $\delta = 0$ to an arbitrary location x where the boundary layer thickness is δ. The result is

$$\frac{\delta^2}{2} = \frac{6\mu}{\rho U} x$$

or

$$\delta = 3.46 \sqrt{\frac{\nu x}{U}} \tag{4}$$

Note that this approximate result (i.e., the velocity profile is not actually the simple straight line we assumed) compares favorably with the (much more laborious to obtain) Blasius result given by Eq. 9.8.

The wall shear stress can also be obtained by combining Eqs. 1, 3, and 4 to give

$$\tau_w = 0.289 U^{3/2} \sqrt{\frac{\rho\mu}{x}} \tag{Ans}$$

Again this approximate result is close (within 13%) to the Blasius value of τ_w given by Eq. 9.11.

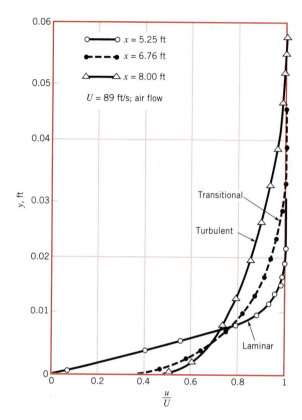

■ **FIGURE 9.9** Typical boundary layer profiles on a flat plate for laminar, transitional, and turbulent flow (Ref. 1).

9.2.4 Transition from Laminar to Turbulent Flow

The analytical results given in Section 9.2.2 are restricted to laminar boundary layer flows along a flat plate with zero pressure gradient. They agree quite well with experimental results up to the point where the boundary layer flow becomes turbulent, which will occur for any free stream velocity and any fluid provided the plate is long enough. This is true because the parameter that governs the transition to turbulent flow is the Reynolds number—in this case the Reynolds number based on the distance from the leading edge of the plate, $Re_x = Ux/\nu$.

The value of the Reynolds number at the transition location is a rather complex function of various parameters involved, including the roughness of the surface, the curvature of the surface (e.g., a flat plate or a sphere), and some measure of the disturbances in the flow outside the boundary layer. On a flat plate with a sharp leading edge in a typical air stream, the transition takes place at a distance x from the leading edge given by $Re_{xcr} = 2 \times 10^5$ to 3×10^6. Unless otherwise stated, we will use $Re_{xcr} = 5 \times 10^5$ in our calculations.

Transition from laminar to turbulent flow also involves a noticeable change in the shape of the boundary layer velocity profile. Typical profiles obtained in the neighborhood of the transition location are indicated in Fig. 9.9. Relative to laminar profiles, turbulent profiles are flatter, have a larger velocity gradient at the wall, and produce a larger boundary layer thickness.

EXAMPLE 9.3 ■ Boundary Layer Transition

GIVEN A fluid flows steadily past a flat plate with a velocity of $U = 10$ ft/s.

FIND At approximately what location will the boundary layer become turbulent, and how thick is the boundary layer at that point if the fluid is water at 60°F, standard air, or glycerin at 68°F?

SOLUTION

For any fluid, the laminar boundary layer thickness is found from Eq. 9.8 as

$$\delta = 5 \sqrt{\frac{\nu x}{U}}$$

The boundary layer remains laminar up to

$$x_{cr} = \frac{\nu \text{Re}_{xcr}}{U}$$

Thus, if we assume $\text{Re}_{xcr} = 5 \times 10^5$ we obtain

$$x_{cr} = \frac{5 \times 10^5}{10 \text{ ft/s}} \nu = 5 \times 10^4 \nu$$

and

$$\delta_{cr} \equiv \delta|_{x=x_{cr}} = 5 \left[\frac{\nu}{10} (5 \times 10^4 \, \nu) \right]^{1/2} = 354 \, \nu$$

where ν is in ft²/s and x_{cr} and δ_{cr} are in feet. The values of the kinematic viscosity obtained from Tables 1.4 to 1.6 are listed in Table E9.3 along with the corresponding x_{cr} and δ_{cr}.

■ **TABLE E9.3**

Fluid	ν (ft²/s)	x_{cr} (ft)	δ_{cr} (ft)
Water	1.21×10^{-5}	0.605	0.00428
Air	1.57×10^{-4}	7.85	0.0556
Glycerin	1.28×10^{-2}	640.0	4.53

(Ans)

COMMENT   Laminar flow can be maintained on a longer portion of the plate if the viscosity is increased. However, the boundary layer flow eventually becomes turbulent, provided the plate is long enough. Similarly, the boundary layer thickness is greater if the viscosity is increased.

9.2.5   Turbulent Boundary Layer Flow

The structure of turbulent boundary layer flow is very complex, random, and irregular. It shares many of the characteristics described for turbulent pipe flow in Section 8.3. In particular, the velocity at any given location in the flow is unsteady in a random fashion. The flow can be thought of as a jumbled mix of intertwined eddies (or swirls) of different sizes (diameters and angular velocities). The various fluid quantities involved (i.e., mass, momentum, energy) are convected downstream in the free-stream direction as in a laminar boundary layer. For turbulent flow they are also convected across the boundary layer (in the direction perpendicular to the plate) by the random transport of finite-sized fluid particles associated with the turbulent eddies. There is considerable mixing involved with these finite-sized eddies—considerably more than is associated with the mixing found in laminar flow where it is confined to the molecular scale. Consequently, the shear force for turbulent boundary layer flow is considerably greater than it is for laminar boundary layer flow (see Section 8.3.2).

There are no "exact" solutions for turbulent boundary layer flow. As discussed in Section 9.2.2, it is possible to solve the Prandtl boundary layer equations for laminar flow past a flat plate to obtain the Blasius solution. Since there is no precise expression for the shear stress in turbulent flow (see Section 8.3), solutions are not available for turbulent flow. Thus, it is necessary to use some empirical relationship for the wall shear stress and corresponding drag coefficient.

In general, the drag coefficient for a flat plate of length ℓ, $C_{Df} = \mathcal{D}_f / (\frac{1}{2} \rho U^2 A)$, is a function of the Reynolds number, Re_ℓ, and the relative roughness, ε/ℓ. The results of numerous experiments covering a wide range of the parameters of interest are shown in Fig. 9.10. For laminar boundary layer flow the drag coefficient is a function of only the Reynolds number—surface roughness is not important. This is similar to laminar flow in a pipe. However, for turbulent flow, the surface roughness does affect the shear stress and, hence, the drag coefficient. This is similar to turbulent pipe flow. Values of the roughness, ε, for different materials can be obtained from Table 8.1.

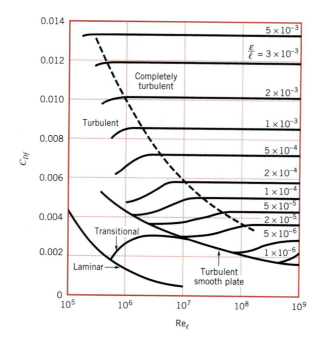

■ **FIGURE 9.10** Friction drag coefficient for a flat plate parallel to the upstream flow (Ref. 12, with permission).

The drag coefficient diagram of Fig. 9.10 (boundary layer flow) shares many characteristics in common with the familiar Moody diagram (pipe flow) of Fig. 8.10, even though the mechanisms governing the flow are quite different. Fully developed horizontal pipe flow is governed by a balance between pressure forces and viscous forces. The fluid inertia remains constant throughout the flow. Boundary layer flow on a horizontal flat plate is governed by a balance between inertia effects and viscous forces. The pressure remains constant throughout the flow.

It is often convenient to have an equation for the drag coefficient as a function of the Reynolds number and relative roughness rather than the graphical representation given in Fig. 9.10. Although there is not one equation valid for the entire Re_ℓ–ε/ℓ range, the equations presented in Table 9.1 do work well for the conditions indicated.

■ **TABLE 9.1**
Empirical Equations for the Flat Plate Drag Coefficient (Ref. 1)

Equation	Flow Conditions
$C_{Df} = 1.328/(\text{Re}_\ell)^{0.5}$	Laminar flow
$C_{Df} = 0.455/(\log \text{Re}_\ell)^{2.58} - 1700/\text{Re}_\ell$	Transitional with $\text{Re}_{xcr} = 5 \times 10^5$
$C_{Df} = 0.455/(\log \text{Re}_\ell)^{2.58}$	Turbulent, smooth plate
$C_{Df} = [1.89 - 1.62 \log(\varepsilon/\ell)]^{-2.5}$	Completely turbulent

EXAMPLE 9.4   Drag on a Flat Plate

GIVEN The water ski shown in Fig. E9.4a moves through 70 °F water with a velocity U.

FIND Estimate the drag caused by the shear stress on the bottom of the ski for $0 < U < 30$ ft/s.

SOLUTION

Clearly the ski is not a flat plate, and it is not aligned exactly parallel to the upstream flow. However, we can obtain a reasonable approximation to the shear force using flat plate results. That is, the friction drag, $\mathcal{D}_f$, caused by the shear stress on the bottom of the ski (the wall shear stress) can be determined as

$$\mathcal{D}_f = \tfrac{1}{2}\rho U^2 \ell b C_{Df}$$

With $A = \ell b = 4\text{ ft} \times 0.5\text{ ft} = 2\text{ ft}^2$, $\rho = 1.94\text{ slugs/ft}^3$, and $\mu = 2.04 \times 10^{-5}\text{ lb·s/ft}^2$ (see Table B.1) we obtain

$$\mathcal{D}_f = \tfrac{1}{2}(1.94\text{ slugs/ft}^3)(2.0\text{ ft}^2)U^2 C_{Df}$$

$$= 1.94\, U^2 C_{Df} \tag{1}$$

where $\mathcal{D}_f$ and U are in pounds and feet per second, respectively.

The friction coefficient, C_{Df}, can be obtained from Fig. 9.10 or from the appropriate equations given in Table 9.1. As we will see, for this problem, much of the flow lies within the transition regime where both laminar and turbulent portions of the boundary layer flow occupy comparable lengths of the plate. We choose to use the values of C_{Df} from the table.

For the given conditions we obtain

$$\text{Re}_\ell = \frac{\rho U \ell}{\mu} = \frac{(1.94\text{ slugs/ft}^3)(4\text{ ft})U}{2.04 \times 10^{-5}\text{ lb·s/ft}^2} = 3.80 \times 10^5\, U$$

where U is in feet per second. With $U = 10$ ft/s, or $\text{Re}_\ell = 3.80 \times 10^6$, we obtain from Table 9.1 $C_{Df} = 0.455/(\log \text{Re}_\ell)^{2.58} - 1700/\text{Re}_\ell = 0.00308$. From Eq. 1 the corresponding drag is

$$\mathcal{D}_f = 1.94(10)^2(0.00308) = 0.598\text{ lb}$$

By covering the range of upstream velocities of interest we obtain the results shown in Fig. E9.4b.

COMMENTS If $\text{Re} \lesssim 1000$, the results of boundary layer theory are not valid—inertia effects are not dominant enough and the boundary layer is not thin compared with the length of the plate. For our problem this corresponds to $U = 2.63 \times 10^{-3}$ ft/s. For all practical purposes U is greater than this value, and the flow past the ski is of the boundary layer type.

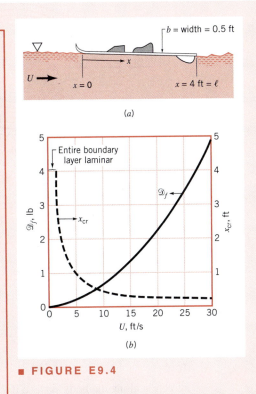

■ FIGURE E9.4

The approximate location of the transition from laminar to turbulent boundary layer flow as defined by $\text{Re}_{cr} = \rho U x_{cr}/\mu = 5 \times 10^5$ is indicated in Fig. E9.4b. Up to $U = 1.31$ ft/s the entire boundary layer is laminar. The fraction of the boundary layer that is laminar decreases as U increases until only the front 0.18 ft is laminar when $U = 30$ ft/s.

For anyone who has water skied, it is clear that it can require considerably more force to be pulled along at 30 ft/s than the 2×4.88 lb $= 9.76$ lb (two skis) indicated in Fig. E9.4b. As is discussed in Section 9.3, the total drag on an object such as a water ski consists of more than just the friction drag. Other components, including pressure drag and wave-making drag, can add considerably to the total resistance.

9.2.6   Effects of Pressure Gradient

The boundary layer discussions in the previous parts of Section 9.2 have dealt with flow along a flat plate in which the pressure is constant throughout the fluid. In general, when a fluid flows past an object other than a flat plate, the pressure field is not uniform. As shown in Fig. 9.5, if the Reynolds number is large, relatively thin boundary layers will develop along the surfaces. Within these layers the component of the pressure gradient in the streamwise direction (i.e., along the body surface) is not zero, although the pressure gradient normal to the surface is negligibly small. That is, if we were to measure the pressure while moving across the boundary layer from the body to the boundary layer edge, we would find that the pressure is essentially constant. However, the pressure does vary in the

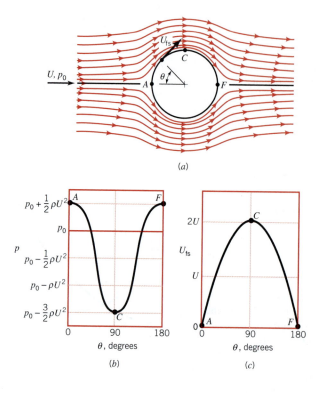

(a)

(b)

(c)

■ **FIGURE 9.11** Inviscid flow past a circular cylinder: (*a*) streamlines for the flow if there were no viscous effects, (*b*) pressure distribution on the cylinder's surface, and (*c*) free-stream velocity on the cylinder's surface.

direction along the body surface if the body is curved. The variation in the *free-stream velocity, U_{fs},* the fluid velocity at the edge of the boundary layer, is the cause of the pressure gradient in the boundary layer. The characteristics of the entire flow (both within and outside of the boundary layer) are often highly dependent on the pressure gradient effects on the fluid within the boundary layer.

For a flat plate parallel to the upstream flow, the upstream velocity (that far ahead of the plate) and the free-stream velocity (that at the edge of the boundary layer) are equal— $U = U_{fs}$. This is a consequence of the negligible thickness of the plate. For bodies of nonzero thickness, these two velocities are different. This can be seen in the flow past a circular cylinder of diameter D. The upstream velocity and pressure are U and p_0, respectively. If the fluid were completely inviscid ($\mu = 0$), the Reynolds number would be infinite ($\text{Re} = \rho UD/\mu = \infty$) and the streamlines would be symmetrical, as are shown in Fig. 9.11a. The fluid velocity along the surface would vary from $U_{fs} = 0$ at the very front and rear of the cylinder (points A and F are stagnation points) to a maximum of $U_{fs} = 2U$ at the top and bottom of the cylinder (point C). This is also indicated in the figure in the margin. The pressure on the surface of the cylinder would be symmetrical about the vertical midplane of the cylinder, reaching a maximum value of $p_0 + \rho U^2/2$ (the stagnation pressure) at both the front and the back of the cylinder, and a minimum of $p_0 - 3\rho U^2/2$ at the top and bottom of the cylinder. The pressure and free-stream velocity distributions are shown in Figs. 9.11b and 9.11c. Because of the absence of viscosity (therefore, $\tau_w = 0$) and the symmetry of the pressure distribution for inviscid flow past a circular cylinder, it is clear that the drag on the cylinder is zero.

Consider large Reynolds number flow of a real (viscous) fluid past a circular cylinder. As discussed in Section 9.1.2, we expect the viscous effects to be confined to thin boundary layers near the surface. This allows the fluid to stick ($\mathbf{V} = 0$) to the surface—a necessary condition for any fluid, provided $\mu \neq 0$. The basic idea of boundary layer theory is that the boundary layer is thin enough so that it does not greatly disturb the flow outside

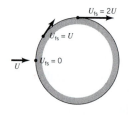

the boundary layer. *Based on this reasoning,* for large Reynolds numbers the flow throughout most of the flow field would be expected to be as is indicated in Fig. 9.11a, the inviscid flow field.

The pressure distribution indicated in Fig. 9.11b is imposed on the boundary layer flow along the surface of the cylinder. In fact, there is negligible pressure variation across the thin boundary layer so that the pressure within the boundary layer is that given by the inviscid flow field. This pressure distribution along the cylinder is such that the stationary fluid at the nose of the cylinder ($U_{fs} = 0$ at $\theta = 0$) is accelerated to its maximum velocity ($U_{fs} = 2U$ at $\theta = 90°$) and then is decelerated back to zero velocity at the rear of the cylinder ($U_{fs} = 0$ at $\theta = 180°$). This is accomplished by a balance between pressure and inertia effects; viscous effects are absent for the inviscid flow outside the boundary layer.

Physically, in the absence of viscous effects, a fluid particle traveling from the front to the back of the cylinder coasts down the "pressure hill" from $\theta = 0$ to $\theta = 90°$ (from point A to C in Fig. 9.11b) and then back up the hill to $\theta = 180°$ (from point C to F) without any loss of energy. There is an exchange between kinetic and pressure energy, but there are no energy losses. The same pressure distribution is imposed on the viscous fluid within the boundary layer. The decrease in pressure in the direction of flow along the front half of the cylinder is termed a *favorable pressure gradient.* The increase in pressure in the direction of flow along the rear half of the cylinder is termed an *adverse pressure gradient.*

Consider a fluid particle within the boundary layer indicated in Fig. 9.12. In its attempt to flow from A to F it experiences the same pressure distribution as the particles in the free stream immediately outside the boundary layer—the inviscid flow field pressure. However, because of the viscous effects involved, the particle in the boundary layer experiences a loss of energy as it flows along. This loss means that the particle does not have enough energy to coast all of the way up the pressure hill (from C to F) and to reach point F at the rear of the cylinder. This kinetic energy deficit is seen in the velocity profile detail at point C, shown in Fig. 9.12a. Because of friction, the boundary layer fluid cannot travel from the front to the rear of the cylinder. (This conclusion can also be obtained from the concept that due to viscous effects the particle at C does not have enough momentum to allow it to coast up the pressure hill to F.)

Thus, the fluid flows against the increasing pressure as far as it can, at which point the boundary layer separates from (lifts off) the surface, as indicated by the figure in the margin. This *boundary layer separation* is indicated in Fig. 9.12a. Typical velocity profiles at representative locations along the surface are shown in Fig. 9.12b. At the separation location (profile D), the velocity gradient at the wall and the wall shear stress are zero. Beyond that location (from D to E) there is reverse flow in the boundary layer.

As is indicated in Fig. 9.12c, because of boundary layer separation, the average pressure on the rear half of the cylinder is considerably less than that on the front half. Thus, a large pressure drag is developed, even though (because of small viscosity) the viscous shear drag may be quite small.

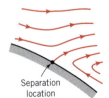

Separation
location

F l u i d s   i n   t h e   N e w s

Increasing truck mpg A large portion of the aerodynamic drag on semis (tractor-trailer rigs) is a result of the low pressure on the flat back end of the trailer. Researchers have recently developed a drag-reducing attachment that could reduce fuel costs on these big rigs by 10 percent. The device consists of a set of flat plates (attached to the rear of the trailer) that fold out into a box shape, thereby making the originally flat rear of the trailer a somewhat more "aerodynamic" shape. Based on thorough wind tunnel testing and actual tests conducted with a prototype design used in a series of cross-country runs, it is estimated that trucks using the device could save approximately $4,000 a year in fuel costs. ∎

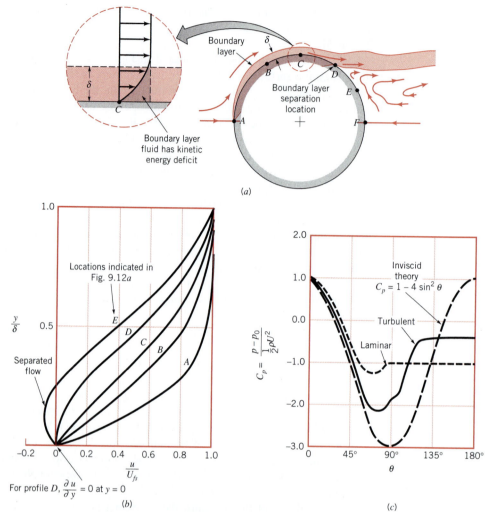

■ **FIGURE 9.12** **Boundary layer characteristics on a circular cylinder: (*a*) boundary layer separation location, (*b*) typical boundary layer velocity profiles at various locations on the cylinder, and (*c*) surface pressure distributions for inviscid flow and boundary layer flow.**

V9.4 Snow drifts

The location of separation, the width of the wake region behind the object, and the pressure distribution on the surface depend on the nature of the boundary layer flow. Compared with a laminar boundary layer, a turbulent boundary layer flow has more kinetic energy and momentum associated with it because (1) as is indicated in Fig. 9.9, the velocity profile is fuller, more nearly like the ideal uniform profile, and (2) there can be considerable energy associated with the swirling, random components of the velocity that do not appear in the time-averaged *x* component of velocity. Thus, as is indicated in Fig. 9.12*c*, the turbulent boundary layer can flow farther around the cylinder (farther up the pressure hill) before it separates than the laminar boundary layer.

9.3 Drag

As discussed in Section 9.1, any object moving through a fluid will experience a drag, $\mathcal{D}$— a net force in the direction of flow due to the pressure and shear forces on the surface of the object. This net force, a combination of flow direction components of the normal and tangential forces on the body, can be determined by use of Eqs. 9.1 and 9.2, provided the distributions of pressure, p, and wall shear stress, τ_w, are known. Only in very rare instances can these distributions be determined analytically.

Most of the information pertaining to drag on objects is a result of numerous experiments with wind tunnels, water tunnels, towing tanks, and other ingenious devices used to measure the drag on scale models. Typically, the result for a given shaped object is a drag coefficient, C_D, where

$$C_D = \frac{\mathcal{D}}{\frac{1}{2}\rho U^2 A}$$
(9.22)

and C_D is a function of other dimensionless parameters such as Reynolds number, Re, Mach number, Ma, Froude number, Fr, and relative roughness of the surface, ε/ℓ. That is,

$$C_D = \phi(\text{shape, Re, Ma, Fr, } \varepsilon/\ell)$$

9.3.1   Friction Drag

Friction drag, $\mathcal{D}_f$, is that part of the drag that is due directly to the shear stress, τ_w, on the object. It is a function of not only the magnitude of the wall shear stress, but also of the orientation of the surface on which it acts. This is indicated by the factor $\tau_w \sin \theta$ in Eq. 9.1. For highly streamlined bodies or for low Reynolds number flow most of the drag may be due to friction.

The friction drag on a flat plate of width b and length ℓ oriented parallel to the upstream flow can be calculated from

$$\mathcal{D}_f = \tfrac{1}{2}\rho U^2 b\ell C_{Df}$$

where C_{Df} is the friction drag coefficient. The value of C_{Df} is given as a function of Reynolds number, $\text{Re}_\ell = \rho U\ell/\mu$, and relative surface roughness, ε/ℓ, in Fig. 9.10 and Table 9.1.

Most objects are not flat plates parallel to the flow; instead, they are curved surfaces along which the pressure varies. The precise determination of the shear stress along the surface of a curved body is quite difficult to obtain. Although approximate results can be obtained by a variety of techniques (Refs. 1, 2), these are outside the scope of this text.

9.3.2   Pressure Drag

Pressure drag, $\mathcal{D}_p$, is that part of the drag that is due directly to the pressure, p, on an object. It is often referred to as *form drag* because of its strong dependency on the shape or form of the object. Pressure drag is a function of the magnitude of the pressure and the orientation of the surface element on which the pressure force acts. For example, the pressure force on either side of a flat plate parallel to the flow may be very large, but it does not contribute to the drag because it acts in the direction normal to the upstream velocity. However, the pressure force on a flat plate normal to the flow provides the entire drag.

As noted previously, for most bodies, there are portions of the surface that are parallel to the upstream velocity, others normal to the upstream velocity, and the majority of which are at some angle in between. The pressure drag can be obtained from Eq. 9.1 provided a detailed description of the pressure distribution and the body shape is given. That is,

$$\mathcal{D}_p = \int p \cos \theta \, dA$$

which can be rewritten in terms of the *pressure drag coefficient, C_{Dp},* as

$$C_{Dp} = \frac{\mathcal{D}_p}{\frac{1}{2}\rho U^2 A} = \frac{\int p \cos \theta \, dA}{\frac{1}{2}\rho U^2 A} = \frac{\int C_p \cos \theta \, dA}{A}$$
(9.23)

Here $C_p = (p - p_0)/(\rho U^2/2)$ is the *pressure coefficient*, where p_0 is a reference pressure. The level of the reference pressure, p_0, does not influence the drag directly because the net pressure force on a body is zero if the pressure is constant (i.e., p_0) on the entire surface.

9.3.3 Drag Coefficient Data and Examples

As discussed in previous sections, the net drag is produced by both pressure and shear stress effects. In most instances these two effects are considered together and an overall drag coefficient, C_D, as defined in Eq. 9.22 is used. There is an abundance of such drag coefficient data available in the literature. In this section we consider a small portion of this information for representative situations. Additional data can be obtained from various sources (Refs. 4, 5).

Shape Dependence. Clearly the drag coefficient for an object depends on the shape of the object, with shapes ranging from those that are streamlined to those that are blunt. The drag on an ellipse with aspect ratio ℓ/D, where D and ℓ are the thickness and length parallel to the flow, illustrates this dependence. The drag coefficient $C_D = \mathcal{D}/(\rho U^2 bD/2)$, based on the frontal area, $A = bD$, where b is the length normal to the flow, is as shown in Fig. 9.13. The more blunt the body, the larger the drag coefficient. With $\ell/D = 0$ (i.e., a flat plate normal to the flow) we obtain the flat plate value of $C_D = 1.9$. With $\ell/D = 1$ the corresponding value for a circular cylinder is obtained. As ℓ/D becomes larger the value of C_D decreases.

For very large aspect ratios ($\ell/D \to \infty$) the ellipse behaves as a flat plate parallel to the flow. For such cases, the friction drag is greater than the pressure drag. For extremely thin bodies (i.e., an ellipse with $\ell/D \to \infty$, a flat plate, or very thin airfoils) it is customary to use the planform area, $A = b\ell$, in defining the drag coefficient. The ellipse drag coefficient based on the planform area, $C_D = \mathcal{D}/(\rho U^2 b\ell/2)$, is also shown in Fig. 9.13. Clearly the drag obtained by using either of these drag coefficients would be the same. They merely represent two different ways to package the same information.

The amount of streamlining can have a considerable effect on the drag. Incredibly, the drag on the two two-dimensional objects drawn to scale in Fig. 9.14 is the same. The width of the wake for the streamlined strut is very thin, on the order of that for the much smaller diameter circular cylinder.

V9.5 Skydiving practice

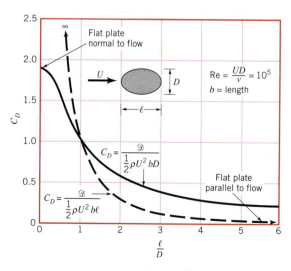

■ **FIGURE 9.13** Drag coefficient for an ellipse with the characteristic area either the frontal area, $A = bD$, or the planform area, $A = b\ell$ (Ref. 4).

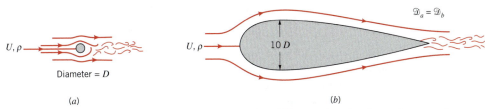

■ **FIGURE 9.14** Two objects of considerably different size that have the same drag force: (*a*) circular cylinder $C_D = 1.2$ and (*b*) streamlined strut $C_D = 0.12$.

Fluids in the News

At 10,240 mpg it doesn't cost much to "fill 'er up" Typical gas consumption for a Formula 1 racer, a sports car, and a sedan is approximately 2 mpg, 15 mpg, and 30 mpg, respectively. Thus, just how did the winning entry in the 2002 Shell Eco-Marathon achieve an incredible 10,240 mpg? To be sure, this vehicle is not as fast as a Formula 1 racer (although the rules require it to average at least 15 mph), and it can't carry as large a load as your family sedan can (the vehicle has barely enough room for the driver). However, by using a number of clever engineering design considerations, this amazing fuel efficiency was obtained. The type (and number) of tires, the appropriate engine power and weight, the specific chassis design, and the design of the body shell are all important and interrelated considerations. To reduce *drag*, the aerodynamic shape of the high-efficiency vehicle was given special attention through theoretical considerations and wind tunnel model testing. The result is an amazing vehicle that can travel a long distance without hearing the usual "fill 'er up." (See Problem 9.46.) ■

Reynolds Number Dependence. Another parameter on which the drag coefficient can be very dependent is the Reynolds number. Low Reynolds number flows (Re < 1) are governed by a balance between viscous and pressure forces. Inertia effects are negligibly small. In such instances the drag is expected to be a function of the upstream velocity, U, the body size, ℓ, and the viscosity, μ. That is,

$$\mathcal{D} = f(U, \ell, \mu)$$

From dimensional considerations (see Section 7.7.1)

$$\mathcal{D} = C\mu\ell U \qquad \textbf{(9.24)}$$

where the value of the constant C depends on the shape of the body. If we put Eq. 9.24 into dimensionless form using the standard definition of the drag coefficient, $C_D = \mathcal{D}/\frac{1}{2}\rho U^2 A$, we obtain

$$C_D = \frac{\text{constant}}{\text{Re}}$$

where Re = $\rho U \ell / \mu$. For a sphere it can be shown that $C_D = 24/\text{Re}$, where $\ell = D$, the sphere diameter. For most objects, the low Reynolds number flow results are valid up to a Reynolds number of about 1.

EXAMPLE 9.5 | Low Reynolds Number Flow Drag

GIVEN A small grain of sand diameter $D = 0.10$ mm and specific gravity $SG = 2.3$ settles to the bottom of a lake after having been stirred up by a passing boat.

FIND Determine how fast it falls through the still water.

SOLUTION

A free-body diagram of the particle (relative to the moving particle) is shown in Fig. E9.5a. The particle moves downward with a constant velocity U that is governed by a balance between the weight of the particle, $\mathcal{W}$, the buoyancy force of the surrounding water, F_B, and the drag of the water on the particle, $\mathcal{D}$.

From the free-body diagram we obtain

$$\mathcal{W} = \mathcal{D} + F_B$$

where

$$\mathcal{W} = \gamma_{sand}\forall = SG\,\gamma_{H_2O}\frac{\pi}{6}D^3 \qquad (1)$$

and

$$F_B = \gamma_{H_2O}\forall = \gamma_{H_2O}\frac{\pi}{6}D^3 \qquad (2)$$

We assume (because of the smallness of the object) that the flow will be creeping flow (Re < 1) with $C_D = 24/Re$ so that

$$\mathcal{D} = \frac{1}{2}\rho_{H_2O}U^2\frac{\pi}{4}D^2C_D = \frac{1}{2}\rho_{H_2O}U^2\frac{\pi}{4}D^2\left(\frac{24}{\rho_{H_2O}UD/\mu_{H_2O}}\right)$$

or

$$\mathcal{D} = 3\pi\mu_{H_2O}UD \qquad (3)$$

We must eventually check to determine if this assumption is valid or not. Equation 3 is called Stokes law in honor of G. G. Stokes, a British mathematician and a physicist. By combining Eqs. 1, 2, and 3, we obtain

$$SG\,\gamma_{H_2O}\frac{\pi}{6}D^3 = 3\pi\mu_{H_2O}UD + \gamma_{H_2O}\frac{\pi}{6}D^3$$

or, since $\gamma = \rho g$,

$$U = \frac{(SG-1)\rho_{H_2O}\,gD^2}{18\mu} \qquad (4)$$

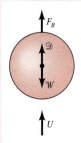

From Table 1.5 for water at 15.6° C we obtain $\rho_{H_2O} = 999$ kg/m³ and $\mu_{H_2O} = 1.12 \times 10^{-3}$ N·s/m². Thus, from Eq. 4 we obtain

$$U = \frac{(2.3-1)(999\text{ kg/m}^3)(9.81\text{ m/s}^2)(0.10\times10^{-3}\text{ m})^2}{18(1.12\times10^{-3}\text{ N·s/m}^2)}$$

or

$$U = 6.32 \times 10^{-3} \text{ m/s} \qquad \textbf{(Ans)}$$

Since

$$Re = \frac{\rho DU}{\mu} = \frac{(999\text{ kg/m}^3)(0.10\times10^{-3}\text{ m})(0.00632\text{ m/s})}{1.12\times10^{-3}\text{ N·s/m}^2}$$

$$= 0.564$$

we see that Re < 1, and the form of the drag coefficient used is valid.

COMMENTS By repeating the calculations for various particle diameters, D, the results shown in Fig. E9.5b are obtained. Note that very small particles fall extremely slowly. Thus, it can take considerable time for silt to settle to the bottom of a river or lake.

Note that if the density of the particle were the same as the surrounding fluid (i.e., $SG = 1$), from Eq. 4 we would obtain $U = 0$. This is reasonable since the particle would be neutrally buoyant and there would be no force to overcome the motion-induced drag. Note also that we have assumed that the particle falls at its steady terminal velocity. That is, we have neglected the acceleration of the particle from rest to its terminal velocity. Since the terminal velocity is small, this acceleration time is quite small. For faster objects (such as a free-falling skydiver) it may be important to consider the acceleration portion of the fall.

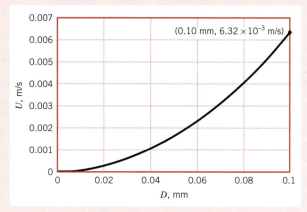

▪ **FIGURE E9.5a**

▪ **FIGURE E9.5b**

V9.6 Oscillating sign

Moderate Reynolds number flows tend to take on a boundary layer flow structure. For such flows past streamlined bodies, the drag coefficient tends to decrease slightly with Reynolds number. The $C_D \sim Re^{-1/2}$ dependence for a laminar boundary layer on a flat plate (see Table 9.1) is such an example. Moderate Reynolds number flows past blunt bodies generally produce drag coefficients that are relatively constant. The C_D values for the spheres and circular cylinders shown in Fig. 9.15a indicate this character in the range $10^3 < Re < 10^5$.

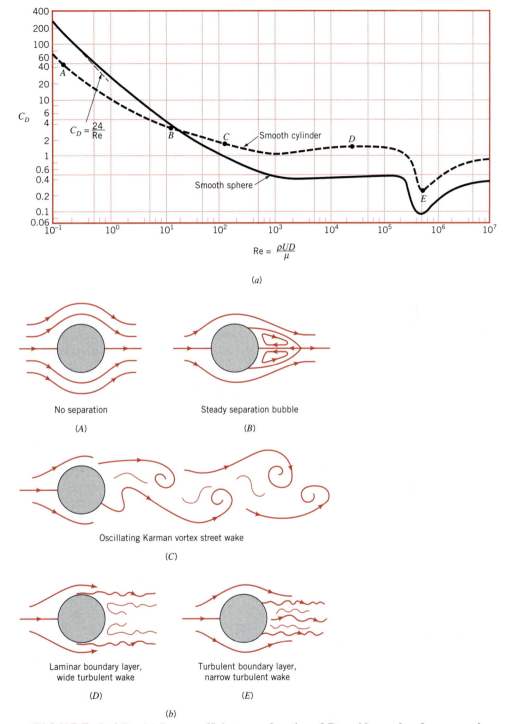

■ **FIGURE 9.15** (a) Drag coefficient as a function of Reynolds number for a smooth circular cylinder and a smooth sphere. (b) Typical flow patterns for flow past a circular cylinder at various Reynolds numbers as indicated in (a).

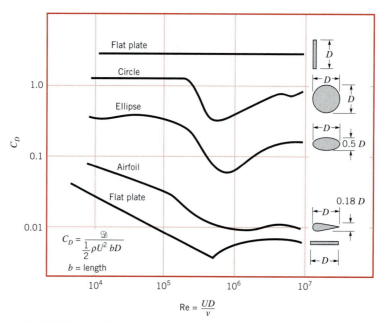

■ **FIGURE 9.16** Character of the drag coefficient as a function of the Reynolds number for objects with various degrees of streamlining, from a flat plate normal to the upstream flow to a flat plate parallel to the flow (two-dimensional flow) (Ref. 4).

The structure of the flow field at selected Reynolds numbers indicated in Fig. 9.15a is shown in Fig. 9.15b. For a given object there is a wide variety of flow situations, depending on the Reynolds number involved. The curious reader is strongly encouraged to study the many beautiful photographs and videos of these (and other) flow situations found in Refs. 6 and 19.

For many shapes there is a sudden change in the character of the drag coefficient when the boundary layer becomes turbulent. This is illustrated in Fig. 9.10 for the flat plate and in Fig. 9.15 for the sphere and the circular cylinder. The Reynolds number at which this transition takes place is a function of the shape of the body.

For streamlined bodies, the drag coefficient increases when the boundary layer becomes turbulent because most of the drag is due to the shear force, which is greater for turbulent flow than for laminar flow. However, the drag coefficient for a relatively blunt object, such as a cylinder or sphere, actually decreases when the boundary layer becomes turbulent. As discussed in Section 9.2.6, a turbulent boundary layer can travel further along the surface into the adverse pressure gradient on the rear portion of the cylinder before separation occurs. The result is a thinner wake and smaller pressure drag for turbulent boundary layer flow. This is indicated in Fig. 9.15 by the sudden decrease in C_D for $10^5 < \text{Re} < 10^6$.

For extremely blunt bodies, such as a flat plate perpendicular to the flow, the flow separates at the edge of the plate regardless of the nature of the boundary layer flow. Thus, the drag coefficient shows very little dependence on the Reynolds number.

The drag coefficients for a series of two-dimensional bodies of varying bluntness are given as a function of Reynolds number in Fig. 9.16. The characteristics described earlier are evident.

EXAMPLE 9.6 ■ Terminal Velocity of a Falling Object

GIVEN Hail is produced by the repeated rising and falling of ice particles in the updraft of a thunderstorm, as is indicated in Fig. E9.6a. When the hail becomes large enough, the aerodynamic drag from the updraft can no longer support the weight of the hail, and it falls from the storm cloud.

FIND Estimate the velocity, U, of the updraft needed to make $D = 1.5$-in.-diameter (i.e., "golf ball-sized") hail.

SOLUTION

As discussed in Example 9.5, for steady-state conditions a force balance on an object falling through a fluid gives

$$\mathcal{W} = \mathcal{D} + F_B$$

where $F_B = \gamma_{air}\mathcal{V}$ is the buoyant force of the air on the particle, $\mathcal{W} = \gamma_{ice}\mathcal{V}$ is the particle weight, and $\mathcal{D}$ is the aerodynamic drag. This equation can be rewritten as

$$\tfrac{1}{2}\rho_{air}U^2 \frac{\pi}{4}D^2 C_D = \mathcal{W} - F_B \qquad (1)$$

With $\mathcal{V} = \pi D^3/6$ and because $\gamma_{ice} \gg \gamma_{air}$ (i.e., $\mathcal{W} \gg F_B$), Eq. 1 can be simplified to

$$U = \left(\frac{4}{3}\frac{\rho_{ice}}{\rho_{air}}\frac{gD}{C_D}\right)^{1/2} \qquad (2)$$

By using $\rho_{ice} = 1.84$ slugs/ft^3, $\rho_{air} = 2.38 \times 10^{-3}$ slugs/ft^3, and $D = 1.5$ in. $= 0.125$ ft, Eq. 2 becomes

$$U = \left[\frac{4(1.84 \text{ slugs/ft}^3)(32.2 \text{ ft/s}^2)(0.125 \text{ ft})}{3(2.38 \times 10^{-3} \text{ slugs/ft}^3)C_D}\right]^{1/2}$$

or

$$U = \frac{64.5}{\sqrt{C_D}} \qquad (3)$$

where U is in feet per second. To determine U, we must know C_D. Unfortunately, C_D is a function of the Reynolds number (see Fig. 9.15), which is not known unless U is known. Thus, we must use an iterative technique similar to that done with the Moody chart for certain types of pipe flow problems (see Section 8.5).

From Fig. 9.15 we expect that C_D is on the order of 0.5. Thus, we assume $C_D = 0.5$ and from Eq. 3 obtain

$$U = \frac{64.5}{\sqrt{0.5}} = 91.2 \text{ ft/s}$$

The corresponding Reynolds number (assuming $\nu = 1.57 \times 10^{-4}$ ft^2/s) is

$$Re = \frac{UD}{\nu} = \frac{91.2 \text{ ft/s} (0.125 \text{ ft})}{1.57 \times 10^{-4} \text{ ft}^2/\text{s}} = 7.26 \times 10^4$$

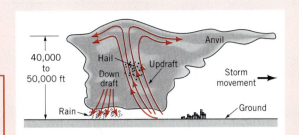

■ **FIGURE E9.6a**

For this value of Re we obtain $C_D = 0.5$ from Fig. 9.15. Thus, our assumed value of $C_D = 0.5$ was correct. The corresponding value of U is

$$U = 91.2 \text{ ft/s} = 62.2 \text{ mph} \qquad \textbf{(Ans)}$$

COMMENTS By repeating the calculations for various altitudes, z, above sea level (using the properties of the U.S. Standard Atmosphere given in **Appendix C**), the results shown in Fig. E9.6b are obtained. Because of the decease in density with altitude, the hail falls even faster through the upper portions of the storm than when it hits the ground.

Clearly, an airplane flying through such an updraft would feel its effects (even if it were able to dodge the hail). As seen from Eq. 2, the larger the hail, the stronger the necessary updraft. Hailstones greater than 6 in. in diameter have been reported. In reality, a hailstone is seldom spherical and often not smooth. However, the calculated updraft velocities are in agreement with measured values.

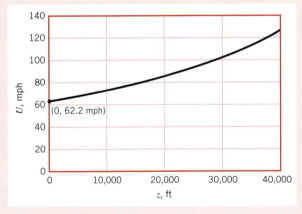

■ **FIGURE E9.6b**

Compressibility Effects. If the velocity of the object is sufficiently large, compressibility effects become important and the drag coefficient becomes a function of the Mach number, $Ma = U/c$, where c is the speed of sound in the fluid. For low Mach numbers, $Ma < 0.5$ or so, compressibility effects are unimportant and the drag coefficient is essentially independent of Ma. However, for larger Mach number flows, the drag coefficient can be strongly dependent on Ma.

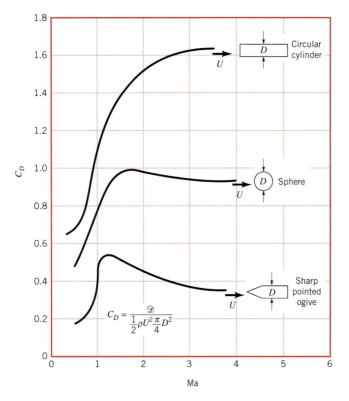

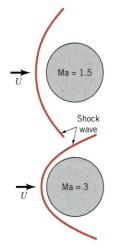

For most objects, values of C_D increase dramatically in the vicinity of Ma = 1 (i.e., sonic flow). This change in character, indicated by Fig. 9.17, is due to the existence of shock waves as indicated by the figure in the margin. Shock waves are extremely narrow regions in the flow field across which the flow parameters change in a nearly discontinuous manner. Shock waves, which cannot exist in subsonic flows, provide a mechanism for the generation of drag that is not present in the relatively low-speed subsonic flows. More information on these important topics can be found in standard texts about compressible flow and aerodynamics (Refs. 7, 8, 18).

Surface Roughness. In general, for streamlined bodies, the drag increases with increasing surface roughness. Great care is taken to design the surfaces of airplane wings to be as smooth as possible, as protruding rivets or screw heads can cause a considerable increase in the drag. However, for an extremely blunt body, such as a flat plate normal to the flow, the drag is independent of the surface roughness, as the shear stress is not in the upstream flow direction and contributes nothing to the drag.

For blunt bodies such as a circular cylinder or sphere, an increase in surface roughness can actually cause a decrease in the drag. This is illustrated for a sphere in Fig. 9.18. As discussed in Section 9.2.6, when the Reynolds number reaches the critical value (Re = 3×10^5 for a smooth sphere), the boundary layer becomes turbulent and the wake region behind the sphere becomes considerably narrower than if it were laminar (see Figs. 9.12 and 9.15). The result is a considerable drop in pressure drag with a slight increase in friction drag, combining to give a smaller overall drag (and C_D).

The boundary layer can be tripped into turbulence at a smaller Reynolds number by using a rough-surfaced sphere. For example, the critical Reynolds number for a golf ball is approximately Re = 4×10^4. In the range $4 \times 10^4 < $ Re $ < 4 \times 10^5$, the drag on the standard rough (i.e., dimpled) golf ball is considerably less ($C_{Drough}/C_{Dsmooth} \approx 0.25/0.5 = 0.5$)

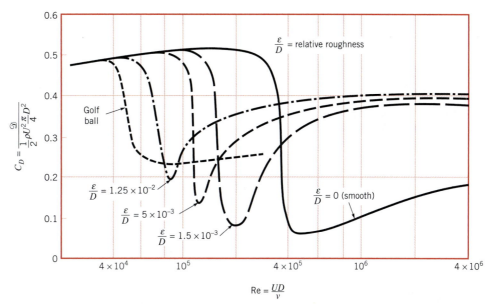

■ **FIGURE 9.18**   **The effect of surface roughness on the drag coefficient of a sphere in the Reynolds number range for which the laminar boundary layer becomes turbulent (Ref. 4).**

than for the smooth ball. As is shown in Example 9.7, this is precisely the Reynolds number range for well-hit golf balls—hence, a reason for dimples on golf balls. The Reynolds number range for well-hit table tennis balls is less than $Re = 4 \times 10^4$. Thus, table tennis balls are smooth.

EXAMPLE 9.7   Effect of Surface Roughness

GIVEN   A well-hit golf ball (diameter $D = 1.69$ in., weight $\mathcal{W} = 0.0992$ lb) can travel at $U = 200$ ft/s as it leaves the tee. A well-hit table tennis ball (diameter $D = 1.50$ in., weight $\mathcal{W} = 0.00551$ lb) can travel at $U = 60$ ft/s as it leaves the paddle.

FIND   Determine the drag on a standard golf ball, a smooth golf ball, and a table tennis ball for the conditions given. Also determine the deceleration of each ball for these conditions.

SOLUTION

For either ball, the drag can be obtained from

$$\mathcal{D} = \frac{1}{2}\rho U^2 \frac{\pi}{4} D^2 C_D \qquad (1)$$

where the drag coefficient, C_D, is given in Fig. 9.18 as a function of the Reynolds number and surface roughness. For the golf ball in standard air

$$Re = \frac{UD}{\nu} = \frac{(200 \text{ ft/s})(1.69/12 \text{ ft})}{1.57 \times 10^{-4} \text{ ft}^2/\text{s}} = 1.79 \times 10^5$$

while for the table tennis ball

$$Re = \frac{UD}{\nu} = \frac{(60 \text{ ft/s})(1.50/12 \text{ ft})}{1.57 \times 10^{-4} \text{ ft}^2/\text{s}} = 4.78 \times 10^4$$

The corresponding drag coefficients are $C_D = 0.25$ for the standard golf ball, $C_D = 0.51$ for the smooth golf ball, and $C_D = 0.50$ for the table tennis ball. Hence, from Eq. 1 for the standard golf ball

$$\mathcal{D} = \frac{1}{2}(2.38 \times 10^{-3} \text{ slugs/ft}^3)(200 \text{ ft/s})^2 \frac{\pi}{4}\left(\frac{1.69}{12} \text{ ft}\right)^2(0.25)$$

$$= 0.185 \text{ lb} \qquad \text{(Ans)}$$

for the smooth golf ball

$$\mathcal{D} = \frac{1}{2}(2.38 \times 10^{-3} \text{ slugs/ft}^3)(200 \text{ ft/s})^2 \frac{\pi}{4}\left(\frac{1.69}{12} \text{ ft}\right)^2(0.51)$$

$$= 0.378 \text{ lb} \qquad \text{(Ans)}$$

and for the table tennis ball

$$\mathcal{D} = \frac{1}{2}(2.38 \times 10^{-3} \text{ slugs/ft}^3)(60 \text{ ft/s})^2 \frac{\pi}{4}\left(\frac{1.50}{12}\text{ ft}\right)^2 (0.50)$$

$$= 0.0263 \text{ lb} \qquad \textbf{(Ans)}$$

The corresponding decelerations are $a = \mathcal{D}/m = g\mathcal{D}/W$, where m is the mass of the ball. Thus, the deceleration relative to the acceleration of gravity, a/g (i.e., the number of g's deceleration), is $a/g = \mathcal{D}/W$ or

$$\frac{a}{g} = \frac{0.185 \text{ lb}}{0.0992 \text{ lb}} = 1.86 \text{ for the standard golf ball} \quad \textbf{(Ans)}$$

$$\frac{a}{g} = \frac{0.378 \text{ lb}}{0.0992 \text{ lb}} = 3.81 \text{ for the smooth golf ball} \quad \textbf{(Ans)}$$

and

$$\frac{a}{g} = \frac{0.0263 \text{ lb}}{0.00551 \text{ lb}} = 4.77 \text{ for the table tennis ball} \quad \textbf{(Ans)}$$

COMMENTS   Note that there is a considerably smaller deceleration for the rough golf ball than for the smooth one. Because of its much larger drag-to-mass ratio, the table tennis ball slows down relatively quickly and does not travel as far as the golf ball. (Note that with $U = 60$ ft/s the standard golf ball has a drag of $\mathcal{D} = 0.0200$ lb and a deceleration of $a/g = 0.202$, considerably less than the $a/g = 4.77$ of the table tennis ball. Conversely, a table tennis ball hit from a tee at 200 ft/s would decelerate at a rate of $a = 1740$ ft/s^2, or $a/g = 54.1$. It would not travel nearly as far as the golf ball.)

By repeating the above calculations, the drag as a function of speed for both a standard golf ball and a smooth golf ball is shown in Fig. E9.7

The Reynolds number range for which a rough golf ball has smaller drag than a smooth one (i.e., 4×10^4 to 3.6×10^5) corresponds to a flight velocity range of $45 < U < 400$ ft/s. This is comfortably within the range of most golfers. (The fastest tee shot by top professional golfers is approximately 280 ft/s.) As discussed in Section 9.4.2, the dimples (roughness) on a golf ball also help produce a lift (due to the spin of the ball) that allows the ball to travel farther than a smooth ball.

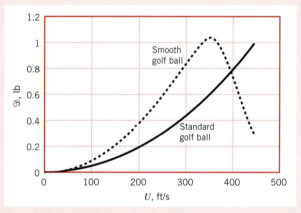

■ **FIGURE E9.7**

Froude Number Effects.   Another parameter on which the drag coefficient may be strongly dependent is the Froude number, $\text{Fr} = U/\sqrt{g\ell}$. As is discussed in Chapter 10, the Froude number is a ratio of the free-stream speed to a typical wave speed on the interface of two fluids, such as the surface of the ocean. An object moving on the surface, such as a ship, often produces waves that require a source of energy to generate. This energy comes from the ship and is manifest as a drag. [Recall that the rate of energy production (power) equals speed times force.] The nature of the waves produced often depends on the Froude number of the flow and the shape of the object—the waves generated by a water skier "plowing"

V9.7 Jet ski

through the water at a low speed (low Fr) are different from those generated by the skier "planing" along the surface at a high speed (larger Fr).

Composite Body Drag. Approximate drag calculations for a complex body can often be obtained by treating the body as a composite collection of its various parts.

EXAMPLE 9.8   Drag on a Composite Body

GIVEN A 60-mph (i.e., 88-fps) wind blows past the water tower shown in Fig. E9.8a.

FIND Estimate the moment (torque), M, needed at the base to keep the tower from tipping over.

SOLUTION

We treat the water tower as a sphere resting on a circular cylinder and assume that the total drag is the sum of the drag from these parts. The free-body diagram of the tower is shown in Fig. E9.8b. By summing moments about the base of the tower, we obtain

$$M = \mathscr{D}_s\left(b + \frac{D_s}{2}\right) + \mathscr{D}_c\left(\frac{b}{2}\right) \qquad (1)$$

where

$$\mathscr{D}_s = \frac{1}{2}\rho U^2 \frac{\pi}{4} D_s^2 C_{Ds} \qquad (2)$$

and

$$\mathscr{D}_c = \frac{1}{2}\rho U^2 b D_c C_{Dc} \qquad (3)$$

are the drag on the sphere and cylinder, respectively. For standard atmospheric conditions, the Reynolds numbers are

$$\mathrm{Re}_s = \frac{U D_s}{\nu} = \frac{(88\ \text{ft/s})(40\ \text{ft})}{1.57 \times 10^{-4}\ \text{ft}^2/\text{s}} = 2.24 \times 10^7$$

and

$$\mathrm{Re}_c = \frac{U D_c}{\nu} = \frac{(88\ \text{ft/s})(15\ \text{ft})}{1.57 \times 10^{-4}\ \text{ft}^2/\text{s}} = 8.41 \times 10^6$$

The corresponding drag coefficients, C_{Ds} and C_{Dc}, can be approximated from Fig. 9.15 as

$$C_{Ds} \approx 0.3 \quad \text{and} \quad C_{Dc} \approx 0.7$$

Note that the value of C_{Ds} was obtained by an extrapolation of the given data to Reynolds numbers beyond those given (a potentially dangerous practice!). From Eqs. 2 and 3 we obtain

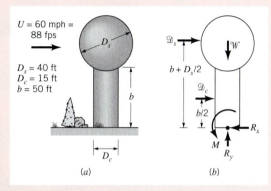

■ FIGURE E9.8

$$\mathscr{D}_s = 0.5(2.38 \times 10^{-3}\ \text{slugs/ft}^3)(88\ \text{ft/s})^2 \frac{\pi}{4}(40\ \text{ft})^2(0.3)$$

$$= 3470\ \text{lb}$$

and

$$\mathscr{D}_c = 0.5(2.38 \times 10^{-3}\ \text{slugs/ft}^3)(88\ \text{ft/s})^2(50\ \text{ft} \times 15\ \text{ft})(0.7)$$

$$= 4840\ \text{lb}$$

From Eq. 1 the corresponding moment needed to prevent the tower from tipping is

$$M = 3470\ \text{lb}\left(50\ \text{ft} + \frac{40}{2}\ \text{ft}\right) + 4840\ \text{lb}\left(\frac{50}{2}\ \text{ft}\right)$$

$$= 3.64 \times 10^5\ \text{ft·lb} \qquad \textbf{(Ans)}$$

COMMENT The above result is only an estimate because (a) the wind is probably not uniform from the top of the tower to the ground, (b) the tower is not exactly a combination of a smooth sphere and a circular cylinder, (c) the cylinder is not of infinite length, (d) there will be some interaction between the flow past the cylinder and that past the sphere so that the net drag is not exactly the sum of the two, and (e) a drag coefficient value was obtained by extrapolation of the given data. However, such approximate results are often quite accurate.

V9.8 Drag on a truck

Drag coefficient information for a very wide range of objects is available in the literature. Some of this information is given in Figs. 9.19, 9.20, and 9.21 for a variety of two- and three-dimensional, natural and man-made objects. Recall that a drag coefficient of unity is equivalent to the drag produced by the dynamic pressure acting on an area of size A. That is, $\mathcal{D} = \frac{1}{2}\rho U^2 A C_D = \frac{1}{2}\rho U^2 A$ if $C_D = 1$. Typical nonstreamlined objects have drag coefficients on this order.

Shape	Reference area A (b = length)	Drag coefficient $C_D = \dfrac{\mathcal{D}}{\frac{1}{2}\rho U^2 A}$	Reynolds number $Re = \rho U D/\mu$
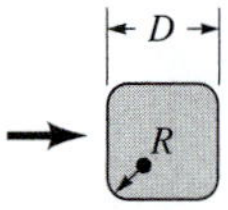 Square rod with rounded corners	$A = bD$	$\begin{array}{c\|c} R/D & C_D \\ \hline 0 & 2.2 \\ 0.02 & 2.0 \\ 0.17 & 1.2 \\ 0.33 & 1.0 \end{array}$	$Re = 10^5$
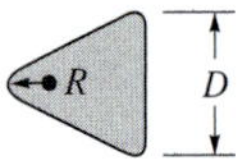 Rounded equilateral triangle	$A = bD$	$\begin{array}{c\|c c} R/D & \rightarrow & \leftarrow \\ \hline 0 & 1.4 & 2.1 \\ 0.02 & 1.2 & 2.0 \\ 0.08 & 1.3 & 1.9 \\ 0.25 & 1.1 & 1.3 \end{array}$	$Re = 10^5$
 Semicircular shell	$A = bD$	$\rightarrow$ 2.3 $\leftarrow$ 1.1	$Re = 2 \times 10^4$
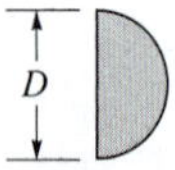 Semicircular cylinder	$A = bD$	$\rightarrow$ 2.15 $\leftarrow$ 1.15	$Re > 10^4$
 T-beam	$A = bD$	$\rightarrow$ 1.80 $\leftarrow$ 1.65	$Re > 10^4$
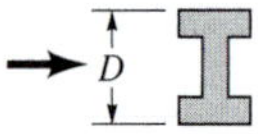 I-beam	$A = bD$	2.05	$Re > 10^4$
 Angle	$A = bD$	$\rightarrow$ 1.98 $\leftarrow$ 1.82	$Re > 10^4$
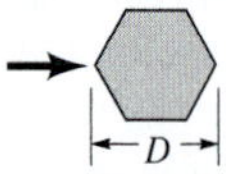 Hexagon	$A = bD$	1.0	$Re > 10^4$
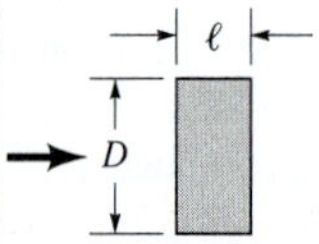 Rectangle	$A = bD$	$\begin{array}{c\|c} \ell/D & C_D \\ \hline \leq 0.1 & 1.9 \\ 0.5 & 2.5 \\ 0.65 & 2.9 \\ 1.0 & 2.2 \\ 2.0 & 1.6 \\ 3.0 & 1.3 \end{array}$	$Re = 10^5$

■ **FIGURE 9.19** Typical drag coefficients for regular two-dimensional objects (Refs. 4 and 5).

Shape		Reference area A	Drag coefficient C_D	Reynolds number $Re = \rho U D/\mu$
	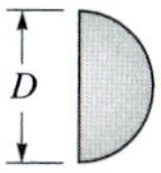 Solid hemisphere	$A = \dfrac{\pi}{4}D^2$	→ 1.17 ← 0.42	$Re > 10^4$
	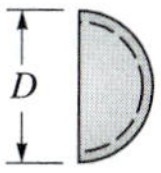 Hollow hemisphere	$A = \dfrac{\pi}{4}D^2$	→ 1.42 ← 0.38	$Re > 10^4$
	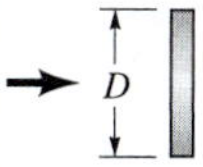 Thin disk	$A = \dfrac{\pi}{4}D^2$	1.1	$Re > 10^3$
	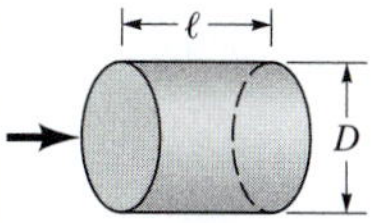 Circular rod parallel to flow	$A = \dfrac{\pi}{4}D^2$	ℓ/D C_D 0.5 1.1 1.0 0.93 2.0 0.83 4.0 0.85	$Re > 10^5$
	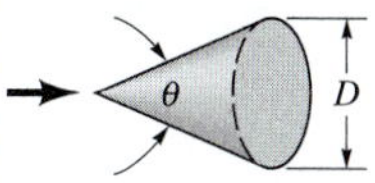 Cone	$A = \dfrac{\pi}{4}D^2$	θ, degrees C_D 10 0.30 30 0.55 60 0.80 90 1.15	$Re > 10^4$
	Cube	$A = D^2$	1.05	$Re > 10^4$
	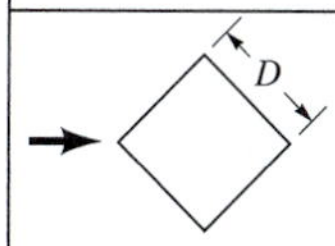 Cube	$A = D^2$	0.80	$Re > 10^4$
	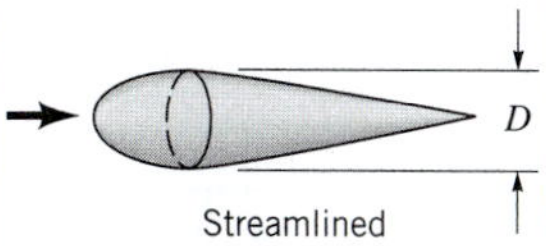 Streamlined body	$A = \dfrac{\pi}{4}D^2$	0.04	$Re > 10^5$

■ **FIGURE 9.20** Typical drag coefficients for regular three-dimensional objects (Ref. 4).

Shape		Reference area	Drag coefficient C_D		
	 Parachute	Frontal area $A = \frac{\pi}{4}D^2$	1.4		
	Porous parabolic dish	Frontal area $A = \frac{\pi}{4}D^2$	Porosity	0	0.2	0.5

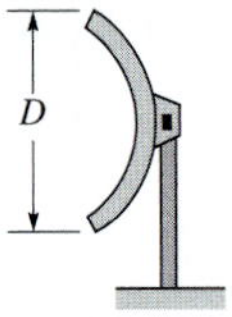

Inside the porous parabolic dish cell:

Porosity	0	0.2	0.5
→	1.42	1.20	0.82
←	0.95	0.90	0.80

Porosity = open area/total area

Shape		Reference area	Drag coefficient C_D
	 Average person	Standing Sitting Crouching	$C_D A = 9\ \text{ft}^2$ $C_D A = 6\ \text{ft}^2$ $C_D A = 2.5\ \text{ft}^2$
	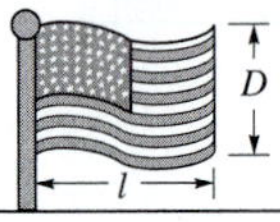 Fluttering flag	$A = \ell D$	$\begin{array}{c\|c} \ell/D & C_D \\ \hline 1 & 0.07 \\ 2 & 0.12 \\ 3 & 0.15 \end{array}$
	 Empire State Building	Frontal area	1.4
 Six-car passenger train		Frontal area	1.8
Bikes	Upright commuter	$A = 5.5\ \text{ft}^2$	1.1
	 Racing	$A = 3.9\ \text{ft}^2$	0.88
	 Drafting	$A = 3.9\ \text{ft}^2$	0.50
	 Streamlined	$A = 5.0\ \text{ft}^2$	0.12
Tractor-trailer tucks	Standard	Frontal area	0.96
	With fairing	Frontal area	0.76
	With fairing and gap seal	Frontal area	0.70
	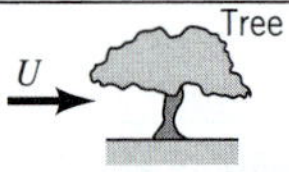 $U = 10$ m/s $U = 20$ m/s $U = 30$ m/s	Frontal area	0.43 0.26 0.20
	 Dolphin	Wetted area	0.0036 at $Re = 6 \times 10^6$ (flat plate has $C_{Df} = 0.0031$)
	 Large birds	Frontal area	0.40

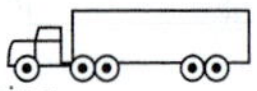

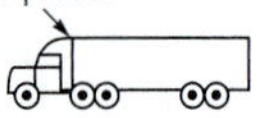

■ **FIGURE 9.21** **Typical drag coefficients for objects of interest (Refs. 4, 5, 11, and 14).**

9.4   Lift

As indicated in Section 9.1, any object moving through a fluid will experience a net force of the fluid on the object. For symmetrical objects, this force will be in the direction of the free stream—a drag, $\mathcal{D}$. If the object is not symmetrical (or if it does not produce a symmetrical flow field, such as the flow around a rotating sphere), there may also be a force normal to the free stream—a lift, $\mathcal{L}$.

9.4.1   Surface Pressure Distribution

The lift can be determined from Eq. 9.2 if the distributions of pressure and wall shear stress around the entire body are known. As indicated in Section 9.1, such data are usually not known. Typically, the lift is given in terms of the lift coefficient

$$C_L = \frac{\mathcal{L}}{\frac{1}{2}\,\rho U^2 A} \tag{9.25}$$

which is obtained from experiments, advanced analysis, or numerical considerations.

The most important parameter that affects the lift coefficient is the shape of the object. Considerable effort has gone into designing optimally shaped lift-producing devices. We will emphasize the effect of the shape on lift—the effects of the other dimensionless parameters can be found in the literature (Refs. 9, 10, 18).

Most common lift-generating devices (i.e., airfoils, fans, spoilers on cars) operate in the large Reynolds number range in which the flow has a boundary layer character, with viscous effects confined to the boundary layers and wake regions. For such cases the wall shear stress, τ_w, contributes little to the lift. Most of the lift comes from the surface pressure distribution.

A typical device designed to produce lift does so by generating a pressure distribution that is different on the top and bottom surfaces. For large Reynolds number flows these pressure distributions are usually directly proportional to the dynamic pressure, $\rho U^2/2$, with viscous effects being of secondary importance. Hence, as indicated by the figure in the margin, for a given airfoil the lift is proportional to the square of the airspeed. Two airfoils used to produce lift are indicated in Fig. 9.22. Clearly the symmetrical one cannot produce lift unless the angle of attack, α, is nonzero. Because of the asymmetry of the nonsymmetrical airfoil, the pressure distributions on the upper and lower surfaces are different, and a lift is produced even with $\alpha = 0$.

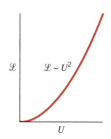

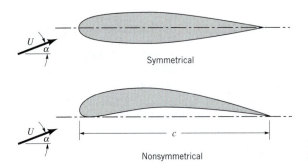

■ **FIGURE 9.22** **Symmetrical and nonsymmetrical airfoils.**

Because most airfoils are thin, it is customary to use the planform area, $A = bc$, in the definition of the lift coefficient. Here b is the length of the airfoil and c is the *chord length*—the length from the leading edge to the trailing edge as indicated in Fig. 9.22. Typical lift coefficients so defined are on the order of unity (see Fig. 9.23). That is, the lift force is on the order of the dynamic pressure times the planform area of the wing, $\mathcal{L} \approx (\rho U^2/2)A$. The *wing loading,* defined as the average lift per unit area of the wing, $\mathcal{L}/A$, therefore, increases with speed. For example, the wing loading of the 1903 Wright Flyer aircraft was 1.5 lb/ft^2, whereas for the present-day Boeing 747 aircraft it is 150 lb/ft^2. The wing loading for a bumble bee is approximately 1 lb/ft^2 (Ref. 11).

In many lift-generating devices the important quantity is the ratio of the lift to drag developed, $\mathcal{L}/\mathcal{D} = C_L/C_D$. Such information is often presented in terms of C_L/C_D versus α, as shown in Fig. 9.23a, or in a *lift-drag polar* of C_L versus C_D with α as a parameter, as

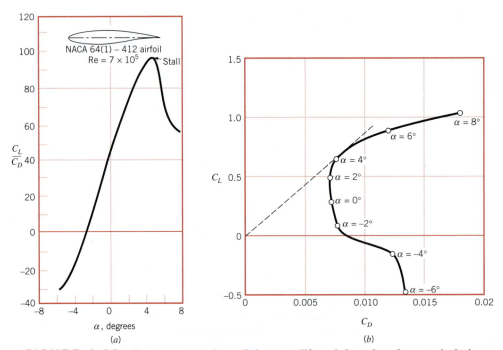

■ **FIGURE 9.23** **Two representations of the same lift and drag data for a typical airfoil: (a) lift-to-drag ratio as a function of angle of attack, with the onset of boundary layer separation on the upper surface indicated by the occurrence of stall, and (b) the lift and drag polar diagram with the angle of attack indicated (Ref. 17).**

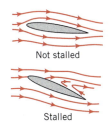

Not stalled

Stalled

shown in Fig. 9.23*b*. The most efficient angle of attack (i.e., largest C_L/C_D) can be found by drawing a line tangent to the $C_L - C_D$ curve from the origin, as shown in Fig. 9.23*b*.

Although viscous effects and the wall shear stress contribute little to the direct generation of lift, they play an extremely important role in the design and use of lifting devices. This is because of the viscosity-induced boundary layer separation that can occur on nonstreamlined bodies such as airfoils that have too large an angle of attack. Up to a certain point, the lift coefficient increases rather steadily with the angle of attack. If α is too large, the boundary layer on the upper surface separates, the flow over the wing develops a wide, turbulent wake region, the lift decreases, and the drag increases. This condition, indicated in the figure in the margin, is termed *stall.* Such conditions are extremely dangerous if they occur while the airplane is flying at a low altitude where there is not sufficient time and altitude to recover from the stall.

F l u i d s   i n   t h e   N e w s

Bats feel turbulence Researchers have discovered that at certain locations on the wings of bats, there are special touch-sensing cells with a tiny hair poking out of the center of the cell. These cells, which are very sensitive to air flowing across the wing surface, can apparently detect turbulence in the flow over the wing. If these hairs are removed the bats fly well in a straight line, but when maneuvering to avoid obstacles, their elevation control is erratic. When the hairs grow back, the bats regain their complete flying skills. It is proposed that these touch-sensing cells are used to detect turbulence on the wing surface and thereby tell bats when to adjust the angle of attack and curvature of their wings in order to avoid stalling out in midair. ■

As indicated earlier, the lift and drag on an airfoil can be altered by changing the angle of attack. This actually represents a change in the shape of the object. Other shape changes can be used to alter the lift and drag when desirable. In modern airplanes it is common to utilize leading edge and trailing edge flaps as shown in Fig. 9.24. To generate

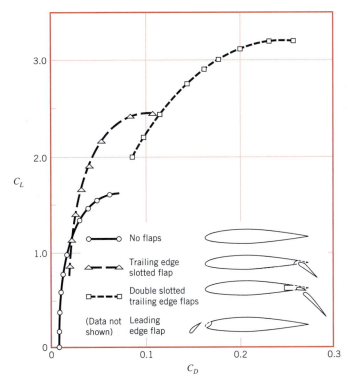

■ **FIGURE 9.24**
Typical lift and drag alterations possible with the use of various types of flap designs (Ref. 15).

the necessary lift during the relatively low-speed landing and takeoff procedures, the airfoil shape is altered by extending special flaps on the front and/or rear portions of the wing. Use of the flaps considerably enhances the lift, although it is at the expense of an increase in the drag (the airfoil is in a "dirty" configuration). This increase in drag is not of much concern during landing and takeoff operations—the decrease in landing or takeoff speed is more important than is a temporary increase in drag. During normal flight with the flaps retracted (the "clean" configuration), the drag is relatively small, and the needed lift force is achieved with the smaller lift coefficient and the larger dynamic pressure (higher speed).

A wide variety of lift and drag information for airfoils can be found in standard aerodynamics books (Refs. 9, 10, 18).

EXAMPLE 9.9   Lift and Power for Human Powered Flight

GIVEN   In 1977 the *Gossamer Condor*, shown in Fig. E9.9a, won the Kremer prize by being the first human-powered aircraft to complete a prescribed figure-eight course around two turning points 0.5 mi apart (Ref. 16). The following data pertain to this aircraft:

$$\text{flight speed} = U = 15 \text{ ft/s}$$
$$\text{wing size} = b = 96 \text{ ft}, c = 7.5 \text{ ft (average)}$$
$$\text{weight (including pilot)} = W = 210 \text{ lb}$$
$$\text{drag coefficient} = C_D = 0.046 \text{ (based on planform area)}$$
$$\text{power train efficiency} = \eta = \text{power to overcome drag/pilot power} = 0.8$$

FIND   Determine

(a)   the lift coefficient, C_L, and

(b)   the power, $\mathcal{P}$, required by the pilot.

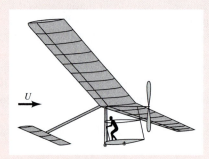

■ **FIGURE E9.9a**

SOLUTION

(a)   For steady flight conditions the lift must be exactly balanced by the weight, or

$$W = \mathcal{L} = \tfrac{1}{2}\rho U^2 A C_L$$

Thus,

$$C_L = \frac{2W}{\rho U^2 A}$$

where $A = bc = 96 \text{ ft} \times 7.5 \text{ ft} = 720 \text{ ft}^2$, $W = 210$ lb, and $\rho = 2.38 \times 10^{-3}$ slugs/ft^3 for standard air. This gives

$$C_L = \frac{2(210 \text{ lb})}{(2.38 \times 10^{-3} \text{ slugs/ft}^3)(15 \text{ ft/s})^2(720 \text{ ft}^2)}$$
$$= 1.09 \qquad \textbf{(Ans)}$$

a reasonable number. The overall lift-to-drag ratio for the aircraft is $C_L/C_D = 1.09/0.046 = 23.7$.

(b)   The product of the power that the pilot supplies and the power train efficiency equals the useful power needed to overcome the drag, $\mathcal{D}$. That is,

$$\eta\mathcal{P} = \mathcal{D}U$$

where

$$\mathcal{D} = \tfrac{1}{2}\rho U^2 A C_D$$

Thus,

$$\mathcal{P} = \frac{\mathcal{D}U}{\eta} = \frac{\tfrac{1}{2}\rho U^2 A C_D U}{\eta} = \frac{\rho A C_D U^3}{2\eta} \qquad \textbf{(1)}$$

or

$$\mathcal{P} = \frac{(2.38 \times 10^{-3} \text{ slugs/ft}^3)(720 \text{ ft}^2)(0.046)(15 \text{ ft/s})^3}{2(0.8)}$$

$$\mathcal{P} = 166 \text{ ft·lb/s}\left(\frac{1 \text{ hp}}{550 \text{ ft·lb/s}}\right) = 0.302 \text{ hp} \qquad \textbf{(Ans)}$$

COMMENT   This power level is obtainable by a well-conditioned athlete (as is indicated by the fact that the flight was completed successfully). Note that only 80% of the pilot's power (i.e., $0.8 \times 0.302 = 0.242$ hp, which corresponds to a drag of $\mathcal{D} = 8.86$ lb) is needed to force the aircraft through the air. The other 20% is lost because of the power train inefficiency.

By repeating the calculations for various flight speeds, the results shown in Fig. E9.9b are obtained. Note from Eq. 1 that for a constant drag coefficient, the power required increases as U^3—a doubling of the speed to 30 ft/s would require an eight-fold increase in power (i.e., 2.42 hp, well beyond the range of any human).

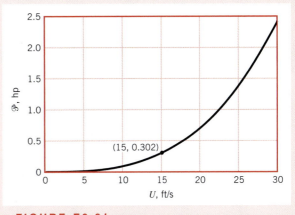

(15, 0.302)

U, ft/s

■ **FIGURE E9.9b**

9.4.2 Circulation

Consider the flow past a finite length airfoil, as indicated in Fig. 9.25. For lift-generating conditions the average pressure on the lower surface is greater than that on the upper surface. Near the tips of the wing this pressure difference will cause some of the fluid to attempt to migrate from the lower to the upper surface, as indicated in Fig. 9.25b. At the same time, this fluid is swept downstream, forming a *trailing vortex* (swirl) from each wing tip (see Fig. 4.3).

V9.9 Wing tip vortices

The trailing vortices from the right and left wing tips are connected by the *bound vortex* along the length of the wing. It is this vortex that generates the *circulation* that produces the lift. The combined vortex system (the bound vortex and the trailing vortices) is termed a horseshoe vortex. The strength of the trailing vortices (which is equal to the strength of the bound vortex) is proportional to the lift generated. Large aircraft (for example, a Boeing 747) can generate very strong trailing vortices that persist for a long time before viscous effects finally cause them to die out. Such vortices are strong enough to flip smaller aircraft out of control if they follow too closely behind the large aircraft.

F l u i d s   i n   t h e   N e w s

Why winglets? Winglets, those upward turning ends of airplane wings, boost the performance by reducing drag. This is accomplished by reducing the strength of the *wingtip vortices* formed by the difference between the high pressure on the lower surface of the wing and the low pressure on the upper surface of the wing. These vortices represent an energy loss and an increase in drag. In essence, the winglet provides an effective increase in the aspect ratio of the wing without extending the wingspan. Winglets come in a variety of styles—the Airbus A320 has a very small upper and lower winglet; the Boeing 747-400 has a conventional, vertical upper winglet; and the Boeing Business Jet (a derivative of the Boeing 737) has an eight-foot winglet with a curving transition from wing to winglet. Since the airflow around the winglet is quite complicated, the winglets must be carefully designed and tested for each aircraft. In the past winglets were more likely to be retrofitted to existing wings, but new airplanes are being designed with winglets from the start. Unlike tailfins on cars, winglets really do work. (See Problem 9.58.) ■

As indicated earlier, the generation of lift is directly related to the production of circulation or vortex flow around the object. A nonsymmetric airfoil, by design, generates its own prescribed amount of swirl and lift. A symmetrical object, such as a circular cylinder or sphere, which normally provides no lift, can generate swirl and lift if it rotates.

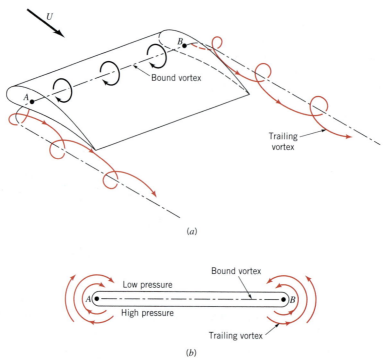

(a)

(b)

■ **FIGURE 9.25** Flow past a finite length wing: (*a*) the horseshoe vortex system produced by the bound vortex and the trailing vortices and (*b*) the leakage of air around the wing tips produces the trailing vortices.

As discussed in Section 6.6.2, inviscid flow past a circular cylinder has the symmetrical flow pattern indicated in Fig. 9.26*a*. By symmetry the lift and drag are zero. However, if the cylinder is rotated about its axis in a stationary real ($\mu \neq 0$) fluid, the rotation will drag some of the fluid around, producing circulation about the cylinder as in Fig. 9.26*b*. When this circulation is combined with an ideal, uniform upstream flow, the flow pattern indicated in Fig. 9.26*c* is obtained. The flow is no longer symmetrical about the horizontal plane through the center of the cylinder; the average pressure is greater on the lower half of the cylinder than on the upper half, and a lift is generated. This effect is called the *Magnus effect*. A similar lift is generated on a rotating sphere. It accounts for the various types of pitches in baseball (i.e., curve ball, floater, sinker), the ability of a soccer player to hook the ball, and the hook, slice, or lift of a golf ball.

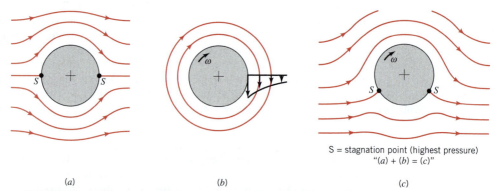

(a)   (b)   (c)

■ **FIGURE 9.26** Inviscid flow past a circular cylinder: (*a*) uniform upstream flow without circulation, (*b*) free vortex at the center of the cylinder, and (*c*) combination of free vortex and uniform flow past a circular cylinder giving nonsymmetric flow and a lift.

9.5 Chapter Summary and Study Guide

In this chapter the flow past objects is discussed. It is shown how the pressure and shear stress distributions on the surface of an object produce the net lift and drag forces on the object.

The character of flow past an object is a function of the Reynolds number. For large Reynolds number flows a thin boundary layer forms on the surface. Properties of this boundary layer flow are discussed. These include the boundary layer thickness, whether the flow is laminar or turbulent, and the wall shear stress exerted on the object. In addition, boundary layer separation and its relationship to the pressure gradient are considered.

The drag, which contains portions due to friction (viscous) effects and pressure effects, is written in terms of the dimensionless drag coefficient. It is shown how the drag coefficient is a function of shape, with objects ranging from very blunt to very streamlined. Other parameters affecting the drag coefficient include the Reynolds number, Froude number, Mach number, and surface roughness.

The lift is written in terms of the dimensionless lift coefficient, which is strongly dependent on the shape of the object. Variation of the lift coefficient with shape is illustrated by the variation of an airfoil's lift coefficient with angle of attack.

The following checklist provides a study guide for this chapter. When your study of the entire chapter and end-of-chapter exercises has been completed you should be able to

drag
lift
lift coefficient
drag coefficient
wake region
boundary layer
boundary layer thickness
laminar boundary layer
transition
turbulent boundary layer
free-stream velocity
favorable pressure gradient
adverse pressure gradient
boundary layer separation
friction drag
pressure drag
stall
circulation

- write out meanings of the terms listed here in the margin and understand each of the related concepts. These terms are particularly important and are set in color and bold type in the text.
- determine the lift and drag on an object from the given pressure and shear stress distributions on the object.
- for flow past a flat plate, calculate the boundary layer thickness, the wall shear stress, the friction drag, and determine whether the flow is laminar or turbulent.
- explain the concept of the pressure gradient and its relationship to boundary layer separation.
- for a given object, obtain the drag coefficient from appropriate tables, figures, or equations and calculate the drag on the object.
- explain why golf balls have dimples.
- for a given object, obtain the lift coefficient from appropriate figures and calculate the lift on the object.

References

1. Schlichting, H., *Boundary Layer Theory,* Eighth Edition, McGraw-Hill, New York, 2000.
2. Rosenhead, L., *Laminar Boundary Layers,* Oxford University Press, London, 1963.
3. White, F. M., *Viscous Fluid Flow,* Third Edition, McGraw-Hill, New York, 2006.
4. Blevins, R. D., *Applied Fluid Dynamics Handbook,* Van Nostrand Reinhold, New York, 1984.
5. Hoerner, S. F., *Fluid-Dynamic Drag,* published by the author, Library of Congress No. 64,19666, 1965.
6. Van Dyke, M., *An Album of Fluid Motion,* Parabolic Press, Stanford, Calif., 1982.
7. Thompson, P. A., *Compressible-Fluid Dynamics,* McGraw-Hill, New York, 1972.
8. Zucrow, M. J., and Hoffman, J. D., *Gas Dynamics, Vol. I,* Wiley, New York, 1976.

9. Shevell, R. S., *Fundamentals of Flight,* Second Edition, Prentice Hall, Englewood Cliffs, NJ, 1989.

10. Kuethe, A. M. and Chow, C. Y., *Foundations of Aerodynamics, Bases of Aerodynamics Design,* Fourth Edition, Wiley, 1986.

11. Vogel, J., *Life in Moving Fluids,* Second Edition, Princeton University Press, New Jersey, 1996.

12. White, F. M., *Fluid Mechanics,* Fifth Edition, McGraw-Hill, New York, 2003.

13. Vennard, J. K., and Street, R. L., *Elementary Fluid Mechanics,* Seventh Edition, Wiley, New York, 1995.

14. Gross, A. C., Kyle, C. R., and Malewicki, D. J., The Aerodynamics of Human Powered Land Vehicles, *Scientific American,* Vol. 249, No. 6, 1983.

15. Abbott, I. H., and Von Doenhoff, A. E., *Theory of Wing Sections,* Dover Publications, New York, 1959.

16. MacReady, P. B., "Flight on 0.33 Horsepower: The Gossamer Condor," *Proc. AIAA 14th Annual Meeting* (Paper No. 78-308), Washington, DC, 1978.

17. Abbott, I. H., von Doenhoff, A. E. and Stivers, L. S., Summary of Airfoil Data, NACA Report No. 824, Langley Field, Va., 1945.

18. Anderson, J. D., *Fundamentals of Aerodynamics,* Fourth Edition, McGraw-Hill, New York, 2007.

19. Homsy, G. M., et al., *Multimedia Fluid Mechanics,* CD-ROM, Cambridge University Press, New York, 2000.

Review Problems

Go to Appendix F for a set of review problems with answers. Detailed solutions can be found in *Student Solution Manual for a Brief Introduction to Fluid Michanics,* by Young et al. (© 2006 John Wiley and Sons, Inc.)

Problems

Note: Unless otherwise indicated use the values of fluid properties found in the tables on the inside of the front cover. Problems designated with an (*) are intended to be solved with the aid of a programmable calculator or a computer. Problems designated with a (†) are "open-ended" problems and require critical thinking in that to work them one must make various assumptions and provide the necessary data. There is not a unique answer to these problems.

Answers to the even-numbered problems are listed at the end of the book. Access to the videos that accompany problems can be obtained through the book's web site, www.wiley.com/college/young. The lab-type problems, FE problems, and FlowLab problems can also be accessed on this web site.

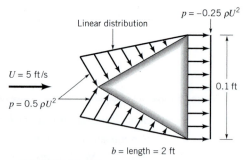

■ FIGURE P9.1

Section 9.1 General External Flow Characteristics

9.1 Assume that water flowing past the equilateral triangular bar shown in Fig. P9.1 produces the pressure distributions indicated. Determine the lift and drag on the bar and the corresponding lift and drag coefficients (based on frontal area). Neglect shear forces.

9.2 Fluid flows past the two-dimensional bar shown in Fig. P9.2. The pressures on the ends of the bar are as shown, and the average shear stress on the top and bottom of the bar is τ_{avg}. Assume that the drag due to pressure is equal to the drag due to viscous effects. (a) Determine τ_{avg} in terms of the dynamic pressure, $\rho U^2/2$. (b) Determine the drag coefficient for this object.

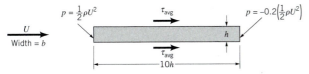

■ **FIGURE P9.2**

9.3   A 0.10-m-diameter circular cylinder moves through air with a speed U. The pressure distribution on the cylinder's surface is approximated by the three straight-line segments shown in Fig. P9.3. Determine the drag on the cylinder. Neglect shear forces.

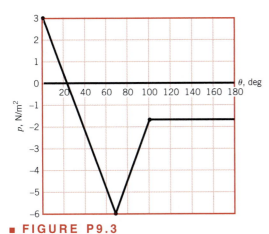

■ **FIGURE P9.3**

9.4   Repeat Problem 9.1 if the object is a cone (made by rotating the equilateral triangle about the horizontal axis through its tip) rather than a triangular bar.

†9.5   Estimate the Reynolds numbers associated with the following objects moving through air: **(a)** a snowflake, settling to the ground, **(b)** a mosquito, **(c)** the space shuttle, **(d)** you walking.

Section 9.2   Boundary Layer Characteristics
(also see Lab Problems 9.61 and 9.62)

9.6   A 17-ft-long kayak moves with a speed of 5 ft/s (see **Video V9.2**). Would a boundary layer type flow be developed along the sides of the boat? Explain.

9.7   A viscous fluid flows past a flat plate such that the boundary layer thickness at a distance x_0 from the leading edge is δ_0. Determine the boundary layer thickness at distances $4x_0$, $25x_0$, and $100x_0$ from the leading edge. Assume laminar flow.

9.8   **(a)** A viscous fluid flows past a flat plate such that the boundary layer thickness at a distance 1.3 m from the leading edge is 12 mm. Determine the boundary layer thickness at distances of 0.20, 2.0, and 20 m from the leading edge. Assume laminar flow. **(b)** If the upstream velocity of the flow in part (a) is $U = 1.5$ m/s, determine the kinematic viscosity of the fluid.

9.9   Air enters a square duct through a 1-ft opening as shown in Fig. P9.9. Because the boundary layer displacement thickness increases in the direction of flow, it is necessary to increase the cross-sectional size of the duct if a constant $U = 2$ ft/s velocity is to be maintained outside the boundary layer. Plot a graph of the duct size, d, as a function of x for $0 \leq x \leq 10$ ft if U is to remain constant. Assume laminar flow.

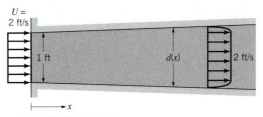

■ **FIGURE P9.9**

9.10   A smooth flat plate of length $\ell = 6$ m and width $b = 4$ m is placed in water with an upstream velocity of $U = 0.5$ m/s. Determine the boundary layer thickness and the wall shear stress at the center and the trailing edge of the plate. Assume a laminar boundary layer.

9.11   An atmospheric boundary layer is formed when the wind blows over the earth's surface. Typically, such velocity profiles can be written as a power law: $u = ay^n$, where the constants a and n depend on the roughness of the terrain. As indicated in Fig. P9.11, typical values are $n = 0.40$ for urban areas, $n = 0.28$ for woodland or suburban areas, and $n = 0.16$ for flat open country. **(a)** If the velocity is 20 ft/s at the bottom of the sail on your boat ($y = 4$ ft), what is the velocity at the top of the mast ($y = 30$ ft)? **(b)** If the average velocity is 10 mph on the 10th floor of an urban building, what is the average velocity on the 60th floor?

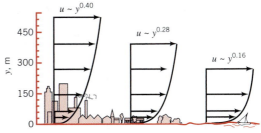

■ **FIGURE P9.11**

9.12   A 30-story office building (each story is 12 ft tall) is built in a suburban industrial park. Plot the dynamic pressure, $\rho u^2/2$, as a function of elevation if the wind blows at hurricane strength (75 mph) at the top of the building. Use the atmospheric boundary layer information of Problem 9.11.

†9.13   If the boundary layer on the hood of your car behaves as one on a flat plate, estimate how far from the front edge of the hood the boundary layer becomes turbulent. How thick is the boundary layer at this location?

9.14 A laminar boundary layer velocity profile is approximated by a sine wave with $u/U = \sin[\pi(y/\delta)/2]$ for $y \leq \delta$, and $u = U$ for $y > \delta$. **(a)** Show that this profile satisfies the appropriate boundary conditions. **(b)** Use the momentum integral equation to determine the boundary layer thickness, $\delta = \delta(x)$.

9.15 A laminar boundary layer velocity profile is approximated by the two straight-line segments indicated in Fig. P9.15. Use the momentum integral equation to determine the boundary layer thickness, $\delta = \delta(x)$, and wall shear stress, $\tau_w = \tau_w(x)$. Compare these results with those in Eqs. 9.8 and 9.11.

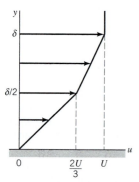

■ **FIGURE P9.15**

*9.16 For a fluid of specific gravity $SG = 0.86$ flowing past a flat plate with an upstream velocity of $U = 5$ m/s, the wall shear stress on a flat plate was determined to be as indicated in the following table. Use the momentum integral equation to determine the boundary layer momentum thickness, $\Theta = \Theta(x)$. Assume $\Theta = 0$ at the leading edge, $x = 0$.

x (m)	τ_w N/m^2
0	—
0.2	13.4
0.4	9.25
0.6	7.68
0.8	6.51
1.0	5.89
1.2	6.57
1.4	6.75
1.6	6.23
1.8	5.92
2.0	5.26

9.17 A plate is oriented parallel to the free stream as is indicated in Fig. 9.17. If the boundary layer flow is laminar, determine the ratio of the drag for case *a* to that for case *b*. Explain your answer physically.

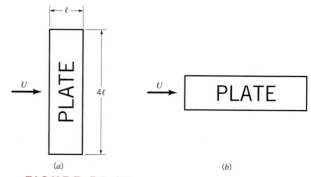

■ **FIGURE P9.17**

9.18 If the drag on one side of a flat plate parallel to the upstream flow is $\mathcal{D}$ when the upstream velocity is U, what will the drag be when the upstream velocity is $2U$; or $U/2$? Assume laminar flow.

Section 9.3 Drag

9.19 It is often assumed that "sharp objects can cut through the air better than blunt ones." Based on this assumption, the drag on the object shown in Fig. P9.19 should be less when the wind blows from right to left than when it blows from left to right. Experiments show that the opposite is true. Explain.

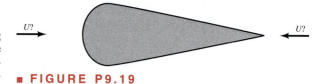

■ **FIGURE P9.19**

9.20 The 5×10^{-6} kg dandelion seed shown in Fig. P9.20 settles through the air with a constant speed of 0.15 m/s. Determine the drag coefficient for this object.

■ **FIGURE P9.20**

9.21 As shown in **Video V9.2** and Fig. P9.21*a*, a kayak is a relatively streamlined object. As a first approximation in calculating the drag on a kayak, assume that the kayak acts as if it were a smooth flat plate 17 ft long and 2 ft wide. Determine the drag as a function of speed and compare your results with measured values given in Fig. P9.21*b*. Comment on reasons why the two sets of values may differ.

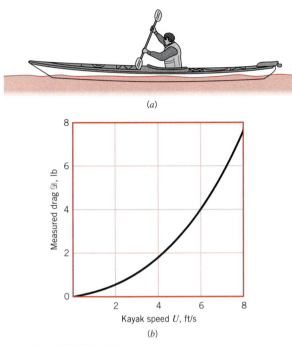

(a)

(b)

■ **FIGURE P9.21**

9.22   Compare the rise velocity of an $\frac{1}{8}$-in.-diameter air bubble in water to the fall velocity of an $\frac{1}{8}$-in.-diameter water drop in air. Assume each to behave as a solid sphere.

9.23   The large, newly planted tree shown in Fig. P9.23 is kept from tipping over in a wind by use of a rope as shown. It is

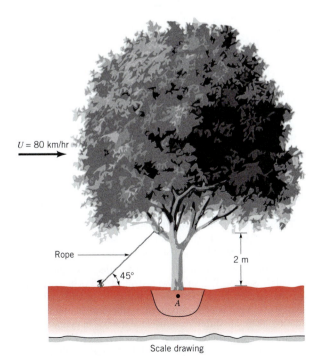

Scale drawing

■ **FIGURE P9.23**

assumed that the sandy soil cannot support any moment about the center of the soil ball, point A. Estimate the tension in the rope if the wind is 80 km/hr. See Fig. 9.21 for drag coefficient data.

9.24   A 38.1-mm-diameter, 0.0245-N table tennis ball is released from the bottom of a swimming pool. With what velocity does it rise to the surface? Assume it has reached its terminal velocity.

†9.25   How fast will a toy balloon filled with helium rise through still air? List all of your assumptions.

9.26   The structure shown in Fig. P9.26 consists of three cylindrical support posts to which an elliptical flat-plate sign is attached. Estimate the drag on the structure when a 50-mph wind blows against it.

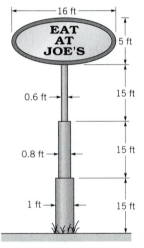

■ **FIGURE P9.26**

9.27   Determine the moment needed at the base of a 30-m-tall, 0.12-m-diameter flag pole to keep it in place in a 20-m/s wind.

9.28   Repeat Problem 9.27 if a 2 × 2.5-m flag is attached to the top of the pole. See Fig. 9.21 for drag coefficient data for flags.

9.29   If for a given vehicle it takes 20 hp to overcome aerodynamic drag while being driven at 55 mph, estimate the horsepower required at 65 mph.

9.30   (See Fluids in the News article titled "Armstrong's aerodynamic bike and suit," Section 9.1.) By appropriate streamlining, the amount of power needed to peddle a bike can be lowered. How much must the drag coefficient for a bike and rider be reduced if the power a bike racer expends while riding 13 m/s is to be reduced by 10 watts? Assume the cross-sectional area of the bike and rider is 0.36 m^2.

9.31   On a day without any wind, your car consumes x gallons of gasoline when you drive at a constant speed, U, from point A to point B and back to point A. Assume that you repeat the journey, driving at the same speed, on another day when there is a steady wind blowing from B to A. Would you expect your fuel consumption to be less than, equal to, or greater than *x* gallons for this windy round trip? Support your answer with appropriate analysis.

†**9.32** Estimate the wind velocity necessary to knock over a garbage can. List your assumptions.

9.33 A 25-ton (50,000-lb) truck coasts down a steep 7% mountain grade without brakes, as shown in Fig. P9.33. The truck's ultimate steady-state speed, *V*, is determined by a balance among weight, rolling resistance, and aerodynamic drag. Assume that the rolling resistance for a truck on concrete is 1.2% of the weight and the drag coefficient is 0.96 for a truck without an air deflector, but 0.70 if it has an air deflector (see **Video V9.8**). Determine *V* for these two situations.

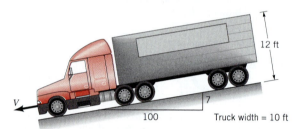

■ **FIGURE P9.33**

9.34 Estimate the velocity with which you would contact the ground if you jumped from an airplane at an altitude of 5000 ft and **(a)** air resistance is negligible, **(b)** air resistance is important, but you forgot your parachute, or **(c)** you use a 25-ft-diameter parachute.

*∗**9.35** The helium-filled balloon shown in Fig. P9.35 is to be used as a wind speed indicator. The specific weight of the helium is $\gamma = 0.011 \text{ lb/ft}^3$, the weight of the balloon material is 0.20 lb, and the weight of the anchoring cable is negligible. Plot a graph of θ as a funtion of *U* for $1 \leq U \leq 50$ mph. Would this be an effective device over the range of *U* indicated? Explain.

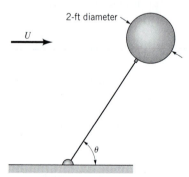

■ **FIGURE P9.35**

9.36 Explain how the drag on a given smokestack could be the same in a 2-mph wind as in a 4-mph wind. Assume the values of ρ and μ are the same for each case.

9.37 A 22 × 34-in. speed limit sign is supported on a 3-in.-wide, 5-ft-long pole. Estimate the bending moment in the

pole at ground level when a 30-mph wind blows against the sign. (See **Video V9.6.**) List any assumptions used in your calculations.

†**9.38** Estimate the wind force on your hand when you hold it out your car window while driving 55 mph. Repeat your calculations if you were to hold your hand out of the window of an airplane flying 550 mph.

9.39 A 4-mm-diameter meteor of specific gravity 2.9 has a speed of 6 km/s at an altitude of 50,000 m where the air density is $1.03 \times 10^{-3} \text{ kg/m}^3$. If the drag coefficient at this large Mach number condition is 1.5, determine the deceleration of the meteor.

9.40 A 30-ft-tall tower is constructed of equal 1-ft segments as is indicated in Fig. P9.40. Each of the four sides is similar. Estimate the drag on the tower when a 75-mph wind blows against it.

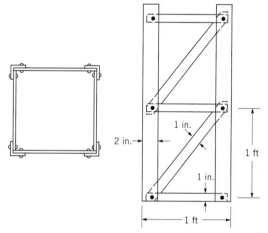

■ **FIGURE P9.40**

9.41 The United Nations Building in New York is approximately 87.5 m wide and 154 m tall. **(a)** Determine the drag on this building if the drag coefficient is 1.3 and the wind speed is a uniform 20 m/s. **(b)** Repeat your calculations if the velocity profile against the building is a typical profile for an urban area (see Problem 9.11) and the wind speed half-way up the building is 20 m/s.

9.42 A hot air balloon roughly spherical in shape has a volume of 70,000 ft³ and a weight of 500 lb (including passengers, basket, balloon fabric, etc.). If the outside air temperature is 80 °F and the temperature within the balloon is 165 °F, estimate the rate at which it will rise under steady-state conditions if the atmospheric pressure is 14.7 psi.

9.43 (See Fluids in the News article titled "Dimpled baseball bats," Section 9.3.3.) How fast must a 3.5-in.-diameter, dimpled baseball bat move through the air in order to take advantage of drag reduction produced by the dimples on the bat. Although there are differences, assume the bat (a cylinder) acts the same as a golf ball in terms of how the dimples affect the transition from a laminar to a turbulent boundary layer.

9.44 A strong wind can blow a golf ball off the tee by pivoting it about point 1 as shown in Fig. P9.44. Determine the wind speed necessary to do this.

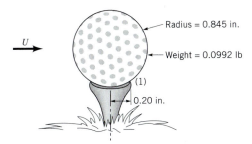

■ FIGURE P9.44

9.45 An airplane tows a banner that is $b = 0.8$ m tall and $\ell = 25$ m long at a speed of 150 km/hr. If the drag coefficient based on the area $b\ell$ is $C_D = 0.06$, estimate the power required to tow the banner. Compare the drag force on the banner with that on a rigid flat plate of the same size. Which has the larger drag force and why?

9.46 (See Fluids in the News article titled "At 10,240 mpg it doesn't cost much to 'fill 'er up,' " Section 9.3.3.) **(a)** Determine the power it takes to overcome aerodynamic drag on a small (6 ft^2 cross section), streamlined ($C_D = 0.12$) vehicle traveling 15 mph. **(b)** Compare the power calculated in part **(a)** with that for a large (36 ft^2 cross-sectional area), non-streamlined ($C_D = 0.48$) SUV traveling 65 mph on the interstate.

9.47 Estimate the power needed to overcome the aerodynamic drag of a person who runs at a rate of 100 yds in 10 s in still air. Repeat the calculations if the race is run into a 20-mph headwind.

9.48 As shown in **Video V9.5** and Fig. P9.48, a vertical wind tunnel can be used for skydiving practice. Estimate the vertical wind speed needed if a 150-lb person is to be able to "float" motionless when the person **(a)** curls up as in a crouching position or **(b)** lies flat. See Fig. 9.21 for appropriate drag coefficient data.

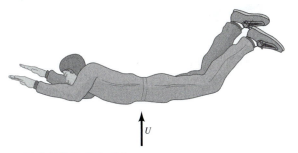

■ FIGURE P9.48

9.49 The wings of old airplanes are often strengthened by the use of wires that provided cross-bracing as shown in Fig. P9.49. If the drag coefficient for the wings was 0.020 (based on the planform area), determine the ratio of the drag from the wire bracing to that from the wings.

Speed: 70 mph
Wing area: 148 ft^2
Wire: length = 160 ft
diameter = 0.05 in.

■ FIGURE P9.49

Section 9.4 Lift

9.50 When the 0.9-lb box kite shown in Fig. P9.50 is flown in a 20-ft/s wind, the tension in the string, which is at a 30° angle relative to the ground, is 3.0 lb. **(a)** Determine the lift and drag coefficients for the kite based on the frontal area of 6.0 ft^2. **(b)** If the wind speed increased to 30 ft/s, would the kite rise or fall? That is, would the 30° angle shown in the figure increase or decrease? Assume the lift and drag coefficients remain the same. Support your answer with appropriate calculations.

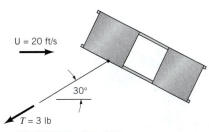

U = 20 ft/s

30°

T = 3 lb

■ FIGURE P9.50

9.51 A Piper Cub airplane has a gross weight of 1750 lb, a cruising speed of 115 mph, and a wing area of 179 ft^2. Determine the lift coefficient of this airplane for these conditions.

9.52 As shown in **Video V9.9** and Fig. P9.52, a spoiler is used on race cars to produce a negative lift, thereby giving a better tractive force. The lift coefficient for the airfoil shown is $C_L = 1.1$, and the coefficient of friction between the wheels and the pavement is 0.6. At a speed of 200 mph, by how much would use of the spoiler increase the maximum tractive force that could

b = spoiler length = 4 ft

Spoiler   1.5 ft

200 mph

■ FIGURE P9.52

be generated between the wheels and ground? Assume the air speed past the spoiler equals the car speed and that the airfoil acts directly over the drive wheels.

9.53 If the takeoff speed of a particular airplane is 120 mi/hr at sea level, what will it be at Denver (elevation 5000 ft)? Use properties of the U.S. Standard Atmosphere.

9.54 A Boeing 747 aircraft weighing 580,000 lb when loaded with fuel and 100 passengers takes off with an air speed of 140 mph. With the same configuration (i.e., angle of attack, flap settings), what is the takeoff speed if it is loaded with 372 passengers? Assume each passenger with luggage weighs 200 lb.

9.55 (a) Show that for unpowered flight (for which the lift, drag, and weight forces are in equilibrium) the glide slope angle, θ, is given by $\tan \theta = C_D/C_L$. (b) If the lift coefficient for a Boeing 757 aircraft is 16 times greater than its drag coefficient, can it glide from an altitude of 30,000 ft to an airport 60 miles away if it loses power from its engines? Explain.

9.56 (See Fluids in the News article titled "Learning from nature," Section 9.4.1.) As indicated in Fig. P9.56, birds can significantly alter their body shape and increase their planform area, A, by spreading their wing and tail feathers, thereby reducing their flight speed. If during landing the planform area is increased by 50% and the lift coefficient increased by 30% while all other parameters are held constant, by what percent is the flight speed reduced?

■ **FIGURE P9.56**

9.57 The landing speed of an airplane such as the space shuttle is dependent on the air density. (See **Video V9.1.**) By what percent must the landing speed be increased on a day when

the temperature is 110° F compared to a day when it is 50° F? Assume the atmospheric pressure remains constant.

9.58 (See Fluids in the News article titled "Why winglets?," Section 9.4.2.) It is estimated that by installing appropriately designed winglets on a certain airplane the drag coefficient will be reduced by 5%. For the same engine thrust, by what percent will the aircraft speed be increased by use of the winglets?

9.59 Over the years there has been a dramatic increase in the flight speed (U), altitude (h), weight ($\mathcal{W}$), and wing loading ($\mathcal{W}/A$ = weight divided by wing area) of aircraft. Use data given in the following table to determine the lift coefficient for each of the aircraft listed.

Aircraft	Year	$\mathcal{W}$ (lb)	U (mph)	$\mathcal{W}/A$ (lb/ft²)	h (ft)
Wright Flyer	1903	750	35	1.5	0
Douglas DC-3	1935	25,000	180	25.0	10,000
Douglas DC-6	1947	105,000	315	72.0	15,000
Boeing 747	1970	800,000	570	150.0	30,000

9.60 When air flows past the airfoil shown in Fig. P9.60, the velocity just outside the boundary layer, u, is as indicated. Estimate the lift coefficient for these conditions.

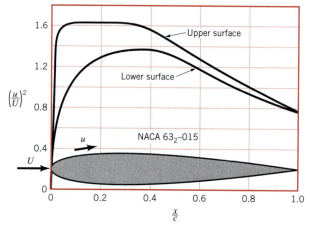

■ **FIGURE P9.60**

Lab Problems

9.61 This problem involves measuring the boundary layer profile on a flat plate. To proceed with this problem, go to the book's web site, www.wiley.com/college/young.

9.62 This problem involves measuring the pressure distribution on a circular cylinder. To proceed with this problem, go to the book's web site, www.wiley.com/college/young.

FlowLab Problems

*9.63 This FlowLab problem involves simulation of flow past an airfoil and investigation of the surface pressure distribution as a function of angle of attack. To proceed with this problem, go to the book's web site, www.wiley.com/college/young.

*9.64 This FlowLab problem involves investigation of the effects of angle-of-attack on lift and drag for flow past an airfoil. To proceed with this problem, go to the book's web site, www.wiley.com/college/young.

*9.65 This FlowLab problem involves simulating the effects of altitude on the lift and drag of an airfoil. To proceed with this problem, go to the book's web site, www.wiley.com/college/young.

*9.66 This FlowLab problem involves comparison between inviscid and viscous flows past an airfoil. To proceed with this problem, go to the book's web site, www.wiley.com/college/young.

*9.67 This FlowLab problem involves simulating the pressure distribution for flow past a cylinder and investigating the differences between inviscid and viscous flows. To proceed with this problem, go to the book's web site, www.wiley.com/college/young.

*9.68 This FlowLab problem involves comparing CFD predictions and theoretical values of the drag coefficient of flow past a cylinder. To proceed with this problem, go to the book's web site, www.wiley.com/college/young.

*9.69 This FlowLab problem involves simulating the unsteady flow past a cylinder. To proceed with this problem, go to the book's web site, www.wiley.com/college/young.

FE Exam Problems

Sample FE (Fundamentals of Engineering) exam questions for fluid mechanics are provided on the book's web site, www.wiley.com/college/young.

Hydraulic jump: *Under certain conditions, when water flows in an open channel the depth of the water may increase considerably over a short distance along the channel. This phenomenon is termed a hydraulic jump (water flow from left to right).* (Photograph by Bruce Munson.)

Open-Channel Flow

LEARNING OBJECTIVES

After completing this chapter, you should be able to:

- discuss the general characteristics of open-channel flow.
- use a specific energy diagram.
- apply appropriate equations to analyze open-channel flow with uniform depth.
- calculate key properties of a hydraulic jump.
- determine flowrates based on open-channel flow-measuring devices.

O*pen-channel flow* involves the flow of a liquid in a channel or conduit that is not completely filled. A free surface exists between the flowing fluid (usually water) and fluid above it (usually the atmosphere). The main driving force for such flows is the fluid weight—gravity forces the fluid to flow downhill. Most open-channel flow results are based on correlations obtained from model and full-scale experiments. Additional information can be gained from various analytical and numerical efforts.

10.1   General Characteristics of Open-Channel Flow

An open-channel flow is classified as *uniform flow* (UF) if the depth of flow does not vary along the channel ($dy/dx = 0$, where y is the fluid depth and x is the distance along the channel). Conversely, it is *nonuniform flow* or *varied flow* if the depth varies with distance ($dy/dx \neq 0$). Nonuniform flows are further classified as *rapidly varying flow* (RVF) if the flow depth changes considerably over a relatively short distance, $dy/dx \sim 1$. *Gradually varying flows* (GVF) are those in which the flow depth changes slowly with distance along the channel, $dy/dx \ll 1$. Examples of these types of flow are illustrated in Fig. 10.1.

As for any flow geometry, open-channel flow may be *laminar, transitional,* or *turbulent,* depending on various conditions involved. Which type of flow occurs depends on the Reynolds number, Re $= \rho V R_h / \mu$, where V is the average velocity of the fluid and R_h is the hydraulic radius of the channel (see Section 10.4). Since most open-channel flows involve water (which has a fairly small viscosity) and have relatively large characteristic lengths, most open-channel flows have large Reynolds numbers. Thus, it is rare to have laminar open-channel flows.

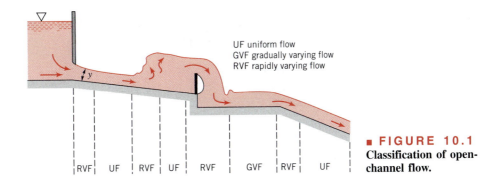

UF uniform flow
GVF gradually varying flow
RVF rapidly varying flow

| RVF | UF | RVF | UF | RVF | GVF | RVF | UF |

■ **FIGURE 10.1**
Classification of open-channel flow.

Open-channel flows involve a free surface that can deform from its undisturbed relatively flat configuration to form waves. The character of an open-channel flow may depend strongly on how fast the fluid is flowing relative to how fast a typical wave moves relative to the fluid. The dimensionless parameter that describes this behavior is termed the Froude number, $\text{Fr} = V/(gy)^{1/2}$, where y is the fluid depth. The special case of a flow with a Froude number of unity, $\text{Fr} = 1$, is termed a ***critical flow.*** If the Froude number is less than 1, the flow is ***subcritical*** (or *tranquil*). A flow with the Froude number greater than 1 is termed ***supercritical*** (or *rapid*).

10.2   Surface Waves

The distinguishing feature of flows involving a free surface (as in open-channel flows) is the opportunity for the free surface to distort into various shapes. The surface of a lake or the ocean is seldom "smooth as a mirror." It is usually distorted into ever-changing patterns associated with surface waves.

F l u i d s   i n   t h e   N e w s

Highest tides in the world The *unsteady*, periodic ocean tides are a result of the slightly variable gravitational forces of the moon and the sun felt at any location on the surface of the spinning Earth. At most ocean shores the tides are only a few feet. The greatest difference between high and low tide ever recorded, 53.38 ft, occurred in the Bay of Fundy, in Canada. The extremely high tides in the Bay of Fundy are a result of two factors. First, the gradual tapering (width) and shallowing (depth) of the bay constricts the tidal flow into the bay from its mouth on the Atlantic Ocean. Second, the precise size of the bay which tapers from 62 miles wide and 600 ft deep at its mouth to mud flats 180 miles upstream at the end of the bay causes a resonance effect. The time it takes water to flood the length of the Bay of Fundy (its natural rhythm) is nearly identical to the time between high and low tides in the ocean. The closeness of these two natural rhythms sets up a resonance effect, enhancing the tides, similar to the wave motion produced by sloshing water back and forth in a bathtub. ■

10.2.1   Wave Speed

Consider the situation illustrated in Fig. 10.2a in which a single elementary wave of small height, δy, is produced on the surface of a channel by suddenly moving the initially stationary end wall with speed δV. The water in the channel was stationary at the initial time, $t = 0$. A stationary observer will observe a single wave move down the channel with a ***wave***

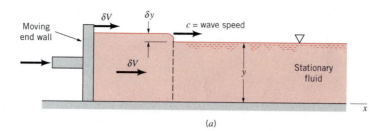

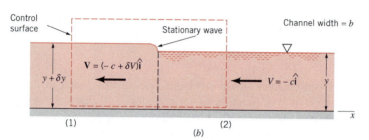

■ FIGURE 10.2
(*a*) **Production of a single elementary wave in a channel as seen by a stationary observer.** (*b*) **Wave as seen by an observer moving with a speed equal to the wave speed.**

speed c, with no fluid motion ahead of the wave and a fluid velocity of δV behind the wave. The motion is unsteady for such an observer. For an observer moving along the channel with speed c, the flow will appear steady as shown in Fig. 10.2*b*. To this observer, the fluid velocity will be $\mathbf{V} = -c\hat{\mathbf{i}}$ on the observer's right and $\mathbf{V} = (-c + \delta V)\hat{\mathbf{i}}$ to the left of the observer.

The relationship between the various parameters involved for this flow can be obtained by application of the continuity and momentum equations to the control volume shown in Fig. 10.2*b* as follows. With the assumption of uniform one-dimensional flow, the continuity equation (Eq. 5.10) becomes

$$-cyb = (-c + \delta V)(y + \delta y)b$$

where b is the channel width. This simplifies to

$$c = \frac{(y + \delta y)\,\delta V}{\delta y}$$

or in the limit of small amplitude waves with $\delta y \ll y$

$$c = y\,\frac{\delta V}{\delta y} \tag{10.1}$$

Similarly, the momentum equation (Eq. 5.17) is

$$\tfrac{1}{2}\gamma y^2 b - \tfrac{1}{2}\gamma(y + \delta y)^2 b = \rho b c y[(c - \delta V) - c]$$

where we have written the mass flowrate as $\dot{m} = \rho b c y$ and have assumed that the pressure variation is hydrostatic within the fluid. That is, the pressure forces on the channel across sections (1) and (2) are $F_1 = \gamma y_{c1} A_1 = \gamma(y + \delta y)^2 b/2$ and $F_2 = \gamma y_{c2} A_2 = \gamma y^2 b/2$, respectively. If we again impose the assumption of small amplitude waves [that is, $(\delta y)^2 \ll y\,\delta y$], the momentum equation reduces to

$$\frac{\delta V}{\delta y} = \frac{g}{c} \tag{10.2}$$

Combination of Eqs. 10.1 and 10.2 gives the wave speed as

$$c = \sqrt{gy} \tag{10.3}$$

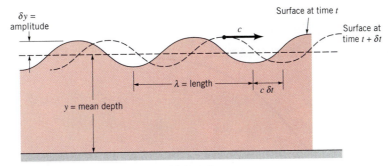

■ **FIGURE 10.3**   **Sinusoidal surface wave.**

The speed of a small amplitude solitary wave as indicated in Fig. 10.2 is proportional to the square root of the fluid depth, y, and independent of the wave amplitude, δy.

A more general description of wave motion can be obtained by considering continuous (not solitary) waves of sinusoidal shape as is shown in Fig. 10.3. An advanced analysis of such sinusoidal surface waves of small amplitude shows that the wave speed varies with both the wavelength, λ, and fluid depth, y, as (Ref. 1)

$$c = \left[\frac{g\lambda}{2\pi} \tanh\left(\frac{2\pi y}{\lambda} \right) \right]^{1/2} \tag{10.4}$$

where $\tanh(2\pi y/\lambda)$ is the hyperbolic tangent of the argument $2\pi y/\lambda$. This result is plotted in Fig. 10.4. For conditions for which the water depth is much greater than the wavelength ($y \gg \lambda$, as in the ocean), the wave speed is independent of y and given by

$$c = \sqrt{\frac{g\lambda}{2\pi}}$$

However, if the fluid layer is shallow ($y \ll \lambda$, as often happens in open channels), the wave speed is given by $c = (gy)^{1/2}$, as derived for the solitary wave in Fig. 10.2. These two limiting cases are shown in Fig. 10.4.

F l u i d s   i n   t h e   N e w s

Tsunami, the nonstorm wave A tsunami, often miscalled a "tidal wave," is a wave produced by a disturbance (for example, an earthquake, volcanic eruption, or meteorite impact) that vertically displaces the water column. Tsunamis are characterized as shallow-water waves, with long periods, very long wavelengths, and extremely large wave speeds. For example, the waves of the great December 2005, Indian Ocean tsunami traveled with speeds to 500–1000 m/s. Typically, these waves were of small amplitude in deep water far from land. Satellite radar measured the wave height less than 1 m in these areas. However, as the waves approached shore and moved into shallower water, they slowed down considerably and reached heights up to 30 m. Because the rate at which a wave loses its energy is inversely related to its wavelength, tsunamis, with their wavelengths on the order of 100 km, not only travel at high speeds, they also travel great distances with minimal energy loss. The furthest reported death from the Indian Ocean tsunami occurred approximately 8000 km from the epicenter of the earthquake that produced it. (See Problem 10.5.) ■

10.2.2   Froude Number Effects

Consider an elementary wave traveling on the surface of a fluid as is shown in Fig. 10.2a. If the fluid layer is stationary, the wave moves to the right with speed c relative to the fluid

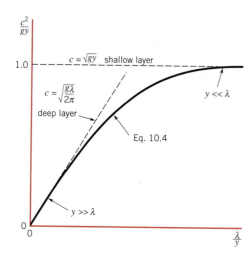

■ **FIGURE 10.4** **Wave speed as a function of wavelength.**

V10.1 Bicycle through a puddle

and the stationary observer. If the fluid is flowing to the left with velocity $V < c$, the wave (which travels with speed c relative to the fluid) will travel to the right with a speed of $c - V$ relative to a fixed observer. If the fluid flows to the left with $V = c$, the wave will remain stationary, but if $V > c$ the wave will be washed to the left with speed $V - c$.

These ideas can be expressed in dimensionless form by use of the ***Froude number,*** $\mathrm{Fr} = V/(gy)^{1/2}$, where we take the characteristic length to be the fluid depth, y. Thus, the Froude number, $\mathrm{Fr} = V/(gy)^{1/2} = V/c$, is the ratio of the fluid velocity to the wave speed.

The following characteristics are observed when a wave is produced on the surface of a moving stream, as happens when a rock is thrown into a river. If the stream is not flowing, the wave spreads equally in all directions. If the stream is nearly stationary or moving in a tranquil manner (i.e., $V < c$), the wave can move upstream. Upstream locations are said to be in hydraulic communication with the downstream locations. That is, an observer upstream of a disturbance can tell that there has been a disturbance on the surface because that disturbance can propagate upstream to the observer. Such conditions, $V < c$ or $\mathrm{Fr} < 1$, are termed subcritical.

However, if the stream is moving rapidly so that the flow velocity is greater than the wave speed (i.e., $V > c$), no upstream communication with downstream locations is possible. Any disturbance on the surface downstream from the observer will be washed further downstream. Such conditions, $V > c$ or $\mathrm{Fr} > 1$, are termed supercritical. For the special case of $V = c$ or $\mathrm{Fr} = 1$, the upstream propagating wave remains stationary and the flow is termed critical.

10.3 Energy Considerations

A typical segment of an open-channel flow is shown in Fig. 10.5. The slope of the channel bottom (or *bottom slope*), $S_0 = (z_1 - z_2)/\ell$, is assumed constant over the segment shown. The fluid depths and velocities are y_1, y_2, V_1, and V_2 as indicated.

With the assumption of a uniform velocity profile across any section of the channel, the one-dimensional energy equation for this flow (Eq. 5.57) becomes

$$\frac{p_1}{\gamma} + \frac{V_1^2}{2g} + z_1 = \frac{p_2}{\gamma} + \frac{V_2^2}{2g} + z_2 + h_L \tag{10.5}$$

where h_L is the head loss due to viscous effects between sections (1) and (2) and $z_1 - z_2 = S_0\ell$. Since the pressure is essentially hydrostatic at any cross section, we find that $p_1/\gamma = y_1$ and

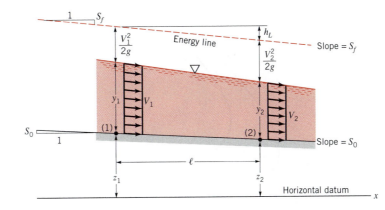

■ **FIGURE 10.5**
Typical open-channel geometry.

$p_2/\gamma = y_2$ so that Eq. 10.5 becomes

$$y_1 + \frac{V_1^2}{2g} + S_0\ell = y_2 + \frac{V_2^2}{2g} + h_L \tag{10.6}$$

We write the head loss in terms of the slope of the energy line, $S_f = h_L/\ell$ (often termed the *friction slope*), as indicated in Fig. 10.5. Recall from Chapter 3 that the energy line is located a distance z (the elevation from some datum to the channel bottom) plus the pressure head (p/γ) plus the velocity head $(V^2/2g)$ above the datum.

10.3.1 Specific Energy

The concept of the *specific energy, E,* defined as

$$E = y + \frac{V^2}{2g} \tag{10.7}$$

is often useful in open-channel flow considerations. The energy equation, Eq. 10.6, can be written in terms of E as

$$E_1 = E_2 + (S_f - S_0)\ell \tag{10.8}$$

If we consider a simple channel whose cross-sectional shape is a rectangle of width b, the specific energy can be written in terms of the flowrate per unit width, $q = Q/b = Vyb/b = Vy$, as

$$E = y + \frac{q^2}{2gy^2} \tag{10.9}$$

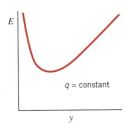

which is illustrated by the figure in the margin.

For a given channel of constant width, the value of q remains constant along the channel, although the depth, y, may vary. To gain insight into the flow processes involved, we consider the *specific energy diagram,* a graph of $y = y(E)$ with q fixed, as shown in Fig. 10.6. The relationship between the specific energy, E, the flow depth, y, and the velocity head, $V^2/2g$, as given by Eq. 10.7, is indicated in the figure.

For given q and E, Eq. 10.9 is a cubic equation $[y^3 - Ey^2 + (q^2/2g) = 0]$ with three solutions, y_{sup}, y_{sub}, and y_{neg}. If the specific energy is large enough (i.e., $E > E_{min}$, where E_{min} is a function of q), two of the solutions are positive and the other, y_{neg}, is negative. The negative root, represented by the curved dashed line in Fig. 10.6, has no physical meaning and can be ignored. Thus, for a given flowrate and specific energy there are two possible depths. These two depths are termed *alternate depths.*

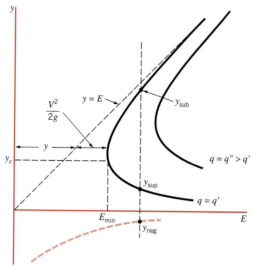

■ **FIGURE 10.6**   Specific energy diagram.

It can be shown that critical conditions (Fr = 1) occur at the location of E_{min}, where $E_{min} = 3y_c/2$ and $y_c = (q^2/g)^{1/3}$. Because the layer is deeper and the velocity smaller for the upper part of the specific energy diagram (compared with the conditions at E_{min}), such flows are subcritical (Fr < 1). Conversely, flows for the lower part of the diagram are supercritical. Thus, for a given flowrate, q, if $E > E_{min}$ there are two possible depths of flow, one subcritical and the other supercritical.

EXAMPLE 10.1   Specific Energy Diagram—Quantitative

GIVEN   Water flows up a 0.5-ft-tall ramp in a constant width rectangular channel at a rate $q = 5.75$ ft^2/s as is shown in Fig. E10.1a. (For now disregard the "bump.") The upstream depth is 2.3 ft and viscous effects are negligible.

FIND   Determine the elevation of the water surface downstream of the ramp, $y_2 + z_2$.

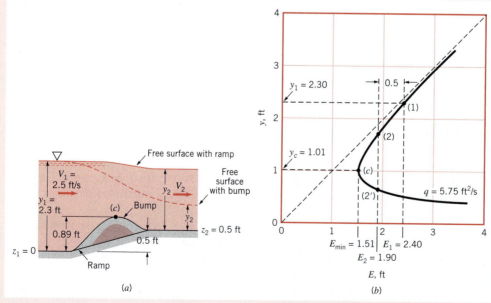

■ **FIGURE E10.1**

SOLUTION

With $S_0\ell = z_1 - z_2$ and $h_L = 0$, conservation of energy (Eq. 10.6 which, under these conditions, is actually the Bernoulli equation) requires that

$$y_1 + \frac{V_1^2}{2g} + z_1 = y_2 + \frac{V_2^2}{2g} + z_2$$

For the conditions given ($z_1 = 0$, $z_2 = 0.5$ ft, $y_1 = 2.3$ ft, and $V_1 = q/y_1 = 2.5$ ft/s), this becomes

$$1.90 = y_2 + \frac{V_2^2}{64.4} \tag{1}$$

where V_2 and y_2 are in ft/s and feet, respectively. The continuity equation provides the second equation

$$y_2 V_2 = y_1 V_1$$

or

$$y_2 V_2 = 5.75 \text{ ft}^2/\text{s} \tag{2}$$

Equations 1 and 2 can be combined to give

$$y_2^3 - 1.90 y_2^2 + 0.513 = 0$$

which has solutions

$$y_2 = 1.72 \text{ ft}, \quad y_2 = 0.638 \text{ ft}, \quad \text{or} \quad y_2 = -0.466 \text{ ft}$$

Note that two of these solutions are physically realistic, but the negative solution is meaningless. This is consistent with the previous discussions concerning the specific energy (recall the three roots indicated in Fig. 10.6). The corresponding elevations of the free surface are either

$$y_2 + z_2 = 1.72 \text{ ft} + 0.50 \text{ ft} = 2.22 \text{ ft}$$

or

$$y_2 + z_2 = 0.638 \text{ ft} + 0.50 \text{ ft} = 1.14 \text{ ft}$$

The question is which of these two flows is to be expected? This can be answered by use of the specific energy diagram obtained from Eq. 10.9, which for this problem is

$$E = y + \frac{0.513}{y^2}$$

where E and y are in feet. The diagram is shown in Fig. E10.1b. The upstream condition corresponds to subcritical flow; the downstream condition is either subcritical or supercritical, corresponding to points 2 or 2′. Note that since $E_1 = E_2 + (z_2 - z_1) = E_2 + 0.5$ ft, it follows that the downstream conditions are located 0.5 ft to the left of the upstream conditions on the diagram.

With a constant width channel, the value of q remains the same for any location along the channel. That is, all points for the flow from (1) to (2) or (2′) must lie along the $q = 5.75$ ft^2/s curve shown. Any deviation from this curve would imply either a change in q or a relaxation of the one-dimensional flow assumption. To stay on the curve and go from (1) around the critical point (point c) to point (2′) would require a reduction in specific energy to $E_{\min}$. As is seen from Fig. E10.1a, this would require a specified elevation (bump) in the channel bottom so that critical conditions would occur above this bump. The height of this bump can be obtained from the energy equation (Eq. 10.8) written between points (1) and (c) with $S_f = 0$ (no viscous effects) and $S_0\ell = z_1 - z_c$. That is, $E_1 = E_{\min} - z_1 + z_c$. In particular, since $E_1 = y_1 + 0.513/y_1^2 = 2.40$ ft and $E_{\min} = 3y_c/2 = 3(q^2/g)^{1/3}/2 = 1.51$ ft, the top of this bump would need to be $z_c - z_1 = E_1 - E_{\min} = 2.40 \text{ ft} - 1.51 \text{ ft} = 0.89$ ft above the channel bottom at section (1). The flow could then accelerate to supercritical conditions (Fr$_{2'} > 1$) as is shown by the free surface represented by the dashed line in Fig. E10.1a.

Since the actual elevation change (a ramp) shown in Fig. E10.1a does not contain a bump, the downstream conditions will correspond to the subcritical flow denoted by (2), not the

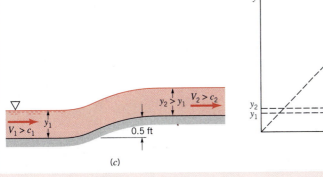

(c)

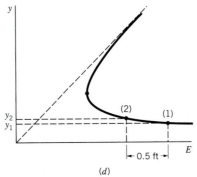

(d)

■ **FIGURE E10.1**   (*Continued*)

supercritical condition $(2')$. Without a bump on the channel bottom, the state $(2')$ is inaccessible from the upstream condition state (1). Such considerations are often termed the *accessibility of flow regimes*. Thus, the surface elevation is

$$y_2 + z_2 = 2.22 \text{ ft} \qquad \textbf{(Ans)}$$

COMMENTS Note that since $y_1 + z_1 = 2.30 \text{ ft}$ and $y_2 + z_2 = 2.22 \text{ ft}$, the elevation of the free surface decreases as it goes across the ramp.

If the flow conditions upstream of the ramp were supercritical, the free-surface elevation and fluid depth would increase as the fluid flows up the ramp. This is indicated in Fig. E10.1c along with the corresponding specific energy diagram, as is shown in Fig. E10.1d. For this case the flow starts at (1) on the lower (supercritical) branch of the specific energy curve and ends at (2) on the same branch with $y_2 > y_1$. Since both y and z increase from (1) to (2), the surface elevation, $y + z$, also increases. Thus, flow up a ramp is different for subcritical than it is for supercritical conditions.

10.4 Uniform Depth Channel Flow

V10.2 Merging channels

Many channels are designed to carry fluid at a uniform depth all along their length. *Uniform depth flow* ($dy/dx = 0$, or $y_1 = y_2$ and $V_1 = V_2$) can be accomplished by adjusting the bottom slope, S_0, so that it precisely equals the slope of the energy line, S_f. That is, $S_0 = S_f$ (see Eq. 10.8). From an energy point of view, uniform depth flow is achieved by a balance between the potential energy lost by the fluid as it coasts downhill and the energy that is dissipated by viscous effects (head loss) associated with shear stresses throughout the fluid.

10.4.1 Uniform Flow Approximations

We consider fluid flowing in an open channel of constant cross-sectional size and shape such that the depth of flow remains constant as indicated in Fig. 10.7. The area of the section is A and the **wetted perimeter** (i.e., the length of the perimeter of the cross section in contact with the fluid) is P. Reasonable analytical results can be obtained by assuming a uniform velocity profile, V, and a constant wall shear stress, τ_w, along the wetted perimeter.

10.4.2 The Chezy and Manning Equations

Under the assumptions of steady, uniform flow, the x component of the momentum equation (Eq. 5.17) applied to the control volume indicated in Fig. 10.8 simply reduces to

$$\Sigma F_x = \rho Q(V_2 - V_1) = 0$$

since $V_1 = V_2$. There is no acceleration of the fluid, and the momentum flux across section (1) is the same as that across section (2). The flow is governed by a simple balance between the forces in the direction of the flow. Thus, $\Sigma F_x = 0$, or

$$F_1 - F_2 - \tau_w P\ell + \mathcal{W} \sin \theta = 0 \qquad \textbf{(10.10)}$$

where F_1 and F_2 are the hydrostatic pressure forces across either end of the control volume. Because the flow is at a uniform depth ($y_1 = y_2$), it follows that $F_1 = F_2$ so that these

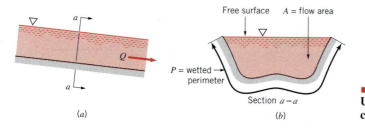

(a)

Free surface   A = flow area

P = wetted perimeter

Section a–a

(b)

■ **FIGURE 10.7**
Uniform flow in an open channel.

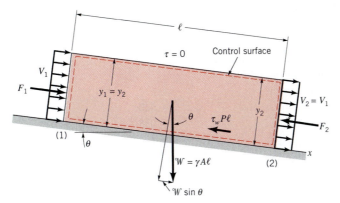

■ **FIGURE 10.8**
Control volume for uniform flow in an open channel.

two forces do not contribute to the force balance. The term $\mathcal{W} \sin \theta$ is the component of the fluid weight that acts down the slope, and $\tau_w P\ell$ is the shear force on the fluid, acting up the slope as a result of the interaction of the water and the channel's wetted perimeter. Thus, Eq. 10.10 becomes

$$\tau_w = \frac{\mathcal{W}\sin\theta}{P\ell} = \frac{\mathcal{W}S_0}{P\ell}$$

where we have used the approximation that $\sin\theta \approx \tan\theta = S_0$, since the bottom slope is typically very small (i.e., $S_0 \ll 1$). Because $\mathcal{W} = \gamma A\ell$ and the *hydraulic radius* is defined as $R_h = A/P$, the force balance equation becomes

$$\tau_w = \frac{\gamma A\ell S_0}{P\ell} = \gamma R_h S_0 \tag{10.11}$$

Most open-channel flows are turbulent (rather than laminar), and the wall shear stress is nearly proportional to the dynamic pressure, $\rho V^2/2$, and independent of the viscosity. That is,

$$\tau_w = K\rho\frac{V^2}{2}$$

where K is a constant dependent upon the roughness of the channel. In such situations, Eq. 10.11 becomes

$$K\rho\frac{V^2}{2} = \gamma R_h S_0$$

or

$$V = C\sqrt{R_h S_0} \tag{10.12}$$

where the constant C is termed the Chezy coefficient and Eq. 10.12 is termed the *Chezy equation.*

From a series of experiments it was found that the slope dependence of Eq. 10.12 ($V \sim S_0^{1/2}$) is reasonable, but that the dependence on the hydraulic radius is more nearly $V \sim R_h^{2/3}$ rather than $V \sim R_h^{1/2}$. Thus, the following somewhat modified equation for open-channel flow is used to more accurately describe the R_h dependence:

$$V = \frac{R_h^{2/3}S_0^{1/2}}{n} \tag{10.13}$$

Equation 10.13 is termed the *Manning equation,* and the parameter n is the *Manning resistance coefficient.* Its value is dependent on the surface material of the channel's wetted

■ **TABLE 10.1**
Values of the Manning Coefficient, n (Ref. 5)

Wetted Perimeter	n
A. Natural channels	
Clean and straight	0.030
Sluggish with deep pools	0.040
Major rivers	0.035
B. Floodplains	
Pasture, farmland	0.035
Light brush	0.050
Heavy brush	0.075
Trees	0.15
C. Excavated earth channels	
Clean	0.022
Gravelly	0.025
Weedy	0.030
Stony, cobbles	0.035
D. Artificially lined channels	
Glass	0.010
Brass	0.011
Steel, smooth	0.012
Steel, painted	0.014
Steel, riveted	0.015
Cast iron	0.013
Concrete, finished	0.012
Concrete, unfinished	0.014
Planed wood	0.012
Clay tile	0.014
Brickwork	0.015
Asphalt	0.016
Corrugated metal	0.022
Rubble masonry	0.025

V10.3 Uniform
channel flow

perimeter and is obtained from experiments. It is not dimensionless, having the units of $s/m^{1/3}$ or $s/ft^{1/3}$.

Typical values of the Manning coefficient are indicated in Table 10.1. As expected, the rougher the wetted perimeter, the larger the value of n. The values of n were developed for SI units. Standard practice is to use the same value of n when using the BG system of units, and to insert a conversion factor into the equation.

Thus, uniform flow in an open channel is obtained from the Manning equation written as

$$V = \frac{\kappa}{n} R_h^{2/3} S_0^{1/2}$$

(10.14)

and

$$Q = \frac{\kappa}{n} A R_h^{2/3} S_0^{1/2}$$

(10.15)

where $\kappa = 1$ if SI units are used, and $\kappa = 1.49$ if BG units are used. Thus, by using R_h in meters, A in m^2, and $\kappa = 1$, the average velocity is meters per second and the flowrate m^3/s. By using R_h in feet, A in ft^2, and $\kappa = 1.49$, the average velocity is feet per second and the flowrate ft^3/s.

F l u i d s   i n   t h e   N e w s

Done without GPS or lasers Two thousand years before the invention of such tools as the GPS or laser surveying equipment, Roman engineers were able to design and construct structures that made a lasting contribution to Western civilization. For example, one of the best surviving examples of Roman aqueduct construction is the Pont du Gard, an aqueduct that spans the Gardon River near Nîmes, France. This aqueduct is part of a circuitous, 50-km-long open channel that transported water to Rome from a spring located 20 km from Rome. The spring is only 14.6 m above the point of delivery, giving an average *bottom slope* of only $S_0 = 3 \times 10^{-4}$. It is obvious that to carry out such a project, the Roman understanding of hydraulics, surveying, and construction was well advanced. (See Problem 10.18.) ∎

10.4.3   Uniform Depth Examples

A variety of interesting and useful results can be obtained from the Manning equation. The following examples illustrate some of the typical considerations.

Determination of the flowrate for a given channel with flow at a given depth (often termed the *normal flowrate* or *normal depth*) is obtained from a straightforward calculation as shown in Example 10.2.

EXAMPLE 10.2   Uniform Flow—Determine Flow Rate

GIVEN Water flows in the canal of trapezoidal cross section shown in Fig. E10.2a. The bottom drops 1.4 ft per 1000 ft of length, and the canal is lined with new finished concrete.

FIND Determine

(a) the flowrate and

(b) the Froude number for this flow.

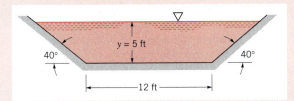

■ **FIGURE E10.2a**

SOLUTION

(a) From Eq. 10.15,

$$Q = \frac{1.49}{n} A R_h^{2/3} S_0^{1/2} \qquad (1)$$

where we have used $\kappa = 1.49$, since the dimensions are given in BG units. For a depth of $y = 5$ ft, the flow area is

$$A = 12 \text{ ft (5 ft)} + 5 \text{ ft} \left(\frac{5}{\tan 40°} \text{ ft} \right) = 89.8 \text{ ft}^2$$

so that with a wetted perimeter of $P = 12 \text{ ft} + 2(5/\sin 40° \text{ ft}) = 27.6$ ft, the hydraulic radius is determined to be $R_h = A/P = 3.25$ ft. Note that even though the channel is quite wide (the free-surface width is 23.9 ft), the hydraulic radius is only 3.25 ft, which is less than the depth.

Thus, with $S_0 = 1.4 \text{ ft}/1000 \text{ ft} = 0.0014$, Eq. 1 becomes

$$Q = \frac{1.49}{n} (89.8 \text{ ft}^2)(3.25 \text{ ft})^{2/3}(0.0014)^{1/2} = \frac{10.98}{n}$$

where Q is in ft^3/s.

From Table 10.1, we obtain $n = 0.012$ for finished concrete. Thus,

$$Q = \frac{10.98}{0.012} = 915 \text{ cfs} \qquad \textbf{(Ans)}$$

COMMENT The corresponding average velocity, $V = Q/A$, is 10.2 ft/s. It does not take a very steep slope ($S_0 = 0.0014$ or $\theta = \tan^{-1}(0.0014) = 0.080°$) for this velocity.

By repeating the calculations for various surface types (i.e., various Manning coefficient values) the results shown in Fig. E10.2b are obtained. Note that the increased roughness causes a decrease in the flowrate. This is an indication that for the turbulent flows involved, the wall shear stress increases with surface roughness. [For water at 50 °F, the Reynolds number based on the 3.25-ft hydraulic radius of the channel is $Re = R_h V/\nu = $ 3.25 ft $(10.2$ ft/s$)/(1.41 \times 10^{-5}$ ft^2/s$) = 2.35 \times 10^6$, well into the turbulent regime.]

(b) The Froude number based on the maximum depth for the flow can be determined from $Fr = V/(gy)^{1/2}$. For the finished concrete case,

$$Fr = \frac{10.2 \text{ ft/s}}{(32.2 \text{ ft/s}^2 \times 5 \text{ ft})^{1/2}} = 0.804 \quad \textbf{(Ans)}$$

The flow is subcritical.

COMMENT The same results would be obtained for the channel if its size were given in meters. We would use the same value of n but set $\kappa = 1$ for this SI units situation.

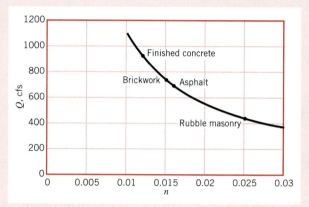

■ **FIGURE E10.2b**

Fluids in the News

Plumbing the Everglades Because of all of the economic development that has occurred in southern Florida, the natural drainage pattern of that area has been greatly altered during the past century. Previously there was a vast network of surface flow southward from the Orlando area, to Lake Okeechobee, through the Everglades, and out to the Gulf of Mexico. Currently a vast amount of freshwater from Lake Okeechobee and surrounding waterways (1.7 billion gallons per day) is sluiced into the ocean for flood control, bypassing the Everglades. A new long-term Comprehensive Everglades Restoration Plan is being implemented to restore, preserve, and protect the south Florida ecosystem. Included in the plan is the use of numerous aquifer-storage-and-recovery systems that will recharge the ecosystem. In addition, surface water reservoirs using artificial wetlands will clean agricultural runoff. In an attempt to improve the historical flow from north to south, old levees will be removed, parts of the Tamiami Trail causeway will be altered, and stored water will be redirected through miles of new pipes and rebuilt *canals*. Strictly speaking, the Everglades will not be "restored." However, by 2030, 1.6 million acres of national parkland will have cleaner water and more of it. (See Problem 10.30.) ■

In some instances a trial-and-error or iteration method must be used to solve for the dependent variable. This is often encountered when the flowrate, channel slope, and channel material are given, and the flow depth is to be determined as illustrated in the following example.

EXAMPLE 10.3 | Uniform Flow—Determine Flow Depth

GIVEN Water flows in the channel shown in Fig. E10.2a at a rate of $Q = 10.0$ m^3/s. The canal lining is weedy.

FIND Determine the depth of the flow.

SOLUTION

In this instance neither the flow area nor the hydraulic radius is known, although they can be written in terms of the depth, y. Since the flowrate is given in m^3/s, we will solve this problem using SI units. Hence, the bottom width is (12 ft)(1 m/3.281 ft) = 3.66 m and the area is

$$A = y\left(\frac{y}{\tan 40°}\right) + 3.66y = 1.19y^2 + 3.66y$$

where A and y are in square meters and meters, respectively. Also, the wetted perimeter is

$$P = 3.66 + 2\left(\frac{y}{\sin 40°}\right) = 3.11y + 3.66$$

so that

$$R_h = \frac{A}{P} = \frac{1.19y^2 + 3.66y}{3.11y + 3.66}$$

where R_h and y are in meters. Thus, with $n = 0.030$ (from Table 10.1), Eq. 10.15 can be written as

$$Q = 10 = \frac{\kappa}{n} AR_h^{2/3} S_0^{1/2}$$

$$= \frac{1.0}{0.030}(1.19y^2 + 3.66y)\left(\frac{1.19y^2 + 3.66y}{3.11y + 3.66}\right)^{2/3}$$

$$\times (0.0014)^{1/2}$$

which can be rearranged into the form

$$(1.19y^2 + 3.66y)^5 - 515(3.11y + 3.66)^2 = 0 \quad \textbf{(1)}$$

where y is in meters. The solution of Eq. 1 can be obtained easily by use of a simple root-finding numerical technique or by trial-and-error methods. The only physically meaningful root of Eq. 1 (i.e., a positive, real number) gives the solution for the normal flow depth at this flowrate as

$$y = 1.50 \text{ m} \quad \textbf{(Ans)}$$

COMMENT   By repeating the calculations for various flowrates, the results shown in Fig. E10.3 are obtained. Note that the water depth is not linearly related to the flowrate. That is, if the flowrate is doubled, the depth is not doubled.

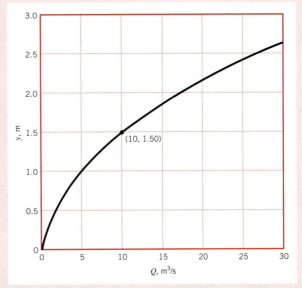

■ **FIGURE E10.3**

In many man-made channels the surface roughness (and hence the Manning coefficient) varies along the wetted perimeter of the channel. In such cases the channel cross section can be divided into N subsections, each with its own wetted perimeter, P_i, and A_i, and Manning coefficient, n_i. The P_i values do not include the imaginary boundaries between the different subsections. The total flowrate is assumed to be the sum of the flowrates through each section. This technique is illustrated by Example 10.4.

EXAMPLE 10.4 Uniform Flow—Variable Roughness

GIVEN   Water flows along the drainage canal having the properties shown in Fig. E10.4a. The bottom slope is $S_0 =$ 1 ft/500 ft = 0.002.

FIND   Estimate the flowrate.

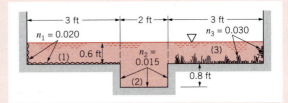

■ **FIGURE E10.4a**

SOLUTION

We divide the cross section into three subsections as is indicated in Fig. E10.4a and write the flowrate as $Q = Q_1 + Q_2 + Q_3$, where for each section

$$Q_i = \frac{1.49}{n_i} A_i R_{hi}^{2/3} S_0^{1/2}$$

The appropriate values of A_i, P_i, R_{hi}, and n_i are listed in Table E10.4. Note that the imaginary portions of the perimeters between sections (denoted by the dashed lines in Fig. E10. 4a) are not included in the P_i. That is, for section (2)

$$A_2 = 2 \text{ ft } (0.8 + 0.6) \text{ ft} = 2.8 \text{ ft}^2$$

and

$$P_2 = 2 \text{ ft} + 2(0.8 \text{ ft}) = 3.6 \text{ ft}$$

so that

$$R_{h_2} = \frac{A_2}{P_2} = \frac{2.8 \text{ ft}^2}{3.6 \text{ ft}} = 0.778 \text{ ft}$$

■ **TABLE E10.4**

i	A_i (ft^2)	P_i (ft)	R_{hi} (ft)	n_i
1	1.8	3.6	0.500	0.020
2	2.8	3.6	0.778	0.015
3	1.8	3.6	0.500	0.030

Thus, the total flowrate is

$$Q = Q_1 + Q_2 + Q_3 = 1.49(0.002)^{1/2}$$
$$\times \left[\frac{(1.8 \text{ ft}^2)(0.500 \text{ ft})^{2/3}}{0.020} + \frac{(2.8 \text{ ft}^2)(0.778 \text{ ft})^{2/3}}{0.015} \right.$$
$$+ \left. \frac{(1.8 \text{ ft}^2)(0.500 \text{ ft})^{2/3}}{0.030} \right]$$

or

$$Q = 16.8 \text{ ft}^3/\text{s} \qquad \text{(Ans)}$$

COMMENTS If the entire channel cross section were considered as one flow area, then $A = A_1 + A_2 + A_3 = 6.4 \text{ ft}^2$ and

$P = P_1 + P_2 + P_3 = 10.8 \text{ ft}$, or $R_h = A/P = 6.4 \text{ ft}^2/10.8 \text{ ft} = 0.593 \text{ ft}$. The flowrate is given by Eq. 10.15, which can be written as

$$Q = \frac{1.49}{n_{eff}} AR_h^{2/3} S_0^{1/2}$$

where n_{eff} is the effective value of n for this channel. With $Q = 16.8 \text{ ft}^3/\text{s}$ as determined earlier, the value of n_{eff} is found to be

$$n_{eff} = \frac{1.49 AR_h^{2/3} S_0^{1/2}}{Q}$$
$$= \frac{1.49(6.4)(0.593)^{2/3}(0.002)^{1/2}}{16.8} = 0.0179$$

As expected, the effective roughness (Manning n) is between the minimum ($n_2 = 0.015$) and maximum ($n_3 = 0.030$) values for the individual subsections.

By repeating the calculations for various depths, y, the results shown in Fig. E10.4b are obtained. Note that there are two distinct portions of the graph—one when the water is contained entirely within the main, center channel ($y < 0.8$ ft); the other when the water overflows into the side portions of the channel ($y > 0.8$ ft).

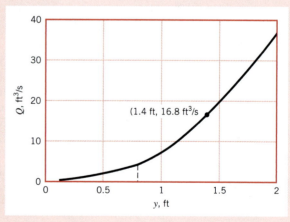

■ **FIGURE E10.4b**

One type of problem often encountered in open-channel flows is that of determining the *best hydraulic cross section* defined as the section of the minimum area for a given flowrate, Q, slope, S_0, and roughness coefficient, n. Example 10.5 illustrates the concept of the best hydraulic cross section for rectangular channels.

EXAMPLE 10.5 | Uniform Flow—Best Hydraulic Cross Section

GIVEN Water flows uniformly in a rectangular channel of width b and depth y.

FIND Determine the aspect ratio, b/y, for the best hydraulic cross section.

SOLUTION

The uniform flow is given by Eq. 10.15 as

$$Q = \frac{\kappa}{n} A R_h^{2/3} S_0^{1/2} \qquad (1)$$

where $A = by$ and $P = b + 2y$, so that $R_h = A/P = by/(b + 2y)$. We rewrite the hydraulic radius in terms of A as

$$R_h = \frac{A}{(2y + b)} = \frac{A}{(2y + A/y)} = \frac{Ay}{(2y^2 + A)}$$

so that Eq. 1 becomes

$$Q = \frac{\kappa}{n} A \left(\frac{Ay}{2y^2 + A} \right)^{2/3} S_0^{1/2}$$

This can be rearranged to give

$$A^{5/2} y = K(2y^2 + A) \qquad (2)$$

where $K = (nQ/\kappa S_0^{1/2})^{3/2}$ is a constant. The best hydraulic section is the one that gives the minimum A for all y. That is, $dA/dy = 0$. By differentiating Eq. 2 with respect to y, we obtain

$$\frac{5}{2} A^{3/2} \frac{dA}{dy} y + A^{5/2} = K \left(4y + \frac{dA}{dy} \right)$$

which, with $dA/dy = 0$, reduces to

$$A^{5/2} = 4Ky \qquad (3)$$

With $K = A^{5/2} y/(2y^2 + A)$ from Eq. 2, Eq. 3 can be written in the form

$$A^{5/2} = \frac{4A^{5/2} y^2}{(2y^2 + A)}$$

which simplifies to $y = (A/2)^{1/2}$. Thus, because $A = by$, the best hydraulic cross section for a rectangular shape has a width b and a depth

$$y = \left(\frac{A}{2} \right)^{1/2} = \left(\frac{by}{2} \right)^{1/2}$$

or

$$2y^2 = by$$

That is, the rectangle with the best hydraulic cross section is twice as wide as it is deep, or

$$b/y = 2 \qquad \text{(Ans)}$$

COMMENTS A rectangular channel with $b/y = 2$ will give the smallest area (and smallest wetted perimeter) for a given flowrate. Conversely, for a given area, the largest flowrate in a rectangular channel will occur when $b/y = 2$. For $A = by = $ constant, if $y \to 0$ then $b \to \infty$, and the flowrate is small because of the large wetted perimeter $P = b + 2y \to \infty$. The maximum Q occurs when $y = b/2$. However, as seen in Fig. E10.5a, the maximum represented by this optimal configuration is a rather weak one. For example, for aspect ratios between 1 and 4, the flowrate is within 96%

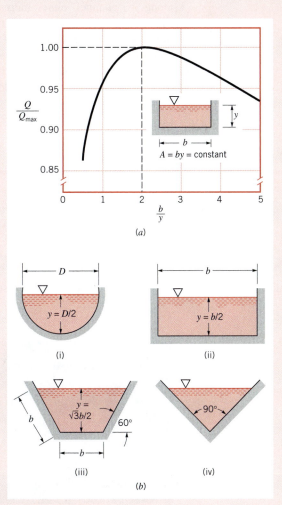

(a)

(b)

■ **FIGURE E10.5**

of the maximum flowrate obtained with the same area and $b/y = 2$.

An alternate but equivalent method to obtain the aforementioned answer is to use the fact that $dR_h/dy = 0$, which follows from Eq. 1 using $dQ/dy = 0$ (constant flowrate) and $dA/dy = 0$ (best hydraulic cross section has minimum area). Differentiation

of $R_h = Ay/(2y^2 + A)$ with constant A gives $b/y = 2$ when $dR_h/dy = 0$.

The best hydraulic cross section can be calculated for other shapes in a similar fashion. The results (given here without proof) for circular, rectangular, trapezoidal (with 60° sides), and triangular shapes are shown in Fig. E10.5b.

10.5   Gradually Varied Flow

In many situations the flow in an open channel is not of uniform depth (y = constant) along the channel. This can occur because of several reasons: The bottom slope is not constant, the cross-sectional shape and area vary in the flow direction, or there is some obstruction across a portion of the channel. Such flows are classified as gradually varying flows if $dy/dx \ll 1$.

If the bottom slope and the energy line slope are not equal, the flow depth will vary along the channel, either increasing or decreasing in the flow direction. In such cases $dy/dx \neq 0$, $dV/dx \neq 0$ and the right-hand side of Eq. 10.10 is not zero. Physically, the difference between the component of weight and the shear forces in the direction of flow produces a change in the fluid momentum that requires a change in velocity and, from continuity considerations, a change in depth. Whether the depth increases or decreases depends on various parameters of the flow, with a variety of surface profile configurations [flow depth as a function of distance, $y = y(x)$] possible (Refs. 4, 7).

10.6   Rapidly Varied Flow

V10.4 Erosion in a channel

In many open channels, flow depth changes occur over a relatively short distance so that $dy/dx \sim 1$. Such *rapidly varied flow* conditions are often quite complex and difficult to analyze in a precise fashion. Fortunately, many useful approximate results can be obtained by using a simple one-dimensional model along with appropriate experimentally determined coefficients when necessary. This section discusses several of these flows.

The hydraulic jump is one such case. As indicated in Fig. 10.9, the flow may change from a relatively shallow, high-speed condition into a relatively deep, low-speed condition

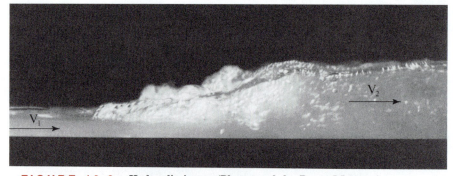

■ **FIGURE 10.9**   **Hydraulic jump. (Photograph by Bruce Munson.)**

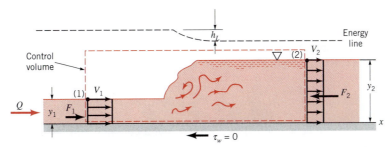

■ FIGURE 10.10 **Hydraulic jump geometry.**

within a horizontal distance of just a few channel depths. Also, many open-channel flow-measuring devices are based on principles associated with rapidly varied flows. Among these devices are broad-crested weirs, sharp-crested weirs, critical flow flumes, and sluice gates. The operation of such devices is discussed in the following sections.

10.6.1 The Hydraulic Jump

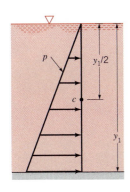

V10.5 Hydraulic jump in a river

Observations of flows in open channels show that under certain conditions it is possible that the fluid depth will change very rapidly over a short length of the channel without any change in the channel configuration. Such changes in depth can be approximated as a discontinuity in the free-surface elevation ($dy/dx = \infty$). Physically, this near discontinuity, called a *hydraulic jump,* may result when there is a conflict between the upstream and the downstream influences that control a particular section (or reach) of a channel.

The simplest type of hydraulic jump occurs in a horizontal, rectangular channel as indicated in Fig. 10.10. Although the flow within the jump itself is extremely complex and agitated, it is reasonable to assume that the flow at sections (1) and (2) is nearly uniform, steady, and one dimensional. In addition, we neglect any wall shear stresses, τ_w, within the relatively short segment between these two sections. Under these conditions the x component of the momentum equation (Eq. 5.17) for the control volume indicated can be written as

$$F_1 - F_2 = \rho Q(V_2 - V_1) = \rho V_1 y_1 b(V_2 - V_1)$$

where, as indicated by the figure in the margin, the pressure force at either section is hydrostatic. That is, $F_1 = p_{c1}A_1 = \gamma y_1^2 b/2$ and $F_2 = p_{c2}A_2 = \gamma y_2^2 b/2$, where for the rectangular channel cross sections, $p_{c1} = \gamma y_1/2$ and $p_{c2} = \gamma y_2/2$ are the pressures at the centroids of the channel cross sections and b is the channel width. Thus, the momentum equation becomes

$$\frac{y_1^2}{2} - \frac{y_2^2}{2} = \frac{V_1 y_1}{g}(V_2 - V_1) \tag{10.16}$$

In addition to the momentum equation, we have the conservation of mass equation (Eq. 5.11)

$$y_1 b V_1 = y_2 b V_2 = Q \tag{10.17}$$

and the energy equation (Eq. 5.57)

$$y_1 + \frac{V_1^2}{2g} = y_2 + \frac{V_2^2}{2g} + h_L \tag{10.18}$$

The head loss, h_L, in Eq. 10.18 is due to the violent turbulent mixing and dissipation that occur within the jump itself. We have neglected any head loss due to wall shear stresses.

Clearly Eqs. 10.16, 10.17, and 10.18 have a solution $y_1 = y_2$, $V_1 = V_2$, and $h_L = 0$. This represents the trivial case of no jump. Because these are nonlinear equations, it may be possible that more than one solution exists. The other solutions can be obtained as follows. By combining Eqs. 10.16 and 10.17 to eliminate V_2 we obtain

$$\frac{y_1^2}{2} - \frac{y_2^2}{2} = \frac{V_1 y_1}{g}\left(\frac{V_1 y_1}{y_2} - V_1\right) = \frac{V_1^2 y_1}{g y_2}(y_1 - y_2)$$

which can be simplified by factoring out a common nonzero factor $y_1 - y_2$ from each side to give

$$\left(\frac{y_2}{y_1}\right)^2 + \left(\frac{y_2}{y_1}\right) - 2\,\mathrm{Fr}_1^2 = 0$$

where $\mathrm{Fr}_1 = V_1/\sqrt{gy_1}$ is the upstream Froude number. By using the quadratic formula we obtain

$$\frac{y_2}{y_1} = \frac{1}{2}\left(-1 \pm \sqrt{1 + 8\mathrm{Fr}_1^2}\right)$$

Clearly the solution with the minus sign is not possible (it would give a negative y_2/y_1). Thus,

$$\frac{y_2}{y_1} = \frac{1}{2}\left(-1 + \sqrt{1 + 8\mathrm{Fr}_1^2}\right) \tag{10.19}$$

This depth ratio, y_2/y_1, across the hydraulic jump is shown as a function of the upstream Froude number in Fig. 10.11. The portion of the curve for $\mathrm{Fr}_1 < 1$ is dashed in recognition of the fact that to have a hydraulic jump the flow must be supercritical. That is, the solution as given in Eq. 10.19 must be restricted to $\mathrm{Fr}_1 \geq 1$, for which $y_2/y_1 \geq 1$. This can be shown by consideration of the energy equation, Eq. 10.18, as follows. The dimensionless head loss, h_L/y_1, can be obtained from Eq. 10.18 as

$$\frac{h_L}{y_1} = 1 - \frac{y_2}{y_1} + \frac{\mathrm{Fr}_1^2}{2}\left[1 - \left(\frac{y_1}{y_2}\right)^2\right] \tag{10.20}$$

V10.6 Hydraulic jump in a sink

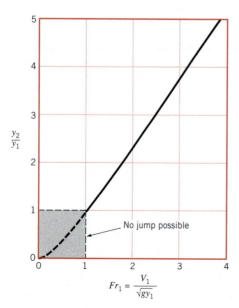

■ **FIGURE 10.11**   Depth ratio across a hydraulic jump as a function of upstream Froude number.

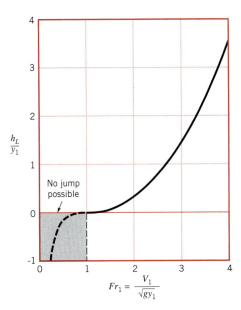

where, for given values of Fr_1, the values of y_2/y_1 are obtained from Eq. 10.19. As indicated in Fig. 10.12, the head loss is negative if $Fr_1 < 1$. Since negative head losses are not possible (viscous effects dissipate energy, they cannot create energy; see Section 5.3), it is not possible to produce a hydraulic jump with $Fr_1 < 1$. The head loss across the jump is indicated by the lowering of the energy line shown in Fig. 10.10.

F l u i d s   i n   t h e   N e w s

Grand Canyon rapids building Virtually all of the rapids in the Grand Canyon were formed by rock debris carried into the Colorado River from side canyons. Severe storms wash large amounts of sediment into the river, building debris fans that narrow the river. This debris forms crude dams which back up the river to form quiet pools above the rapids. Water exiting the pool through the narrowed channel can reach supercritical conditions and produce *hydraulic jumps* downstream. Since the configuration of the jumps is a function of the flowrate, the difficulty in running the rapids can change from day to day. Also, rapids change over

the years as debris is added to or removed from the rapids. For example, Crystal Rapid, one of the notorious rafting stretches of the river, changed very little between the first photos of 1890 and those of 1966. However, a debris flow from a severe winter storm in 1966 greatly constricted the river. Within a few minutes the configuration of Crystal Rapid was completely changed. The new, immature rapid was again drastically changed by a flood in 1983. While Crystal Rapid is now considered full grown, it will undoubtedly change again, perhaps in 100 or 1000 years. (See Problem 10.40.) ■

EXAMPLE 10.6   Hydraulic Jump

GIVEN Water on the horizontal apron of the 100-ft-wide spillway shown in Fig. E10.6a has a depth of 0.60 ft and a velocity of 18 ft/s.

FIND Determine

(a) the depth, y_2, after the jump, the Froude numbers before and after the jump, Fr_1 and Fr_2, and

(b) the power dissipated, $\mathcal{P}_d$, within the jump.

SOLUTION

(a) Conditions across the jump are determined by the upstream Froude number

$$Fr_1 = \frac{V_1}{\sqrt{gy_1}} = \frac{18 \text{ ft/s}}{[(32.2 \text{ ft/s}^2)(0.60 \text{ ft})]^{1/2}} = 4.10 \text{ (Ans)}$$

Thus, the upstream flow is supercritical, and it is possible to generate a hydraulic jump as sketched.

From Eq. 10.19 we obtain the depth ratio across the jump as

$$\frac{y_2}{y_1} = \frac{1}{2}(-1 + \sqrt{1 + 8Fr_1^2})$$

$$= \frac{1}{2}[-1 + \sqrt{1 + 8(4.10)^2}] = 5.32$$

or

$$y_2 = 5.32 \,(0.60 \text{ ft}) = 3.19 \text{ ft} \qquad \text{(Ans)}$$

Since $Q_1 = Q_2$, or $V_2 = (y_1 V_1)/y_2 = 0.60$ ft$(18 \text{ ft/s})/3.19$ ft $= 3.39$ ft/s, it follows that

$$Fr_2 = \frac{V_2}{\sqrt{gy_2}} = \frac{3.39 \text{ ft/s}}{[(32.2 \text{ ft/s}^2)(3.19 \text{ ft})]^{1/2}} = 0.334 \text{ (Ans)}$$

COMMENT As is true for any hydraulic jump, the flow changes from supercritical to subcritical flow across the jump.

(b) The power (energy per unit time) dissipated, $\mathcal{P}_d$, by viscous effects within the jump can be determined from the head loss as (see Eq. 5.58)

$$\mathcal{P}_d = \gamma Q h_L = \gamma b y_1 V_1 h_L \qquad (1)$$

where h_L is obtained from Eqs. 10.18 or 10.20 as

$$h_L = \left(y_1 + \frac{V_1^2}{2g}\right) - \left(y_2 + \frac{V_2^2}{2g}\right) = \left[0.60 \text{ ft} + \frac{(18.0 \text{ ft/s})^2}{2(32.2 \text{ ft/s}^2)}\right]$$

$$- \left[3.19 \text{ ft} + \frac{(3.39 \text{ ft/s})^2}{2(32.2 \text{ ft/s}^2)}\right]$$

or

$$h_L = 2.26 \text{ ft}$$

Thus, from Eq. 1,

$$\mathcal{P}_d = (62.4 \text{ lb/ft}^3)(100 \text{ ft})(0.60 \text{ ft})(18.0 \text{ ft/s})(2.26 \text{ ft})$$
$$= 1.52 \times 10^5 \text{ ft·lb/s}$$

or

$$\mathcal{P}_d = \frac{1.52 \times 10^5 \text{ ft·lb/s}}{550[(\text{ft·lb/s})/\text{hp}]} = 277 \text{ hp} \qquad \text{(Ans)}$$

COMMENTS This power, which is dissipated within the highly turbulent motion of the jump, is converted into an increase in water temperature, T. That is, $T_2 > T_1$. Although the

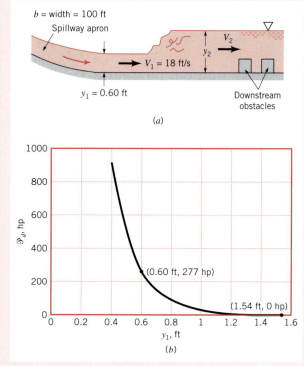

■ **FIGURE E10.6a, b**

power dissipated is considerable, the difference in temperature is not great because the flowrate is quite large.

By repeating the calculations for the given flowrate $Q_1 = A_1 V_1 = b_1 y_1 V_1 = 100$ ft $(0.6 \text{ ft})(18 \text{ ft/s}) = 1080 \text{ ft}^3/\text{s}$ but with various upstream depths, y_1, the results shown in Fig. E10.6b are obtained. Note that a slight change in water depth can produce a considerable change in energy dissipated. Also, if $y_1 > 1.54$ ft the flow is subcritical (Fr$_1 < 1$) and no hydraulic jump can occur.

The hydraulic jump flow process can be illustrated by use of the specific energy concept introduced in Section 10.3 as follows. Equation 10.18 can be written in terms of the specific energy, $E = y + V^2/2g$, as $E_1 = E_2 + h_L$, where $E_1 = y_1 + V_1^2/2g = 5.63$ ft and $E_2 = y_2 + V_2^2/2g = 3.37$ ft. As discussed in Section 10.3, the specific energy diagram for this flow can be obtained by using $V = q/y$, where

$$q = q_1 = q_2 = \frac{Q}{b} = y_1 V_1 = 0.60 \text{ ft } (18.0 \text{ ft/s})$$

$$= 10.8 \text{ ft}^2/\text{s}$$

Thus,

$$E = y + \frac{q^2}{2gy^2} = y + \frac{(10.8 \text{ ft}^2/\text{s})^2}{2(32.2 \text{ ft/s}^2)y^2}$$

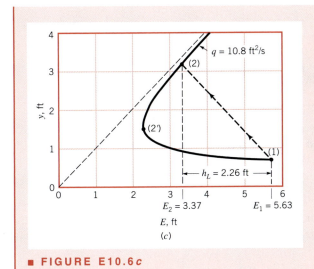

■ FIGURE E10.6c

or

$$E = y + \frac{1.81}{y^2}$$

where y and E are in feet. The resulting specific energy diagram is shown in Fig. E10.6c. Because of the head loss across the jump, the upstream and downstream values of E are different. In going from state (1) to state (2) the fluid does not proceed along the specific energy curve and pass through the critical condition at state 2′. Rather, it jumps from (1) to (2) as is represented by the dashed line in the figure. From a one-dimensional consideration, the jump is a discontinuity. In actuality, the jump is a complex three-dimensional flow incapable of being represented on the one-dimensional specific energy diagram.

10.6.2   Sharp-Crested Weirs

A weir is an obstruction on a channel bottom over which the fluid must flow. It provides a convenient method of determining the flowrate in an open channel in terms of a single depth measurement. A *sharp-crested weir* is essentially a vertical sharp-edged flat plate placed across the channel in a way such that the fluid must flow across the sharp edge and drop into the pool downstream of the weir plate, as shown in Fig. 10.13. The specific shape of the flow area in the plane of the weir plate is used to designate the type of weir (see Fig. 10.14). The complex nature of the flow over a weir makes it impossible to obtain precise analytical expressions for the flow as a function of other parameters, such as the weir height, P_w, the *weir head, H,* the fluid depth upstream, and the geometry of the weir plate (angle θ for triangular weirs or aspect ratio, b/H, for rectangular weirs).

As a first approximation, we assume that the velocity profile upstream of the weir plate is uniform and that the pressure within the nappe (see Fig. 10.13) is atmospheric. In addition, we assume that the fluid flows horizontally over the weir plate with a nonuniform velocity profile, as indicated in Fig. 10.15. With $p_B = 0$ the Bernoulli equation for flow along the arbitrary streamline A–B indicated can be written as

$$\frac{p_A}{\gamma} + \frac{V_1^2}{2g} + z_A = (H + P_w - h) + \frac{u_2^2}{2g} \tag{10.21}$$

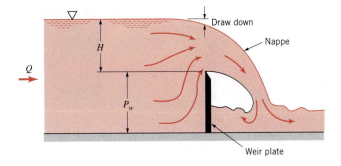

■ FIGURE 10.13
Sharp-crested weir geometry.

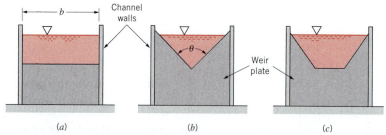

■ **FIGURE 10.14**   Sharp-crested weir plate geometry: (*a*) rectangular, (*b*) triangular, and (*c*) trapezoidal.

where h is the distance that point B is below the free surface. The total head for any particle along the vertical section (1) is the same, $z_A + p_A/\gamma + V_1^2/2g = H + P_w + V_1^2/2g$. Thus, the velocity of the fluid over the weir plate is obtained from Eq. 10.21 as

$$u_2 = \sqrt{2g\left(h + \frac{V_1^2}{2g}\right)}$$

The flowrate can be calculated from

$$Q = \int_{(2)} u_2\, dA = \int_{h=0}^{h=H} u_2 \ell\, dh \tag{10.22}$$

where $\ell = \ell(h)$ is the cross-channel width of a strip of the weir area, as indicated in Fig. 10.15*b*. For a rectangular weir ℓ is constant. For other weirs, such as triangular or circular weirs, the value of ℓ is known as a function of h.

For a rectangular weir (see Fig. 10.14*a*), $\ell = b$, and the flowrate becomes

$$Q = \sqrt{2g}\, b \int_0^H \left(h + \frac{V_1^2}{2g}\right)^{1/2} dh$$

or

$$Q = \frac{2}{3}\sqrt{2g}\, b\left[\left(H + \frac{V_1^2}{2g}\right)^{3/2} - \left(\frac{V_1^2}{2g}\right)^{3/2}\right] \tag{10.23}$$

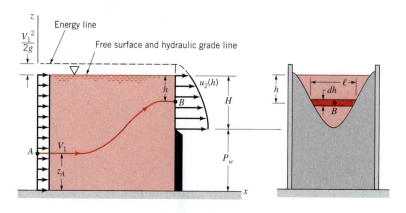

■ **FIGURE 10.15**   Assumed flow structure over a weir.

Equation 10.23 is a rather cumbersome expression that can be simplified by using the fact that with $P_w \gg H$ (as often happens in practical situations) the upstream velocity is negligibly small. That is, $V_1^2/2g \ll H$ and Eq. 10.23 simplifies to the basic rectangular weir equation

$$Q = \tfrac{2}{3} \sqrt{2g} \, b \, H^{3/2} \qquad (10.24)$$

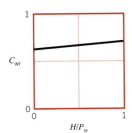

C_{wr}

H/P_w

Because of the numerous approximations made to obtain Eq. 10.24, it is not unexpected that an experimentally determined correction factor must be used to obtain the actual flowrate as a function of weir head. Thus, the final form is

$$Q = C_{\mathrm{wr}} \tfrac{2}{3} \sqrt{2g} \, b \, H^{3/2} \qquad (10.25)$$

where C_{wr} is the rectangular weir coefficient. In most practical situations the following correlation, shown in the figure in the margin, can be used (Refs. 3 and 6):

$$C_{\mathrm{wr}} = 0.611 + 0.075 \left(\frac{H}{P_w}\right) \qquad (10.26)$$

More precise values of C_{wr} can be found in the literature, if needed (Refs. 2, 9).

The triangular sharp-crested weir (see Fig. 10.14b) is often used for flow measurements, particularly for measuring flowrates over a wide range of values. For small flowrates, the head, H, for a rectangular weir would be very small and the flowrate could not be measured accurately. However, with the triangular weir, the flow area decreases as H decreases so that even for small flowrates, reasonable heads are developed. Accurate results can be obtained over a wide range of Q.

The triangular weir equation can be obtained from Eq. 10.22 by using

$$\ell = 2(H - h) \tan\left(\frac{\theta}{2}\right)$$

V10.7 Triangular
weir

where θ is the angle of the vee notch (see Figs. 10.14 and 10.15). After carrying out the integration and again neglecting the upstream velocity ($V_1^2/2g \ll H$), we obtain

$$Q = C_{\mathrm{wt}} \frac{8}{15} \tan\left(\frac{\theta}{2}\right) \sqrt{2g} \, H^{5/2} \qquad (10.27)$$

where the experimentally determined triangular weir coefficient, C_{wt}, is used to account for the real-world effects neglected in the analysis. Typical values of C_{wt} for triangular weirs are in the range of 0.58 to 0.62, as shown in Fig. 10.16.

10.6.3   Broad-Crested Weirs

A *broad-crested weir* is a structure in an open channel that has a horizontal crest above which the fluid pressure may be considered hydrostatic. A typical configuration is shown in Fig. 10.17.

The operation of a broad-crested weir is based on the fact that for an appropriately designed weir, nearly uniform critical flow is achieved in the short reach above the weir block. (If $H/L_w < 0.08$, viscous effects are important, and the flow is subcritical over the weir. However, if $H/L_w > 0.5$ the streamlines are not horizontal.) If the kinetic energy of the upstream flow is negligible, then $V_1^2/2g \ll y_1$ and the upstream specific energy is $E_1 = V_1^2/2g + y_1 \approx y_1$. Observations show that as the flow passes over the weir block, it accelerates and reaches critical conditions, $y_2 = y_c$ and $\mathrm{Fr}_2 = 1$ (i.e., $V_2 = c_2$), corresponding to the nose of the specific energy curve (see Fig. 10.6). The flow does not accelerate to supercritical conditions ($\mathrm{Fr}_2 > 1$).

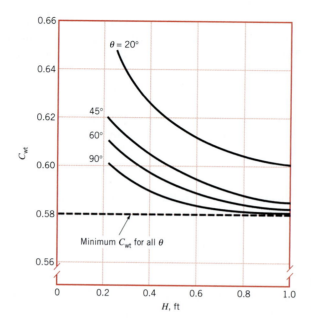

■ **FIGURE 10.16** Weir coefficient for triangular sharp-crested weirs (Ref. 8).

The Bernoulli equation can be applied between point (1) upstream of the weir and point (2) over the weir where the flow is critical to obtain

$$H + P_w + \frac{V_1^2}{2g} = y_c + P_w + \frac{V_c^2}{2g}$$

or, if the upstream velocity head is negligible

$$H - y_c = \frac{(V_c^2 - V_1^2)}{2g} = \frac{V_c^2}{2g}$$

However, because $V_2 = V_c = (gy_c)^{1/2}$, we find that $V_c^2 = gy_c$ so that we obtain

$$H - y_c = \frac{y_c}{2}$$

or

$$y_c = \frac{2H}{3}$$

Thus, the flowrate is

$$Q = by_2 V_2 = by_c V_c = by_c (gy_c)^{1/2} = b\sqrt{g}\, y_c^{3/2}$$

V10.8 Low-head dam

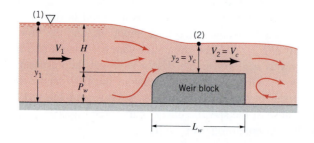

■ **FIGURE 10.17** Broad-crested weir geometry.

or

$$Q = b\sqrt{g}\left(\frac{2}{3}\right)^{3/2} H^{3/2}$$

Again an empirical weir coefficient is used to account for the various real-world effects not included in the aforementioned simplified analysis. That is

$$Q = C_{wb}\, b\sqrt{g}\left(\frac{2}{3}\right)^{3/2} H^{3/2} \tag{10.28}$$

where approximate values of C_{wb}, the broad-crested weir coefficient shown in the figure in the margin, can be obtained from the equation (Ref. 5)

$$C_{wb} = 1.125\left(\frac{1 + H/P_w}{2 + H/P_w}\right)^{1/2} \tag{10.29}$$

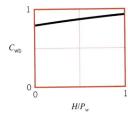

EXAMPLE 10.7   Sharp-Crested and Broad-Crested Weirs

GIVEN Water flows in a rectangular channel of width $b = 2$ m with flowrates between $Q_{min} = 0.02$ m³/s and $Q_{max} = 0.60$ m³/s. This flowrate is to be measured by using **(a)** a rectangular sharp-crested weir, **(b)** a triangular sharp-crested weir with $\theta = 90°$, or **(c)** a broad-crested weir. In all cases the bottom of the flow area over the weir is a distance $P_w = 1$ m above the channel bottom.

FIND Plot a graph of $Q = Q(H)$, flowrate as a function of weir head, for each weir and comment on which weir would be best for this application.

SOLUTION

For the rectangular weir with $P_w = 1$ m, Eqs. 10.25 and 10.26 give

$$Q = C_{wr}\frac{2}{3}\sqrt{2g}\, bH^{3/2}$$

$$= \left(0.611 + 0.075\frac{H}{P_w}\right)\frac{2}{3}\sqrt{2g}\, bH^{3/2}$$

Thus,

$$Q = (0.611 + 0.075H)\frac{2}{3}\sqrt{2(9.81 \text{ m/s}^2)}\,(2\text{ m})\,H^{3/2}$$

or

$$Q = 5.91(0.611 + 0.075H)H^{3/2} \tag{1}$$

where H and Q are in meters and m³/s, respectively. The results from Eq. 1 are plotted in Fig. E10.7.

Similarly, for the triangular weir, Eq. 10.27 gives

$$Q = C_{wt}\frac{8}{15}\tan\left(\frac{\theta}{2}\right)\sqrt{2g}\, H^{5/2}$$

$$= C_{wt}\frac{8}{15}\tan(45°)\,\sqrt{2(9.81 \text{ m/s}^2)}\, H^{5/2}$$

or

$$Q = 2.36C_{wt}\, H^{5/2} \tag{2}$$

where H and Q are in meters and m³/s and C_{wt} is obtained from Fig. 10.16. For example, with $H = 0.20$ m, we find $C_{wt} = 0.60$, or $Q = 2.36\,(0.60)(0.20)^{5/2} = 0.0253$ m³/s. The triangular weir results are also plotted in Fig. E10.7.

For the broad-crested weir, Eqs. 10.28 and 10.29 give

$$Q = C_{wb}\, b\sqrt{g}\left(\frac{2}{3}\right)^{3/2} H^{3/2}$$

$$= 1.125\left(\frac{1 + H/P_w}{2 + H/P_w}\right)^{1/2} b\sqrt{g}\left(\frac{2}{3}\right)^{3/2} H^{3/2}$$

Thus, with $P_w = 1$ m

$$Q = 1.125\left(\frac{1 + H}{2 + H}\right)^{1/2}(2\text{ m})\,\sqrt{9.81 \text{ m/s}^2}\left(\frac{2}{3}\right)^{3/2} H^{3/2}$$

or

$$Q = 3.84\left(\frac{1 + H}{2 + H}\right)^{1/2} H^{3/2} \tag{3}$$

where, again, H and Q are in meters and m³/s. This result is also plotted iin Fig. E10.7.

COMMENT Although it appears as though any of the three weirs would work well for the upper portion of the flowrate range, neither the rectangular nor the broad-crested weir would be very accurate for small flowrates near $Q = Q_{min}$ because of the small head, H, at these conditions. The triangular weir, however, would allow reasonably large values of H at the lowest flowrates. The corresponding heads with $Q = Q_{min} = 0.02$ m³/s for rectangular, triangular, and broad-crested weirs are 0.0312, 0.182, and 0.0375 m, respectively.

In addition, as discussed in this section, for proper operation the broad-crested weir geometry is restricted to $0.08 < H/L_w < 0.50$, where L_w is the weir block length. From Eq. 3 with $Q_{max} = 0.60$ m³/s, we obtain $H_{max} = 0.349$. Thus, we must have $L_w > H_{max}/0.5 = 0.698$ m to maintain proper critical flow conditions at the largest flowrate in the channel. However, with $Q = Q_{min} = 0.02$ m³/s, we obtain $H_{min} = 0.0375$ m. Thus, we must have $L_w < H_{min}/0.08 = 0.469$ m to ensure that frictional effects are not important. Clearly, these two constraints on the geometry of the weir block, L_w, are incompatible.

A broad-crested weir will not function properly under the wide range of flowrates considered in this example. The sharp-crested triangular weir would be the best of the three types considered, provided the channel can handle the $H_{max} = 0.719$-m head.

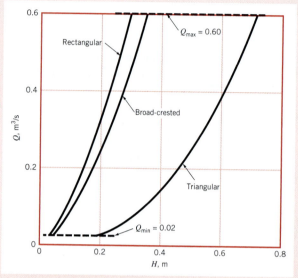

■ **FIGURE E10.7**

10.6.4 Underflow Gates

A variety of ***underflow gate*** structures is available for flowrate control. Three typical types are illustrated in Fig. 10.18.

The flow under a gate is said to be free outflow when the fluid issues as a jet of supercritical flow with a free surface open to the atmosphere as shown in Fig. 10.18. In such cases it is customary to write this flowrate as the product of the distance, a, between the channel bottom and the bottom of the gate times the convenient reference velocity $(2gy_1)^{1/2}$. That is

$$q = C_d a \sqrt{2gy_1} \tag{10.30}$$

where q is the flowrate per unit width. The discharge coefficient, C_d, is a function of the contraction coefficient, $C_c = y_2/a$, and the depth ratio y_1/a. Typical values of the discharge coefficient for free outflow (or free discharge) from a vertical sluice gate are on the order of 0.55 to 0.60 as indicated by the top line in Fig. 10.19 (Ref. 2).

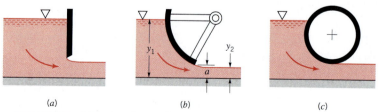

(a)   (b)   (c)

■ **FIGURE 10.18** **Three variations of underflow gates: (*a*) vertical gate, (*b*) radial gate, and (*c*) drum gate.**

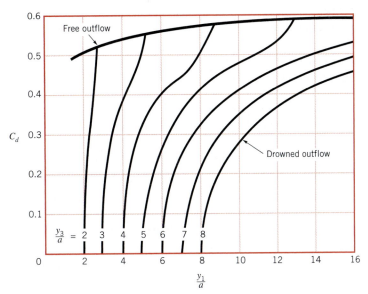

■ **FIGURE 10.19** **Typical discharge coefficients for underflow gates (Ref. 3).**

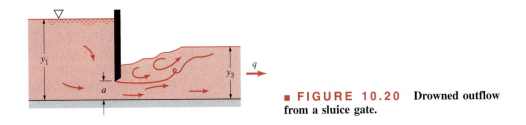

■ **FIGURE 10.20** **Drowned outflow from a sluice gate.**

As indicated in Fig. 10.20, in certain situations the jet of water issuing from under the gate is overlaid by a mass of water that is quite turbulent. Typical values of C_d for these drowned outflow cases are indicated as the series of lower curves in Fig. 10.19.

10.7   Chapter Summary and Study Guide

open-channel flow
critical flow
subcritical flow
supercritical flow
wave speed
Froude number
specific energy
specific energy diagram

This chapter discussed various aspects of flows in an open channel. A typical open channel flow is driven by the component of gravity in the direction of flow. The character of such flows can be a strong function of the Froude number, which is the ratio of the fluid speed to the free surface wave speed. The specific energy diagram is used to provide insight into the flow processes involved in open channel flow.

Uniform depth channel flow is achieved by a balance between the potential energy lost by the fluid as it coasts downhill and the energy dissipated by viscous effects. Alternately, it represents a balance between weight and friction forces. The relationship among the flowrate, the slope of the channel, the geometry of the channel, and the roughness of the channel surfaces is given by the Manning equation. Values of the Manning coefficient used in the Manning equation are dependent on the surface material roughness.

The hydraulic jump is an example of nonuniform depth open channel flow. If the Froude number of a flow is greater than one, the flow is supercritical, and a hydraulic jump may occur. The momentum and mass equations are used to obtain the relationship between the upstream Froude number and the depth ratio across the jump. The energy dissipated in the jump and the head loss can then be determined by use of the energy equation.

The use of weirs to measure the flowrate in an open channel is discussed. The relationships between the flowrate and the weir head are given for both sharp-crested and broad-crested weirs.

The following checklist provides a study guide for this chapter. When your study of the entire chapter and end-of-chapter exercises has been completed you should be able to

- write out meanings of the terms listed here in the margin and understand each of the related concepts. These terms are particularly important and are set in color and bold type in the text.
- determine the Froude number for a given flow and explain the concepts of subcritical, critical, and supercritical flows.
- plot and interpret the specific energy diagram for a given flow.
- use the Manning equation to analyze uniform depth flow in an open channel.
- calculate properties such as the depth ratio and the head loss for a hydraulic jump.
- determine the flowrates over sharp-crested weirs, broad-crested weirs, and under the underflow gates.

References

1. Currie, C. G., and Currie, I. G., *Fundamental Mechanics of Fluids,* Third Edition, Marcel Dekker, New York, 2003.
2. Henderson, F. M., *Open Channel Flow,* Macmillan, New York, 1966.
3. Rouse, H., *Elementary Fluid Mechanics,* Wiley, New York, 1946.
4. French, R. H., *Open Channel Hydraulics,* McGraw-Hill, New York, 1985.
5. Chow, V. T., *Open Channel Hydraulics,* McGraw-Hill, New York, 1959.
6. Blevins, R. D., *Applied Fluid Dynamics Handbook,* Van Nostrand Reinhold, New York, 1984.
7. Vennard, J. K., and Street, R. L., *Elementary Fluid Mechanics,* Seventh Edition, Wiley, New York, 1995.
8. Lenz, A. T., "Viscosity and Surface Tension Effects on V-Notch Weir Coefficients," *Transactions of the American Society of Chemical Engineers,* Vol. 108, 759–820, 1943.
9. Spitzer, D. W., editor, *Flow Measurement: Practical Guides for Measurement and Control,* Instrument Society of America, Research Triangle Park, N. C., 1991.
10. Wallet, A., and Ruellan, F., *Houille Blanche,* Vol. 5, 1950.

Review Problems

Go to Appendix F for a set of review problems with answers. Detailed solutions can be found in *Student Solution Manual for a Brief Introduction to Fluid Mechanics,* by Young et al. (© 2006 John Wiley and Sons, Inc.).

Problems

Note: Unless otherwise indicated, use the values of fluid properties found in the tables on the inside of the front cover. Problems designated with an (*) are intended to be solved with the aid of a programmable calculator or a computer.

Answers to the even-numbered problems are listed at the end of the book. Access to the videos that accompany problems can be obtained through the book's web site, www.wiley.com/college/young. The lab-type problems and FE problems can also be accessed on this web site.

Section 10.2 Surface Waves

10.1 The flowrate in a 10-ft-wide, 2-ft-deep river is $Q = 190$ cfs. Is the flow subcritical or supercritical?

10.2 Consider waves made by dropping objects (one after another from a fixed location) into a stream of depth y that is moving with speed V as shown in Fig. P10.2 (see **Video V9.1**). The circular wave crests that are produced travel with speed $c = (gy)^{1/2}$ relative to the moving water. Thus, as the circular waves are washed downstream, their diameters increase and the center of each circle is fixed relative to the moving water. **(a)** Show that if the flow is supercritical, lines tangent to the waves generate a wedge of half-angle $\alpha/2 = \arcsin(1/Fr)$, where $Fr = V/(gy)^{1/2}$ is the Froude number. **(b)** Discuss what happens to the wave pattern when the flow is subcritical, $Fr < 1$.

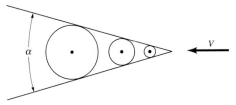

■ **FIGURE P10.2**

10.3 Waves on the surface of a tank are observed to travel at a speed of 2 m/s. How fast would these waves travel if **(a)** the tank were in an elevator accelerating upward at a rate of 4 m/s², **(b)** the tank accelerated horizontally at a rate of 9.81 m/s², and **(c)** the tank were aboard the orbiting space shuttle. Explain.

10.4 Observations at a shallow sandy beach show that even though the waves several hundred yards out from the shore are not parallel to the beach, the waves often "break" on the beach

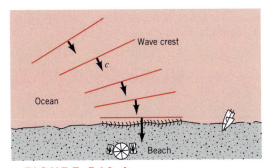

■ **FIGURE P10.4**

nearly parallel to the shore as is indicated in Fig. P10.4. Explain this behavior based on the wave speed $c = (gy)^{1/2}$.

10.5 (See Fluids in the News article titled "Tsunami, the nonstorm wave," Section 10.2.1.) An earthquake causes a shift in the ocean floor that produces a tsunami with a wavelength of 100 km. How fast will this wave travel across the ocean surface if the ocean depth is 3000 m?

Section 10.3 Energy Considerations

10.6 Water flows in a rectangular channel with a flowrate per unit width of $q = 2.5$ m²/s. Plot the specific energy diagram for this flow. Determine the two possible depths of flow if $E = 2.5$ m.

*10.7 Water flows in a rectangular channel with a specific energy of $E = 5$ ft. If the flowrate per unit width is $q = 30$ ft²/s, determine the two possible flow depths and the corresponding Froude numbers. Plot the specific energy diagram for this flow. Repeat the problem for $E = 1, 2, 3,$ and 4 ft.

10.8 Water flows in a rectangular channel at a rate of $q = 20$ cfs/ft. When a Pitot tube is placed in the stream, water in the tube rises to a level of 4.5 ft above the channel bottom. Determine the two possible flow depths in the channel. Illustrate this flow on a specific energy diagram.

10.9 Water flows in a 10-ft-wide rectangular channel with a flowrate of $Q = 60$ ft³/s and an upstream depth of $y_1 = 2$ ft as shown in Fig. P10.9. Determine the flow depth and the surface elevation at section (2).

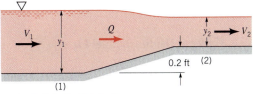

■ **FIGURE P10.9**

10.10 Repeat Problem 10.9 if the upstream depth is $y_1 = 0.5$ ft.

10.11 Water in a rectangular channel flows into a gradual contraction section as indicated in Fig. P10.11. If the flowrate is

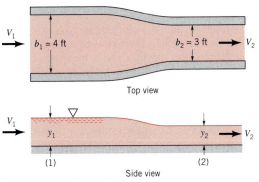

Top view

Side view

■ **FIGURE P10.11**

$Q = 25$ ft^3/s and the upstream depth is $y_1 = 2$ ft, determine the downstream depth, y_2.

10.12   Repeat Problem 10.11 if the upstream depth is $y_1 = 0.5$ ft. Assume that there are no losses between sections (1) and (2).

10.13   Water flows in a horizontal, rectangular channel with an initial depth of 2 ft and an initial velocity of 12 ft/s. Determine the depth downstream if losses are negligible. Note that there may be more than one solution. Repeat the problem if the initial depth remains the same, but the initial velocity is 6 ft/s.

*10.14   Water flows over the bump in the bottom of the rectangular channel shown in Fig. P10.14 with a flowrate per unit width of $q = 4$ m^2/s. The channel bottom contour is given by $z_B = 0.2e^{-x^2}$, where z_B and x are in meters. The water depth far upstream of the bump is $y_1 = 2$ m. Plot a graph of the water depth, $y = y(x)$, and the surface elevation, $z = z(x)$, for -4 m $\leq x \leq 4$ m. Assume one-dimensional flow.

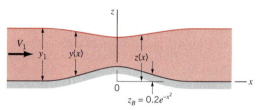

■ FIGURE P10.14

10.15   Determine the maximum depth in a 3-m-wide rectangular channel if the flow is to be supercritical with a flowrate of $Q = 60$ m^3/s.

Section 10.4 Uniform Depth Channel Flow—Determine Flowrate

10.16   Water flows in a 10-ft-wide rectangular channel with a flowrate of 200 cfs and a depth of 3 ft. If the slope is 0.005, determine the Manning coefficient, n, and the average shear stress at the sides and bottom of the channel.

10.17   The following data are obtained for a particular reach of the Provo River in Utah: $A = 183$ ft^2, free-surface width $= 55$ ft, average depth $= 3.3$ ft, $R_h = 3.22$ ft, $V = 6.56$ ft/s, length of reach $= 116$ ft, and elevation drop of reach $= 1.04$ ft. Determine **(a)** the average shear stress on the wetted perimeter, **(b)** the Manning coefficient, n, and **(c)** the Froude number of the flow.

10.18   (See Fluids in the News article titled "Done without a GPS or lasers," Section 10.4.2.) Determine the number of gallons of water delivered per day by a rubble masonary, 1.2-m-wide aqueduct laid on an average slope of 14.6 m per 50 km if the water depth is 1.8 m.

10.19   An open channel of square cross section had a flowrate of 80 ft^3/s when first used. After extended use, the channel became half-filled with silt. Determine the flowrate for this silted condition. Assume the Manning coefficient is the same for all the surfaces.

10.20   The great Kings River flume in Fresno County, California, was used from 1890 to 1923 to carry logs from an elevation of 4500 ft where trees were cut to an elevation of 300 ft at the railhead. The flume was 54 miles long, constructed of wood, and had a V cross section as indicated in Fig. P10.20. It is claimed that logs would travel the length of the flume in 15 hr. Do you agree with this claim? Provide appropriate calculations to support your answer.

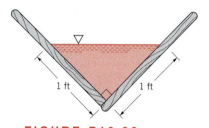

■ FIGURE P10.20

10.21   Water flows in a 2-m-diameter finished concrete pipe so that it is completely full and the pressure is constant all along the pipe. If the slope is $S_0 = 0.005$, determine the flowrate by using open-channel flow methods. Compare this result with that obtained using the pipe flow methods of Chapter 8.

10.22   Because of neglect, an irrigation canal has become weedy and the maximum flowrate possible is only 90% of the desired flowrate. Would removing the weeds, thus making the surface gravel, allow the canal to carry the desired flowrate? Support your answer with appropriate calculations.

10.23   Water flows in a channel with an equilateral triangle cross section as is shown in Fig. P10.23. Let Q_{full} denote the flowrate when $y = h$. By what percent is Q_{full} less than Q when $y = h - \delta y$, where $\delta y \ll h$? That is, placing a lid on this channel reduces the flowrate by what percent?

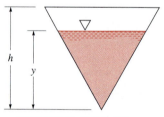

■ FIGURE P10.23

*10.24   Water flows in the fiberglass ($n = 0.014$) triangular channel with a round bottom shown in Fig. P10.24. The channel slope is 0.1 m/90 m. Plot a graph of flowrate as a function of water depth for $0 \leq y \leq 0.50$ m with bottom radii of $r = 0$, 0.25, 0.50, 0.75, and 1.0 m.

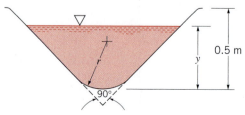

■ FIGURE P10.24

10.25 The smooth concrete-lined channel shown in Fig. P10.25 is built on a slope of 2 m/km. Determine the flowrate if the depth is $y = 1.5$ m.

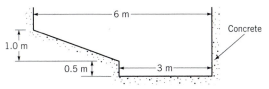

■ **FIGURE P10.25**

***10.26** The cross section of a creek valley is shown in Fig. P10.26. Plot a graph of flowrate as a function of depth, y, for $0 \le y \le 10$ ft. The slope is 5 ft/mi.

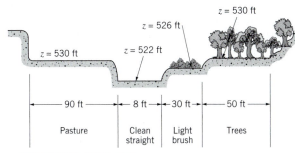

■ **FIGURE P10.26**

10.27 Repeat Problem 10.25 if the surfaces are smooth concrete as indicated except for the diagonal surface, which is gravelly with $n = 0.025$.

10.28 Determine the flowrate for the symmetrical channel shown in Fig. P10.28 if the bottom is smooth concrete and the sides are weedy. The bottom slope is $S_0 = 0.001$.

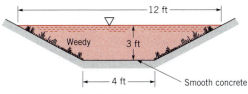

■ **FIGURE P10.28**

Section 10.4 Uniform Depth Channel Flow— Determine Depth or Size

10.29 Rainwater runoff from a 200×500-ft parking lot is to drain through a circular concrete pipe that is laid on a slope of 5 ft/mi. Determine the pipe diameter if it is to be full with a steady rainfall of 1.5 in./hr.

10.30 (See Fluids in the News article titled "Plumbing the Everglades," Section 10.4.3.) The canal shown in Fig. P10.30 is to be widened so that it can carry twice the amount of water. Determine the additional width, L, required if all other parameters (i.e., flow depth, bottom slope, surface material, side slope) are to remain the same.

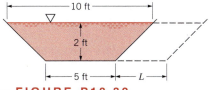

■ **FIGURE P10.30**

10.31 A rectangular, unfinished 28-ft-wide concrete channel is laid on a slope of 8 ft/mi. Determine the flow depth and Froude number of the flow if the flowrate is 400 ft³/s.

10.32 An engineer is to design a channel lined with planed wood to carry water at a flowrate of 2 m³/s on a slope of 10 m/800 m. The channel cross section can be either a 90° triangle or a rectangle with a cross section twice as wide as its depth. Which would require less wood and by what percent?

10.33 Two canals join to form a larger canal as shown in **Video V10.2** and Fig. P10.33. Each of the three rectangular canals is lined with the same material and has the same bottom slope. The water depth in each is to be 2 m. Determine the width of the merged canal, b. Explain physically (i.e., without using any equations) why it is expected that the width of the merged canal is less than the combined widths of the two original canals (i.e., $b < 4$ m $+ 8$ m $= 12$ m).

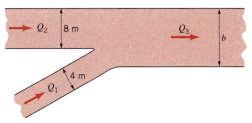

■ **FIGURE P10.33**

10.34 A smooth steel water slide at an amusement park is of semicircular cross section with a diameter of 2.5 ft. The slide descends a vertical distance of 35 ft in its 420-ft length. If pumps supply water to the slide at a rate of 6 cfs, determine the depth of flow. Neglect the effects of the curves and bends of the slide.

10.35 Determine the flow depth for the channel shown in Fig. P10.25 if the flowrate is 15 m³/s.

Section 10.4 Uniform Depth Channel Flow— Determine Slope

10.36 Water flows in a river with a speed of 3 ft/s. The river is a clean, straight natural channel, 400 ft wide with a nearly uniform 3-ft depth. Is the slope of this river greater than or less than the average slope of the Mississippi River, which drops a distance of 1475 ft in its 2552-mi length? Support your answer with appropriate calculations.

10.37 To prevent weeds from growing in a clean earthen-lined canal, it is recommended that the velocity be no less than 2.5 ft/s. For the symmetrical canal shown in Fig. P10.37, determine the minimum slope needed.

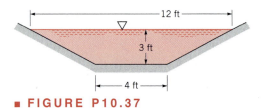

■ **FIGURE P10.37**

10.38 The smooth concrete-lined symmetrical channel shown in **Video V10.3** and Fig. P10.37 carries silt-laden water. If the velocity must be 4.0 ft/s to prevent the silt from settling out (and eventually clogging the channel), determine the minimum slope needed.

10.39 The symmetrical channel shown in Fig. P10.37 is dug in sandy loam soil with $n = 0.020$. For such surface material it is recommended that to prevent scouring of the surface the average velocity be no more than 1.75 ft/s. Determine the maximum slope allowed.

Section 10.6.1 The Hydraulic Jump (also see Lab Problems 10.59 and 10.60)

10.40 (See Fluids in the News article titled "Grand Canyon rapids building," Section 10.6.1.) During the flood of 1983, a large hydraulic jump formed at "Crystal Hole" rapid on the Colorado River. People rafting the river at that time report "entering the rapid at almost 30 mph, hitting a 20-ft-tall wall of water, and exiting at about 10 mph." Is this information (i.e., upstream and downstream velocities and change in depth) consistent with the principles of a hydraulic jump? Show calculations to support your answer.

10.41 A hydraulic jump at the base of a spillway of a dam is such that the depths upstream and downstream of the jump are 0.90 and 3.6 m, respectively (see **Video V10.5**). If the spillway is 10 m wide, what is the flowrate over the spillway?

10.42 Under appropriate conditions, water flowing from a faucet, onto a flat plate, and over the edge of the plate can produce a circular hydraulic jump as shown in Fig. P10.42 and **Video V10.6.** Consider a situation where a jump forms 3.0 in. from the center of the plate with depths upstream and downstream of the jump of 0.05 and 0.20 in., respectively. Determine the flowrate from the faucet.

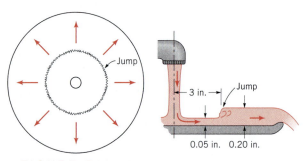

■ **FIGURE P10.42**

10.43 Water flows in a 2-ft-wide rectangular channel at a rate of 10 ft³/s. If the water depth downstream of a hydraulic jump is 2.5 ft, determine **(a)** the water depth upstream of the jump, **(b)** the upstream and downstream Froude numbers, and **(c)** the head loss across the jump.

10.44 A hydraulic jump at the base of a spillway of a dam is such that the depths upstream and downstream of the jump are 0.90 and 3.6 ft, respectively (see **Video V10.5**). If the spillway is 10 ft wide, what is the flowrate upstream of the jump; downstream of the jump?

10.45 Determine the head loss and power dissipated by the hydraulic jump of Problem 10.41.

10.46 Water flows in a rectangular channel at a depth of $y = 1$ ft and a velocity of $V = 20$ ft/s. When a gate is suddenly placed across the end of the channel, a wave (a moving hydraulic jump) travels upstream with velocity V_w as is indicated in Fig. P10.46. Determine V_w. Note that this is an unsteady problem for a stationary observer. However, for an observer moving to the left with velocity V_w, the flow appears as a steady hydraulic jump.

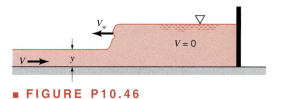

■ **FIGURE P10.46**

Section 10.6 Weirs (also see Lab Problems 10.57 and 10.58)

10.47 Water flows over a 5-ft-wide, rectangular sharp-crested weir that is $P_w = 4.5$ ft tall. If the depth upstream is 5 ft, determine the flowrate.

10.48 A rectangular sharp-crested weir is used to measure the flowrate in a channel of width 10 ft. It is desired to have the channel flow depth be 6 ft when the flowrate is 50 cfs. Determine the height, P_w, of the weir plate.

10.49 Water flows from a storage tank, over two triangular weirs, and into two irrigation channels as shown in **Video V10.7** and Fig. P10.49. The head for each weir is 0.4 ft, and the flowrate in the channel fed by the 90° V-notch weir is to be twice the flowrate in the other channel. Determine the angle θ for the second weir.

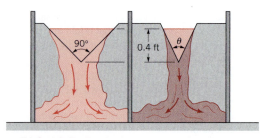

■ **FIGURE P10.49**

10.50 Water flows over a broad-crested weir that has a width of 4 m and a height of $P_w = 1.5$ m. The free-surface well upstream of the weir is at a height of 0.5 m above the surface of the weir. Determine the flowrate in the channel and the minimum depth of the water above the weir block.

10.51 Determine the flowrate per unit width, q, over a broad-crested weir that is 2.0 m tall if the head, H, is 0.50 m.

10.52 Determine the head, H, required to allow a flowrate of 300 m^3/hr over a sharp-crested triangular weir with $\theta = 60°$.

10.53 Water flows in a rectangular channel of width $b = 20$ ft at a rate of 100 ft^3/s. The flowrate is to be measured by using either a rectangular weir of height $P_w = 4$ ft or a triangular ($\theta = 90°$) sharp-crested weir. Determine the head, H, necessary. If measurement of the head is accurate to only ± 0.04 ft, determine the accuracy of the measured flowrate expected for each of the weirs. Which weir would be the most accurate? Explain.

10.54 A water-level regulator (not shown) maintains a depth of 2.0 m downstream from a 50-ft-wide drum gate as

shown in Fig. P10.54. Plot a graph of flowrate, Q, as a function of water depth upstream of the gate, y_1, for $2.0 \le y_1 \le 5.0$ m.

10.55 Water flows under a sluice gate in a channel of 10-ft width. If the upstream depth remains constant at 5 ft, plot a graph of flowrate as a function of the distance between the gate and the channel bottom as the gate is slowly opened. Assume free outflow.

10.56 Water flows over a triangular weir as shown in Fig. P10.56a and **Video V10.7.** It is proposed that in order to increase the flowrate, Q, for a given head, H, the triangular weir should be changed to a trapezoidal weir as shown in Fig. P10.56b. **(a)** Derive an equation for the flowrate as a function of the head for the trapezoidal weir. Neglect the upstream velocity head and assume the weir coefficient is 0.60, independent of H. **(b)** Use the equation obtained in part **(a)** to show that when $b \ll H$ the trapezoidal weir functions as if it were a triangular weir. Similarly, show that when $b \gg H$ the trapezoidal weir functions as if it were a rectangular weir.

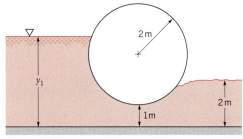

■ **FIGURE P10.54**

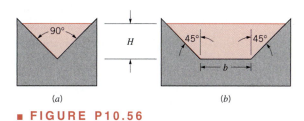

■ **FIGURE P10.56**

Lab Problems

10.57 This problem involves the calibration of a triangular weir. To proceed with this problem, go to the book's web site, www.wiley.com/college/young.

10.58 This problem involves the calibration of a rectangular weir. To proceed with this problem, go to the book's website, www.wiley.com/college/young.

10.59 This problem involves the depth ratio across a hydraulic jump. To proceed with this problem, go to the book's web site, www.wiley.com/college/young.

10.60 This problem involves the head loss across a hydraulic jump. To proceed with this problem, go to the book's web site, www.wiley.com/college/young.

FE Exam Problems

Sample FE (Fundamentals of Engineering) exam questions for fluid mechanics are provided on the book's web site, www.wiley.com/college/young.

CHAPTER 11

Turbine blades on the rotor of a centrifugal compressor used in an automobile turbocharger. (Photograph by Bruce Munson.)

Turbomachines

LEARNING OBJECTIVES

After completing this chapter, you should be able to:

- construct appropriate velocity triangles.
- calculate parameters related to centrifugal pumps.
- apply appropriate equations and principles to predict pump performance.
- calculate shaft torque and power for turbines.
- use the concept of specific speed to select the appropriate pump or turbine.

Pumps and turbines (sometimes called *fluid machines*) occur in a wide variety of configurations. In general, pumps add energy to the fluid—they do work on the fluid; turbines extract energy from the fluid—the fluid does work on them. The term "pump" will be used to generically refer to all pumping machines, including *pumps, fans, blowers,* and *compressors.* Fluid machines can be divided into two main categories: *positive displacement machines* (denoted as the static type) and ***turbomachines*** (denoted as the dynamic type). The majority of this chapter deals with turbomachines.

Positive displacement machines force a fluid into or out of a chamber by changing the volume of the chamber. The pressures developed and the work done are a result of essentially static forces rather than dynamic effects. Typical examples include the common tire pump (shown by the figure in the margin) used to fill bicycle tires, the human heart, and the gear pump.

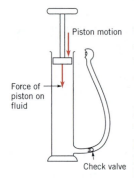

Piston motion

Force of piston on fluid

Check valve

Turbomachines, however, involve a collection of blades, buckets, flow channels, or passages arranged around an axis of rotation to form a rotor. Rotation of the rotor produces dynamic effects that either add energy to the fluid or remove energy from the fluid. Examples of turbomachine-type pumps include simple window fans, propellers on ships or airplanes, squirrel-cage fans on home furnaces, and compressors in automobile turbochargers. Examples of turbines include the turbine portion of gas turbine engines on aircraft, steam turbines used to drive generators at electrical generation stations, and the small, high-speed air turbines that power dentist drills.

11.1   Introduction

Turbomachines are mechanical devices that either extract energy from a fluid (turbine) or add energy to a fluid (pump) as a result of dynamic interactions between the device and the fluid.

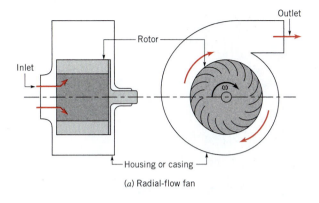

(a) Radial-flow fan

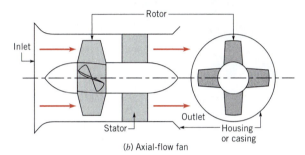

(b) Axial-flow fan

■ **FIGURE 11.1** (a) **A radial-flow turbomachine.** (b) **An axial-flow turbomachine.**

Turbomachines contain blades, airfoils, "buckets," flow channels, or passages attached to a rotating shaft. Energy is either supplied to the rotating shaft (by a motor, for example) and transferred to the fluid by the blades (a pump) or transferred from the fluid to the blades and made available at the rotating shaft as shaft power (a turbine). The fluid used can be either a gas (as with a window fan or a gas turbine engine) or a liquid (as with the water pump on a car or a turbine at a hydroelectric power plant).

Many turbomachines contain some type of housing or casing that surrounds the rotating blades or rotor, thus forming an internal flow passageway through which the fluid flows (see Fig. 11.1). Others, such as a windmill or a window fan, are unducted. Some turbomachines include stationary blades or vanes in addition to rotor blades.

Turbomachines are classified as *axial-flow, mixed-flow,* or *radial-flow* machines depending on the predominant direction of the fluid motion relative to the rotor's axis (see Fig. 11.1). For an axial-flow machine, the fluid maintains a significant axial-flow direction component from the inlet to outlet of the rotor. A radial-flow machine involves a substantial radial-flow component at the rotor inlet, exit, or both. In mixed-flow machines there are significant radial- and axial-flow velocity components for the flow through the rotor row.

11.2 Basic Energy Considerations

An understanding of the work transfer in turbomachines can be obtained by considering the basic operation of a household fan (pump) and a windmill (turbine). Although the actual flows in such devices are very complex (i.e., three-dimensional and unsteady), the essential phenomena can be illustrated by use of simplified considerations and velocity triangles.

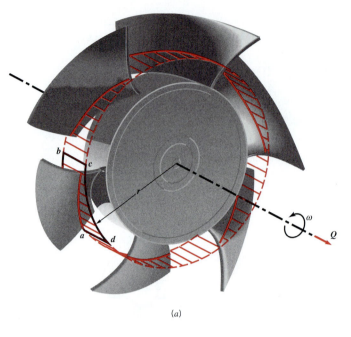

(a)

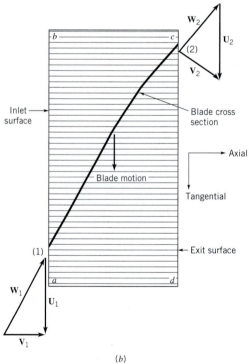

(b)

Consider a fan blade driven at constant angular velocity, ω, by a motor as is shown in Fig. 11.2a. We denote the blade speed as $U = \omega r$, where r is the radial distance from the axis of the fan. The absolute fluid velocity (that seen by a person sitting stationary at the table on which the fan rests) is denoted **V,** and the relative velocity (that seen by a person riding on the moving fan blade) is denoted **W.** As shown by the figure in the margin,

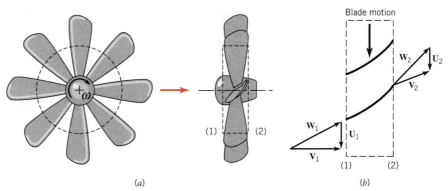

■ **FIGURE 11.3**   **Idealized flow through a windmill: (*a*) windmill blade geometry and (*b*) absolute velocity, V; relative velocity, W; and blade velocity, U; at the inlet and exit of the windmill blade section.**

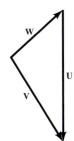

the actual (absolute) fluid velocity is the vector sum of the relative velocity and the blade velocity

$$\mathbf{V} = \mathbf{W} + \mathbf{U} \tag{11.1}$$

A simplified sketch of the fluid velocity as it "enters" and "exits" the fan at radius r is shown in Fig. 11.2*b*. The shaded surface labeled *a-b-c-d* is a portion of the cylindrical surface (including a "slice" through the blade) shown in Fig. 11.2*a*. We assume for simplicity that the flow moves smoothly along the blade so that relative to the moving blade the velocity is parallel to the leading and trailing edges (points 1 and 2) of the blade. For now we assume that the fluid enters and leaves the fan at the same distance from the axis of rotation; thus, $U_1 = U_2 = \omega r$.

With this information we can construct the ***velocity triangles*** shown in Fig. 11.2*b*. Note that this view is looking radially toward the axis of rotation. The motion of the blade is down; the motion of the incoming air is assumed to be directed along the axis of rotation. The important concept to grasp from this sketch is that the fan blade (because of its shape and motion) "pushes" the fluid, causing it to change direction. The absolute velocity vector, **V,** is turned during its flow across the blade from section (1) to section (2). Initially the fluid had no component of absolute velocity in the direction of the motion of the blade, the θ (or tangential) direction. When the fluid leaves the blade, this tangential component of absolute velocity is nonzero. For this to occur, the blade must push on the fluid in the tangential direction. That is, the blade exerts a tangential force component on the fluid in the direction of the motion of the blade. This tangential force component and the blade motion are in the same direction—the blade does work on the fluid. This device is a pump.

However, consider the windmill shown in Fig. 11.3*a*. Rather than the rotor being driven by a motor, it is rotated in the opposite direction (compared to the fan in Fig. 11.2) by the wind blowing through the rotor. We again note that because of the blade shape and motion, the absolute velocity vectors at sections (1) and (2), V_1 and V_2, have different directions. For this to happen, the blades must have pushed up on the fluid—opposite to the direction of their motion. Alternatively, because of equal and opposite forces (action/reaction) the fluid must have pushed on the blades in the direction of their motion—the fluid does work on the blades. This extraction of energy from the fluid is the purpose of a turbine.

V11.1 Windmills

EXAMPLE 11.1   Basic Difference Between a Pump and a Turbine

GIVEN The rotor shown in Fig. E11.1a rotates at a constant angular velocity of $\omega = 100$ rad/s. Although the fluid initially approaches the rotor in an axial direction, the flow across the blades is primarily radial (see Fig. 11.1a). Measurements indicate that the absolute velocity at the inlet and outlet are $V_1 = 12$ m/s and $V_2 = 25$ m/s, respectively.

FIND Is this device a pump or a turbine?

SOLUTION

To answer this question, we need to know if the tangential component of the force of the blade on the fluid is in the direction of the blade motion (a pump) or opposite to it (a turbine). We assume that the blades are tangent to the incoming relative velocity and that the relative flow leaving the rotor is tangent to

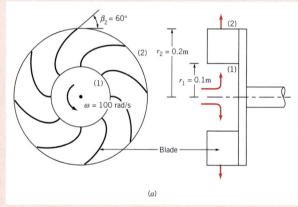

■ **FIGURE E11.1a**

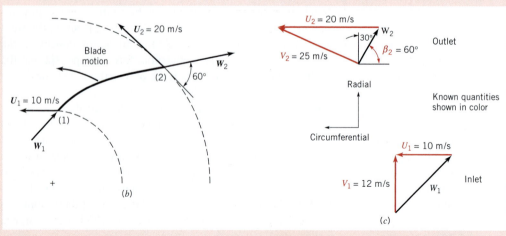

■ **FIGURE E11.1b, c**

the blades as shown in Fig. E11.1*b*. We can also calculate the inlet and outlet blade speeds as

$$U_1 = \omega r_1 = (100 \text{ rad/s})(0.1 \text{ m}) = 10 \text{ m/s}$$

and

$$U_2 = \omega r_2 = (100 \text{ rad/s})(0.2 \text{ m}) = 20 \text{ m/s}$$

With the known absolute fluid velocity and blade velocity at the inlet, we can draw the velocity triangle (the graphical representation of Eq. 11.1) at that location as shown in Fig. E11.1*c*. Note that we have assumed that the absolute flow at the blade row inlet is radial (i.e., the direction of V_1 is radial). At the outlet we know the blade velocity, U_2, the outlet speed, V_2, and the relative velocity direction, β_2 (because of the blade geometry). Therefore, we can graphically (or trigonometrically) construct the outlet velocity triangle as shown in the figure. By comparing the velocity triangles at the inlet and outlet, it can be seen that as the fluid flows across the blade row, the absolute velocity vector turns in the direction of the blade motion. At the inlet there is no component of absolute velocity in the direction of rotation; at the outlet this component is not zero. That is, the blade pushes the fluid in the direction of the blade motion, thereby doing work on the fluid, adding energy to it.

This device is a pump.　　　　**(Ans)**

COMMENT　　On the other hand, by reversing the direction of flow from larger to smaller radii, this device can become a radial-flow turbine. In this case (Fig. E11.1*d*) the flow direction is reversed (compared to that in Figs. E. 11.1*a*, *b*, and *c*) and the velocity triangles are as indicated. Stationary vanes around the perimeter of the rotor would be needed to achieve V_1 as shown. Note that the component of the absolute velocity, **V**, in the direction of the blade motion is smaller at the outlet than at the inlet. The blade must push against the fluid in the direction opposite the motion of the blade to cause this. Hence (by equal and opposite forces), the fluid pushes against the blade in the direction of blade motion, thereby doing work on the blade. There is a transfer of work from the fluid to the blade—a turbine operation.

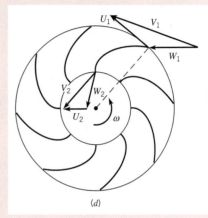

(*d*)

■ **FIGURE E11.1*d***

11.3　Basic Angular Momentum Considerations

Since turbomachines involve the rotation of an impeller or a rotor about a central axis, it is appropriate to discuss their performance in terms of torque and angular momentum.

Recall that work can be written as force times distance or as torque times angular displacement. Hence, if the shaft torque (the torque that the shaft applies to the rotor) and the rotation of the rotor are in the same direction, energy is transferred from the shaft to the rotor and from the rotor to the fluid—the machine is a pump. Conversely, if the torque exerted by the shaft on the rotor is opposite to the direction of rotation, the energy transfer is from the fluid to the rotor—a turbine. The amount of shaft torque (and hence shaft work) can be obtained from the moment-of-momentum equation derived formally in Section 5.2.3 and discussed as follows.

Consider a fluid particle traveling through the rotor in the radial-flow machine shown in Figs. E11.1*a*, *b*, and *c*. For now, assume that the particle enters the rotor with a radial velocity only (i.e., no "swirl"). After being acted upon by the rotor blades during its passage from the inlet [section (1)] to the outlet [section (2)], the particle exits with radial (r) and circumferential (θ) components of velocity. Thus, the particle enters with no angular momentum about the rotor axis of rotation but leaves with nonzero angular momentum about that axis. (Recall that the axial component of angular momentum

V11.2 Self-propelled lawn sprinkler

for a particle is its mass times the distance from the axis times the θ component of absolute velocity.)

In a turbomachine a series of particles (a continuum) passes through the rotor. Thus, the moment-of-momentum equation applied to a control volume as derived in Section 5.2.3 is valid. For steady flow Eq. 5.21 gives

$$\sum (\mathbf{r} \times \mathbf{F}) = \sum (\mathbf{r} \times \mathbf{V})_{\text{out}} \rho_{\text{out}} A_{\text{out}} V_{\text{out}} - \sum (\mathbf{r} \times \mathbf{V})_{\text{in}} \rho_{\text{in}} A_{\text{in}} V_{\text{in}}$$

Recall that the left-hand side of this equation represents the sum of the external torques (moments) acting on the contents of the control volume, and the right-hand side is the net rate of flow of moment-of-momentum (*angular momentum*) through the control surface.

The axial component of this equation applied to the one-dimensional simplification of flow through a turbomachine rotor with section (1) as the inlet and section (2) as the outlet results in

$$T_{\text{shaft}} = -\dot{m}_1 (r_1 V_{\theta 1}) + \dot{m}_2 (r_2 V_{\theta 2}) \tag{11.2}$$

where T_{shaft} is the *shaft torque* applied to the contents of the control volume. The "$-$" is associated with mass flowrate into the control volume, and the "$+$" is used with the outflow. The sign of the V_{θ} component depends on the direction of V_{θ} and the blade motion, U. If V_{θ} and U are in the same direction, then V_{θ} is positive. The sign of the torque exerted by the shaft on the rotor, T_{shaft}, is positive if T_{shaft} is in the same direction as rotation, and negative otherwise.

As seen from Eq. 11.2, the shaft torque is directly proportional to the mass flowrate, $\dot{m} = \rho Q$. (It takes considerably more torque and power to pump water than to pump air with the same volume flowrate.) The torque also depends on the tangential component of the absolute velocity, V_{θ}. Equation 11.2 is often called the *Euler turbomachine equation.*

Also recall that the *shaft power,* $\dot{W}_{shaft}$, is related to the shaft torque and angular velocity by

$$\dot{W}_{\text{shaft}} = T_{\text{shaft}} \, \omega \tag{11.3}$$

By combining Eqs. 11.2 and 11.3 and using the fact that $U = \omega r$, we obtain

$$\dot{W}_{\text{shaft}} = -\dot{m}_1 (U_1 V_{\theta 1}) + \dot{m}_2 (U_2 V_{\theta 2}) \tag{11.4}$$

Again, the value of V_{θ} is positive when V_{θ} and U are in the same direction and negative otherwise. Also, $\dot{W}_{\text{shaft}}$ is positive when the shaft torque and ω are in the same direction and negative otherwise. Thus, $\dot{W}_{\text{shaft}}$ is positive when power is supplied to the contents of the control volume (pumps) and negative otherwise (turbines). This outcome is consistent with the sign convention involving the work term in the energy equation considered in Chapter 5 (see Eq. 5.44).

Finally, in terms of work per unit mass, $w_{\text{shaft}} = \dot{W}_{\text{shaft}}/\dot{m}$, we obtain

$$w_{\text{shaft}} = -U_1 V_{\theta 1} + U_2 V_{\theta 2} \tag{11.5}$$

where we have used the fact that by conservation of mass, $\dot{m}_1 = \dot{m}_2$. Equations 11.3, 11.4, and 11.5 are the basic governing equations for pumps or turbines whether the machines are radial-, mixed-, or axial-flow devices and for compressible and incompressible flows. Note that neither the axial nor the radial component of velocity enters into the specific work (work per unit mass) equation.

1948 Buick Dynaflow started it Prior to 1948 almost all cars had manual transmissions which required the use of a clutch pedal to shift gears. The 1948 Buick Dynaflow was the first automatic transmission to use the hydraulic *torque* converter and was the model for present-day automatic transmissions. Currently, in the U.S. over 84% of the cars have automatic transmissions. The torque converter replaces the clutch found on manual shift vehicles and allows the engine to continue running when the vehicle comes to a stop. In principle, but certainly not in detail or complexity, operation of a torque converter is similar to blowing air from a *fan* onto another fan which is unplugged. One can hold the *blade* of the unplugged fan and keep it from turning, but as soon as it is let go, it will begin to speed up until it comes close to the speed of the powered fan. The torque converter uses transmission *fluid* (not air) and consists of a *pump* (the powered fan) driven by the engine drive shaft, a *turbine* (the unplugged fan) connected to the input shaft of the transmission, and a *stator* (absent in the fan model) to efficiently direct the flow between the pump and turbine. ■

11.4   The Centrifugal Pump

V11.3 Windshield washer pump

One of the most common radial-flow turbomachines is the ***centrifugal pump.*** This type of pump has two main components: an *impeller* attached to a rotating shaft and a stationary *casing, housing,* or *volute* enclosing the impeller. The impeller consists of a number of blades (usually curved) arranged in a regular pattern around the shaft. A sketch showing the essential features of a centrifugal pump is shown in Fig. 11.4. As the impeller rotates, fluid is sucked in through the *eye* of the casing and flows radially outward. Energy is added to the fluid by the rotating blades, and both pressure and absolute velocity are increased as the fluid flows from the eye to the periphery of the blades. For the simplest type of centrifugal pump, the fluid discharges directly into a volute-shaped casing. The casing shape is designed to reduce the velocity as the fluid leaves the impeller, and this decrease in kinetic energy is converted into an increase in pressure. The volute-shaped casing, with its increasing area in the direction of flow, is used to produce an essentially uniform velocity distribution as the fluid moves around the casing into the discharge opening. For large centrifugal pumps, a different design is often used in which diffuser guide vanes surround the impeller.

11.4.1   Theoretical Considerations

Although flow through a pump is very complex (unsteady and three-dimensional), the basic theory of operation of a centrifugal pump can be developed by considering the time averaged, steady, one-dimensional flow of the fluid as it passes between the inlet and the outlet

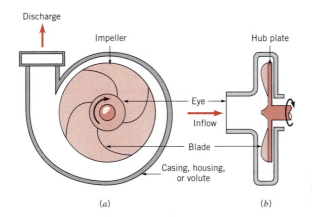

Discharge

Impeller

Hub plate

Eye

Inflow

Blade

Casing, housing, or volute

(a)

(b)

■ **FIGURE 11.4**   **Schematic diagram of basic elements of a centrifugal pump.**

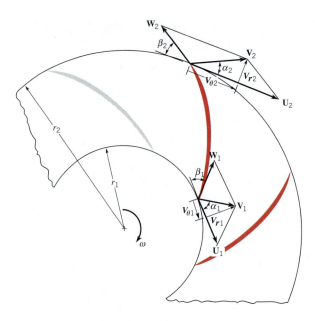

sections of the impeller as the blades rotate. As shown in Fig. 11.5, for a typical blade passage, the absolute velocity, $\mathbf{V}_1$, of the fluid entering the passage is the vector sum of the velocity of the blade, $\mathbf{U}_1$, rotating in a circular path with angular velocity ω, and the relative velocity, $\mathbf{W}_1$, within the blade passage so that $\mathbf{V}_1 = \mathbf{W}_1 + \mathbf{U}_1$. Similarly, at the exit $\mathbf{V}_2 = \mathbf{W}_2 + \mathbf{U}_2$. Note that $U_1 = r_1\omega$ and $U_2 = r_2\omega$. Fluid velocities are taken to be average velocities over the inlet and exit sections of the blade passage. The relationship between the various velocities is shown graphically in Fig. 11.5.

As discussed in Section 11.3, the moment-of-momentum equation indicates that the shaft torque, T_{shaft}, required to rotate the pump impeller is given by Eq. 11.2 applied to a pump with $\dot{m}_1 = \dot{m}_2 = \dot{m}$. That is,

$$T_{\text{shaft}} = \dot{m}(r_2 V_{\theta 2} - r_1 V_{\theta 1}) = \rho Q(r_2 V_{\theta 2} - r_1 V_{\theta 1}) \tag{11.6}$$

where $V_{\theta 1}$ and $V_{\theta 2}$ are the tangential components of the absolute velocities, $\mathbf{V}_1$ and $\mathbf{V}_2$ (see Fig. 11.5).

For a rotating shaft, the power transferred is given by $\dot{W}_{\text{shaft}} = T_{\text{shaft}}\omega$ and therefore from Eq. 11.6

$$\dot{W}_{\text{shaft}} = \rho Q\omega(r_2 V_{\theta 2} - r_1 V_{\theta 1})$$

Since $U_1 = r_1\omega$ and $U_2 = r_2\omega$ we obtain

$$\dot{W}_{\text{shaft}} = \rho Q(U_2 V_{\theta 2} - U_1 V_{\theta 1}) \tag{11.7}$$

Equation 11.7 shows how the power supplied to the shaft of the pump is transferred to the flowing fluid. It also follows that the shaft power per unit mass of flowing fluid is

$$w_{\text{shaft}} = \frac{\dot{W}_{\text{shaft}}}{\rho Q} = U_2 V_{\theta 2} - U_1 V_{\theta 1} \tag{11.8}$$

Recall from Section 5.3.3 that the energy equation is often written in terms of heads—velocity head, pressure head, and elevation head. The head that a pump adds to the fluid is an important parameter. The ideal or maximum head rise possible, h_i, is found from

$$\dot{W}_{\text{shaft}} = \rho g Q h_i$$

which is obtained from Eq. 5.57 by setting head loss (h_L) equal to zero and multiplying by the weight flowrate, $\rho g Q$. Combining this result with Eq. 11.8 we get

$$h_i = \frac{1}{g}(U_2 V_{\theta 2} - U_1 V_{\theta 1}) \tag{11.9}$$

This ideal head rise, h_i, is the amount of energy per unit weight of fluid added to the fluid by the pump. The actual head rise realized by the fluid is less than the ideal amount by the head loss suffered.

An appropriate relationship between the flowrate and the pump ideal head rise can be obtained as follows. Often the fluid has no tangential component of velocity $V_{\theta 1}$, or *swirl*, as it enters the impeller; that is, the angle between the absolute velocity and the tangential direction is 90° ($\alpha_1 = 90°$ in Fig. 11.5). In this case, Eq. 11.19 reduces to

$$h_i = \frac{U_2 V_{\theta 2}}{g} \tag{11.10}$$

From Fig. 11.5

$$\cot \beta_2 = \frac{U_2 - V_{\theta 2}}{V_{r2}}$$

so that Eq. 11.10 can be expressed as

$$h_i = \frac{U_2^2}{g} - \frac{U_2 V_{r2} \cot \beta_2}{g} \tag{11.11}$$

The flowrate, Q, is related to the radial component of the absolute velocity through the equation

$$Q = 2\pi r_2 b_2 V_{r2} \tag{11.12}$$

where b_2 is the impeller blade height at the radius r_2. Thus, combining Eqs. 11.11 and 11.12 yields

$$h_i = \frac{U_2^2}{g} - \frac{U_2 \cot \beta_2}{2\pi r_2 b_2 g} Q \tag{11.13}$$

This equation shows that the ideal or maximum head rise for a centrifugal pump varies linearly with Q for a given blade geometry and angular velocity. For actual pumps, the blade angle β_2 falls in the range of 15–35°, with a normal range of $20° < \beta_2 < 25°$, and with $15° < \beta_1 < 50°$ (Ref. 1). Blades with $\beta_2 < 90°$ are called *backward curved*, whereas blades with $\beta_2 > 90°$ are called *forward curved*. Pumps are not usually designed with forward curved vanes since such pumps tend to suffer unstable flow conditions.

Figure 11.6 shows the ideal head versus flowrate curve (Eq. 11.13) for a centrifugal pump with backward curved vanes ($\beta_2 < 90°$). Since there are simplifying assumptions (i.e., zero losses) associated with the equation for h_i, we would expect that the actual rise in head of the fluid, h_a, would be less than the ideal head rise, and this is indeed the case. As shown in Fig. 11.6, the h_a versus Q curve lies below the ideal head–rise curve and shows a nonlinear variation with Q. Differences between the two curves (as represented by the shaded areas between the curves) arise from several sources. These differences include hydraulic losses due to fluid skin friction in the blade passages, which vary as Q^2, and other losses due to such factors as flow separation, impeller blade-casing clearance flows, and other three-dimensional flow effects. Near the design flowrate, some of these other losses are minimized.

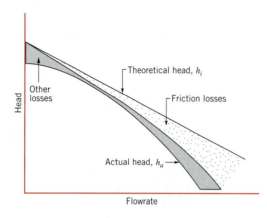

■ **FIGURE 11.6** Effect of losses on the pump head–flowrate curve.

EXAMPLE 11.2 — Centrifugal Pump Performance Based on Inlet/Outlet Velocities

GIVEN Water is pumped at the rate of 1400 gpm through a centrifugal pump operating at a speed of 1750 rpm. The impeller has a uniform blade height, b, of 2 in. with $r_1 = 1.9$ in. and $r_2 = 7.0$ in., and the exit blade angle β_2 is 23° (see Fig. 11.5). Assume ideal flow conditions and that the tangential velocity component, $V_{\theta 1}$, of the water entering the blade is zero ($\alpha_1 = 90°$).

FIND Determine

(a) the tangential velocity component, $V_{\theta 2}$, at the exit,

(b) the ideal head rise, h_i, and

(c) the power, $\dot{W}_{\text{shaft}}$, transferred to the fluid.

SOLUTION

(a) At the exit the velocity diagram is as shown in Fig. 11.5, where $\mathbf{V}_2$ is the absolute velocity of the fluid, $\mathbf{W}_2$ is the relative velocity, and $\mathbf{U}_2$ is the tip velocity of the impeller with

$$U_2 = r_2\omega = (7/12\ \text{ft})(2\pi\ \text{rad/rev})\frac{(1750\ \text{rpm})}{(60\ \text{s/min})}$$

$$= 107\ \text{ft/s}$$

Since the flowrate is given, it follows that $Q = 2\pi r_2 b_2 V_{r2}$

or

$$V_{r2} = \frac{Q}{2\pi r_2 b_2}$$

$$= \frac{1400\ \text{gpm}}{(7.48\ \text{gal/ft}^3)(60\ \text{s/min})(2\pi)(7/12\ \text{ft})(2/12\ \text{ft})}$$

$$= 5.11\ \text{ft/s}$$

From Fig. 11.5, we see that

$$\cot\beta_2 = \frac{U_2 - V_{\theta 2}}{V_{r2}}$$

so that

$$V_{\theta 2} = U_2 - V_{r2}\cot\beta_2$$

$$= (107 - 5.11 \cot 23°)\ \text{ft/s}$$

$$= 95.0\ \text{ft/s} \qquad \textbf{(Ans)}$$

(b) From Eq. 11.10 the ideal head rise is given by

$$h_i = \frac{U_2 V_{\theta 2}}{g} = \frac{(107\ \text{ft/s})(95.0\ \text{ft/s})}{32.2\ \text{ft/s}^2}$$

$$= 316\ \text{ft} \qquad \textbf{(Ans)}$$

Alternatively, from Eq. 11.11, the ideal head rise is

$$h_i = \frac{U_2^2}{g} - \frac{U_2 V_{r2}\cot\beta_2}{g}$$

$$= \frac{(107\ \text{ft/s})^2}{32.2\ \text{ft/s}^2} - \frac{(107\ \text{ft/s})(5.11\ \text{ft/s})\cot 23°}{32.2\ \text{ft/s}^2}$$

$$= 316\ \text{ft} \qquad \textbf{(Ans)}$$

(c) From Eq. 11.7, with $V_{\theta 1} = 0$, the power transferred to the fluid is given by the equation

$$\dot{W}_{\text{shaft}} = \rho Q U_2 V_{\theta 2}$$

$$= \frac{(1.94\ \text{slugs/ft}^3)(1400\ \text{gpm})(107\ \text{ft/s})(95.0\ \text{ft/s})}{[1(\text{slug·ft/s}^2)/\text{lb}](7.48\ \text{gal/ft}^3)(60\ \text{s/min})}$$

$$= 61{,}500\ \text{ft·lb/s}\ (1\ \text{hp/550 ft·lb/s}) = 112\ \text{hp} \qquad \textbf{(Ans)}$$

COMMENT Note that the ideal head rise and the power transferred are related through the relationship

$$\dot{W}_{\text{shaft}} = \rho g Q h_i$$

It should be emphasized that results given in the previous equation involve the ideal head rise. The actual headrise performance characteristics of a pump are usually determined by experimental measurements obtained in a testing laboratory. The actual head rise is always less than the ideal head rise for a specific flowrate because of the loss of available energy associated with actual flows. Also, it is important to note that even if actual values of U_2 and V_{r2} are used in Eq. 11.11, the ideal head rise is calculated. The only idealization used in this example problem is that the exit flow angle is identical to the blade angle at the exit. If the actual exit flow angle was made available in this example, it could have been used in Eq. 11.11 to calculate the ideal head rise.

The pump power, $\dot{W}_{shaft}$, is the actual power required to achieve a blade speed of 107 ft/s, a flowrate of 1400 gpm, and the tangential velocity, $V_{\theta 2}$, associated with this example. If pump losses could somehow be reduced to zero (every pump designer's dream), the actual and ideal head rise would have been identical at 316 ft. As is, the ideal head rise is 316 ft and the actual head rise something less.

11.4.2 Pump Performance Characteristics

Centrifugal pump design is a highly developed field, with much known about pump theory and design procedures (see, e.g., Refs. 1, 2, 3, 4, 17). However, due to the general complexity of flow through a centrifugal pump, the actual performance of the pump cannot be accurately predicted on a completely theoretical basis as indicated by the data of Fig. 11.6. Actual pump performance is determined experimentally through tests on the pump. From these tests, pump characteristics are determined and presented as *pump performance curves.* It is this information that is most helpful to the engineer responsible for incorporating pumps into a given flow system.

The actual head rise, h_a, gained by fluid flowing through a pump can be determined with an experimental arrangement of the type shown in Fig. 11.7, using the energy equation (Eq. 5.57 with $h_a = h_s - h_L$ where h_s is the shaft work head and is identical to h_i, and h_L is the pump head loss)

$$h_a = \frac{p_2 - p_1}{\gamma} + z_2 - z_1 + \frac{V_2^2 - V_1^2}{2g} \qquad (11.14)$$

with sections (1) and (2) at the pump inlet and exit, respectively. The head, h_a, is the same as h_p used with the energy equation, Eq. 5.57, where h_p is interpreted to be the net head rise actually gained by the fluid flowing through the pump, that is, $h_a = h_p = h_s - h_L$. Typically, the differences in elevations and velocities are small so that

$$h_a \approx \frac{p_2 - p_1}{\gamma} \qquad (11.15)$$

The power, $\mathscr{P}_f$, gained by the fluid is given by the equation

$$\mathscr{P}_f = \gamma Q h_a \qquad (11.16)$$

In addition to the head or power added to the fluid, the *overall efficiency,* η, is of interest, where

$$\eta = \frac{\text{power gained by the fluid}}{\text{shaft power driving the pump}} = \frac{\mathscr{P}_f}{\dot{W}_{shaft}}$$

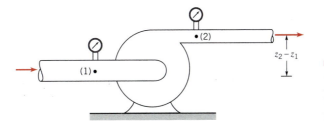

■ FIGURE 11.7 Typical experimental arrangement for determining the head rise gained by a fluid flowing through a pump.

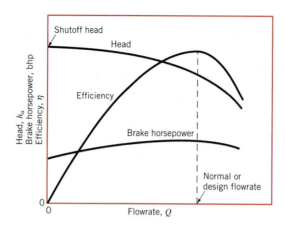

■ **FIGURE 11.8** Typical performance characteristics for a centrifugal pump of a given size operating at a constant impeller speed.

The denominator of this relationship represents the total power applied to the shaft of the pump and is often referred to as *brake horsepower* (bhp). Thus,

$$\eta = \frac{\gamma Q h_a / 550}{\text{bhp}} \tag{11.17}$$

The overall pump efficiency is affected by the *hydraulic losses* in the pump and by the *mechanical losses* in the bearings and seals.

Performance characteristics for a given pump geometry and operating speed are usually given in the form of plots of h_a, η, and bhp versus Q (commonly referred to as *capacity*) as illustrated in Fig. 11.8. Actually, only two curves are needed, as h_a, η, and bhp are related through Eq. 11.17. For convenience, all three curves are usually provided. The head developed by the pump at zero discharge is called the shutoff head, and it represents the rise in pressure head across the pump with the discharge valve closed. Since there is no flow with the valve closed, the related efficiency is zero, and the power supplied by the pump (bhp at $Q = 0$) is simply dissipated as heat. Although centrifugal pumps can be operated for short periods of time with the discharge valve closed, damage will occur due to overheating and large mechanical stress with any extended operation with the valve closed.

As can be seen from Fig. 11.8, as the discharge is increased from zero the brake horsepower increases, with a subsequent fall as the maximum discharge is approached. As noted previously, with h_a and bhp known, the efficiency can be calculated. As shown in Fig. 11.8, the efficiency is a function of the flowrate and reaches a maximum value at some particular value of the flowrate, commonly referred to as the *normal* or *design* flowrate or capacity for the pump. The points on the various curves corresponding to the maximum efficiency are denoted as the *best efficiency points* (BEP). It is apparent that when selecting a pump for a particular application, it is usually desirable to have the pump operate near its maximum efficiency. Thus, performance curves of the type shown in Fig. 11.8 are very important to the engineer responsible for the selection of pumps for a particular flow system. Matching the pump to a particular flow system is discussed in Section 11.4.3.

Pump performance characteristics are also presented in charts of the type shown in Fig. 11.9. Since impellers with different diameters may be used in a given casing, performance characteristics for several impeller diameters can be provided with corresponding lines of constant efficiency and brake horsepower as illustrated in Fig. 11.9. Thus, the same information can be obtained from this type of graph as from the curves shown in Fig. 11.8.

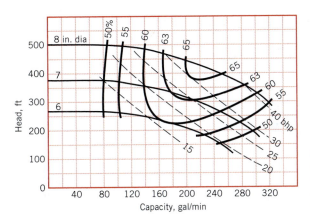

■ **FIGURE 11.9** Typical performance curves for a two-stage centrifugal pump operating at 3500 rpm. Data given for three different impeller diameters.

11.4.3   System Characteristics and Pump Selection

A typical flow system in which a pump is used is shown in Fig. 11.10. The energy equation applied between points (1) and (2) indicates that

$$h_p = z_2 - z_1 + \Sigma h_L \tag{11.18}$$

where h_p is the actual head gained by the fluid from the pump, and Σh_L represents all friction losses in the pipe and minor losses for pipe fittings and valves. From our study of pipe flow, we know that typically h_L varies approximately as the flowrate squared; that is, $h_L \propto Q^2$ (see Section 8.4). Thus, Eq. 11.18 can be written in the form

$$h_p = z_2 - z_1 + KQ^2 \tag{11.19}$$

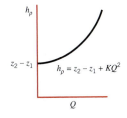

where K depends on the pipe sizes and lengths, friction factors, and minor loss coefficients. Equation 11.19, shown graphically in the figure in the margin, is the ***system equation*** and shows how the actual head gained by the fluid from the pump is related to the system parameters. In this case the parameters include the change in elevation head, $z_2 - z_1$, and the losses due to friction as expressed by KQ^2. Each flow system has its own specific system equation. If the flow is laminar, the frictional losses will be proportional to Q rather than Q^2 (see Section 8.2).

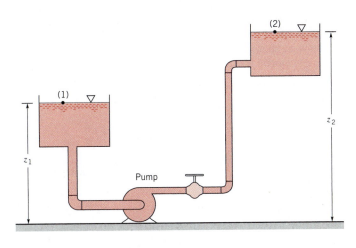

■ **FIGURE 11.10** Typical flow system.

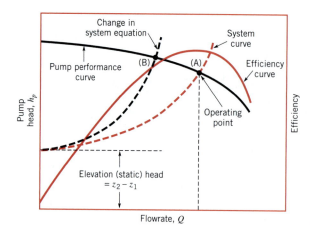

■ **FIGURE 11.11**
Utilization of the system curve and the pump performance curve to obtain the operating point for the system.

There is also a unique relationship between the actual pump head gained by the fluid and the flowrate, which is governed by the pump design (as indicated by the pump performance curve). To select a pump for a particular application, it is necessary to utilize both the *system curve,* as determined by the system equation, and the pump performance curve. If both curves are plotted on the same graph, as illustrated in Fig. 11.11, their intersection (point A) represents the operating point for the system. That is, this point gives the head and flowrate that satisfies both the system equation and the pump equation. On the same graph the pump efficiency is shown. Ideally, we want the operating point to be near the best efficiency point (BEP) for the pump. For a given pump, it is clear that as the system equation changes, the operating point will shift. For example, if the pipe friction increases due to pipe wall fouling, the system changes, resulting in the operating point A shifting to point B in Fig. 11.11 with a reduction in flowrate and efficiency. The following example shows how the system and pump characteristics can be used to decide if a particular pump is suitable for a given application.

EXAMPLE 11.3   Use of Pump Performance Curves

GIVEN Water is to be pumped from one large, open tank to a second large, open tank as shown in Fig. E11.3a. The pipe diameter throughout is 6 in. and the total length of the pipe between the pipe entrance and exit is 200 ft. Minor loss coefficients for the entrance, exit, and the elbow are shown, and the friction factor for the pipe can be assumed constant and equal to 0.02. A certain centrifugal pump having the performance characteristics shown in Fig. E11.3b is suggested as a good pump for this flow system.

FIND With this pump, what would be the flowrate between the tanks? Do you think this pump would be a good choice?

SOLUTION

Application of the energy equation between the two free surfaces, points (1) and (2) as indicated, gives

$$\frac{p_1}{\gamma} + \frac{V_1^2}{2g} + z_1 + h_p = \frac{p_2}{\gamma} + \frac{V_2^2}{2g} + z_2$$

$$+ f\frac{\ell}{D}\frac{V^2}{2g} + \sum K_L \frac{V^2}{2g} \quad (1)$$

Thus, with $p_1 = p_2 = 0$, $V_1 = V_2 = 0$, $\Delta z = z_2 - z_1 = 10$ ft, $f = 0.02$, $D = 6/12$ ft, and $\ell = 200$ ft, Eq. 1 becomes

$$h_p = 10 + \left[0.02\,\frac{(200\ \text{ft})}{(6/12\ \text{ft})} \right.$$

$$\left. + (0.5 + 1.5 + 1.0) \right] \frac{V^2}{2(32.2\ \text{ft/s}^2)} \quad (2)$$

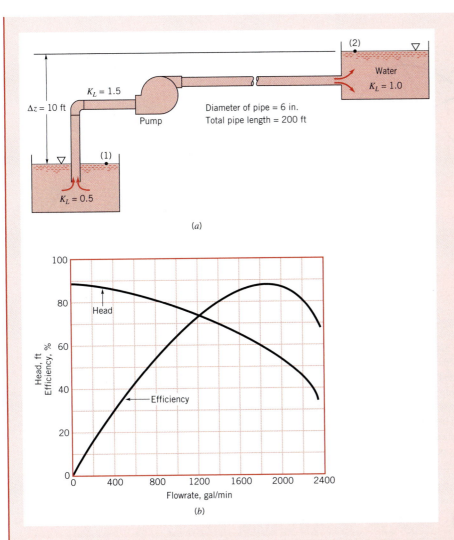

■ **FIGURE E11.3a, b**

where the given minor loss coefficients have been used. Since

$$V = \frac{Q}{A} = \frac{Q(\text{ft}^3/\text{s})}{(\pi/4)(6/12 \text{ ft})^2}$$

Eq. 2 can be expressed as

$$h_p = 10 + 4.43\, Q^2 \qquad (3)$$

where Q is in ft^3/s, or with Q in gallons per minute.

$$h_p = 10 + 2.20 \times 10^{-5} Q^2 \qquad (4)$$

Equation 3 or 4 represents the system equation for this particular flow system and reveals how much actual head the fluid will need to gain from the pump to maintain a certain flowrate. Performance data shown in Fig. E11.3b indicate the actual head the fluid will gain from this particular pump when it operates at a certain flowrate. Thus, when Eq. 4 is plotted on the same graph with performance data, the intersection of the two curves represents the operating point for the pump and the system.

This combination is shown in Fig. E11.3c with the intersection (as obtained graphically) occurring at

$$Q = 1600 \text{ gal/min} \qquad \textbf{(Ans)}$$

with the corresponding actual head gained equal to 66.5 ft.

Another concern is whether the pump is operating efficiently at the operating point. As can be seen from Fig. E11.3c, although this is not peak efficiency, which is about 86%, it is close (about 84%). Thus, this pump would be a satisfactory choice, assuming the 1600 gal/min flowrate is at or near the desired flowrate.

The amount of pump head needed at the pump shaft is 66.5 ft/0.84 = 79.2 ft. The power needed to drive the pump is

$$\dot{W}_{\text{shaft}} = \frac{\gamma Q h_a}{\eta}$$

$$= \frac{(62.4 \text{ lb/ft}^3)[(1600 \text{ gal/min})/(7.48 \text{ gal/ft}^3)(60 \text{ s/min})](66.5 \text{ ft})}{0.84}$$

$$= 17{,}600 \text{ ft} \cdot \text{lb/s} = 32.0 \text{ hp}$$

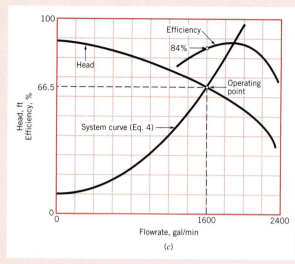

■ **FIGURE E11.3c (*Continued*)**

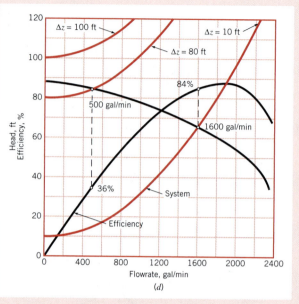

■ **FIGURE E11.3*d***

COMMENT By repeating the calculations for $\Delta z = z_2 - z_1 = 80$ ft and 100 ft (rather than the given 10 ft), the results shown in Fig. E11.3d are obtained. Although the given pump could be used with $\Delta z = 80$ ft (provided that the 500 gal/min flowrate produced is acceptable), it would not be an ideal pump for this application since its efficiency would be only 36 percent. Energy could be saved by using a different pump with a performance curve that more nearly matches the new system requirements (i.e., higher efficiency at the operating condition). On the other hand, the given pump would not work at all for $\Delta z = 100$ ft since its maximum head ($h_p = 88$ ft when $Q = 0$) is not enough to lift the water 100 ft, let alone overcome head losses. This is shown in Fig. E11.3d by the fact that for $\Delta z = 100$ ft the system curve and the pump performance curve do not intersect.

F l u i d s   i n   t h e   N e w s

Space Shuttle fuel pumps The fuel *pump* of your car engine is vital to its operation. Similarly, the fuels (liquid hydrogen and oxygen) of each Space Shuttle main engine (there are three per shuttle) rely on multistage *turbopumps* to get from storage tanks to main combustors. High pressures are utilized throughout the pumps to avoid cavitation. The pumps, some *centrifugal* and some *axial*, are driven by axial-flow, *multi-stage turbines*. Pump speeds are as high as 35,360 rpm. The liquid oxygen is pumped from 100 to 7420 psia, the liquid hydrogen from 30 to 6515 psia. Liquid hydrogen and oxygen flowrates of about 17,200 gpm and 6100 gpm, respectively, are achieved. These pumps could empty your home swimming pool in seconds. The hydrogen goes from −423 °F in storage to +6000 °F in the combustion chamber! ■

11.5   Dimensionless Parameters and Similarity Laws

As discussed in Chapter 7, dimensional analysis is particularly useful in the planning and execution of experiments. Since the characteristics of pumps are usually determined experimentally, it is expected that dimensional analysis and similitude considerations will prove to be useful in the study and documentation of these characteristics.

From the previous section we know that the principal, dependent pump variables are the actual head rise, h_a, shaft power, W_{shaft}, and efficiency, η. We expect that these variables will depend on the geometrical configuration, which can be represented by some characteristic diameter, D, other pertinent lengths, ℓ_i, and surface roughness, ε. In addition, the other important variables are flowrate, Q, the pump shaft rotational speed, ω, fluid viscosity, μ, and fluid density, ρ. We will only consider incompressible fluids presently, so compressibility effects need not concern us yet. Thus, any one of the dependent variables h_a, W_{shaft}, and η can be expressed as

$$\text{dependent variable} = f(D, \ell_i, \varepsilon, Q, \omega, \mu, \rho)$$

and a straightforward application of dimensional analysis leads to

$$\text{dependent pi term} = \phi\left(\frac{\ell_i}{D}, \frac{\varepsilon}{D}, \frac{Q}{\omega D^3}, \frac{\rho\omega D^2}{\mu}\right) \tag{11.20}$$

The dependent pi term involving the head is usually expressed as $C_H = gh_a/\omega^2 D^2$, where gh_a is the actual head rise in terms of energy per unit mass, rather than simply h_a, which is energy per unit weight. This dimensionless parameter is called the ***head rise coefficient.*** The dependent pi term involving the shaft power is expressed as $C_{\mathscr{P}} = W_{shaft}/\rho\omega^3 D^5$, and this standard dimensionless parameter is termed the ***power coefficient.*** The power appearing in this dimensionless parameter is commonly based on the shaft (brake) horsepower, bhp, so that in BG units, $W_{shaft} = 550 \times (\text{bhp})$. The rotational speed, ω, which appears in these dimensionless groups is expressed in rad/s. The final dependent pi term is the efficiency, η, which is already dimensionless. Thus, in terms of dimensionless parameters the performance characteristics are expressed as

$$C_H = \frac{gh_a}{\omega^2 D^2} = \phi_1\left(\frac{\ell_i}{D}, \frac{\varepsilon}{D}, \frac{Q}{\omega D^3}, \frac{\rho\omega D^2}{\mu}\right)$$

$$C_{\mathscr{P}} = \frac{\dot{W}_{shaft}}{\rho\omega^3 D^5} = \phi_2\left(\frac{\ell_i}{D}, \frac{\varepsilon}{D}, \frac{Q}{\omega D^3}, \frac{\rho\omega D^2}{\mu}\right)$$

$$\eta = \frac{\rho g Q h_a}{\dot{W}_{shaft}} = \phi_3\left(\frac{\ell_i}{D}, \frac{\varepsilon}{D}, \frac{Q}{\omega D^3}, \frac{\rho\omega D^2}{\mu}\right)$$

The last pi term in each of the aforementioned equations is a form of the Reynolds number that represents the relative influence of viscous effects. When the pump flow involves high Reynolds numbers, as is usually the case, experience has shown that the effect of the Reynolds number can be neglected. For simplicity, the relative roughness, ε/D, can also be neglected in pumps, as the highly irregular shape of the pump chamber is usually the dominant geometric factor rather than the surface roughness. Thus, with these simplifications and for *geometrically similar* pumps (all pertinent dimensions, ℓ_i, scaled by a common length scale), the dependent pi terms are functions of only $Q/\omega D^3$, so that

$$\frac{gh_a}{\omega^2 D^2} = \phi_1\left(\frac{Q}{\omega D^3}\right) \tag{11.21}$$

$$\frac{\dot{W}_{shaft}}{\rho\omega^3 D^5} = \phi_2\left(\frac{Q}{\omega D^3}\right) \tag{11.22}$$

$$\eta = \phi_3\left(\frac{Q}{\omega D^3}\right) \tag{11.23}$$

The dimensionless parameter $C_Q = Q/\omega D^3$ is called the ***flow coefficient.*** These three equations provide the desired similarity relationships among a family of geometrically similar pumps. If two pumps from the family are operated at the same value of flow coefficient

$$\left(\frac{Q}{\omega D^3}\right)_1 = \left(\frac{Q}{\omega D^3}\right)_2 \tag{11.24}$$

it then follows that

$$\left(\frac{gh_a}{\omega^2 D^2}\right)_1 = \left(\frac{gh_a}{\omega^2 D^2}\right)_2 \tag{11.25}$$

$$\left(\frac{\dot{W}_{\text{shaft}}}{\rho\omega^3 D^5}\right)_1 = \left(\frac{\dot{W}_{\text{shaft}}}{\rho\omega^3 D^5}\right)_2 \tag{11.26}$$

$$\eta_1 = \eta_2 \tag{11.27}$$

where subscripts 1 and 2 refer to any two pumps from the family of geometrically similar pumps.

With these so-called ***pump scaling laws*** it is possible to experimentally determine the performance characteristics of one pump in the laboratory and then use these data to predict the corresponding characteristics for other pumps within the family under different operating conditions. Figure 11.12a shows some typical curves obtained for a centrifugal pump. Figure 11.12b shows the results plotted in terms of the dimensionless coefficients, C_Q, C_H, $C_{\mathscr{P}}$, and η. From these curves the performance of different-sized, geometrically similar pumps can be predicted, as can the effect of changing speeds on the performance of the pump from which the curves were obtained. It is to be noted that the efficiency, η, is related to the other coefficients through the relationship $\eta = C_Q C_H C_{\mathscr{P}}^{-1}$. This follows directly from the definition of η.

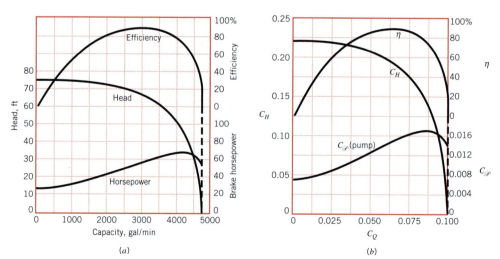

■ **FIGURE 11.12** **Typical performance data for a centrifugal pump: (a) characteristic curves for a 12-in. centrifugal pump operating at 1000 rpm and (b) dimensionless characteristic curves. (Data from Ref. 5, used with permission.)**

EXAMPLE 11.4   Use of Pump Scaling Laws

GIVEN An 8-in.-diameter centrifugal pump operating at 1200 rpm is geometrically similar to the 12-in.-diameter pump having the performance characteristics of Figs. 11.12a and 11.12b while operating at 1000 rpm. The working fluid is water at 60° F.

FIND For peak efficiency, predict the discharge, actual head rise, and shaft horsepower for this smaller pump.

SOLUTION

As is indicated by Eq. 11.23, for a given efficiency the flow coefficient has the same value for a given family of pumps. From Fig. 11.12b we see that at peak efficiency $C_Q = 0.0625$. Thus, for the 8-in. pump

$$Q = C_Q \omega D^3$$
$$= (0.0625)(1200/60 \text{ rev/s})(2\pi \text{ rad/rev})(8/12 \text{ ft})^3$$
$$Q = 2.33 \text{ ft}^3/\text{s} \qquad \textbf{(Ans)}$$

or in terms of gpm

$$Q = (2.33 \text{ ft}^3/\text{s})(7.48 \text{ gal/ft}^3)(60 \text{ s/min})$$
$$= 1046 \text{ gpm} \qquad \textbf{(Ans)}$$

The actual head rise and the shaft horsepower can be determined in a similar manner since at peak efficiency $C_H = 0.19$ and $C_\mathcal{P} = 0.014$, so that with $\omega = 1200 \text{ rev/min}$ (1 min/60 s) × $(2\pi \text{ rad/rev}) = 126 \text{ rad/s}$

$$h_a = \frac{C_H \omega^2 D^2}{g} = \frac{(0.19)(126 \text{ rad/s})^2(8/12 \text{ ft})^2}{32.2 \text{ ft/s}^2} = 41.6 \text{ ft } \textbf{(Ans)}$$

Also,

$$\dot{W}_{shaft} = C_\mathcal{P} \rho \omega^3 D^5$$
$$= (0.014)(1.94 \text{ slugs/ft}^3)(126 \text{ rad/s})^3(8/12 \text{ ft})^5$$
$$= 7150 \text{ ft·lb/s}$$
$$\dot{W}_{shaft} = \frac{7150 \text{ ft·lb/s}}{550 \text{ ft·lb/s/hp}} = 13.0 \text{ hp} \qquad \textbf{(Ans)}$$

COMMENT The last result gives the shaft horsepower, which is the power supplied to the pump shaft. The power actually gained by the fluid is equal to $\gamma Q h_a$, which in this example is

$$\mathcal{P}_f = \gamma Q h_a = (62.4 \text{ lb/ft}^3)(2.33 \text{ ft}^3/\text{s})(41.6 \text{ ft}) = 6050 \text{ ft·lb/s}$$

Thus, the efficiency, η, is

$$\eta = \frac{\mathcal{P}_f}{\dot{W}_{shaft}} = \frac{6050}{7150} = 85\%$$

which checks with the efficiency curve of Fig. 11.12b.

11.5.1   Specific Speed

A useful pi term can be obtained by eliminating diameter D between the flow coefficient and the head rise coefficient. This is accomplished by raising the flow coefficient to an appropriate exponent (1/2) and dividing this result by the head coefficient raised to another appropriate exponent (3/4) so that

$$\frac{(Q/\omega D^3)^{1/2}}{(g h_a/\omega^2 D^2)^{3/4}} = \frac{\omega\sqrt{Q}}{(g h_a)^{3/4}} = N_s \qquad \textbf{(11.28)}$$

The dimensionless parameter N_s is called the *specific speed.* Specific speed varies with the flow coefficient just as the other coefficients and efficiency discussed earlier do. However, for any pump it is customary to specify a value of specific speed at the flow coefficient corresponding to peak efficiency only. For pumps with low Q and high h_a, the specific speed is low compared to a pump with high Q and low h_a. Centrifugal pumps typically are low-capacity, high-head pumps, and therefore have low specific speeds.

Specific speed as defined by Eq. 11.28 is dimensionless and therefore independent of the system of units used in its evaluation as long as a consistent unit system is used. However,

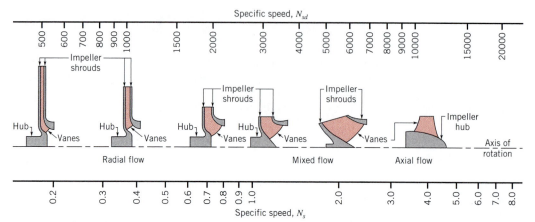

■ **FIGURE 11.13** Variation in specific speed with type of pump (for pumps operating at peak efficiency). (Adapted from Ref. 6, used with permission.)

in the United States a modified, dimensional form of specific speed, N_{sd}, is commonly used, where

$$N_{sd} = \frac{\omega(\text{rpm})\sqrt{Q(\text{gpm})}}{[h_a(\text{ft})]^{3/4}} \qquad (11.29)$$

In this case N_{sd} is said to be expressed in *U.S. customary units.* Typical values of N_{sd} are in the range $500 < N_{sd} < 4000$ for centrifugal pumps. Both N_s and N_{sd} have the same physical meaning, but their magnitudes will differ by a constant conversion factor ($N_{sd} = 2733 N_s$) when ω in Eq. 11.28 is expressed in rad/s.

Each family or class of pumps has a particular range of values of specific speed associated with it. Thus, pumps that have low-capacity, high-head characteristics will have specific speeds that are smaller than pumps that have high-capacity, low-head characteristics. The concept of specific speed is very useful to engineers and designers, since if the required head, flowrate, and speed are specified, it is possible to select an appropriate (most efficient) type of pump for a particular application. As the specific speed, N_{sd}, increases beyond about 2000, the peak efficiency of the purely radial-flow centrifugal pump starts to fall off, and other types of more efficient pump design are preferred. In addition to the centrifugal pump, the *axial-flow* pump is widely used. As discussed in Section 11.6, in an axial-flow pump the direction of the flow is primarily parallel to the rotating shaft rather than radial as in the centrifugal pump. Axial-flow pumps are essentially high-capacity, low-head pumps and therefore have large specific speeds ($N_{sd} > 9000$) compared to centrifugal pumps. *Mixed-flow* pumps combine features of both radial-flow and axial-flow pumps and have intermediate values of specific speed. Figure 11.13 illustrates how the specific speed changes as the configuration of the pump changes from centrifugal or radial to axial.

11.6   Axial-Flow and Mixed-Flow Pumps

As noted previously, centrifugal pumps are radial-flow machines that operate most efficiently for applications requiring high heads at relatively low flowrates. This head–flowrate combination typically yields specific speeds (N_{sd}) that are less than approximately 4000.

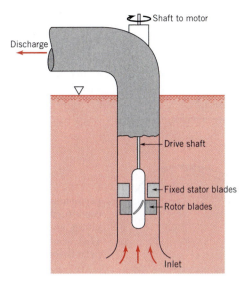

For many applications, such as those associated with drainage and irrigation, high flowrates at low heads are required and centrifugal pumps are not suitable. In this case, axial-flow pumps are commonly used. This type of pump consists essentially of a propeller confined within a cylindrical casing. Axial-flow pumps are often called *propeller pumps*. For this type of pump the flow is primarily in the axial direction (parallel to the axis of rotation of the shaft), as opposed to the radial flow found in the centrifugal pump. Whereas the head developed by a centrifugal pump includes a contribution due to centrifugal action, the head developed by an axial-flow pump is due primarily to the tangential force exerted by the rotor blades on the fluid. A schematic of an axial-flow pump arranged for vertical operation is shown in Fig. 11.14. The rotor is connected to a motor through a shaft, and as it rotates (usually at a relatively high speed) the fluid is sucked in through the inlet. Typically the fluid discharges through a row of fixed stator (guide) vanes used to straighten the flow leaving the rotor. Some axial-flow pumps also have inlet guide vanes upstream of the rotor row, and some are multistage in which pairs (*stages*) of rotating blades (*rotor blades*) and fixed vanes (*stator blades*) are arranged in series. Axial-flow pumps usually have specific speeds (N_{sd}) in excess of 9000.

The definitions and broad concepts that were developed for centrifugal pumps are also applicable to axial-flow pumps. The actual flow characteristics, however, are quite different. In Fig. 11.15 typical head, power, and efficiency characteristics are compared for a centrifugal pump and an axial-flow pump. It is noted that at design capacity (maximum efficiency) the head and brake horsepower are the same for the two pumps selected, but as the flowrate decreases, the power input to the centrifugal pump falls to 180 hp at shutoff, whereas for the axial-flow pump the power input increases to 520 hp at shutoff. This characteristic of the axial-flow pump can cause overloading of the drive motor if the flowrate is reduced significantly from the design capacity. It is also noted that the head curve for the axial-flow pump is much steeper than that for the centrifugal pump. Thus, with axial-flow pumps there will be a large change in head with a small change in the flowrate, whereas for the centrifugal pump, with its relatively flat head curve, there will be only a small change in head with large changes in the flowrate. It is further observed from Fig. 11.15 that, except at design capacity, the efficiency of the axial-flow pump is lower than that of the centrifugal pump. To improve operating characteristics, some axial-flow pumps are constructed with adjustable blades.

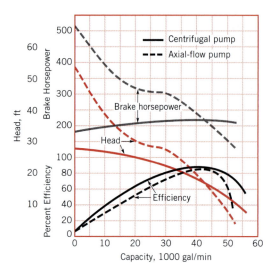

■ **FIGURE 11.15** **Comparison of performance characteristics for a centrifugal pump and an axial-flow pump, each rated 42,000 gal/min at a 17-ft head. (Data from Ref. 7, used with permission.)**

For applications requiring specific speeds intermediate to those for centrifugal and axial-flow pumps, mixed-flow pumps have been developed that operate efficiently in the specific speed range $4000 < N_{sd} < 9000$. As the name implies, the flow in a mixed-flow pump has both a radial and an axial component. Figure 11.16 shows some typical data for centrifugal, mixed-flow, and axial-flow pumps, each operating at the same flowrate for peak efficiency. These data indicate that as we proceed from the centrifugal pump to the mixed-flow pump to the axial-flow pump, the specific speed increases, the head decreases, the speed increases, the impeller diameter decreases, and the eye diameter increases. These general trends are commonly found when these three types of pumps are compared.

The dimensionless parameters and scaling relationships developed in the previous sections apply to all three types of pumps—centrifugal, mixed flow, and axial flow—since the dimensional analysis used is not restricted to a particular type of pump. Additional information about pumps can be found in Refs. 1, 7, 8, 9, and 10.

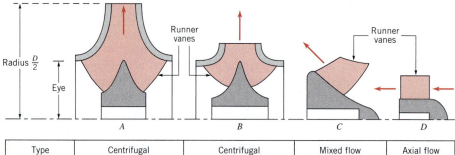

Type	Centrifugal	Centrifugal	Mixed flow	Axial flow
N_{sd}	1,250	2,200	6,200	13,500
Gal/min	2,400	2,400	2,400	2,400
Head, ft	70	48	33	20
Rpm	870	1,160	1,750	2,600
D, in.	19	12	10	7
D_{eye}/D	0.5	0.7	0.9	1.0

■ **FIGURE 11.16** **Comparison of different types of impellers. Specific speed for centrifugal pumps based on single suction. (Adapted from Ref. 7, used with permission.)**

Mechanical heart assist devices As with any pump, the human heart can suffer various malfunctions and problems during its useful life. Recent developments in artificial heart technology may be able to provide help to those whose pumps have broken down beyond repair. One of the more promising techniques is use of a left-ventricular assist device (LVAD), which supplements a diseased heart. Rather than replacing a diseased heart, an LVAD pump is implanted alongside the heart and works in parallel with the cardiovascular system to assist the pumping function of the heart's left ventricle. (The left ventricle supplies oxygenated blood to the entire body and performs about 80% of the heart's work.) Some LVADs are *positive displacement pumps* that use mechanical means to force a membrane back and forth to "beat" in conjunction with the patient's natural heart. Others use a *centrifugal or axial-flow pump* to provide a continuous flow of blood. The continuous flow devices may take some adjustment on the part of patients, who do not hear a pulse or a heartbeat. Despite advances in artificial heart technology, it is probably still several years before fully implantable, quiet, and reliable devices will be considered for widespread use. ■

11.7   Turbines

As discussed in Section 11.2, turbines are devices that extract energy from a flowing fluid. The geometry of turbines is such that the fluid exerts a torque on the rotor in the direction of its rotation. The shaft power generated is available to drive generators or other devices. In the following sections we discuss mainly the operation of hydraulic turbines (those for which the working fluid is water). Although there are numerous ingenious hydraulic turbine designs, most of these turbines can be classified into two basic types—*impulse turbines* and *reaction turbines.* In general, impulse turbines are high-head, low-flowrate devices, whereas reaction turbines are low-head, high-flowrate devices.

For hydraulic impulse turbines, the pressure drop across the rotor is zero; all of the pressure drop across the turbine stage occurs in the nozzle row. The *Pelton wheel* shown in Fig. 11.17 is a classical example of an impulse turbine. In these machines the total head of the incoming fluid (the sum of the pressure head, velocity head, and elevation head) is converted into a large-velocity head at the exit of the supply nozzle (or nozzles if a multiple nozzle configuration is used). Both the pressure drop across the bucket (blade) and the change in relative speed (i.e., fluid speed relative to the moving bucket) of the fluid across the bucket are negligible. The space surrounding the rotor is not completely filled with fluid. It is the impulse of the individual jets of fluid striking the buckets that generates the torque.

For reaction turbines, however, the rotor is surrounded by a casing (or volute), which is completely filled with the working fluid. There is both a pressure drop and a change in fluid relative speed across the rotor. As shown for the radial-inflow turbine in Fig. 11.18, guide vanes act as nozzles to accelerate the flow and turn it in the appropriate direction as

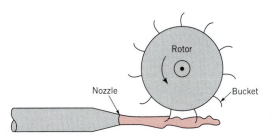

■ **FIGURE 11.17** Schematic diagram of a Pelton wheel turbine.

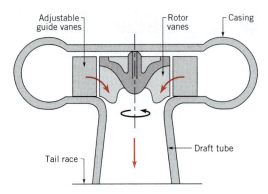

■ **FIGURE 11.18** Schematic diagram of a reaction turbine.

the fluid enters the rotor. Thus, part of the pressure drop occurs across the guide vanes and part occurs across the rotor.

11.7.1 Impulse Turbines

Although there are various types of impulse turbine designs, perhaps the easiest to understand is the *Pelton wheel*. Lester Pelton, an American mining engineer during the California gold-mining days, is responsible for many of the still-used features of this type of turbine. It is most efficient when operated with a large head (e.g., a water source from a lake located significantly above the turbine nozzle), which is converted into a relatively large velocity at the exit of the nozzle (see Fig. 11.17).

V11.4 Pelton wheel lawn sprinkler

As shown in Fig. 11.19, for a Pelton wheel a high-speed jet of water strikes the buckets and is deflected. The water enters and leaves the control volume surrounding the wheel as free jets (atmospheric pressure). In addition, a person riding on the bucket would note that the speed of the water does not change as it slides across the buckets (assuming viscous effects are negligible). That is, the magnitude of the relative velocity does not change, but its direction does. The change in direction of the velocity of the fluid jet causes a torque on the rotor, resulting in a power output from the turbine.

Design of the optimum, complex shape of the buckets to obtain maximum power output is a very difficult matter. Ideally, the radial component of velocity is zero. (In practice there often is a small but negligible radial component.) In addition, the buckets would ideally turn the relative velocity vector through a 180° turn, but physical constraints dictate that β, the angle of the exit edge of the blade, is less than 180°. Thus, the fluid leaves with an axial component of velocity as shown in Fig. 11.20.

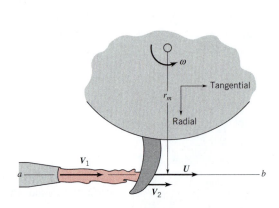

■ **FIGURE 11.19** Ideal fluid velocities for a Pelton wheel turbine.

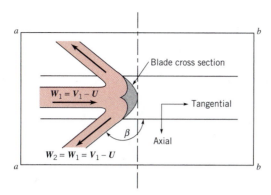

■ **FIGURE 11.20**   **Flow as viewed by an observer riding on the Pelton wheel—relative velocities.**

The inlet and exit velocity triangles at the arithmetic mean radius, r_m, are assumed to be as shown in Fig. 11.21. To calculate the torque and power, we must know the tangential components of the absolute velocities at the inlet and exit. (Recall from the discussion in Section 11.3 that neither the radial nor the axial components of velocity enter into the torque or power equations.) From Fig. 11.21 we see that

$$V_{\theta 1} = V_1 = W_1 + U \tag{11.30}$$

and

$$V_{\theta 2} = W_2 \cos \beta + U \tag{11.31}$$

Thus, with the assumption that $W_1 = W_2$ (i.e., the relative speed of the fluid does not change as it is deflected by the buckets), we can combine Eqs. 11.30 and 11.31 to obtain

$$V_{\theta 2} - V_{\theta 1} = (U - V_1)(1 - \cos \beta) \tag{11.32}$$

This change in tangential component of velocity combined with the torque and power equations developed in Section 11.3 (i.e., Eqs. 11.2 and 11.4) gives

$$T_{\text{shaft}} = \dot{m} r_m (U - V_1)(1 - \cos \beta)$$

where $\dot{m} = \rho Q$ is the mass flowrate through the turbine.
Thus, since $U = \omega r_m$

$$\dot{W}_{\text{shaft}} = T_{\text{shaft}} \omega = \dot{m} U (U - V_1)(1 - \cos \beta) \tag{11.33}$$

These results are plotted in Fig. 11.22 along with typical experimental results. Note that $V_1 > U$ (i.e., the jet impacts the bucket), and $\dot{W}_{\text{shaft}} < 0$ (i.e., the turbine extracts power from the fluid).

Several interesting points can be noted from the aforementioned results. First, the power is a function of β. However, a typical value of $\beta = 165°$ (rather than the optimum 180°) results

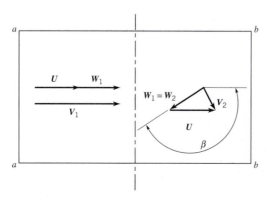

■ **FIGURE 11.21**   **Inlet and exit velocity triangles for a Pelton wheel turbine.**

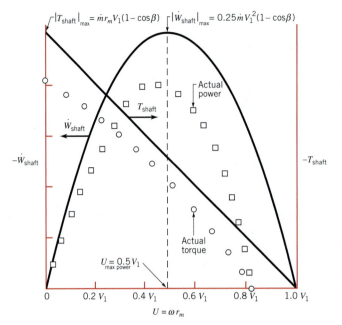

■ **FIGURE 11.22**
Typical theoretical and experimental power and torque for a Pelton wheel turbine as a function of bucket speed.

in a relatively small (less than 2%) reduction in power since $1 - \cos 165 = 1.966$, compared to $1 - \cos 180 = 2$. Second, although the torque is maximum when the wheel is stopped ($U = 0$), there is no power under this condition—to extract power one needs force and motion. However, the power output is a maximum when

$$U_{\text{max power}} = \frac{V_1}{2} \tag{11.34}$$

This can be shown by using Eq. 11.33 and solving for U that gives $d\dot{W}_{\text{shaft}}/dU = 0$. A bucket speed of one-half the speed of the fluid coming from the nozzle gives the maximum power. Third, the maximum speed occurs when $T_{\text{shaft}} = 0$ (i.e., the load is completely removed from the turbine, as would happen if the shaft connecting the turbine to the generator were to break and frictional torques were negligible). For this case $U = \omega R = V_1$, the turbine is "free wheeling," and the water simply passes across the rotor without putting any force on the buckets.

EXAMPLE 11.5 | Pelton Wheel Turbine Characteristics

GIVEN Water to drive a Pelton wheel is supplied through a pipe from a lake as indicated in Fig. E11.5a. The head loss due to friction in the pipe is important, but minor losses can be neglected.

FIND (a) Determine the nozzle diameter, D_1, that will give the maximum power output.

(b) Determine the maximum power and the angular velocity of the rotor at the conditions found in part (a).

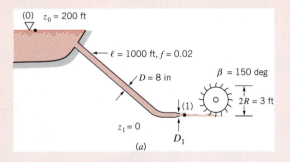

■ **FIGURE E11.5a**

SOLUTION

(a) As indicated by Eq. 11.33, the power output depends on the flowrate, $Q = \dot{m}/\rho$, and the jet speed at the nozzle exit, V_1, both of which depend on the diameter of the nozzle, D_1, and the head loss associated with the supply pipe. That is

$$\dot{W}_{shaft} = \rho Q U (U - V_1)(1 - \cos\beta) \qquad (1)$$

The nozzle exit speed, V_1, can be obtained by applying the energy equation (Eq. 5.57) between a point on the lake surface (where $V_0 = p_0 = 0$) and the nozzle outlet (where $z_1 = p_1 = 0$) to give

$$z_0 = \frac{V_1^2}{2g} + h_L \qquad (2)$$

where the head loss is given in terms of the friction factor, f, as (see Eq. 8.18)

$$h_L = f\frac{\ell}{D}\frac{V^2}{2g}$$

The speed, V, of the fluid in the pipe of diameter D is obtained from the continuity equation

$$V = \frac{A_1 V_1}{A} = \left(\frac{D_1}{D}\right)^2 V_1$$

We have neglected minor losses associated with the pipe entrance and the nozzle. With the given data, Eq. 2 becomes

$$z_0 = \left[1 + f\frac{\ell}{D}\left(\frac{D_1}{D}\right)^4\right]\frac{V_1^2}{2g} \qquad (3)$$

or

$$V_1 = \left[\frac{2gz_0}{1 + f\frac{\ell}{D}\left(\frac{D_1}{D}\right)^4}\right]^{1/2}$$

$$= \left[\frac{2(32.2 \text{ ft/s}^2)(200 \text{ ft})}{1 + 0.02\left(\frac{1000 \text{ ft}}{8/12 \text{ ft}}\right)\left(\frac{D_1}{8/12}\right)^4}\right]^{1/2}$$

$$= \frac{113.5}{\sqrt{1 + 152 D_1^4}} \qquad (4)$$

where D_1 is in feet.

By combining Eqs. 1 and 4 and using $Q = \pi D_1^2 V_1/4$ we obtain the power as a function of D_1 and U as

$$\dot{W}_{shaft} = \frac{323\, U D_1^2}{\sqrt{1 + 152 D_1^4}}\left[U - \frac{113.5}{\sqrt{1 + 152 D_1^4}}\right] \qquad (5)$$

where U is in feet per second and $\dot{W}_{shaft}$ is in ft·lb/s. These results are plotted as a function of U for various values of D_1 in Fig. 11.5b.

As shown by Eq. 11.34, the maximum power (in terms of its variation with U) occurs when $U = V_1/2$, which, when used with Eqs. 4 and 5, gives

$$\dot{W}_{shaft} = -\frac{1.04 \times 10^6 D_1^2}{(1 + 152 D_1^4)^{3/2}} \qquad (6)$$

The maximum power possible occurs when $d\dot{W}_{shaft}/dD_1 = 0$, which according to Eq. 6 can be found as

$$\frac{d\dot{W}_{shaft}}{dD_1} = -1.04 \times 10^6\left[\frac{2 D_1}{(1 + 152 D_1^4)^{3/2}}\right.$$

$$\left. -\left(\frac{3}{2}\right)\frac{4(152) D_1^5}{(1 + 152 D_1^4)^{5/2}}\right] = 0$$

or

$$304 D_1^4 = 1$$

Thus, the nozzle diameter for maximum power output is

$$D_1 = 0.239 \text{ ft} \qquad \textbf{(Ans)}$$

(b) The corresponding maximum power can be determined from Eq. 6 as

$$\dot{W}_{shaft} = -\frac{1.04 \times 10^6 (0.239)^2}{[1 + 152(0.239)^4]^{3/2}} = -3.25 \times 10^4 \text{ ft·lb/s}$$

or

$$\dot{W}_{shaft} = -3.25 \times 10^4 \text{ ft·lb/s} \times \frac{1 \text{ hp}}{550 \text{ ft·lb/s}}$$

$$= -59.0 \text{ hp} \qquad \textbf{(Ans)}$$

The rotor speed at the maximum power condition can be obtained from

$$U = \omega R = \frac{V_1}{2}$$

where V_1 is given by Eq. 4. Thus,

$$\omega = \frac{V_1}{2R} = \frac{\dfrac{113.5}{\sqrt{1 + 152(0.239)^4}} \text{ ft/s}}{2\left(\dfrac{3}{2} \text{ ft}\right)}$$

$$= 30.9 \text{ rad/s} \times 1 \text{ rev}/2\pi \text{ rad} \times 60 \text{ s/min}$$

$$= 295 \text{ rpm} \qquad \textbf{(Ans)}$$

COMMENT The reason that an optimum diameter nozzle exists can be explained as follows. A larger diameter nozzle will allow a larger flowrate, but will produce a smaller jet velocity because of the head loss within the supply side. A smaller diameter nozzle will reduce the flowrate but will produce a larger jet velocity. Since the power depends on a product combination of flowrate and jet velocity (see Eq. 1), there is an optimum-diameter nozzle that gives the maximum power.

These results can be generalized (i.e., without regard to the specific parameter values of this problem) by considering Eqs. 1 and 3 and the condition that $U = V_1/2$ to obtain

$$\dot{W}_{\text{shaft}}|_{U=V_1/2} = -\frac{\pi}{16}\rho(1 - \cos\beta)$$

$$\times (2gz_0)^{3/2} D_1^2 \bigg/ \left(1 + f\frac{\ell}{D^5} D_1^4\right)^{3/2}$$

By setting $d\dot{W}_{\text{shaft}}/dD_1 = 0$, it can be shown (see Problem 11.33) that the maximum power occurs when

$$D_1 = D \bigg/ \left(2f\frac{\ell}{D}\right)^{1/4}$$

which gives the same results obtained earlier for the specific parameters of the example problem. Note that the optimum condition depends only on the friction factor and the length to diameter ratio of the supply pipe. What happens if the supply pipe is frictionless or of essentially zero length?

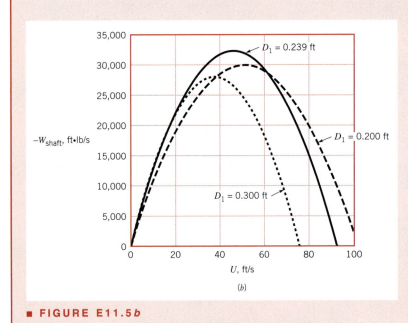

■ **FIGURE E11.5b**

V11.5 Dental drill

A second type of impulse turbine that is widely used (most often with gas as the working fluid) is indicated in Fig. E11.6a. A circumferential series of fluid jets strikes the rotating blades, which, as with the Pelton wheel, alter both the direction and the magnitude of the absolute velocity. As with the Pelton wheel, the inlet and exit pressures (i.e., on either side of the rotor) are equal, and the magnitude of the relative velocity is unchanged as the fluid slides across the blades (if frictional effects are negligible).

EXAMPLE 11.6   Non-Pelton Wheel Impulse Turbine (Dentist Drill)

GIVEN An air turbine used to drive the high-speed drill used by your dentist is shown in Fig. E11.6a. Air exiting from the upstream nozzle holes forces the turbine blades to move in the direction shown. The turbine rotor speed is 300,000 rpm, the tangential component of velocity out of the nozzle is twice the blade speed, and the tangential component of the absolute velocity out of the rotor is zero.

FIND Estimate the shaft energy per unit mass of air flowing through the turbine.

SOLUTION

We use the fixed, nondeforming control volume that includes the turbine rotor and the fluid in the rotor blade passages at an instant of time (see Fig. E11.6b). The only torque acting on this control volume is the shaft torque. For simplicity we analyze this problem using an arithmetic mean radius, r_m, where

$$r_m = \frac{1}{2}(r_0 + r_i)$$

A sketch of the velocity triangles at the rotor entrance and exit is shown in Fig. E11.6c.

Application of Eq. 11.5 (a form of the moment-of-momentum equation) gives

$$w_{\text{shaft}} = -U_1 V_{\theta 1} + U_2 V_{\theta 2} \qquad (1)$$

where w_{shaft} is shaft energy per unit of mass flowing through the turbine. From the problem statement, $V_{\theta 1} = 2U$ and $V_{\theta 2} = 0$, where

$$
\begin{aligned}
U = \omega r_m &= (300{,}000 \text{ rev/min})(1 \text{ min/60 s})(2\pi \text{ rad/rev}) \\
&\quad \times (0.168 \text{ in.} + 0.133 \text{ in.})/2(12 \text{ in./ft}) \qquad (2) \\
&= 394 \text{ ft/s}
\end{aligned}
$$

is the mean-radius blade velocity. Thus, Eq. 1 becomes

$$
\begin{aligned}
w_{\text{shaft}} = -U_1 V_{\theta 1} = -2U^2 &= -2(394 \text{ ft/s})^2 \\
&= -310{,}000 \text{ ft}^2/\text{s}^2 \\
&= (-310{,}000 \text{ ft}^2/\text{s}^2)(1 \text{ lb/slug} \cdot \text{ft/s}^2) \\
&= -310{,}000 \text{ ft} \cdot \text{lb/slug} \qquad \textbf{(Ans)}
\end{aligned}
$$

COMMENT   For each slug of air passing through the turbine there is 310,000 ft·lb of energy available at the shaft to drive the drill. However, because of fluid friction, the actual amount of energy given up by each slug of air will be greater than the amount available at the shaft. How much greater depends on the efficiency of the fluid-mechanical energy transfer between the fluid and the turbine blades.

Recall that the shaft power is given by $\dot{W}_{\text{shaft}} = \dot{m} w_{\text{shaft}}$. Hence, to determine the power we need to know the mass flowrate, $\dot{m}$, which depends on the size and number of the nozzles. Although the energy per unit mass is large (i.e., 310,000 ft·lb/slug), the flowrate is small, so the power is not "large."

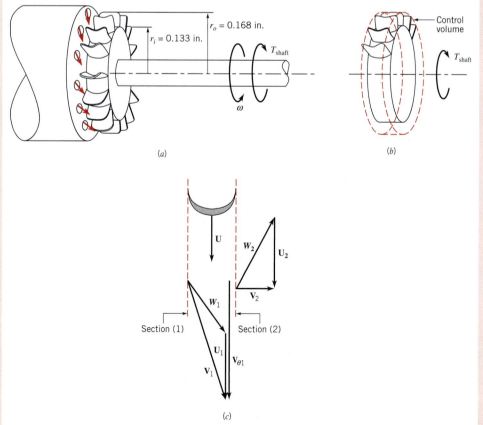

(a)

(b)

(c)

11.7.2 Reaction Turbines

As indicated in the previous section, impulse turbines are best suited (i.e., most efficient) for lower flowrate and higher head operations. Reaction turbines, however, are best suited for higher flowrate and lower head situations such as are often encountered in hydroelectric power plants associated with a dammed river, for example.

In a reaction turbine the working fluid completely fills the passageways through which it flows (unlike an impulse turbine, which contains one or more individual unconfined jets of fluid). The angular momentum, pressure, and velocity of the fluid decrease as it flows through the turbine rotor—the turbine rotor extracts energy from the fluid.

As with pumps, turbines are manufactured in a variety of configurations—radial-flow, mixed-flow, and axial-flow. Typical radial- and mixed-flow hydraulic turbines are called *Francis turbines*, named after James Francis, an American engineer. At very low heads the most efficient type of turbine is the axial-flow or propeller turbine. The *Kaplan turbine*, named after Victor Kaplan, a German professor, is an efficient axial-flow hydraulic turbine with adjustable blades. Cross sections of these different turbine types are shown in Fig. 11.23.

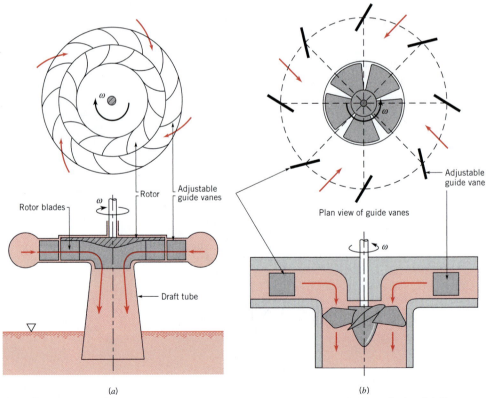

■ **FIGURE 11.23** (*a*) **Typical radial-flow Francis turbine and (*b*) typical axial-flow Kaplan turbine.**

F l u i d s   i n   t h e   N e w s

World's Largest Hydroelectric Complex When completed, the hydroelectric plants associated with the Three Gorges Dam in China will be the largest assembly of turbines in the world. For the first phase of the project, the dam will be 87 m high and the turbine complex will consist of 14 Francis turbines of 700 MW each, giving a total of 9800 MW. They will initially operate at a low head compared to the second phase when the dam will be its maximum 185 m high. In the second phase, 12 more 700-MW units will be added on the opposite bank, bringing the total to 18,200 MW (i.e., 24.4 million horsepower!). The Three Gorges complex will then contain the largest hydroelectric plant in the world, well ahead of Brazil's 12,600-MW Itaipu installation. When operating at its maximum capacity, the turbines will produce up to one-ninth of China's electrical output. ∎

As shown in Fig. 11.23*a,* flow across the rotor blades of a radial-inflow turbine has a major component in the radial direction. Inlet guide vanes (which may be adjusted to allow optimum performance) direct the water into the rotor with a tangential component of velocity. The absolute velocity of the water leaving the rotor is essentially without tangential velocity. Hence, the rotor decreases the angular momentum of the fluid, the fluid exerts a torque on the rotor in the direction of rotation, and the rotor extracts energy from the fluid. The Euler turbomachine equation (Eq. 11.2) and the corresponding power equation (Eq. 11.4) are equally valid for this turbine as they are for the centrifugal pump discussed in Section 11.4.

As shown in Fig. 11.23*b,* for an axial-flow Kaplan turbine, the fluid flows through the inlet guide vanes and achieves a tangential velocity in a vortex (swirl) motion before it reaches the rotor. Flow across the rotor contains a major axial component. Both the inlet guide vanes and the turbine blades can be adjusted by changing their setting angles to produce the best match (optimum output) for the specific operating conditions. For example, the operating head available may change from season to season and/or the flowrate through the rotor may vary.

As with pumps, incompressible flow turbine performance is often specified in terms of appropriate dimensionless parameters. The flow coefficient, $C_Q = Q/\omega D^3$, the head coefficient, $C_H = gh_T/\omega^2 D^2$, and the power coefficient, $C_{\mathcal{P}} = W_{\text{shaft}}/\rho\omega^3 D^5$, are defined in the same way for pumps and turbines. However, turbine efficiency, η, is the inverse of pump efficiency. That is, the efficiency is the ratio of the shaft power output to the power available in the flowing fluid, or

$$\eta = \frac{\dot{W}_{\text{shaft}}}{\rho g Q h_T}$$

For geometrically similar turbines and for negligible Reynolds number and surface roughness difference effects, the relationships between the dimensionless parameters are given functionally by that shown in Eqs. 11.21, 22, and 23. That is,

$$C_H = \phi_1(C_Q), \qquad C_{\mathcal{P}} = \phi_2(C_Q), \qquad \text{and} \qquad \eta = \phi_3(C_Q)$$

where the functions ϕ_1, ϕ_2, and ϕ_3 are dependent on the type of turbine involved. Also, for turbines the efficiency, η, is related to the other coefficients according to $\eta = C_{\mathcal{P}}/C_H C_Q$.

As indicated earlier, the design engineer has a variety of turbine types available for any given application. It is necessary to determine which type of turbine would best fit the job (i.e., be most efficient) before detailed design work is attempted. As with pumps, the use of a specific speed parameter can help provide this information. For hydraulic turbines, the rotor diameter D is eliminated between the flow coefficient and the power coefficient to obtain the *power specific speed, N'_s*, where

$$N'_s = \frac{\omega \sqrt{\dot{W}_{\text{shaft}}/\rho}}{(gh_T)^{5/4}}$$

We use the more common, but not dimensionless, definition of specific speed

$$N'_{sd} = \frac{\omega(\text{rpm}) \sqrt{\dot{W}_{\text{shaft}}/(\text{bhp})}}{[h_T(\text{ft})]^{5/4}} \tag{11.35}$$

That is, N'_{sd} is calculated with angular velocity, ω, in rpm; shaft power, $\dot{W}_{\text{shaft}}$, in brake horsepower; and head, h_T, in feet. Optimum turbine efficiency (for large turbines) as a function of specific speed is indicated in Fig. 11.24. Also shown are representative rotor and casing cross sections. Note that impulse turbines are best at low specific speeds; that is, when operating with larger heads and small flowrate. The other extreme is axial-flow turbines, which are the most efficient type if the head is low and if the flowrate is large. For intermediate values of specific speeds, radial- and mixed-flow turbines offer the best performance.

Data shown in Fig. 11.24 are meant only to provide a guide for turbine-type selection. The actual turbine efficiency for a given turbine depends very strongly on the detailed design of the turbine. Considerable analysis, testing, and experience are needed to produce an efficient turbine. However, data of Fig. 11.24 are representative. Much additional information can be found in the literature (Refs. 11, 17).

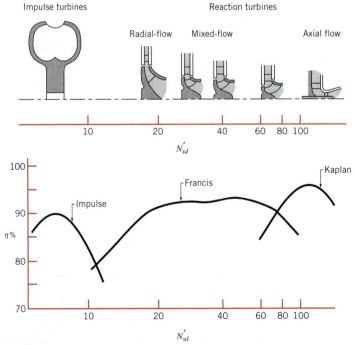

■ **FIGURE 11.24** Typical turbine cross sections and maximum efficiencies as a function of specific speed.

EXAMPLE 11.7	Use of Specific Speed to Select Turbine Type

GIVEN A hydraulic turbine is to operate at an angular velocity of 6 rev/s, a flowrate of 10 ft³/s, and a head of 20 ft.

FIND What type of turbine should be selected?

SOLUTION

The most efficient type of turbine to use can be obtained by calculating the specific speed, N'_{sd}, and using the information of Fig. 11.24. To use the dimensional form of the specific speed indicated in Fig. 11.24 we must convert the given data into the appropriate units. For the rotor speed we get

$$\omega = 6 \text{ rev/s} \times 60 \text{ s/min} = 360 \text{ rpm}$$

To estimate the shaft power, we assume all of the available head is converted into power (i.e., $h_T = z = 20$ ft) and multiply this amount by an assumed efficiency (94%).

$$\dot{W}_{shaft} = \gamma Q z \eta = (62.4 \text{ lb/ft}^3)(10 \text{ ft}^3/\text{s})\left[\frac{20 \text{ ft}(0.94)}{550 \text{ ft} \cdot \text{lb/s} \cdot \text{hp}}\right]$$

$$\dot{W}_{shaft} = 21.3 \text{ hp}$$

Thus for this turbine,

$$N'_{sd} = \frac{\omega \sqrt{\dot{W}_{shaft}}}{(h_T)^{5/4}} = \frac{(360 \text{ rpm})\sqrt{21.3 \text{ hp}}}{(20 \text{ ft})^{5/4}} = 39.3$$

According to the information of Fig. 11.24,

> A mixed-flow Francis turbine would probably give the highest efficiency and an assumed efficiency of 0.94 is appropriate.   **(Ans)**

COMMENT What would happen if we wished to use a Pelton wheel for this application? Note that with only a 20-ft head, the maximum jet velocity, V_1, obtainable (neglecting viscous effects) would be

$$V_1 = \sqrt{2 gz} = \sqrt{2 \times 32.2 \text{ ft/s}^2 \times 20 \text{ ft}} = 35.9 \text{ ft/s}$$

As shown by Eq. 11.34, for maximum efficiency of a Pelton wheel the jet velocity is ideally two times the blade velocity. Thus, $V_1 = 2\omega R$, or the wheel diameter, $D = 2R$, is

$$D = \frac{V_1}{\omega} = \frac{35.9 \text{ ft/s}}{(6 \text{ rev/s} \times 2\pi \text{ rad/rev})} = 0.952 \text{ ft}$$

To obtain a flowrate of $Q = 10$ ft³/s at a velocity of $V_1 = 39.5$ ft/s, the jet diameter, d_1, must be given by

$$Q = \frac{\pi}{4} d_1^2 V_1$$

or

$$d_1 = \left[\frac{4Q}{\pi V_1}\right]^{1/2} = \left[\frac{4(10 \text{ ft}^3/\text{s})}{\pi(35.9 \text{ ft/s})}\right]^{1/2} = 0.596 \text{ ft}$$

A Pelton wheel with a diameter of $D = 0.952$ ft supplied with water through a nozzle of diameter $d_1 = 0.596$ ft is not a practical design. Typically $d_1 \ll D$ (see Fig. 11.17). By using multiple jets it would be possible to reduce the jet diameter. However, even with eight jets, the jet diameter would be 0.211 ft, which is still too large (relative to the wheel diameter) to be practical. Hence, the aforementioned calculations reinforce the results presented in Fig. 11.24—a Pelton wheel would not be practical for this application. If the flowrate were considerably smaller, the specific speed could be reduced to the range where a Pelton wheel would be the type to use (rather than a mixed-flow reaction turbine).

11.8   Compressible Flow Turbomachines

Compressible flow turbomachines are in many ways similar to the incompressible flow pumps and turbines described in previous portions of this chapter. The main difference is that the density of the fluid (a gas or vapor) changes significantly from the inlet to the outlet of the compressible flow machines. This added feature has interesting consequences (e.g., shock waves), benefits (e.g., large pressure changes), and complications (e.g., blade cooling). The photograph in Fig. 11.25 shows a typical centrifugal compressor used in an automobile turbocharger. Exhaust gases from the engine drive the turbine, which powers the compressor so that more air can be forced into the engine.

An in-depth understanding of compressible flow turbomachines requires the mastery of various thermodynamic concepts, as well as information provided earlier about basic

■ **FIGURE 11.25**   Photograph of the rotor from an automobile turbocharger.

energy considerations (Section 11.2) and basic angular momentum considerations (Section 11.3). The interested reader is encouraged to read some of the references available for further information (Refs. 12, 13, 14, 15, 16, and 17).

11.9   Chapter Summary and Study Guide

turbomachine
axial-, mixed-,
*   and radial-flow*
velocity triangle
angular
*   momentum*
shaft torque
Euler turboma-
*   chine equation*
shaft power
centrifugal pump
pump performance
*   curve*
overall efficiency
system equation
head rise
*   coefficient*
power coefficient
flow coefficient
pump scaling
*   laws*
specific speed
impulse turbine
reaction turbine
Pelton wheel

This chapter discussed various aspects of turbomachine analysis and design. The concepts of angular momentum and torque are key to understanding how such pumps and turbines operate.

The shaft torque required to change the axial component of angular momentum of a fluid as it flows through a pump or turbine is described in terms of the inlet and outlet velocity triangles diagrams. Such diagrams indicate the relationship among absolute, relative, and blade velocities.

Performance characteristics for centrifugal pumps are discussed. Standard dimensionless pump parameters, similarity laws, and the concept of specific speed are presented for use in pump analysis. It is shown how to use the pump performance curve and the system curve for proper pump selection. A brief discussion of axial-flow and mixed-flow pumps is given.

The analysis of impulse turbines is provided, with emphasis on the Pelton wheel turbine. For impulse turbines there is negligible pressure difference across the blade; the torque is a result of the impulse of the fluid jet striking the blade. Radial-flow and axial-flow reaction turbines are also discussed.

The following checklist provides a study guide for this chapter. When your study of the entire chapter and end-of-chapter exercises has been completed you should be able to

■ write out meanings of the terms listed here in the margin and understand each of the related concepts. These terms are particularly important and are set in color and bold type in the text.

■ draw appropriate velocity triangles for given pump or turbine configurations.

■ calculate the shaft torque, shaft power, and pump head for a given centrifugal pump configuration.

- use the pump performance curve and the system curve to predict the pump performance for a given system.
- predict the performance characteristics for one pump based on the performance of another pump of the same family using the pump scaling laws.
- use the specific speed to determine whether a radial flow, mixed flow, or axial flow pump would be most appropriate for a given situation.
- calculate the shaft torque and shaft power for an impulse turbine of a given configuration.
- calculate the shaft torque and shaft power for a given reaction turbine.
- use the specific speed to determine whether an impulse or a reaction turbine would be most appropriate for a given situation.

References

1. Stepanoff, H. J., *Centrifugal and Axial Flow Pumps,* 2nd Ed., Wiley, New York, 1957.
2. Shepherd, D. G., *Principles of Turbomachinery,* Macmillan, New York, 1956.
3. Wislicenus, G. F., *Preliminary Design of Turbopumps and Related Machinery,* NASA Reference Publication 1170, 1986.
4. Neumann, B., *The Interaction Between Geometry and Performance of a Centrifugal Pump,* Mechanical Engineering Publications Limited, London, 1991.
5. Rouse, H., *Elementary Mechanics of Fluids,* Wiley, New York, 1946.
6. Hydraulic Institute, *Hydraulic Institute Pump Standards,* CD-ROM, Hydraulic Institute, Cleveland, Ohio, 2005.
7. Kristal, F. A., and Annett, F. A., *Pumps: Types, Selection, Installation, Operation, and Maintenance,* McGraw-Hill, New York, 1953.
8. Garay, P. N., *Pump Application Desk Book,* 3rd Ed., Fairmont Press, Lilburn, Georgia, 1996.
9. Karassick, I. J., et al., *Pump Handbook,* 3rd Ed., McGraw-Hill, New York, 2000.
10. Moody, L. F., and Zowski, T., "Hydraulic Machinery," in *Handbook of Applied Hydraulics,* 3rd Ed., by C. V. Davis and K. E. Sorensen, McGraw-Hill, New York, 1969.
11. Balje, O. E., *Turbomachines: A Guide to Design, Selection, and Theory,* Wiley, New York, 1981.
12. Bathie, W. W., *Fundamentals of Gas Turbines,* Wiley, New York, 1984.
13. Boyce, M. P., *Gas Turbine Engineering Handbook,* 3rd Ed., Gulf Publishing, Houston, 2006.
14. Cohen, H., Rogers, G. F. C., and Saravanamuttoo, H. I. H., *Gas Turbine Theory,* 3rd Ed., Longman Scientific & Technical, Essex, UK, and Wiley, New York, 1987.
15. Johnson, I. A., and Bullock, R. D., Eds., *Aerodynamic Design of Axial-Flow Compressors,* NASA SP-36, National Aeronautics and Space Administration, Washington, 1965.
16. Glassman, A. J., Ed., *Turbine Design and Application,* Vol. 3, NASA SP-290, National Aeronautics and Space Administration, Washington, 1975.
17. Johnson, R. W., Ed. *The Handbook of Fluid Dynamics,* CRC Press, New York, 1998.

Review Problems

Go to Appendix F for a set of review problems with answers. Detailed solutions can be found in *Student Solution Manual for a Brief Introduction to Fluid Mechanics,* by Young et al. (© 2006 John Wiley and Sons, Inc.).

Problems

Section 11.3 Basic Angular Momentum Considerations

11.1 Water flows through a rotating sprinkler arm as shown in Fig. P11.1 and **Video V11.2.** Determine the flowrate if the angular velocity is 150 rpm. Friction is negligible. Is this a turbine or a pump? What is the maximum angular velocity for this flowrate?

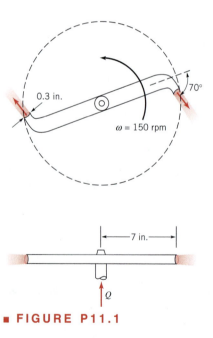

■ **FIGURE P11.1**

11.2 Water flows axially up the shaft and out through the two sprinkler arms as sketched in Fig. P11.1 and as shown in **Video V11.2.** With the help of the moment-of-momentum equation explain why, at a threshold amount of water flow, the sprinkler arms begin to rotate. What happens when the flowrate increases above this threshold amount?

11.3 Uniform horizontal sheets of water of 3-mm thickness issue from the slits on the rotating manifold shown in Fig. P11.3. The velocity relative to the arm is a constant 3 m/s along each slit. Determine the torque needed to hold the

manifold stationary. What would the angular velocity of the manifold be if the resisting torque is negligible?

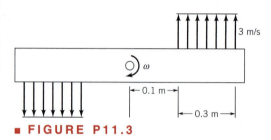

■ **FIGURE P11.3**

11.4 Air (assumed incompressible) flows across the rotor shown in Fig. P11.4 such that the magnitude of the absolute velocity increases from 15 m/s to 25 m/s. Measurements indicate that the absolute velocity at the inlet is in the direction shown. Determine the direction of the absolute velocity at the outlet if the fluid puts no torque on the rotor. Is the rotation CW or CCW? Is this device a pump or a turbine?

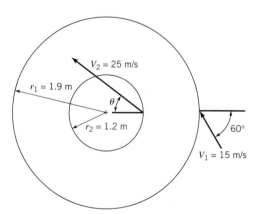

■ **FIGURE P11.4**

11.5 At a given radial location, a 15 ft/s wind against a windmill (see **Video V11.1**) results in the upstream (1) and downstream (2) velocity triangles shown in Fig. P11.5. Sketch an appropriate blade section at that radial location and determine the energy transferred per unit mass of fluid.

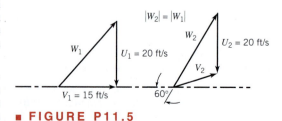

■ **FIGURE P11.5**

Section 11.4 The Centrifugal Pump

11.6 A centrifugal pump impeller is rotating at 1200 rpm in the direction shown in Fig. P11.6. The flow enters parallel to the axis of rotation and leaves at an angle of 30° to the radial direction. The absolute exit velocity, V_2, is 90 ft/s. **(a)** Draw the velocity triangle for the impeller exit flow. **(b)** Estimate the torque necessary to turn the impeller if the fluid density is 2.0 slugs/ft^3. What will the impeller rotation speed become if the shaft breaks?

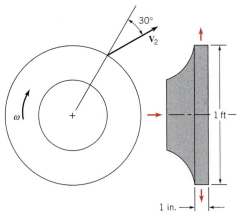

■ **FIGURE P11.6**

11.7 A centrifugal radial water pump has the dimensions shown in Fig. P11.7. The volume rate of flow is 0.25 ft^3/s, and the absolute inlet velocity is directed radially outward. The angular velocity of the impeller is 960 rpm. The exit velocity as seen from a coordinate system attached to the impeller can be assumed to be tangent to the vane at its trailing edge. Calculate the power required to drive the pump.

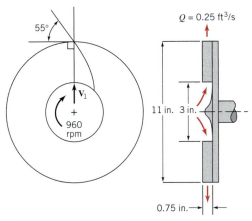

■ **FIGURE P11.7**

11.8 Water is pumped with a centrifugal pump, and measurements made on the pump indicate that for a flowrate of 240 gpm the required input power is 6 hp. For a pump efficiency of 62%, what is the actual head rise of the water being pumped?

11.9 The performance characteristics of a certain centrifugal pump having a 9-in.-diameter impeller and operating at 1750 rpm are determined using an experimental setup similar to that shown in Fig. 11.7. The following data were obtained during a series of tests in which $z_2 - z_1 = 0$, $V_2 = V_1$, and the fluid was water.

Q (gpm)	20	40	60	80	100	120	140
$p_2 - p_1$ (psi)	40.2	40.1	38.1	36.2	33.5	30.1	25.8
Power input (hp)	1.58	2.27	2.67	2.95	3.19	3.49	4.00

Based on these data, show or plot how the actual head rise, h_a, and the pump efficiency, η, vary with the flowrate. What is the design flowrate for this pump?

11.10 The centrifugal pump shown in Fig. P11.10 is not self-priming. That is, if the water is drained from the pump and pipe as shown in Fig. P11.10a, the pump will not draw the water into the pump and start pumping when the pump is turned on. However, if the pump is primed (i.e., filled with water as in Fig. P11.10b), the pump does start pumping water when turned on. Explain this behavior.

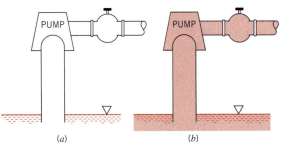

■ **FIGURE P11.10**

11.11 Due to fouling of the pipe wall, the friction factor for the pipe of Example 11.3 increases from 0.02 to 0.03. Determine the new flowrate, assuming all other conditions remain the same. What is the pump efficiency at this new flowrate? Explain how a line valve could be used to vary the flowrate through the pipe of Example 11.3. Would it be better to place the valve upstream or downstream of the pump? Why?

11.12 A centrifugal pump having a head–capacity relationship given by the equation $h_a = 180 - 6.10 \times 10^{-4} Q^2$, with h_a in feet when Q is in gpm, is to be used with a system similar to that shown in Fig. 11.10. For $z_2 - z_1 = 50$ ft, what is the expected flowrate if the total length of a constant-diameter pipe is 600 ft and the fluid is water? Assume the pipe diameter to be 4 in. and the friction factor to be equal to 0.02. Neglect all minor losses.

11.13 A centrifugal pump having a 6-in.-diameter impeller and the characteristics shown in Fig. 11.9 is to be used to pump gasoline through 4000 ft of commercial steel 3-in.-diameter pipe. The pipe connects two reservoirs having open surfaces at the same elevation. Determine the flowrate. Do you think this pump is a good choice? Explain.

11.14 A centrifugal pump having the characteristics shown in Example 11.3 is used to pump water between two large open tanks through 100 ft of 8-in.-diameter pipe. The pipeline contains four regular flanged 90° elbows, a check valve, and a fully open globe valve. Other minor losses are negligible. Assume the friction factor $f = 0.02$ for the 100-ft section of pipe. If the static head (difference in height of fluid surfaces in the two tanks) is 30 ft, what is the expected flowrate? Do you think this pump is a good choice? Explain.

11.15   In a chemical processing plant a liquid is pumped from an open tank, through a 0.1-m-diameter vertical pipe, and into another open tank as shown in Fig. P11.15a. A valve is located in the pipe, and the minor loss coefficient for the valve as a function of the valve setting is shown in Fig. P11.15b. The pump head–capacity relationship is given by the equation $h_a = 52.0 - 1.01 \times 10^3 \, Q^2$ with h_a in meters when Q is in m³/s. Assume the friction factor $f = 0.02$ for the pipe, and all minor losses, except for the valve, are negligible. The fluid levels in the two tanks can be assumed to remain constant. **(a)** Determine the flowrate with the valve wide open. **(b)** Determine the required valve setting (percent open) to reduce the flowrate by 50%

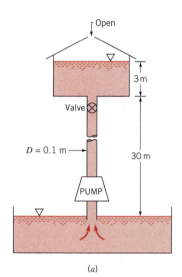

(a)

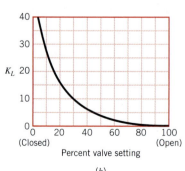

(b)

■ **FIGURE P11.15**

†11.16   Water is pumped between the two tanks described in Example 11.3 once a day, 365 days a year, with each pumping period lasting 2 hr. The water levels in the two tanks remain essentially constant. Estimate the annual cost of the electrical power needed to operate the pump if it were located in your city.

You will have to make a reasonable estimate for the efficiency of the motor used to drive the pump. Due to aging, it can be expected that the overall resistance of the system will increase with time. If the operating point shown in Fig. E11.3c changes to a point where the flowrate has been reduced to 1000 gpm, what will be the new annual cost of operating the pump? Assume the cost of electrical power remains the same.

Section 11.5   Dimensionless Parameters and Similarity Laws

11.17   A small model of a pump is tested in the laboratory and found to have a specific speed, N_{sd}, equal to 1000 when operating at peak efficiency. Predict the discharge of a larger, geometrically similar pump operating at peak efficiency at a speed of 1800 rpm across an actual head rise of 200 ft.

11.18   A centrifugal pump provides a flowrate of 500 gpm when operating at 1760 rpm against a 200-ft head. Determine the pump's flowrate and developed head if the pump speed is increased to 3500 rpm.

11.19   A centrifugal pump with a 12-in.-diameter impeller requires a power input of 60 hp when the flowrate is 3200 gpm against a 60-ft head. The impeller is changed to one with a 10-in. diameter. Determine the expected flowrate, head, and input power if the pump speed remains the same.

11.20   A centrifugal pump has the performance characteristics of the pump with the 6-in.-diameter impeller described in Fig. 11.9. Note that in Fig. 11.9 the pump is operating at 3500 rpm. What is the expected flowrate and head gain if the speed of this pump is reduced to 2800 rpm while operating at peak efficiency?

11.21   A certain axial-flow pump has a specific speed of $N_S = 5.0$. If the pump is expected to deliver 3000 gpm when operating against a 15-ft head, at what speed (rpm) should the pump be run?

11.22   A certain pump is known to have a capacity of 3 m³/s when operating at a speed of 60 rad/s against a head of 20 m. Based on the information in **Fig. 11.13**, would you recommend a radial-flow, mixed-flow, or axial-flow pump?

11.23   Fuel   oil   (sp. wt = 48.0 lb/ft³, viscosity = 2.0 × 10^{-5} lb· s/ft²) is pumped through the piping system of Fig. P11.23 with a velocity of 4.6 ft/s. The pressure 200 ft upstream from the pump is 5 psi. Pipe losses downstream from the pump are negligible, but minor losses are not (minor loss coefficients are given on the figure). **(a)** For a pipe diameter of 2 in. with a relative roughness $\varepsilon/D = 0.001$, determine the head that must

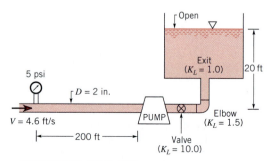

■ **FIGURE P11.23**

be added by the pump. **(b)** For a pump operating speed of 1750 rpm, what type of pump (radial flow, mixed flow, or axial flow) would you recommend for this application?

11.24 A centrifugal pump having an impeller diameter of 1 m is to be constructed so that it will supply a head rise of 200 m at a flowrate of 4.1 m³/s of water when operating at a speed of 1200 rpm. To study the characteristics of this pump, a 1/5 scale, geometrically similar model operated at the same speed is to be tested in the laboratory. Determine the required model discharge and head rise. Assume both model and prototype operate with the same efficiency (and therefore the same flow coefficient).

Section 11.7 Turbines

11.25 A Pelton wheel turbine is illustrated in Fig. P11.25. The radius to the line of action of the tangential reaction force on each vane is 1 ft. Each vane deflects fluid by an angle of 135° as indicated. Assume all of the flow occurs in a horizontal plane. Each of the four jets shown strikes a vane with a velocity of 100 ft/s and a stream diameter of 1 in. The magnitude of velocity of the jet remains constant along the vane surface. **(a)** How much torque is required to hold the wheel stationary? **(b)** How fast will the wheel rotate if shaft torque is negligible, and what practical situation is simulated by this condition?

enters at section (1) (cylindrical cross-section area A_1 at $r_1 = 1.5$ m) at an angle of 100° from the tangential direction and leaves at section (2) (cylindrical cross-section area A_2 at $r_2 = 0.85$ m) at an angle of 50° from the tangential direction. The blade height at sections (1) and (2) is 0.45 m and the volume flowrate through the turbine is 30 m³/s. The runner speed is 130 rpm in the direction shown. Determine the shaft power developed. Is the shaft power greater or less than the power lost by the fluid? Explain.

11.27 A water turbine wheel rotates at the rate of 100 rpm in the direction shown in Fig. P11.27. The inner radius, r_2, of the blade row is 1 ft, and the outer radius, r_1, is 2 ft. The absolute velocity vector at the turbine rotor entrance makes an angle of 20° with the tangential direction. The inlet blade angle is 60° relative to the tangential direction. The blade outlet angle is 120°. The flowrate is 10 ft³/s. For the flow tangent to the rotor blade surface at inlet and outlet, determine an appropriate constant blade height, b, and the corresponding power available at the rotor shaft. Is the shaft power greater or less than the power lost by the fluid? Explain.

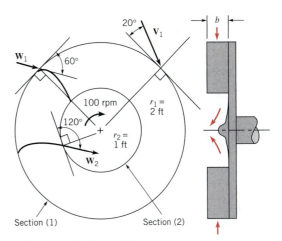

■ **FIGURE P11.27**

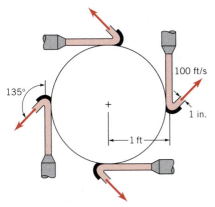

■ **FIGURE P11.25**

11.26 A simplified sketch of a hydraulic turbine runner is shown in Fig. P11.26. Relative to the rotating runner, water

11.28 A sketch of the arithmetic mean radius blade sections of an axial-flow water turbine stage is shown in Fig. P11.28.

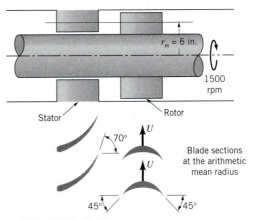

■ **FIGURE P11.28**

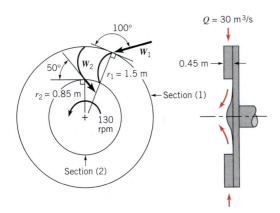

■ **FIGURE P11.26**

The rotor speed is 1500 rpm. **(a)** Sketch and label velocity triangles for the flow entering and leaving the rotor row. Use **V** for absolute velocity, **W** for relative velocity, and **U** for blade velocity. Assume flow enters and leaves each blade row at the blade angles shown. **(b)** Calculate the work per unit mass delivered at the shaft.

11.29 An inward flow radial turbine (see Fig. P11.29) involves a nozzle angle, α_1, of 60° and an inlet rotor tip speed, U_1, of 9 m/s. The ratio of rotor inlet to outlet diameters is 2.0. The radial component of velocity remains constant at 6 m/s through the rotor, and the flow leaving the rotor at section (2) is without angular momentum. If the flowing fluid is water and the stagnation pressure drop across the rotor is 110 kPa, determine the loss of available energy across the rotor and the efficiency involved.

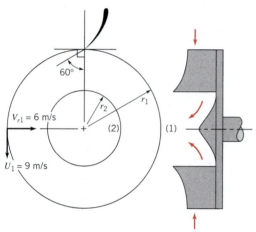

■ **FIGURE P11.29**

11.30 A 10.9-ft-diameter Pelton wheel operates at 500 rpm with a total head just upstream of the nozzle of 5330 ft. Estimate the diameter of the nozzle of the single-nozzle wheel if it develops 25,000 horsepower.

11.31 A small Pelton wheel is used to power an oscillating lawn sprinkler as shown in **Video V11.4** and Fig. P11.31. The arithmetic mean radius of the turbine is 1 in., and the exit angle of the blade is 135° relative to the blade motion. Water is sup-

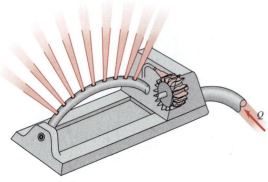

■ **FIGURE P11.31**

plied through a single 0.20-in.-diameter nozzle at a speed of 50 ft/s. Determine the flowrate, the maximum torque developed, and the maximum power developed by this turbine.

11.32 A Pelton wheel has a diameter of 2 m and develops 500 kW when rotating 180 rpm. What is the average force of the water against the blades? If the turbine is operating at maximum efficiency, determine the speed of the water jet from the nozzle and the mass flowrate.

11.33 Water to run a Pelton wheel is supplied by a penstock of length ℓ and diameter D with a friction factor f. If the only losses associated with the flow in the penstock are due to pipe friction, show that the maximum power output of the turbine occurs when the nozzle diameter, D_1, is given by $D_1 = D/(2f\,\ell/D)^{1/4}$.

11.34 A hydraulic turbine operating at 180 rpm with a head of 170 feet develops 20,000 hp. Estimate the power and speed if the turbine were to operate under a head of 190 ft.

11.35 Draft tubes as shown in Fig. P11.35 are often installed at the exit of Kaplan and Francis turbines. Explain why such draft tubes are advantageous.

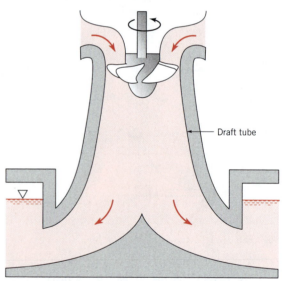

Draft tube

■ **FIGURE P11.35**

11.36 Turbines are to be designed to develop 30,000 hp while operating under a head of 70 ft and an angular velocity of 60 rpm. What type of turbine is best suited for this purpose? Estimate the flowrate needed.

11.37 A 1-m-diameter Pelton wheel rotates at 300 rpm. Which of the following heads (in meters) would be best suited for this turbine: **(a)** 2, **(b)** 5, **(c)** 40, **(d)** 70, or **(e)** 140? Explain.

11.38 Water at 400 psi is available to operate a turbine at 1750 rpm. What type of turbine would you suggest to use if the turbine should have an output of approximately 200 hp?

11.39 A high-speed turbine used to power a dentist's drill is shown in **Video V11.5** and Fig. E11.6. With the conditions stated in Example 11.6, for every slug of air that passes through the turbine there is 310,000 ft·lb of energy available at the shaft to drive the drill. One of the assumptions made to obtain this numerical result is that the tangential component of the absolute velocity out of the rotor is zero. Suppose this assumption were not true (but all other parameter values remain the same). Discuss how and why the value of 310,000 ft·lb/slug would change for these new conditions.

11.40 Test data for the small Francis turbine shown in Fig. P11.40 are given in the table shown. The test was run at a constant 32.8-ft head just upstream of the turbine. The Prony brake on the turbine output shaft was adjusted to give various angular velocities, and the force on the brake arm, F, was recorded. Use the given data to plot curves of torque as a function of angular velocity and turbine efficiency as a function of angular velocity.

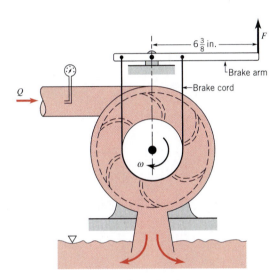

ω (rpm)	Q (ft³/s)	F (lb)
0	0.129	2.63
1000	0.129	2.40
1500	0.129	2.22
1870	0.124	1.91
2170	0.118	1.49
2350	0.0942	0.876
2580	0.0766	0.337
2710	0.068	0.089

■ **FIGURE P11.40**

†11.41 It is possible to generate power using the water from your garden hose to drive a small Pelton wheel turbine. (See **Video V11.4.**) Provide a preliminary design of such a turbine

and estimate the power output expected. List all assumptions and show calculations.

11.42 The device shown in Fig. P11.42 is used to investigate the power produced by a Pelton wheel turbine. Water supplied at a constant flowrate issues from a nozzle and strikes the turbine buckets as indicated. The angular velocity, ω, of the turbine wheel is varied by adjusting the tension on the Prony brake spring, thereby varying the torque, T_{shaft}, applied to the output shaft. This torque can be determined from the measured force, R, needed to keep the brake arm stationary as $T_{shaft} = R\ell$, where ℓ is the moment arm of the brake force.

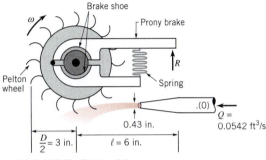

■ **FIGURE P11.42**

Experimentally determined values of ω and R are shown in the following table. Use these results to plot a graph of torque as a function of the angular velocity. On another graph plot the power output, $\dot{W}_{shaft} = T_{shaft}\,\omega$, as a function of the angular velocity. On each of these graphs plot the theoretical curves for this turbine, assuming 100 percent efficiency.

Compare the experimental and theoretical results and discuss some possible reasons for any differences between them.

ω (rpm)	R (lb)
0	2.47
360	1.91
450	1.84
600	1.69
700	1.55
940	1.17
1120	0.89
1480	0.16

FE Exam Problems

Sample FE (Fundamentals of Engineering) exam questions for fluid mechanics are provided on the book's web site, www.wiley.com/college/young.

APPENDICES

Computational Fluid Dynamics and FlowLab

A.1   Introduction

Numerical methods using digital computers are, of course, commonly utilized to solve a wide variety of flow problems. As discussed in Chapter 6, although the differential equations that govern the flow of Newtonian fluids [the Navier–Stokes equations (Eq. 6.120)] were derived many years ago, there are few known analytical solutions to them. However, with the advent of high-speed digital computers it has become possible to obtain approximate numerical solutions to these (and other fluid mechanics) equations for a wide variety of circumstances.

Computational fluid dynamics (CFD) involves replacing the partial differential equations with discretized algebraic equations that approximate the partial differential equations. These equations are then numerically solved to obtain flow field values at the discrete points in space and/or time. Since the Navier–Stokes equations are valid everywhere in the flow field of the fluid continuum, an analytical solution to these equations provides the solution for an infinite number of points in the flow. However, analytical solutions are available for only a limited number of simplified flow geometries. To overcome this limitation, the governing equations can be discretized and put in algebraic form for the computer to solve. The CFD simulation solves for the relevant flow variables only at the discrete points, which make up the grid or mesh of the solution (discussed in more detail below). Interpolation schemes are used to obtain values at non-grid point locations.

CFD can be thought of as a numerical experiment. In a typical fluids experiment, an experimental model is built, measurements of the flow interacting with that model are taken, and the results are analyzed. In CFD, the building of the model is replaced with the formulation of the governing equations and the development of the numerical algorithm. The process of obtaining measurements is replaced with running an algorithm on the computer to simulate the flow interaction. Of course, the analysis of results is common ground to both techniques.

CFD can be classified as a subdiscipline to the study of fluid dynamics. However, it should be pointed out that a thorough coverage of CFD topics is well beyond the scope of this textbook. This appendix highlights some of the more important topics in CFD, but is only intended as a brief introduction. The topics include discretization of the governing equations, grid generation, boundary conditions, application of CFD, and some representative examples. Also included is a section on FlowLab, which is the educational CFD software incorporated with this textbook. FlowLab offers the reader the opportunity to begin using CFD to solve flow problems as well as to reinforce concepts covered in the textbook. For more information, visit the book website to access the FlowLab problems, tutorials, and users guide.

A.2   Discretization

The process of *discretization* involves developing a set of algebraic equations (based on discrete points in the flow domain) to be used in place of the partial differential equations. Of the various discretization techniques available for the numerical solution of the governing differential equations, the following three types are most common: (1) the finite difference method, (2) the finite element (or finite volume) method, and (3) the boundary element method. In each of these methods, the continuous flow field (i.e., velocity or pressure as a function of space and time) is described in terms of discrete (rather than continuous) values at prescribed locations. Through this technique the differential equations are replaced by a set of algebraic equations that can be solved on the computer.

For the *finite element* (or *finite volume*) method, the flow field is broken into a set of small fluid elements (usually triangular areas if the flow is two-dimensional, or small volume elements if the flow is three-dimensional). The conservation equations (i.e., conservation of mass, momentum, and energy) are written in an appropriate form for each element, and the set of resulting algebraic equations for the flow field is solved numerically. The number, size, and shape of elements are dictated in part by the particular flow geometry and flow conditions for the problem at hand. As the number of elements increases (as is necessary for flows with complex boundaries), the number of simultaneous algebraic equations that must be solved increases rapidly. Problems involving one million (or more) grid cells are not uncommon in today's CFD community, particularly for complex three-dimensional geometries. Further information about this method can be found in Refs. 1 and 2.

For the *boundary element method*, the boundary of the flow field (not the entire flow field as in the finite element method) is broken into discrete segments (Ref. 3) and appropriate singularities such as sources, sinks, doublets, and vortices are distributed on these boundary elements. The strengths and type of the singularities are chosen so that the appropriate boundary conditions of the flow are obtained on the boundary elements. For points in the flow field not on the boundary, the flow is calculated by adding the contributions from the various singularities on the boundary. Although the details of this method are rather mathematically sophisticated, it may (depending on the particular problem) require less computational time and space than the finite element method. Typical boundary elements and their associated singularities (vortices) for two-dimensional flow past an airfoil are shown in Fig. A.1. Such use of the boundary element method in aerodynamics is often termed the *panel method* in recognition of the fact that each element plays the role of a panel on the airfoil surface (Ref. 4).

The *finite difference method* for computational fluid dynamics is perhaps the most easily understood and widely used of the three methods listed above. For this method the flow field is dissected into a set of grid points and the continuous functions (velocity, pressure, etc.) are approximated by discrete values of these functions calculated at the grid points. Derivatives of the functions are approximated by using the differences between the function values at local grid points divided by the grid spacing. The standard method for converting the partial differential equations to algebraic equations is through the use of

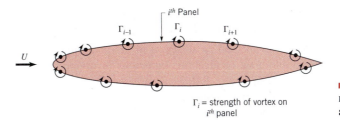

Γ_i = strength of vortex on
i^{th} panel

■ **FIGURE A.1**   **Panel method for flow past an airfoil.**

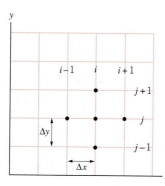

■ **FIGURE A.2**   **Standard rectangular grid.**

Taylor series expansions. (See Ref. 5.) For example, assume a standard rectangular grid is applied to a flow domain as shown in Fig. A.2.

This grid stencil shows five grid points in $x-y$ space with the center point being labeled as i, j. This index notation is used as subscripts on variables to signify location. For example, $u_{i+1, j}$ is the u component of velocity at the first point to the right of the center point i, j. The grid spacing in the i and j directions is given as Δx and Δy, respectively.

To find an algebraic approximation to a first derivative term such as $\partial u/\partial x$ at the i, j grid point, consider a Taylor series expansion written for u at $i + 1$ as

$$u_{i+1, j} = u_{i, j} + \underline{\left(\frac{\partial u}{\partial x}\right)_{i, j} \frac{\Delta x}{1!}} + \left(\frac{\partial^2 u}{\partial x^2}\right)_{i, j} \frac{(\Delta x)^2}{2!} + \left(\frac{\partial^3 u}{\partial x^3}\right)_{i, j} \frac{(\Delta x)^3}{3!} + \cdots \quad \textbf{(A.1)}$$

Solving for the underlined term in the above equation results in the following:

$$\left(\frac{\partial u}{\partial x}\right)_{i, j} = \frac{u_{i+1, j} - u_{i, j}}{\Delta x} + O(\Delta x) \quad \textbf{(A.2)}$$

where $O(\Delta x)$ contains higher order terms proportional to Δx, $(\Delta x)^2$, and so forth. Equation A.2 represents a forward difference equation to approximate the first derivative using values at $i + 1, j$ and i, j along with the grid spacing in the x direction. Obviously in solving for the $\partial u/\partial x$ term we have ignored higher order terms such as the second and third derivatives present in Eq. A.1. This process is termed *truncation* of the Taylor series expansion. The lowest order term that was truncated included $(\Delta x)^2$. Notice that the first derivative term contains Δx. When solving for the first derivative, all terms on the right-hand side were divided by Δx. Therefore, the term $O(\Delta x)$ signifies that this equation has error of "order (Δx)," which is due to the neglected terms in the Taylor series and is called truncation error. Hence, the forward difference is termed first-order accurate.

Thus, we can transform a partial derivative into an algebraic expression involving values of the variable at neighboring grid points. This method of using the Taylor series expansions to obtain discrete algebraic equations is called the finite difference method. Similar procedures can be used to develop approximations termed backward difference and central difference representations of the first derivative. The central difference makes use of both the left and right points (i.e., $i - 1, j$ and $i + 1, j$) and is second-order accurate. In addition, finite difference equations can be developed for the other spatial directions (i.e., $\partial u/\partial y$) as well as for second derivatives $(\partial^2 u/\partial x^2)$, which are also contained in the Navier–Stokes equations (see Ref. 5 for details).

Applying this method to all terms in the governing equations transfers the differential equations into a set of algebraic equations involving the physical variables at the grid points (i.e., $u_{i, j}, p_{i, j}$ for $i = 1, 2, 3, \ldots$ and $j = 1, 2, 3, \ldots$, etc.). This set of equations is then solved by appropriate numerical techniques. The larger the number or grid points used, the larger the number of equations that must be solved.

A student of CFD should realize that the discretization of the continuum governing equations involves the use of algebraic equations that are an approximation to the original partial differential equation. Along with this approximation comes some amount of error. This type of error is termed truncation error because the Taylor series expansion used to represent a derivative is "truncated" at some reasonable point and the higher order terms are ignored. The truncation errors tend to zero as the grid is refined by making Δx and Δy smaller, so grid refinement is one method of reducing this type of error. Another type of unavoidable numerical error is the so-called round-off error. This type of error is due to the limit of the computer on the number of digits it can retain in memory. Engineering students can run into round-off errors from their calculators if they plug values into the equations at an early stage of the solution process. Fortunately, for most CFD cases, if the algorithm is setup properly, round-off errors are usually negligible.

A.3   Grids

CFD computations using the finite difference method provide the flow field at discrete points in the flow domain. The arrangement of these discrete points is termed the *grid* or the *mesh*. The type of grid developed for a given problem can have a significant impact on the numerical simulation, including the accuracy of the solution. The grid must represent the geometry correctly and accurately, since an error in this representation can have a significant effect on the solution.

The grid must also have sufficient grid resolution to capture the relevant flow physics, otherwise they will be lost. This particular requirement is problem dependent. For example, if a flow field has small-scale structures, the grid resolution must be sufficient to capture these structures. It is usually necessary to increase the number of grid points (i.e., use a finer mesh) where large gradients are to be expected, such as in the boundary layer near a solid surface. The same can also be said for the temporal resolution. The time step, Δt, used for unsteady flows must be smaller than the smallest time scale of the flow features being investigated.

Generally, the types of grids fall into two categories: structured and unstructured, depending on whether or not there exists a systematic pattern of connectivity of the grid points with their neighbors. As the name implies, a *structured grid* has some type of regular, coherent structure to the mesh layout that can be defined mathematically. The simplest structured grid is a uniform rectangular grid, as shown in Fig. A.3a. However structured grids

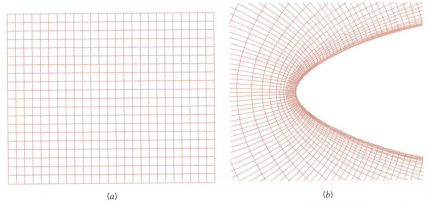

(a)   (b)

■ **FIGURE A.3**   **Structured grids. (*a*) Rectangular grid (*b*) Grid around a parabolic surface.**

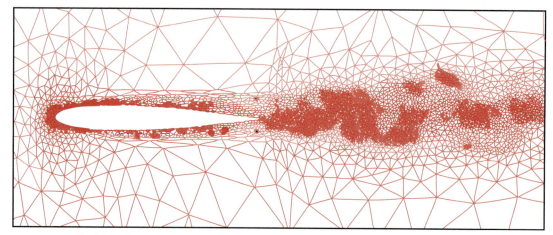

■ **FIGURE A.4**   **Anisotropic adaptive mesh for the calculation of viscous flow over a NACA 0012 airfoil at a Reynolds number of 10,000, Mach number of 0.755, and angle of attack of 1.5°. (From CFD Laboratory, Concordia University, Montreal, Canada. Used by permission.)**

are not restricted to rectangular geometries. Fig. A.3*b* shows a structured grid wrapped around a parabolic surface. Notice that grid points are clustered near the surface (i.e., grid spacing in normal direction increases as one moves away from the surface) to help capture the steep flow gradients found in the boundary layer region. This type of variable grid spacing is used wherever there is a need to increase grid resolution and is termed grid stretching.

For the *unstructured grid*, the grid cell arrangement is irregular and has no systematic pattern. The grid cell geometry usually consists of various-sized triangles for two-dimensional problems and tetrahedrals for three-dimensional grids. An example of an unstructured grid is shown in Fig. A.4. Unlike structured grids, for an unstructured grid each grid cell and the connection information to neighboring cells is defined separately. This produces an increase in the computer code complexity as well as a significant computer storage requirement. The advantage to an unstructured grid is that it can be applied to complex geometries, where structured grids would have severe difficulty. The finite difference method is restricted to structured grids whereas the finite volume (or finite element) method can use either structured or unstructured grids.

Other grids include hybrid, moving, and adaptive grids. A grid that uses a combination of grid elements (rectangles, triangles, etc.) is termed a *hybrid grid.* As the name implies, the *moving grid* is helpful for flows involving a time-dependent geometry. If, for example, the problem involves simulating the flow within a pumping heart or the flow around a flapping wing, a mesh that moves with the geometry is desired. The nature of the adaptive grid lies in its ability to literally adapt itself during the simulation. For this type of grid, while the CFD code is trying to reach a converged solution, the grid will adapt itself to place additional grid resources in regions of high flow gradients. Such a grid is particularly useful when a new problem arises and the user is not quite sure where to refine the grid due to high flow gradients.

A.4   Boundary Conditions

The same governing equations, the Navier–Stokes equations (Eq. 6.120), are valid for all incompressible Newtonian fluid flow problems. Thus, if the same equations are solved for all types of problems, how is it possible to achieve different solutions for different types of flows involving different flow geometries? The answer lies in the *boundary conditions* of the problem. The boundary conditions are what allow the governing equations to differentiate between different flow fields (for example, flow past an automobile and flow past a person running) and produce a solution unique to the given flow geometry.

It is critical to specify the correct boundary conditions so that the CFD simulation is a well-posed problem and is an accurate representation of the physical problem. Poorly defined boundary conditions can ultimately affect the accuracy of the solution. One of the most common boundary conditions used for simulation of viscous flow is the no-slip condition, as discussed in Section 1.6. Thus, for example, for two-dimensional external or internal flows, the x and y components of velocity (u and v) are set to zero at the stationary wall to satisfy the no-slip condition. Other boundary conditions that must be appropriately specified involve inlets, outlets, far-field, wall gradients, etc. It is important to not only select the correct physical boundary condition for the problem, but also to correctly implement this boundary condition into the numerical simulation.

A.5   Basic Representative Examples

A very simple one-dimensional example of the finite difference technique is presented in the following example.

EXAMPLE A.1   Flow from a Tank

GIVEN   A viscous oil flows from a large, open tank and through a long, small-diameter pipe as shown in Fig. EA.1a. At time $t = 0$ the fluid depth is H.

FIND   Use a finite difference technique to determine the liquid depth as a function of time, $h = h(t)$. Compare this result with the exact solution of the governing equation.

SOLUTION

Although this is an unsteady flow (i.e., the deeper the oil, the faster it flows from the tank) we assume that the flow is "quasi-steady" (see Example 3.18) and apply steady flow equations as follows.

As shown by Eq. 6.145, the mean velocity, V, for steady laminar flow in a round pipe of diameter D is given by

$$V = \frac{D^2 \Delta p}{32\mu\ell} \qquad (1)$$

where Δp is the pressure drop over the length ℓ. For this problem the pressure at the bottom of the tank (the inlet of the pipe) is γh and that at the pipe exit is zero. Hence, $\Delta p = \gamma h$ and Eq. 1 becomes

$$V = \frac{D^2 \gamma h}{32\mu\ell} \qquad (2)$$

Conservation of mass requires that the flowrate from the tank, $Q = \pi D^2 V/4$, is related to the rate of change of depth of oil in the tank, dh/dt, by

$$Q = -\frac{\pi}{4} D_T^2 \frac{dh}{dt}$$

where D_T is the tank diameter. Thus,

$$\frac{\pi}{4} D^2 V = -\frac{\pi}{4} D_T^2 \frac{dh}{dt}$$

or

$$V = -\left(\frac{D_T}{D}\right)^2 \frac{dh}{dt} \qquad (3)$$

By combining Eqs. 2 and 3 we obtain

$$\frac{D^2 \gamma h}{32\mu\ell} = -\left(\frac{D_T}{D}\right)^2 \frac{dh}{dt}$$

or

$$\frac{dh}{dt} = -Ch$$

where $C = \gamma D^4/32\mu\ell D_T^2$ is a constant. For simplicity we assume the conditions are such that $C = 1$. Thus, we must solve

$$\frac{dh}{dt} = -h \quad \text{with} \quad h = H \text{ at } t = 0 \qquad (4)$$

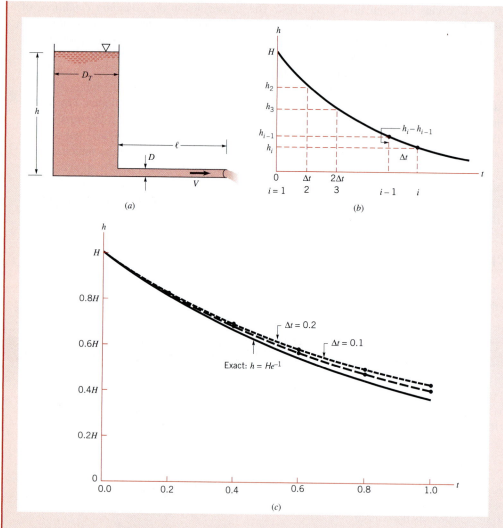

■ **FIGURE EA.1**

The exact solution to Eq. 4 is obtained by separating the variables and integrating to obtain

$$h = He^{-t} \qquad (5)$$

However, assume this solution were not known. The following finite difference technique can be used to obtain an approximate solution.

As shown in Fig. EA.1b, we select discrete points (nodes or grid points) in time and approximate the time derivative of h by the expression

$$\left.\frac{dh}{dt}\right|_{t=t_i} \approx \frac{h_i - h_{i-1}}{\Delta t} \qquad (6)$$

where Δt is the time step between the different node points on the time axis and h_i and h_{i-1} are the approximate values of h at nodes i and $i - 1$. Equation 6 is called the backward-difference approximation to dh/dt. We are free to select whatever value of Δt that we wish. (Although we do not need to space the nodes at equal distances, it is often convenient to do so.) Since the

governing equation (Eq. 4) is an ordinary differential equation, the "grid" for the finite difference method is a one-dimensional grid as shown in Fig. EA.1b rather than a two-dimensional grid (which occurs for partial differential equations) as shown in Fig. EA.2b, or a three-dimensional grid.

Thus, for each value of $i = 2, 3, 4, \ldots$ we can approximate the governing equation, Eq. 4, as

$$\frac{h_i - h_{i-1}}{\Delta t} = -h_i$$

or

$$h_i = \frac{h_{i-1}}{(1 + \Delta t)} \qquad (7)$$

We cannot use Eq. 7 for $i = 1$ since it would involve the nonexisting h_0. Rather we use the initial condition (Eq. 4), which gives

$$h_1 = H$$

The result is the following set of N algebraic equations for the N approximate values of h at times $t_1 = 0$, $t_2 = \Delta t, \ldots, t_N = (N - 1)\Delta t$.

$$h_1 = H$$
$$h_2 = h_1/(1 + \Delta t)$$
$$h_3 = h_2/(1 + \Delta t)$$
$$\vdots \qquad \vdots$$
$$h_N = h_{N-1}/(1 + \Delta t)$$

For most problems the corresponding equations would be more complicated than those just given, and a computer would be used to solve for the h_i. For this problem the solution is simply

$$h_2 = H/(1 + \Delta t)$$
$$h_3 = H/(1 + \Delta t)^2$$
$$\vdots \qquad \vdots$$

or in general

$$h_i = H/(1 + \Delta t)^{i-1}$$

The results for $0 < t < 1$ are shown in Fig. EA.1c. Tabulated values of the depth for $t = 1$ are listed in the table below.

Δt	i for $t = 1$	h_i for $t = 1$
0.2	6	0.4019H
0.1	11	0.3855H
0.01	101	0.3697H
0.001	1001	0.3681H
Exact (Eq. 5)	—	0.3678H

It is seen that the approximate results compare quite favorably with the exact solution given by Eq. 5. It is expected that the finite difference results would more closely approximate the exact results as Δt is decreased since in the limit of $\Delta t \to 0$ the finite difference approximation for the derivatives (Eq. 6) approaches the actual definition of the derivative.

For most CFD problems the governing equations to be solved are partial differential equations [rather than an ordinary differential equation as in the above example (Eq. A.1)] and the finite difference method becomes considerably more involved. The following example illustrates some of the concepts involved.

EXAMPLE A.2   Flow Past a Cylinder

GIVEN Consider steady, incompressible flow of an inviscid fluid past a circular cylinder as shown in Fig. EA.2a. The stream function, ψ, for this flow is governed by the Laplace equation (see Section 6.5)

$$\frac{\partial^2 \psi}{\partial x^2} + \frac{\partial^2 \psi}{\partial y^2} = 0 \tag{1}$$

FIND Describe a simple finite difference technique that can be used to solve this problem.

The exact analytical solution is given in Section 6.6.2.

SOLUTION

The first step is to define a flow domain and set up an appropriate grid for the finite difference scheme. Since we expect the flow field to be symmetrical both above and below and in front of and behind the cylinder, we consider only one-quarter of the entire flow domain as indicated in Fig. EA.2b. We locate the upper boundary and right-hand boundary far enough from the cylinder so that we expect the flow to be essentially uniform at these locations. It is not always clear how far from the object these boundaries must be located. If they are not far enough, the solution obtained will be incorrect because we have imposed artificial, uniform flow conditions at a location where the actual flow is not uniform. If these boundaries are farther than necessary from the object, the flow domain will be larger than necessary and excessive computer time and storage will be required. Experience in solving such problems is invaluable!

Once the flow domain has been selected, an appropriate grid is imposed on this domain (see Fig. EA.2b). Various grid structures can be used. If the grid is too coarse, the numerical solution may not be capable of capturing the fine scale structure of the actual flow field. If the grid is too fine, excessive computer time and storage may be required. Considerable work has gone into forming appropriate grids (Ref. 6). We consider a grid that is uniformly spaced in the x and y directions, as shown in Fig. EA.2b.

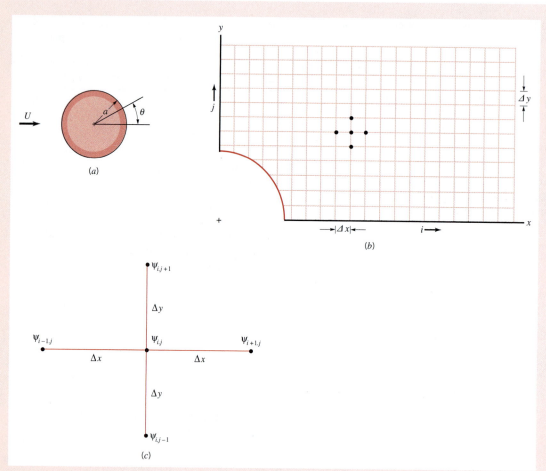

■ **FIGURE EA.2**

As shown in Eq. 6.105, the exact solution to Eq. 1 (in terms of polar coordinates r, θ rather than Cartesian coordinates x, y) is $\psi = Ur(1 - a^2/r^2) \sin \theta$. The finite difference solution approximates these stream function values at a discrete (finite) number of locations (the grid points) as $\psi_{i,j}$, where the i and j indices refer to the corresponding x_i and y_j locations.

The derivatives of ψ can be approximated as follows:

$$\frac{\partial \psi}{\partial x} \approx \frac{1}{\Delta x}(\psi_{i+1,j} - \psi_{i,j})$$

and

$$\frac{\partial \psi}{\partial y} \approx \frac{1}{\Delta y}(\psi_{i,j+1} - \psi_{i,j})$$

This particular approximation is called a forward-difference approximation. Other approximations are possible. By similar reasoning, it is possible to show that the second derivatives of ψ can be written as follows:

$$\frac{\partial^2 \psi}{\partial x^2} \approx \frac{1}{(\Delta x)^2}(\psi_{i+1,j} - 2\psi_{i,j} + \psi_{i-1,j}) \tag{2}$$

and

$$\frac{\partial^2 \psi}{\partial y^2} \approx \frac{1}{(\Delta y)^2}(\psi_{i,j+1} - 2\psi_{i,j} + \psi_{i,j-1}) \tag{3}$$

Thus, by combining Eqs. 1, 2, and 3 we obtain

$$\frac{\partial^2 \psi}{\partial x^2} + \frac{\partial^2 \psi}{\partial y^2} \approx \frac{1}{(\Delta x)^2}(\psi_{i+1,j} + \psi_{i-1,j}) + \frac{1}{(\Delta y)^2}(\psi_{i,j+1}$$
$$+ \psi_{i,j-1}) - 2\left(\frac{1}{(\Delta x)^2} + \frac{1}{(\Delta y)^2}\right)\psi_{i,j} = 0 \tag{4}$$

Equation 4 can be solved for the stream function at x_i and y_j to give

$$\psi_{i,j} = \frac{1}{2[(\Delta x)^2 + (\Delta y)^2]}\left[(\Delta y)^2(\psi_{i+1,j} + \psi_{i-1,j})\right.$$
$$\left. + (\Delta x)^2(\psi_{i,j+1} + \psi_{i,j-1})\right] \tag{5}$$

Note that the value of $\psi_{i,j}$ depends on the values of the stream function at neighboring grid points on either side and above and below the point of interest (see Eq. 5 and Fig. EA. 2c).

To solve the problem (either exactly or by the finite difference technique) it is necessary to specify boundary conditions for points located on the boundary of the flow domain (see Section 6.6.3). For example, we may specify that $\psi = 0$ on the lower boundary of the domain (see Fig. EA.2b) and $\psi = C$, a constant, on the upper boundary of the domain. Appropriate boundary conditions on the two vertical ends of the flow domain can also be specified. Thus, for points interior to the boundary Eq. 5 is valid; similar equations or specified values of $\psi_{i,j}$ are valid for boundary points. The result is an equal number of equations and unknowns, $\psi_{i,j}$, one for every grid point. For this problem, these equations represent a set of linear algebraic equations for $\psi_{i,j}$, the solution of which provides the finite difference approximation for the stream function at discrete grid points in the flow field. Streamlines (lines of constant ψ) can be obtained by interpolating values of $\psi_{i,j}$ between the grid points and "connecting the dots" of $\psi = $ constant. The velocity field can be obtained from the derivatives of the stream function according to Eq. 6.74. That is,

$$u = \frac{\partial \psi}{\partial y} \approx \frac{1}{\Delta y}\left(\psi_{i,j+1} - \psi_{i,j}\right)$$

and

$$v = -\frac{\partial \psi}{\partial x} \approx -\frac{1}{\Delta x}\left(\psi_{i+1,j} - \psi_{i,j}\right)$$

Further details of the finite difference technique can be found in standard references on the topic (Refs. 5, 7, 8). Also, see the completely solved viscous flow CFD problem in Section A6.

The preceding two examples are rather simple because the governing equations are not too complex. A finite difference solution of the more complicated, nonlinear Navier–Stokes equation (Eq. 6.158) requires considerably more effort and insight and larger and faster computers. A typical finite difference grid for a more complex flow, the flow past a turbine blade, is shown in Fig. A.5. Note that the mesh is much finer in regions where large gradients are to be expected (i.e., near the leading and trailing edges of the blade) and more coarse away from the blade.

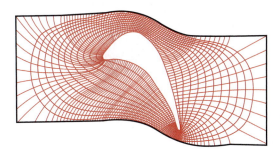

■ **FIGURE A.5** Finite difference grid for flow past a turbine blade. (From Ref. 9, used by permission.)

A.6 Methodology

In general, most applications of CFD take the same basic approach. Some of the differences include problem complexity, available computer resources, available expertise in CFD, and whether a commercially available CFD package is used, or a problem-specific CFD algorithm is developed. In today's market, there are many commercial CFD codes available to solve a wide variety of problems. However, if the intent is to conduct a thorough investigation of a specific fluid flow problem such as in a research environment, it is possible that taking the time to develop a problem-specific algorithm may be most efficient in the long run. The features common to most CFD applications can be summarized in the flowchart shown in Fig. A.6. A complete, detailed CFD solution for a viscous flow obtained by using the steps summarized in the flow chart can be accessed by visiting the book website.

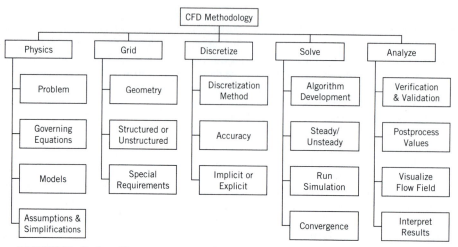

■ **FIGURE A.6** Flow chart of general CFD methodology.

The Algorithm Development box is required only when developing your own CFD code. When using a commercial CFD code, this step is not necessary. This chart represents a generalized methodology to CFD. There are other more complex components that are hidden in the above steps, which are beyond the scope of a brief introduction to CFD.

A.7   Application of CFD

In the early stages of CFD, research and development was primarily driven by the aerospace industry. Today, CFD is still used as a research tool, but it also has found a place in industry as a design tool. There is now a wide variety of industries that make at least some use of CFD, including automotive, industrial, HVAC, naval, civil, chemical, biological, and others. Industries are using CFD as an added engineering tool that complements the experimental and theoretical work in fluid dynamics.

A.7.1   Advantages of CFD

There are many advantages to using CFD for simulation of fluid flow. One of the most important advantages is the realizable savings in time and cost for engineering design. In the past, coming up with a new engineering design meant somewhat of a trial-and-error method of building and testing multiple prototypes prior to finalizing the design. With CFD, many of the issues dealing with fluid flow can be flushed out prior to building the actual prototype. This translates to a significant savings in time and cost. It should be noted that CFD is not meant to replace experimental testing, but rather to work in conjunction with it. Experimental testing will always be a necessary component of engineering design. Other advantages include the ability of CFD to: (1) obtain flow information in regions that would be difficult to test experimentally, (2) simulate real flow conditions, (3) conduct large parametric tests on new designs in a shorter time, and (4) enhance visualization of complex flow phenomena.

A good example of the advantages of CFD is shown in Figure A.7. Researchers use a type of CFD approach called "large-eddy simulation" or LES to simulate the fluid dynamics

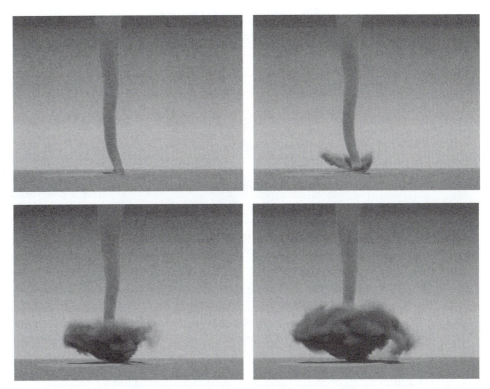

■ FIGURE A.7   Results from a large-eddy simulation showing the visual appearance of the debris and funnel cloud from a simulated medium swirl F3-F4 tornado. The funnel cloud is translating at 15 m/s and is ingesting 1 mm diameter "sand" from the surface as it encounters a debris field. Please visit the book website to access a full animation of this tornado simulation. (Photographs and animation courtesy of Dr. David Lewellen, Ref. 10, and Paul Lewellen, West Virginia University)

of a tornado as it encounters a debris field and begins to pick up sand-sized particles. A full animation of this tornado simulation can be accessed by visiting the book website. The motivation for this work is to investigate whether there are significant differences in the fluid mechanics when debris particles are present. Historically it has been difficult to get comprehensive experimental data throughout a tornado so CFD is helping to shine some light on the complex fluid dynamics involved in such a flow.

A.7.2   Difficulties in CFD

One of the key points that a beginning CFD student should understand is that one cannot treat the computer as a "magic black box" when performing flow simulations. It is quite possible to obtain a fully converged solution for the CFD simulation, but this is no guarantee that the results are physically correct. This is why it is important to have a good understanding of the flow physics and how they are modeled. Any numerical technique (including those discussed above), no matter how simple in concept, contains many hidden subtleties and potential problems. For example, it may seem reasonable that a finer grid would ensure a more accurate numerical solution. While this may be true (as Example A.1), it is not always so straightforward; a variety of stability or convergence problems may occur. In such cases the numerical "solution" obtained may exhibit unreasonable oscillations or the numerical result may "diverge" to an unreasonable (and incorrect) result. Other problems that may arise

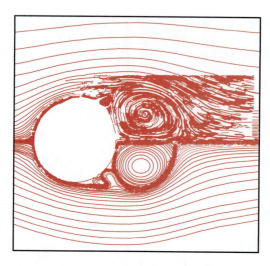

■ **FIGURE A.8**   **Streamlines for flow past a circular cylinder at a short time after the flow was impulsively started. The upper half is a photograph from a flow visualization experiment. The lower half is from a finite difference calculation. (See the photograph at the beginning of Chapter 9.) (From Ref. 9, used by permission.)**

include (but are not limited to): (1) difficulties in dealing with the nonlinear terms of the Navier–Stokes equations, (2) difficulties in modeling or capturing turbulent flows, (3) convergence issues, (4) difficulties in obtaining a quality grid for complex geometries, and (5) managing resources, both time and computational, for complex problems such as unsteady three-dimensional flows.

A.7.3   Verification and Validation

Verification and validation of the simulation are critical steps in the CFD process. This is a necessary requirement for CFD, particularly since it is possible to have a converged solution that is nonphysical. Figure A.8 shows the streamlines for viscous flow past a circular cylinder at a given instant after it was impulsively started from rest. The lower half of the figure represents the results of a finite difference calculation; the upper half of the figure represents the photograph from an experiment of the same flow situation. It is clear that the numerical and experimental results agree quite well. For any CFD simulation, there are several levels of testing that need to be accomplished before one can have confidence in the solution. The most important verification to be performed is grid convergence testing. In its simplest form, it consists of proving that further refinement of the grid (i.e., increasing the number of grid points) does not alter the final solution. When this has been achieved, the solution is termed a grid-independent solution. Other verification factors that need to be investigated include the suitability of the convergence criterion, whether the time step is adequate for the time scale of the problem, and comparison of CFD solutions to existing data, at least for baseline cases. Even when using a commercial CFD code that has been validated on many problems in the past, the CFD practitioner still needs to verify the results through such measures as grid-dependence testing.

A.7.4   Summary

In CFD, there are many different numerical schemes, grid techniques, etc. They all have their advantages and disadvantages. A great deal of care must be used in obtaining approximate numerical solutions to the governing equations of fluid motion. The process is not as simple as the often-heard "just let the computer do it." Remember that CFD is a tool and as such needs to be used appropriately to produce meaningful results. The general field of computational fluid dynamics, in which computers and numerical analysis are combined to solve fluid flow problems, represents an extremely important subject area in advanced fluid mechanics. Considerable progress has been made in the past relatively few years, but much remains to be done. The reader is encouraged to consult some of the available literature.

A.8   FlowLab

The authors of this textbook are working in collaboration with Fluent, Inc., the largest provider of commercial CFD software (www.fluent.com), to offer students the opportunity to use a new CFD tool called FlowLab. FlowLab is designed to be a virtual fluids laboratory to help enhance the educational experience in fluids courses. It uses computational fluid dynamics to help the student grasp various concepts in fluid dynamics and introduces the student to the use of CFD in solving fluid flow problems. Please visit the book website to access FlowLab resources for this textbook.

The motivation behind incorporating FlowLab with a fundamental fluid mechanics textbook is twofold: (1) expose the student to computational fluid dynamics and (2) offer a mechanism for students to conduct experiments in fluid dynamics, numerically in this case. This educational software allows students to reinforce basic concepts covered in class, conduct parametric studies to gain a better understanding of the interaction between geometry, fluid properties, and flow conditions, and provides the student a visualization tool for various flow phenomena.

One of the strengths of FlowLab is the ease-of-use. The CFD simulations are based on previously developed templates which allow the user to start using CFD to solve flow problems without requiring an extensive background in the subject. FlowLab provides the student the opportunity to focus on the results of the simulation rather than the development of the simulation. Typical results showing the developing velocity profile in the entrance region of a pipe are shown in the solution window of Fig. A.9.

New problems have been developed that take advantage of the FlowLab capability of this textbook. Please visit the book website to access these problems (contained in Chapters 7, 8, and 9) as well as a basic tutorial on using FlowLab. The course instructor can provide information on accessing the FlowLab software. Please visit the book website to go to a brief example using FlowLab.

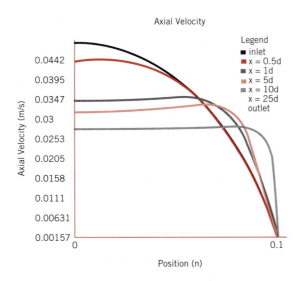

■ **FIGURE A.9** **Entrance flow in a pipe. Velocity profiles as a function of radial position for various locations along the pipe length.**

References

1. Baker, A. J., *Finite Element Computational Fluid Mechanics*, McGraw-Hill, New York, 1983.

2. Carey, G. F., and Oden, J. T., *Finite Elements: Fluid Mechanics*, Prentice-Hall, Englewood Cliffs, N.J., 1986.

3. Brebbia, C. A., and Dominguez, J., *Boundary Elements: An Introductory Course*, McGraw-Hill, New York, 1989.

4. Moran, J., *An Introduction to Theoretical and Computational Aerodynamics*, Wiley, New York, 1984.

5. Anderson, J.D., *Computational Fluid Dynamics: The Basics with Applications*, McGraw-Hill, New York, 1995.

6. Thompson, J. F., Warsi, Z. U. A., and Mastin, C. W., *Numerical Grid Generation: Foundations and Applications*, North-Holland, New York, 1985.

7. Peyret, R., and Taylor, T. D., *Computational Methods for Fluid Flow*, Springer-Verlag, New York, 1983.

8. Tannehill, J. C., Anderson, D. A., and Pletcher, R. H., *Computational Fluid Mechanics and Heat Transfer*, 2nd Ed., Taylor and Francis, Washington, D.C., 1997.

9. Hall, E. J., and Pletcher, R. H., *Simulation of Time Dependent, Compressible Viscous Flow Using Central and Upwind-Biased Finite-Difference Techniques*, Technical Report HTL-52, CFD-22, College of Engineering, Iowa State University, 1990.

10. Lewellen, D. C., Gong, B., and Lewellen, W. S., *Effects of Debris on Near-Surface Tornado Dynamics*, 22nd Conference on Severe Local Storms, Paper 15.5, American Meteorological Society, 2004.

APPENDIX B

Physical Properties of Fluids

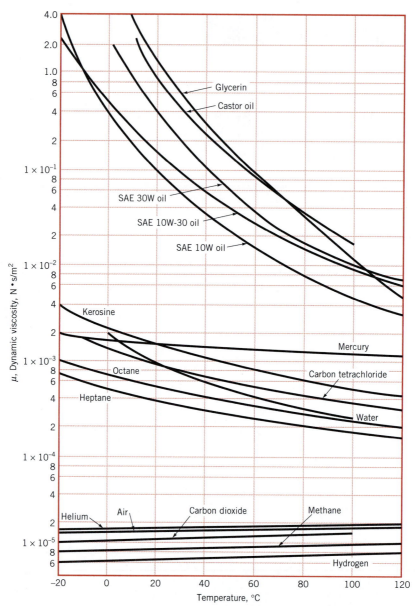

■ **FIGURE B.1** Dynamic (absolute) viscosity of common fluids as a function of temperature. To convert to BG units of lb·s/ft^2 multiply N·s/m^2 by 2.089×10^{-2}. Curves from R. W. Fox and A. T. McDonald, *Introduction to Fluid Mechanics,* Third Edition, Wiley, New York, 1985. Used with permission.

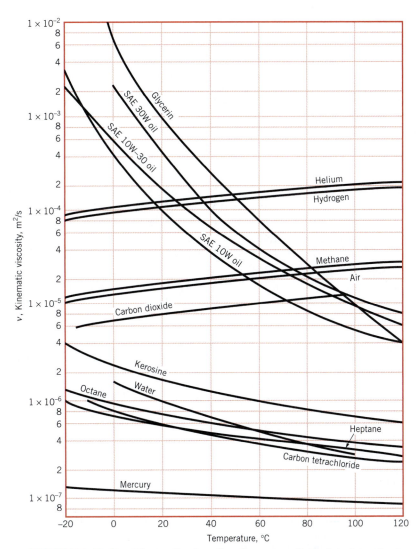

■ **FIGURE B.2** **Kinematic viscosity of common fluids (at atmospheric pressure) as a function of temperature. To convert to BG units of ft²/s multiply m²/s by 10.76. Curves from R. W. Fox and A. T. McDonald,** *Introduction to Fluid Mechanics,* **Third Edition, Wiley, New York, 1985. Used with permission.**

■ **TABLE B.1**
Physical Properties of Water (BG Units)[a]

Temperature (°F)	Density, ρ (slugs/ft³)	Specific Weight,[b] γ (lb/ft³)	Dynamic Viscosity, μ (lb·s/ft²)	Kinematic Viscosity, ν (ft²/s)	Surface Tension,[c] σ (lb/ft)	Vapor Pressure, p_v [lb/in.²(abs)]	Speed of Sound,[d] c (ft/s)
32	1.940	62.42	3.732 E − 5	1.924 E − 5	5.18 E − 3	8.854 E − 2	4603
40	1.940	62.43	3.228 E − 5	1.664 E − 5	5.13 E − 3	1.217 E − 1	4672
50	1.940	62.41	2.730 E − 5	1.407 E − 5	5.09 E − 3	1.781 E − 1	4748
60	1.938	62.37	2.344 E − 5	1.210 E − 5	5.03 E − 3	2.563 E − 1	4814
70	1.936	62.30	2.037 E − 5	1.052 E − 5	4.97 E − 3	3.631 E − 1	4871
80	1.934	62.22	1.791 E − 5	9.262 E − 6	4.91 E − 3	5.069 E − 1	4819
90	1.931	62.11	1.500 E − 5	8.233 E − 6	4.86 E − 3	6.979 E − 1	4960
100	1.927	62.00	1.423 E − 5	7.383 E − 6	4.79 E − 3	9.493 E − 1	4995
120	1.918	61.71	1.164 E − 5	6.067 E − 6	4.67 E − 3	1.692 E + 0	5049
140	1.908	61.38	9.743 E − 6	5.106 E − 6	4.53 E − 3	2.888 E + 0	5091
160	1.896	61.00	8.315 E − 6	4.385 E − 6	4.40 E − 3	4.736 E + 0	5101
180	1.883	60.58	7.207 E − 6	3.827 E − 6	4.26 E − 3	7.507 E + 0	5195
200	1.869	60.12	6.342 E − 6	3.393 E − 6	4.12 E − 3	1.152 E + 1	5089
212	1.860	59.83	5.886 E − 6	3.165 E − 6	4.04 E − 3	1.469 E + 1	5062

[a]Based on data from *Handbook of Chemistry and Physics,* 69th Ed., CRC Press, 1988. Where necessary, values obtained by interpolation.
[b]Density and specific weight are related through the equation $\gamma = \rho g$. For this table, g = 32.174 ft/s².
[c]In contact with air.
[d]From R. D. Blevins, *Applied Fluid Dynamics Handbook,* Van Nostrand Reinhold Co., Inc., New York, 1984.

■ **TABLE B.2**
Physical Properties of Water (SI Units)[a]

Temperature (°C)	Density, ρ (kg/m³)	Specific Weight,[b] γ (kN/m³)	Dynamic Viscosity, μ (N·s/m²)	Kinematic Viscosity, ν (m²/s)	Surface Tension,[c] σ (N/m)	Vapor Pressure, p_v [N/m²(abs)]	Speed of Sound,[d] c (m/s)
0	999.9	9.806	1.787 E − 3	1.787 E − 6	7.56 E − 2	6.105 E + 2	1403
5	1000.0	9.807	1.519 E − 3	1.519 E − 6	7.49 E − 2	8.722 E + 2	1427
10	999.7	9.804	1.307 E − 3	1.307 E − 6	7.42 E − 2	1.228 E + 3	1447
20	998.2	9.789	1.002 E − 3	1.004 E − 6	7.28 E − 2	2.338 E + 3	1481
30	995.7	9.765	7.975 E − 4	8.009 E − 7	7.12 E − 2	4.243 E + 3	1507
40	992.2	9.731	6.529 E − 4	6.580 E − 7	6.96 E − 2	7.376 E + 3	1526
50	988.1	9.690	5.468 E − 4	5.534 E − 7	6.79 E − 2	1.233 E + 4	1541
60	983.2	9.642	4.665 E − 4	4.745 E − 7	6.62 E − 2	1.992 E + 4	1552
70	977.8	9.589	4.042 E − 4	4.134 E − 7	6.44 E − 2	3.116 E + 4	1555
80	971.8	9.530	3.547 E − 4	3.650 E − 7	6.26 E − 2	4.734 E + 4	1555
90	965.3	9.467	3.147 E − 4	3.260 E − 7	6.08 E − 2	7.010 E + 4	1550
100	958.4	9.399	2.818 E − 4	2.940 E − 7	5.89 E − 2	1.013 E + 5	1543

[a]Based on data from *Handbook of Chemistry and Physics,* 69th Ed., CRC Press, 1988.
[b]Density and specific weight are related through the equation $\gamma = \rho g$. For this table, g = 9.807 m/s².
[c]In contact with air.
[d]From R. D. Blevins, *Applied Fluid Dynamics Handbook,* Van Nostrand Reinhold Co., Inc., New York, 1984.

■ **TABLE B.3**
Physical Properties of Air at Standard Atmospheric Pressure (BG Units)[a]

Temperature (°F)	Density, ρ (slugs/ft³)		Specific Weight,[b] γ (lb/ft³)		Dynamic Viscosity, μ (lb·s/ft²)		Kinematic Viscosity, ν (ft²/s)		Specific Heat Ratio, k (–)	Speed of Sound, c (ft/s)
−40	2.939	E − 3	9.456	E − 2	3.29	E − 7	1.12	E − 4	1.401	1004
−20	2.805	E − 3	9.026	E − 2	3.34	E − 7	1.19	E − 4	1.401	1028
0	2.683	E − 3	8.633	E − 2	3.38	E − 7	1.26	E − 4	1.401	1051
10	2.626	E − 3	8.449	E − 2	3.44	E − 7	1.31	E − 4	1.401	1062
20	2.571	E − 3	8.273	E − 2	3.50	E − 7	1.36	E − 4	1.401	1074
30	2.519	E − 3	8.104	E − 2	3.58	E − 7	1.42	E − 4	1.401	1085
40	2.469	E − 3	7.942	E − 2	3.60	E − 7	1.46	E − 4	1.401	1096
50	2.420	E − 3	7.786	E − 2	3.68	E − 7	1.52	E − 4	1.401	1106
60	2.373	E − 3	7.636	E − 2	3.75	E − 7	1.58	E − 4	1.401	1117
70	2.329	E − 3	7.492	E − 2	3.82	E − 7	1.64	E − 4	1.401	1128
80	2.286	E − 3	7.353	E − 2	3.86	E − 7	1.69	E − 4	1.400	1138
90	2.244	E − 3	7.219	E − 2	3.90	E − 7	1.74	E − 4	1.400	1149
100	2.204	E − 3	7.090	E − 2	3.94	E − 7	1.79	E − 4	1.400	1159
120	2.128	E − 3	6.846	E − 2	4.02	E − 7	1.89	E − 4	1.400	1180
140	2.057	E − 3	6.617	E − 2	4.13	E − 7	2.01	E − 4	1.399	1200
160	1.990	E − 3	6.404	E − 2	4.22	E − 7	2.12	E − 4	1.399	1220
180	1.928	E − 3	6.204	E − 2	4.34	E − 7	2.25	E − 4	1.399	1239
200	1.870	E − 3	6.016	E − 2	4.49	E − 7	2.40	E − 4	1.398	1258
300	1.624	E − 3	5.224	E − 2	4.97	E − 7	3.06	E − 4	1.394	1348
400	1.435	E − 3	4.616	E − 2	5.24	E − 7	3.65	E − 4	1.389	1431
500	1.285	E − 3	4.135	E − 2	5.80	E − 7	4.51	E − 4	1.383	1509
750	1.020	E − 3	3.280	E − 2	6.81	E − 7	6.68	E − 4	1.367	1685
1000	8.445	E − 4	2.717	E − 2	7.85	E − 7	9.30	E − 4	1.351	1839
1500	6.291	E − 4	2.024	E − 2	9.50	E − 7	1.51	E − 3	1.329	2114

[a]Based on data from R. D. Blevins, *Applied Fluid Dynamics Handbook,* Van Nostrand Reinhold Co., Inc., New York, 1984.
[b]Density and specific weight are related through the equation $\gamma = \rho g$. For this table $g = 32.174$ ft/s².

■ **TABLE B.4**

Physical Properties of Air at Standard Atmospheric Pressure (SI Units)[a]

Temperature (°C)	Density, ρ (kg/m³)	Specific Weight,[b] γ (N/m³)	Dynamic Viscosity, μ (N·s/m²)	Kinematic Viscosity, ν (m²/s)	Specific Heat Ratio, k (–)	Speed of Sound, c (m/s)
−40	1.514	14.85	1.57 E − 5	1.04 E − 5	1.401	306.2
−20	1.395	13.68	1.63 E − 5	1.17 E − 5	1.401	319.1
0	1.292	12.67	1.71 E − 5	1.32 E − 5	1.401	331.4
5	1.269	12.45	1.73 E − 5	1.36 E − 5	1.401	334.4
10	1.247	12.23	1.76 E − 5	1.41 E − 5	1.401	337.4
15	1.225	12.01	1.80 E − 5	1.47 E − 5	1.401	340.4
20	1.204	11.81	1.82 E − 5	1.51 E − 5	1.401	343.3
25	1.184	11.61	1.85 E − 5	1.56 E − 5	1.401	346.3
30	1.165	11.43	1.86 E − 5	1.60 E − 5	1.400	349.1
40	1.127	11.05	1.87 E − 5	1.66 E − 5	1.400	354.7
50	1.109	10.88	1.95 E − 5	1.76 E − 5	1.400	360.3
60	1.060	10.40	1.97 E − 5	1.86 E − 5	1.399	365.7
70	1.029	10.09	2.03 E − 5	1.97 E − 5	1.399	371.2
80	0.9996	9.803	2.07 E − 5	2.07 E − 5	1.399	376.6
90	0.9721	9.533	2.14 E − 5	2.20 E − 5	1.398	381.7
100	0.9461	9.278	2.17 E − 5	2.29 E − 5	1.397	386.9
200	0.7461	7.317	2.53 E − 5	3.39 E − 5	1.390	434.5
300	0.6159	6.040	2.98 E − 5	4.84 E − 5	1.379	476.3
400	0.5243	5.142	3.32 E − 5	6.34 E − 5	1.368	514.1
500	0.4565	4.477	3.64 E − 5	7.97 E − 5	1.357	548.8
1000	0.2772	2.719	5.04 E − 5	1.82 E − 4	1.321	694.8

[a]Based on data from R. D. Blevins, *Applied Fluid Dynamics Handbook,* Van Nostrand Reinhold Co., Inc., New York, 1984.

[b]Density and specific weight are related through the equation $\gamma = \rho g$. For this table $g = 9.807$ m/s².

APPENDIX C

Properties of the U.S. Standard Atmosphere

■ TABLE C.1

Properties of the U.S. Standard Atmosphere (BG Units)[a]

Altitude (ft)	Temperature (°F)	Acceleration of Gravity, g (ft/s^2)	Pressure, p [lb/in.^{2}(abs)]	Density, ρ (slugs/ft^3)		Dynamic Viscosity, μ (lb·s/ft^2)	
−5,000	76.84	32.189	17.554	2.745	E − 3	3.836	E − 7
0	59.00	32.174	14.696	2.377	E − 3	3.737	E − 7
5,000	47.17	32.159	12.228	2.048	E − 3	3.637	E − 7
10,000	23.36	32.143	10.108	1.756	E − 3	3.534	E − 7
15,000	5.55	32.128	8.297	1.496	E − 3	3.430	E − 7
20,000	−12.26	32.112	6.759	1.267	E − 3	3.324	E − 7
25,000	−30.05	32.097	5.461	1.066	E − 3	3.217	E − 7
30,000	−47.83	32.082	4.373	8.907	E − 4	3.107	E − 7
35,000	−65.61	32.066	3.468	7.382	E − 4	2.995	E − 7
40,000	−69.70	32.051	2.730	5.873	E − 4	2.969	E − 7
45,000	−69.70	32.036	2.149	4.623	E − 4	2.969	E − 7
50,000	−69.70	32.020	1.692	3.639	E − 4	2.969	E − 7
60,000	−69.70	31.990	1.049	2.256	E − 4	2.969	E − 7
70,000	−67.42	31.959	0.651	1.392	E − 4	2.984	E − 7
80,000	−61.98	31.929	0.406	8.571	E − 5	3.018	E − 7
90,000	−56.54	31.897	0.255	5.610	E − 5	3.052	E − 7
100,000	−51.10	31.868	0.162	3.318	E − 5	3.087	E − 7
150,000	19.40	31.717	0.020	3.658	E − 6	3.511	E − 7
200,000	−19.78	31.566	0.003	5.328	E − 7	3.279	E − 7
250,000	−88.77	31.415	0.000	6.458	E − 8	2.846	E − 7

[a]Data abridged from *U.S. Standard Atmosphere,* 1976, U.S. Government Printing Office, Washington, D.C.

■ **TABLE C.2**
Properties of the U.S. Standard Atmosphere (SI Units)[a]

Altitude (m)	Temperature (°C)	Acceleration of Gravity, g (m/s^2)	Pressure, p [N/m^2(abs)]		Density, ρ (kg/m^3)		Dynamic Viscosity, μ (N·s/m^2)	
−1,000	21.50	9.810	1.139	E + 5	1.347	E + 0	1.821	E − 5
0	15.00	9.807	1.013	E + 5	1.225	E + 0	1.789	E − 5
1,000	8.50	9.804	8.988	E + 4	1.112	E + 0	1.758	E − 5
2,000	2.00	9.801	7.950	E + 4	1.007	E + 0	1.726	E − 5
3,000	−4.49	9.797	7.012	E + 4	9.093	E − 1	1.694	E − 5
4,000	−10.98	9.794	6.166	E + 4	8.194	E − 1	1.661	E − 5
5,000	−17.47	9.791	5.405	E + 4	7.364	E − 1	1.628	E − 5
6,000	−23.96	9.788	4.722	E + 4	6.601	E − 1	1.595	E − 5
7,000	−30.45	9.785	4.111	E + 4	5.900	E − 1	1.561	E − 5
8,000	−36.94	9.782	3.565	E + 4	5.258	E − 1	1.527	E − 5
9,000	−43.42	9.779	3.080	E + 4	4.671	E − 1	1.493	E − 5
10,000	−49.90	9.776	2.650	E + 4	4.135	E − 1	1.458	E − 5
15,000	−56.50	9.761	1.211	E + 4	1.948	E − 1	1.422	E − 5
20,000	−56.50	9.745	5.529	E + 3	8.891	E − 2	1.422	E − 5
25,000	−51.60	9.730	2.549	E + 3	4.008	E − 2	1.448	E − 5
30,000	−46.64	9.715	1.197	E + 3	1.841	E − 2	1.475	E − 5
40,000	−22.80	9.684	2.871	E + 2	3.996	E − 3	1.601	E − 5
50,000	−2.50	9.654	7.978	E + 1	1.027	E − 3	1.704	E − 5
60,000	−26.13	9.624	2.196	E + 1	3.097	E − 4	1.584	E − 5
70,000	−53.57	9.594	5.221	E + 0	8.283	E − 5	1.438	E − 5
80,000	−74.51	9.564	1.052	E + 0	1.846	E − 5	1.321	E − 5

[a]Data abridged from *U.S. Standard Atmosphere,* 1975, U.S. Government Printing Office, Washington, D.C.

Reynolds Transport Theorem

D.1 General Reynolds Transport Theorem

In Section 4.4.1 the Reynolds transport theorem (Eq. 4.13) was obtained under the restrictive assumptions of a fixed control volume having uniform properties across the inlets and outlets with the velocity normal to the inlet and outlet areas. This appendix develops a form of the Reynolds transport theorem that is valid for more general flow conditions.

A general, fixed control volume with fluid flowing through it is shown in Fig. D.1. We consider the system to be the fluid within the control volume at the initial time t. A short time later a portion of the fluid (region II) has exited from the control volume and additional fluid (region I, not part of the original system) has entered the control volume.

We consider an extensive fluid property B and seek to determine how the rate of change of B associated with the system is related to the rate of change of B within the control volume at any instant. By repeating the exact steps that we did for the simplified control volume shown in Fig. 4.8, we see that Eq. 4.12 is also valid for the general case, provided that we give the correct interpretation to the terms $\dot{B}_{out}$ and $\dot{B}_{in}$. Thus,

$$\frac{DB_{sys}}{Dt} = \frac{\partial B_{cv}}{\partial t} + \dot{B}_{out} - \dot{B}_{in} \tag{D.1}$$

The term $\dot{B}_{out}$ represents the net flowrate of the property B from the control volume. Its value can be thought of as arising from the addition (integration) of the contributions through each infinitesimal area element of size δA on the portion of the control surface

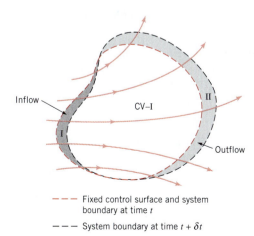

Inflow

CV–I

II

I

Outflow

- - - - Fixed control surface and system
boundary at time t

- - - System boundary at time $t + \delta t$

■ **FIGURE D.1** Control volume and system for flow through an arbitrary, fixed control volume.

477

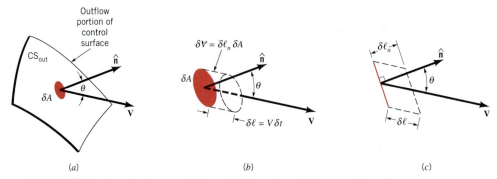

■ **FIGURE D.2** **Outflow across a typical portion of the control surface.**

dividing region II and the control volume. This surface is denoted CS_{out}. As indicated in Fig. D.2, in time δt the volume of fluid that passes across each area element is given by $\delta V = \delta \ell_n \, \delta A$, where $\delta \ell_n = \delta \ell \cos \theta$ is the height (normal to the base, δA) of the small volume element, and θ is the angle between the velocity vector and the outward pointing normal to the surface, $\hat{\mathbf{n}}$. Thus, since $\delta \ell = V \, \delta t$, the amount of the property B carried across the area element δA in the time interval δt is given by

$$\delta B = b\rho \, \delta V = b\rho (V \cos \theta \, \delta t) \, \delta A$$

The rate at which B is carried out of the control volume across the small area element δA, denoted $\delta \dot{B}_{out}$, is

$$\delta \dot{B}_{out} = \lim_{\delta t \to 0} \frac{\rho b \, \delta V}{\delta t} = \lim_{\delta t \to 0} \frac{(\rho b V \cos \theta \, \delta t) \delta A}{\delta t} = \rho b V \cos \theta \, \delta A$$

By integrating over the entire outflow portion of the control surface, CS_{out}, we obtain

$$\dot{B}_{out} = \int_{cs_{out}} d\dot{B}_{out} = \int_{cs_{out}} \rho b V \cos \theta \, dA$$

The quantity $V \cos \theta$ is the component of the velocity normal to the area element δA. From the definition of the dot product, this can be written as $V \cos \theta = \mathbf{V} \cdot \hat{\mathbf{n}}$. Hence, an alternate form of the outflow rate is

$$\dot{B}_{out} = \int_{cs_{out}} \rho b \mathbf{V} \cdot \hat{\mathbf{n}} \, dA \tag{D.2}$$

In a similar fashion, by considering the inflow portion of the control surface, CS_{in}, as shown in Fig. D.3, we find that the inflow rate of B into the control volume is

$$\dot{B}_{in} = -\int_{cs_{in}} \rho b V \cos \theta \, dA = -\int_{cs_{in}} \rho b \mathbf{V} \cdot \hat{\mathbf{n}} \, dA \tag{D.3}$$

Therefore, the net flux (flowrate) of parameter B across the entire control surface is

$$\dot{B}_{out} - \dot{B}_{in} = \int_{cs_{out}} \rho b \mathbf{V} \cdot \hat{\mathbf{n}} \, dA - \left(-\int_{cs_{in}} \rho b \mathbf{V} \cdot \hat{\mathbf{n}} \, dA \right)$$

$$= \int_{cs} \rho b \mathbf{V} \cdot \hat{\mathbf{n}} \, dA \tag{D.4}$$

where the integration is over the entire control surface.

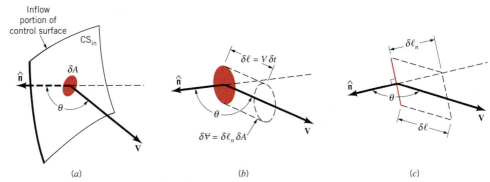

Inflow
portion of
control surface

δA

CS_{in}

$\hat{n}$

θ

V

(a)

$\delta\ell = V\,\delta t$

$\hat{n}$

θ

$\delta\forall = \delta\ell_n\,\delta A$

V

(b)

$\delta\ell_n$

$\hat{n}$

θ

$\delta\ell$

V

(c)

■ **FIGURE D.3**   Inflow across a typical portion of the control surface.

By combining Eqs. D.1 and D.4 we obtain

$$\frac{DB_{sys}}{Dt} = \frac{\partial B_{cv}}{\partial t} + \int_{cs} \rho b \mathbf{V} \cdot \hat{\mathbf{n}}\, dA$$

This can be written in a slightly different form by using $B_{cv} = \int_{cv}\rho b\, d\forall$ so that

$$\frac{DB_{sys}}{Dt} = \frac{\partial}{\partial t}\int_{cv} \rho b\, d\forall + \int_{cs} \rho b\, \mathbf{V} \cdot \hat{\mathbf{n}}\, dA \qquad \textbf{(D.5)}$$

Equation D.5 is the general form of the Reynolds transport theorem for a fixed, nondeforming control volume.

The left side of Eq. D.5 is the time rate of change of an arbitrary extensive parameter of a system. This may represent the rate of change of mass, momentum, energy, or angular momentum of the system, depending on the choice of the parameter B.

The first term on the right side of Eq. D.5 represents the time rate of change of B within the control volume as the fluid flows through it. Recall that b is the amount of B per unit mass so that $\rho b\, d\forall$ is the amount of B in a small volume $d\forall$. Thus, the time derivative of the integral of ρb throughout the control volume is the time rate of change of B within the control volume at a given time.

The last term in Eq. D.5 (an integral over the control surface) represents the net flowrate of the parameter B across the entire control surface. Over a portion of the control surface this property is being carried out of the control volume ($\mathbf{V} \cdot \hat{\mathbf{n}} > 0$); over other portions it is being carried into the control volume ($\mathbf{V} \cdot \hat{\mathbf{n}} < 0$). Over the remainder of the control surface there is no transport of B across the surface since $b\mathbf{V} \cdot \hat{\mathbf{n}} = 0$, because either $b = 0$, $\mathbf{V} = 0$, or $\mathbf{V}$ is parallel to the surface at those locations.

D.2   General Control Volume Equations

In Chapter 5 the simplified form of the Reynolds transport theorem (Eq. 4.14) was used in the derivation of the governing control volume equations involving conservation of mass, linear momentum, moment-of-momentum, and energy. The equations obtained are valid for flows in which the fluid properties (i.e., velocity, pressure) are uniform across the inlet and outlet areas of the control volume and the velocity is normal to these areas. By using the more general form of the Reynolds transport theorem given in Eq. D.5, the governing equations can be written for flows having variable properties across the control surface and velocities that are not normal to the control surface, as shown in Fig. D.1. The resulting equations are given as

a) Conservation of mass equation

$$\frac{\partial}{\partial t} \int_{cv} \rho d\forall + \int_{cs} \rho \mathbf{V} \cdot \hat{\mathbf{n}} \, dA = 0 \tag{D.6}$$

b) Linear momentum equation

$$\frac{\partial}{\partial t} \int_{cv} \mathbf{V}\rho \, d\forall + \int_{cv} \mathbf{V}\rho \mathbf{V} \cdot \hat{\mathbf{n}} \, dA = \sum \mathbf{F}_{\substack{\text{contents of the} \\ \text{control volume}}} \tag{D.7}$$

c) Moment-of-momentum equation

$$\frac{\partial}{\partial t} \int_{cv} (\mathbf{r} \times \mathbf{V})\rho \, d\forall + \int_{cv} (\mathbf{r} \times \mathbf{V})\rho \mathbf{V} \cdot \hat{\mathbf{n}} \, dA = \sum (\mathbf{r} \times \mathbf{F})_{\substack{\text{contents of the} \\ \text{control volume}}} \tag{D.8}$$

d) Energy equation

$$\frac{\partial}{\partial t} \int_{cv} e\rho \, d\forall + \int_{cs} \left(\breve{u} + \frac{p}{\rho} + \frac{V^2}{2} + gz \right) \rho \mathbf{V} \cdot \hat{\mathbf{n}} \, dA = \dot{Q}_{\substack{\text{net} \\ \text{in}}} + \dot{W}_{\substack{\text{shaft} \\ \text{net in}}} \tag{D.9}$$

Equations D.6 through D.9 are very similar to those developed in Chapter 5 under more restrictive conditions (Eqs. 5.5, 5.17, 5.21, and 5.42, respectively). The main difference between these two sets of equations is the way the various flux (or flowrate) terms are treated. The four integral terms of the form $\int_{cs} (\quad) \rho \mathbf{V} \cdot \hat{\mathbf{n}} \, dA$ in Eqs. D.6 through D.9 represent the flux of mass, linear momentum, moment-of-momentum, and energy through the control surface, respectively. If the velocities at the inlets and outlets of the control volume are normal to those areas, and if the fluid properties are uniform across those areas, the integrands of these four terms are constant and the integrals can be written algebraically as $\int_{cs} (\quad) \rho \mathbf{V} \cdot \hat{\mathbf{n}} \, dA = \sum (\quad)_{\text{out}} \dot{m}_{\text{out}} - \sum (\quad)_{\text{in}} \dot{m}_{\text{in}}$, where $\dot{m} = \rho A V$ is the mass flowrate. Note that on the inflow portion of the control surface $\mathbf{V} \cdot \hat{\mathbf{n}} < 0$ and on the outflow portion $\mathbf{V} \cdot \hat{\mathbf{n}} > 0$. The solution to Example 5.5 is repeated here to illustrate the use of the more general form of the linear momentum equation, Eq. D.7.

EXAMPLE D.1   (Repeat of Example 5.5)

GIVEN As shown in Fig. ED.1a, a horizontal jet of water exits a nozzle with a uniform speed of $V_1 = 10$ ft/s, strikes a vane, and is turned through an angle θ.

FIND Determine the anchoring force needed to hold the vane stationary.

SOLUTION

We select a control volume that includes the vane and a portion of the water (see Figs. ED.1b, c) and apply the linear momentum equation to this fixed control volume. The x and z components of Eq. D.7 become

$$\frac{\partial}{\partial t} \overset{\text{0 (flow is steady)}}{\cancel{\int_{cv} u \, \rho \, d\forall}} + \int_{cs} u\rho \mathbf{V} \cdot \hat{\mathbf{n}} \, dA = \sum F_x \tag{1}$$

and

$$\frac{\partial}{\partial t} \overset{\text{0 (flow is steady)}}{\cancel{\int_{cv} w \rho \, d\forall}} + \int_{cs} w\rho \mathbf{V} \cdot \hat{\mathbf{n}} \, dA = \sum F_z \tag{2}$$

where $\mathbf{V} = u\hat{\mathbf{i}} + w\hat{\mathbf{k}}$, and ΣF_x and ΣF_x are the net x and z components of force acting on the contents of the control volume.

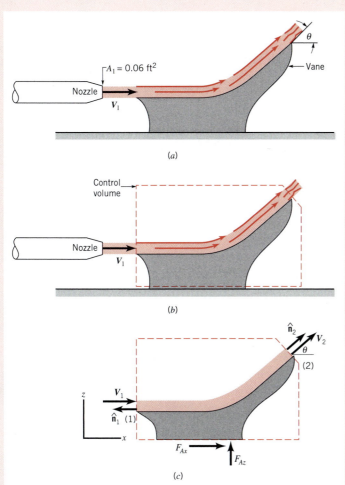

The water enters and leaves the control volume as a free jet at atmospheric pressure. Hence, there is atmospheric pressure surrounding the entire control volume, and the net pressure force on the control volume surface is zero. If we neglect the weight of the water and vane, the only forces applied to the control volume contents are the horizontal and vertical components of the anchoring force, F_{Ax} and F_{Az}, respectively.

The only portions of the control surface across which fluid flows are section 1 (the entrance) where $\mathbf{V} \cdot \hat{\mathbf{n}} = -V_1$ and section 2 (the exit) where $\mathbf{V} \cdot \hat{\mathbf{n}} = +V_2$. (Recall that the unit normal vector is directed out from the control surface.) Also, with negligible gravity and viscous effects, and since $p_1 = p_2$, the speed of the fluid remains constant so that $V_1 = V_2 = 10$ ft/s (see the Bernoulli equation, Eq. 3.6). Hence, at section 1, $u = V_1$, $w = 0$, and at section 2, $u = V_1 \cos \theta$, $w = V_1 \sin \theta$.

By using the above information, Eqs. 1 and 2 can be written as

$$V_1 \rho(-V_1)A_1 + V_1 \cos \theta \rho(V_1)A_2 = F_{Ax} \qquad (3)$$

and

$$(0)\rho(-V_1)A_1 + V_1 \sin \theta \rho(V_1)A_2 = F_{Az} \qquad (4)$$

Note that since the flow is uniform across the inlet and exit, the integrals simply reduce to multiplications. Equations 3 and 4 can be simplified by using conservation of mass, which states that for this incompressible flow $A_1V_1 = A_2V_2$, or $A_1 = A_2$ since $V_1 = V_2$. Thus

$$
\begin{aligned}
F_{Ax} &= -\rho A_1 V_1^2 + \rho A_1 V_1^2 \cos \theta \\
&= -\rho A_1 V_1^2 (1 - \cos \theta)
\end{aligned} \qquad (5)
$$

and

$$F_{Az} = \rho A_1 V_1^2 \sin \theta \qquad (6)$$

With the given data we obtain

$$
\begin{aligned}
F_{Ax} &= -(1.94 \text{ slugs/ft}^3)(0.06 \text{ ft}^2)(10 \text{ ft/s})^2(1 - \cos \theta) \\
&= -11.64(1 - \cos \theta) \text{ slugs·ft/s}^2 \\
&= -11.64(1 - \cos \theta) \text{ lb} \qquad \textbf{(Ans)}
\end{aligned}
$$

and

$$F_{Az} = (1.94 \text{ slugs/ft}^3)(0.06 \text{ ft}^2)(10 \text{ ft/s})^2 \sin \theta$$
$$= 11.64 \sin \theta \text{ lb} \qquad \text{(Ans)}$$

COMMENT   Note that the anchoring force (Eqs. 5, 6) can be written in terms of the mass flowrate, $\dot{m} = \rho A_1 V_1$, as

$$F_{Ax} = -\dot{m} V_1 (1 - \cos \theta)$$

and

$$F_{Az} = \dot{m} V_1 \sin \theta$$

In this example the anchoring force is needed to produce the nonzero net momentum flowrate (mass flowrate times the change in x or z component of velocity) across the control surface.

Comprehensive Table of Conversion Factors[1]

The following tables express the definitions of miscellaneous units of measure as exact numerical multiples of coherent SI units, and provide multiplying factors for converting numbers and miscellaneous units to corresponding new numbers and SI units.

Conversion factors are expressed using computer exponential notation, and an asterisk follows each number that expresses an exact definition. For example, the entry "2.54 E − 2*" expresses the fact that 1 inch = 2.54×10^{-2} meter, exactly by definition. Numbers not followed by an asterisk are only approximate representations of definitions, or are the results of physical measurements. In these tables pound-force is designated as lbf, whereas in the text pound-force is designated as lb.

■ TABLE E.1
Listing by Physical Quantity

To convert from	to	Multiply by
Acceleration		
foot/second^2	meter/second^2	3.048 E − 1*
free fall, standard	meter/second^2	9.806 65 E + 0*
gal (galileo)	meter/second^2	1.00 E − 2*
inch/second^2	meter/second^2	2.54 E − 2*
Area		
acre	meter^2	4.046 856 422 4 E + 3*
are	meter^2	1.00 E + 2*
barn	meter^2	1.00 E − 28*
foot^2	meter^2	9.290 304 E − 2*
hectare	meter^2	1.00 E + 4*
inch^2	meter^2	6.4516 E − 4*
mile^2 (U.S. statute)	meter^2	2.589 988 110 336 E + 6*
section	meter^2	2.589 988 110 336 E + 6*
township	meter^2	9.323 957 2 E + 7
yard^2	meter^2	8.361 273 6 E − 1*
Density		
gram/centimeter^3	kilogram/meter^3	1.00 E + 3*
lbm/inch^3	kilogram/meter^3	2.767 990 5 E + 4

[1]These tables abridged from Mechtly, E. A., *The International System of Units, 2nd Revision*, NASA SP-7012, 1973.

■ **TABLE E.1** (continued)

To convert from	to	Multiply by
lbm/foot^3	kilogram/meter^3	1.601 846 3 E + 1
slug/foot^3	kilogram/meter^3	5.153 79 E + 2

Energy

British thermal unit:		
(IST after 1956)	joule	1.055 056 E + 3
British thermal unit (thermochemical)	joule	1.054 350 E + 3
calorie (International Steam Table)	joule	4.1868 E + 0
calorie (thermochemical)	joule	4.184 E + 0*
calorie (kilogram, International Steam Table)	joule	4.1868 E + 3
calorie (kilogram, thermochemical)	joule	4.184 E + 3*
electron volt	joule	1.602 191 7 E − 19
erg	joule	1.00 E − 7*
foot lbf	joule	1.355 817 9 E + 0
foot poundal	joule	4.214 011 0 E − 2
joule (international of 1948)	joule	1.000 165 E + 0
kilocalorie (International Steam Table)	joule	4.1868 E + 3
kilocalorie (thermochemical)	joule	4.184 E + 3*
kilowatt hour	joule	3.60 E + 6*
watt hour	joule	3.60 E + 3*

Force

dyne	newton	1.00 E − 5*
kilogram force (kgf)	newton	9.806 65 E + 0*
kilopond force	newton	9.806 65 E + 0*
kip	newton	4.448 221 615 260 5 E + 3*
lbf (pound force, avoirdupois)	newton	4.448 221 615 260 5 E + 0*
ounce force (avoirdupois)	newton	2.780 138 5 E − 1
pound force, lbf (avoirdupois)	newton	4.448 221 615 260 5 E + 0*
poundal	newton	1.382 549 543 76 E − 1*

Length

angstrom	meter	1.00 E − 10*
astronomical unit (IAU)	meter	1.496 00 E + 11
cubit	meter	4.572 E − 1*
fathom	meter	1.8288 E + 0*
foot	meter	3.048 E − 1*
furlong	meter	2.011 68 E + 2*
hand	meter	1.016 E − 1*
inch	meter	2.54 E − 2*
league (international nautical)	meter	5.556 E + 3*
light year	meter	9.460 55 E + 15
meter	wavelengths Kr 86	1.650 763 73 E + 6*
micron	meter	1.00 E − 6*
mil	meter	2.54 E − 5*
mile (U.S. statute)	meter	1.609 344 E + 3*
nautical mile (U.S.)	meter	1.852 E + 3*
rod	meter	5.0292 E + 0*
yard	meter	9.144 E − 1*

■ **TABLE E.1** (continued)

To convert from	to	Multiply by
Mass		
carat (metric)	kilogram	2.00 E − 4*
grain	kilogram	6.479 891 E − 5*
gram	kilogram	1.00 E − 3*
ounce mass (avoirdupois)	kilogram	2.834 952 312 5 E − 2*
pound mass, lbm (avoirdupois)	kilogram	4.535 923 7 E − 1*
slug	kilogram	1.459 390 29 E + 1
ton (long)	kilogram	1.016 046 908 8 E + 3*
ton (metric)	kilogram	1.00 E + 3*
ton (short, 2000 pound)	kilogram	9.071 847 4 E + 2*
tonne	kilogram	1.00 E + 3*
Power		
Btu (thermochemical)/second	watt	1.054 350 264 488 E + 3
calorie (thermochemical)/second	watt	4.184 E + 0*
foot lbf/second	watt	1.355 817 9 E + 0
horsepower (550 foot lbf/second)	watt	7.456 998 7 E + 2
kilocalorie (thermochemical)/second	watt	4.184 E + 3*
watt (international of 1948)	watt	1.000 165 E + 0
Pressure		
atmosphere	newton/meter^2	1.013 25 E + 5*
bar	newton/meter^2	1.00 E + 5*
barye	newton/meter^2	1.00 E − 1*
centimeter of mercury (0 °C)	newton/meter^2	1.333 22 E + 3
centimeter of water (4 °C)	newton/meter^2	9.806 38 E + 1
dyne/centimeter^2	newton/meter^2	1.00 E − 1*
foot of water (39.2 °F)	newton/meter^2	2.988 98 E + 3
inch of mercury (32 °F)	newton/meter^2	3.386 389 E + 3
inch of mercury (60 °F)	newton/meter^2	3.376 85 E + 3
inch of water (39.2 °F)	newton/meter^2	2.490 82 E + 2
inch of water (60 °F)	newton/meter^2	2.4884 E + 2
kgf/centimeter^2	newton/meter^2	9.806 65 E + 4*
kgf/meter^2	newton/meter^2	9.806 65 E + 0*
lbf/foot^2	newton/meter^2	4.788 025 8 E + 1
lbf/inch^2 (psi)	newton/meter^2	6.894 757 2 E + 3
millibar	newton/meter^2	1.00 E + 2*
millimeter of mercury (0 °C)	newton/meter^2	1.333 224 E + 2
pascal	newton/meter^2	1.00 E + 0*
psi (lbf/inch^2)	newton/meter^2	6.894 757 2 E + 3
torr (0 °C)	newton/meter^2	1.333 22 E + 2
Speed		
foot/second	meter/second	3.048 E − 1*
inch/second	meter/second	2.54 E − 2*
kilometer/hour	meter/second	2.777 777 8 E − 1
knot (international)	meter/second	5.144 444 444 E − 1
mile/hour (U.S. statute)	meter/second	4.4704 E − 1*

■ **TABLE E.1** (continued)

To convert from	to	Multiply by
Temperature		
Celsius	kelvin	$T_K = T_C + 273.15$
Fahrenheit	kelvin	$T_K = (5/9)(T_F + 459.67)$
Fahrenheit	Celsius	$T_C = (5/9)(T_F - 32)$
Rankine	kelvin	$T_K = (5/9)T_R$
Time		
day (mean solar)	second (mean solar)	8.64　E + 4*
hour (mean solar)	second (mean solar)	3.60　E + 3*
minute (mean solar)	second (mean solar)	6.00　E + 1*
year (calendar)	second (mean solar)	3.1536　E + 7*
Viscosity		
centistoke	meter^2/second	1.00　E − 6*
stoke	meter^2/second	1.00　E − 4*
foot^2/second	meter^2/second	9.290 304　E − 2*
centipoise	newton second/meter^2	1.00　E − 3*
lbm/foot second	newton second/meter^2	1.488 163 9　E + 0
lbf second/foot^2	newton second/meter^2	4.788 025 8　E + 1
poise	newton second/meter^2	1.00　E − 1*
poundal second/foot^2	newton second/meter^2	1.488 163 9　E + 0
slug/foot second	newton second/meter^2	4.788 025 8　E + 1
rhe	meter^2/newton second	1.00　E + 1*
Volume		
acre foot	meter^3	1.233 481 837 547 52　E + 3*
barrel (petroleum, 42 gallons)	meter^3	1.589 873　E − 1
board foot	meter^3	2.359 737 216　E − 3*
bushel (U.S.)	meter^3	3.523 907 016 688　E − 2*
cord	meter^3	3.624 556 3　E + 0
cup	meter^3	2.365 882 365　E − 4*
dram (U.S. fluid)	meter^3	3.696 691 195 312 5　E − 6*
fluid ounce (U.S.)	meter^3	2.957 352 956 25　E − 5*
foot^3	meter^3	2.831 684 659 2　E − 2*
gallon (U.K. liquid)	meter^3	4.546 087　E − 3
gallon (U.S. liquid)	meter^3	3.785 411 784　E − 3*
inch^3	meter^3	1.638 706 4　E − 5*
liter	meter^3	1.00　E − 3*
ounce (U.S. fluid)	meter^3	2.957 352 956 25　E − 5*
peck (U.S.)	meter^3	8.809 767 541 72　E − 3*
pint (U.S. liquid)	meter^3	4.731 764 73　E − 4*
quart (U.S. liquid)	meter^3	9.463 529 5　E − 4
stere	meter^3	1.00　E + 0*
tablespoon	meter^3	1.478 676 478 125　E − 5*
teaspoon	meter^3	4.928 921 593 75　E − 6*
yard^3	meter^3	7.645 548 579 84　E − 1*

Online Appendix List

Answers to Selected Even-Numbered Homework Problems

Chapter 1

1.2 (a) FL^3, ML^4T^{-2}; (b) $F^2L^{-5}T^2$, $M^2L^{-3}T^{-2}$; (c) FT, MLT^{-1}

1.6 Yes

1.10 (a) 4.66×10^4 ft; (b) 5.18×10^{-2} lb/ft^3; (c) 3.12×10^{-3} slugs/ft^3; (d) 2.36×10^{-2} ft · lb/s; (e) 5.17×10^{-6} ft/s

1.12 1.25, 1.25

1.14 1150 kg/m^3, 11.3 kN/m^3

1.16 12.0 kN/m^3; 1.22×10^3 kg/m^3; 1.22

1.20 oxygen

1.22 6.44×10^{-3} slugs/ft^3; 0.622 lb

1.24 2.60×10^{-3} N · s/m^2; μ (blood)/μ(water) = 3.74

1.26 184

1.28 0.552 U/δ N/m^2 acting to left on plate

1.30 $\mathcal{D} = 0.571$ b$\rho\sqrt{\nu\ell U^3}$

1.32 7.22°

1.34 (a) No; (b) Not correct

1.36 31.0%

1.38 286 N

1.40 1.06

1.42 1.78 kg/m^3

1.44 (a) 343 m/s; (b) 1010 m/s; (c) 446 m/s

1.46 31.2 kPa(abs); 4.52 psia

1.48 1.80×10^{-2} ft

1.50 (a) 24.5 deg

Chapter 2

2.2 38.7 psi

2.4 (a) 18.2 ft; (b) 8.73 psi; (c) 21.7 psia

2.8 (a) 58.8 kN/m^2; (b) 442 mm Hg

2.10 4250 ft

2.12 19.3 psia

2.14 103 kPa; 230 mm

2.16 4.67 psi

2.18 0.424 psi

2.20 23.8 ft

2.22 78.4 lb

2.24 0.100 m

2.26 1930 kg/m^3

2.28 (a) 3.46×10^{11} N; (b) Yes

2.30 1350 lb

2.32 33,900 lb; 2.49 ft above base

2.34 $\mathcal{W} = 314$ kN; $R = 497$ kN

2.36 6300 lb

2.38 2.11 m; 941 kN

2.40 1.88 ft

2.42 0.146

2.44 294 kN; 328 kN; yes

2.46 22,500 lb

2.48 7.77×10^9 lb acting 406 ft up from base of dam

2.50 585 lb acting vertically downward along vertical axis of bottle

2.52 2.19 ft^3

2.56 127 lb

2.58 17.2 mm

2.60 158 lb/ft^2

Chapter 3

3.2 (a) $-194(1 + x) - 62.4$ lb/ft^3; (b) 38.6 psi

3.4 -9.99 kPa/m

3.6 $p + \frac{1}{2}\rho V^2 + \rho g_0 z - \frac{1}{2}\rho c z^2 =$ constant

3.8 12.0 kPa; -20.1 kPa

3.10 10.8 lb/ft^2; 102 lb/ft^2

3.12 194 mph

3.14 0.14 in.

3.16 -76.0 lb/ft^2, 88.0 lb/ft^2

3.18	9.80 kPa
3.22	45.7 ft
3.24	4
3.26	$Q = 1.56 D^2$ where $Q \sim$ m^3/s, $D \sim$ m
3.28	0.997 in.
3.30	3.13 ft
3.32	2.94 m
3.34	0.183 ft^3/s
3.36	20.8 lb/s
3.38	1.31 ft
3.40	0.303 ft^3/s; 499 lb/ft^2; 312 lb/ft^2; -187 lb/ft^2
3.42	1.06×10^{-3} m^3/s; 3.02×10^{-3} m^3/s, 0.118 m^3/s
3.44	1.38 ft^3/s
3.46	29.0 ft/s; 7.98 psi; 3.97 ft^3/s
3.48	155 N/m^2
3.50	3.94 m^2/s
3.52	0.0351 m^3/s; 0
3.54	0.191 ft; 0.109 ft
3.56	6.51 m, 25.4 m; 6.51 m, -9.59 m

Chapter 4

4.4	$y = e^{(x^2/2 - x)} - 1$
4.6	$x = -h(u_0/v_0) \ln(1 - y/h)$
4.8	$a_x = 2x^3, a_y = 2x^2y, a_z = x^2 - 2xy$
4.10	(a) 0.5 m/s^2, 1.0 m/s^2; (b) positive
4.12	$a_x = V_0(1 - x/\ell)[ce^{-ct} - (V_0/\ell)(1 - e^{-ct})^2]$; 0.124/s
4.14	0; 100 N/m^2·s
4.16	0; $(225x + 150)\hat{\mathbf{i}}$ m/s^2; $150\hat{\mathbf{i}}$ m/s^2; $375\hat{\mathbf{i}}$ m/s^2
4.18	200 C/s; 100 C/s
4.20	32.2 ft/s^2
4.22	$-25,600$ ft/s^2; -25 ft/s^2
4.24	132 ft^3/s

Chapter 5

5.2	13.8 ft
5.4	2.92 ft/s
5.6	1.70 ft/s
5.8	$h = 0.012x$; 20.8 ft
5.10	45.8 ft/s; 20.4 ft/s
5.12	150 liters/s
5.14	0.510 ft/day
5.16	352 lb to left
5.18	482 N down
5.20	9.27 N
5.22	0
5.24	1120 lb/ft, 531 lb/ft
5.26	0.108 kg
5.28	1.36 kPa

5.30	78.5 kN
5.32	3.97 m
5.34	2.96 lb
5.36	66.6 N
5.38	(a) 231 N · m; (b) 200 N · m; (c) 116 N · m
5.40	(a) 62.4 N · m/kg, (b) 62.4 N · m/kg
5.42	-7.68 MW
5.44	11.4 (N · m)/kg
5.48	0.842
5.50	right to left, 0.32 ft
5.54	566 (ft · lb)/slug
5.56	734 ft
5.58	1.01×10^5 hp
5.60	301 hp
5.62	0.052 m^3/s
5.64	1610 ft · lb/slug, 2.53 hp
5.66	-197 kPa
5.68	28 hp
5.70	1.75 hp
5.72	$R_x = -12,850$ lb, $R_y = 1,540$ lb

Chapter 6

6.2	$12xy^2 \hat{\mathbf{k}}$; No
6.4	(a) 0; (b) $\boldsymbol{\omega} = -(U/2b)\hat{\mathbf{k}}$; (c) $\boldsymbol{\zeta} = -(U/b)\hat{\mathbf{k}}$; (d) $\dot{\gamma} = U/b$
6.6	$-3x^2/2 + x^3/3 + f(y)$
6.8	(a) $v_r = V \sin\theta, v_\theta = V \cos\theta$; (b) $\psi = -Vx + C, \psi = -Vr \cos\theta + C$
6.10	$v = -2y$; (b) 1.41 ft/s
6.14	(a) Yes; (b) Yes, if $A = B$; (c) $y^2 = (B/A)x^2 + C$
6.16	$\psi = U_c y[1 - \frac{1}{3}(y/h)^2] + C$; No ϕ
6.18	(a) $\psi = 2xy$; (b) $q = 2x_i y_i$
6.20	(a) $\psi = -Kr^2/2 + c$; (b) No
6.22	(a) $\partial p/\partial r = (1.60 \times 10^3)r^{-3}$ kPa/m; (b) 184 kPa
6.24	$C = 0.500$ m
6.26	$\sin\theta = (\Gamma/2\pi rU) \ln(r/4)$
6.28	$h^2 = m/2\pi A$
6.30	62.8 mph
6.34	(a) $p_{max} = p_0 + \rho U^2/2$ (at $\theta = 0$ and π); $p_{min} = p_0 - 3\rho U^2/2$ (at $\theta = \pi/2$); (b) $(2r/3a)(1 - a^2/r^2) \sin\theta = 1$
6.36	$y/a \geq 10$
6.38	$\partial p_s/\partial\theta = 4\rho U^2 \sin\theta \cos\theta$; θ falls in range of $\pm 90°$
6.40	(a) $-2x$; (b) $2x^3\hat{\mathbf{i}}$; $2\mu - \rho x^3$
6.42	311 lb/ft^3, 1.35 ft/s
6.44	(a) 0.0111 ft^3/s; (b) 0.00630 in. gap
6.48	$y/b = 1/3$
6.50	$2\pi r_o^3 \mu\omega\ell/(r_o - r_i)$; $-U/b$

6.52 (a) Yes; (b) 57.1 N/m^2 per m
6.56 (b) 1.20 Pa

Chapter 7
7.4 $\omega\sqrt{\ell/g} = \phi(h/\ell)$
7.6 $c\sqrt{\rho h/\sigma} = \phi(\lambda/h)$
7.8 $\mathcal{D}/d_1^2 V^2 \rho = \phi(d_2/d_1, \rho V d_1/\mu)$
7.10 $c\sqrt{\rho/E} = \phi(h/D)$
7.12 (a) $VD\sqrt{\rho/\mathcal{W}} = \phi(b/d, d/D)$;
 (b) $V = \sqrt{2\mathcal{W}b/\pi\rho dD^2}$
7.14 $V = 18.1 \,\Delta\gamma\, d^2/\mu$
7.16 Omit ρ and σ
7.18 Not correct
7.20 8.71×10^{-2} ft/s
7.22 11.0 m/s
7.24 109 mm
7.26 117 lb
7.28 (a) $\Delta p/\rho V^2 = \phi(d/D, \rho VD/\mu)$
 (b) 0.200 mm, 0.500 m/s to 10.0 m/s, 25.0
7.30 0.129 psf
7.32 27.8 psi
7.34 (a) $h/H = \phi(d/H, b/H, \rho g/\gamma_s, V/\sqrt{gH}, \eta)$;
 (b) 5.00 lb/ft^3, 21.2 mph, 8.00 in.
7.36 (a) $\ell_m/h_m = \ell/h$, $b_m/h_m = b/h$, $l_m h_m V_m/\mu_m$
 $= \rho h U/\mu$, $V/U = V_m/U_m$
7.38 (a) $V_m/U_m = V/U$, $V_m D_m/\nu_{sm} = VD/\nu_s$,
 $V_m^2/g_m D_m = V^2/gD$,
 $(\rho - \rho_s)_m/\rho_m = (\rho - \rho_s)/\rho$; (b) no
7.40 (a) 3.58 ft/s; (b) 1.06 ft/s
7.42 (a) $t\sqrt{g/w}$, $\mu/\rho\sqrt{w^3 g}$;
 (b) $t_m/t = 0.316$, no

Chapter 8
8.2 (a) 8.93×10^4; (b) 8.93×10^{-8}
8.4 blue and yellow; green
8.6 17.7 ft
8.8 0.15 lb/ft^2; 0.06 lb/ft^2; 0
8.10 0.0260 lb · s/ft^2; 1.25 lb/ft^2
8.12 3.43 m; 166 kPa
8.14 (a) 12.42 kPa; (b) 15.85 kPa
8.16 0.102 ft
8.20 0.0404
8.22 Yes
8.24 0.0292; 0.0132 ft
8.26 9.00
8.28 9
8.30 Disagree
8.32 440 ft; 0.234 psi; 5.11 hp
8.34 0.211 psi/ft
8.36 1.50 ft^2
8.38 21.0

8.40 Pump; 127 hp
8.42 9
8.44 1012 ft
8.46 0.750 psi
8.48 78.1 psi
8.50 0.710 ft^3/s
8.52 5.73 ft/s
8.54 0.0289 m^3/s
8.56 0.513 ft
8.58 0.491 ft
8.60 0.662 ft
8.62 0.0180 m^3/s
8.66 18.0 m^3/s
8.68 0.115 ft^3/s
8.70 5.77 ft
8.72 0.0936 ft^3/s

Chapter 9
9.2 (a) 0.06 ($\frac{1}{2}\rho U^2$); (b) 2.40
9.4 0.159 lb; 0.833
9.6 Yes
9.8 (a) 4.70 m, 14.8 m, 47.0 m;
 (b) 6.65×10^{-6} m^2/s
9.10 0.0130 m; 0.0716 N/m^2; 0.0183 m;
 0.0506 N/m^2
9.14 $\delta = 4.79\sqrt{\nu x/U}$
9.18 2.83 $\mathcal{D}$; 0.354 $\mathcal{D}$
9.20 2.82
9.22 in water, 0.946 ft/s; in air, 30.2 ft/s
9.24 1.06 m/s
9.26 378 lb
9.28 18,800 N · m
9.30 0.0206
9.34 (a) 567 ft/s; (b) 118 ft/s; (c) 13.5 ft/s
9.38 1.29 lb; 129 lb
9.40 859 lb
9.42 16.8 ft/s
9.44 77.0 ft/s
9.46 (a) 0.0166 hp; (b) 32.4 hp
9.48 (a) 158 mph; (b) 83.2 mph
9.50 (a) 0.840, 0.910; (b) Rise
9.52 405 lb
9.54 146 mph
9.56 28.4 percent
9.58 2.60 percent
9.60 0.206

Chapter 10
10.6 2.45 m, 0.388 m
10.8 1.42 ft; 4.14 ft
10.10 0.528 ft; 0.728 ft

10.12	0.694 ft
10.16	0.0240; 0.585 lb/ft^2
10.18	19.8×10^6 gal/day
10.20	Yes
10.22	Yes
10.28	119 ft^3/s
10.30	5.94 ft
10.32	Same for each
10.34	0.368 ft
10.36	Greater
10.38	0.000505
10.40	Yes
10.42	0.00759 ft^3/s
10.44	153 ft^3/s
10.46	4.36 ft/s
10.48	4.70 ft
10.50	2.05 m^3/s; 0.333 m
10.52	0.406 m
10.56	$Q = C_w[(2/3)\sqrt{2g}\, b\, H^{3/2} + (8/15)\sqrt{2g}\, H^{5/2}]$

Chapter 11

11.4	55.4°; CCW
11.6	(b) 918 ft · lb; 0 rpm
11.8	61.3 ft
11.12	365 gpm
11.14	1740 gal/min
11.18	1000 gpm; 800 ft
11.20	136 gpm, 147 ft
11.22	mixed-flow pump
11.24	0.0328 m^3/s, 8.00 m
11.26	-12.8 MW
11.28	(b) -7020 ft^2/s^2
11.30	0.300 ft
11.32	26,600 N; 37.6 m/s; 707 kg/s
11.34	23,500 hp, 190 rpm
11.36	Francis, 378 ft^3/s
11.38	impulse

Index

493

Index of Fluids Phenomena Videos

Available on www.wiley.com/college/young
Use the registration code included with this new text to access the videos.

 V1.1 Viscous Fluids

 V1.2 No-Slip Condition

 V1.3 Capillary Tube Viscometer

 V1.4 Non-Newtonian Behavior

 V1.5 Floating Razor Blade

 V2.1 Blood Pressure Measurement

 V2.2 Bourdon Gage

 V2.3 Hoover Dam

 V2.4 Pop Bottle

 V2.5 Cartesian Diver

 V2.6 Hydrometer

 V2.7 Stability of a Model Barge

 V3.1 Balancing Ball

 V3.2 Free Vortex

 V3.3 Stagnation Point Flow

 V3.4 Air Speed Indicator

 V3.5 Flow From a Tank

 V3.6 Venturi Channel

 V4.1 Velocity Field

 V4.2 Flow Past a Wing

 V4.3 Flow Types

 V4.4 Jupiter Red Spot

 V4.5 Streamlines

 V4.6 Pathlines

 V5.1 Sink Overflow

 V5.2 Shop Vac Filter

 V5.3 Smokestack Plume Momentum

 V5.4 Force Due to a Water Jet

 V5.5 Rotating Lawn Sprinkler

 V5.6 Impulse-Type Lawn Sprinkler

 V5.7 Energy Transfer

 V5.8 Water Plant Aerator

 V6.1 Shear Deformation

 V6.2 Vortex In a Beaker

 V6.3 Half-Body

 V6.4 Potential Flow

V6.5
No-Slip Boundary Condition

V6.6
Laminar Flow

V6.7
CFD Example

V7.1
Reynolds Number

V7.2
Environmental Models

V7.3
Model of Fish Hatchery Pond

V7.4
Wind Engineering Models

V7.5
Wind Tunnel Train Model

V7.6
River Flow Model

V7.7
Boat Model

V8.1
Laminar/Turbulent Pipe Flow

V8.2
Turbulence in a Bowl

V8.3
Laminar/Turbulent Velocity Profiles

V8.4
Entrance/Exit Flows

V8.5
Car Exhaust System

V8.6
Rotameter

V8.7
Water Meter

V9.1
Space Shuttle Landing

V9.2
Streamlined and Blunt Bodies

V9.3
Laminar/Turbulent Transition

V9.4
Snow Drifts

V9.5
Sky Diving Practice

V9.6
Oscillating Sign

V9.7
Jet Ski

V9.8
Drag on a Truck

V9.9
Wing Tip Vorticies

V10.1
Bicycle Through a Puddle

V10.2
Merging Channels

V10.3
Uniform Channel Flow

V10.4
Erosion in a Channel

V10.5
Hydraulic Jump in a River

V10.6
Hydraulic Jump in a Sink

V10.7
Triangular Weir

V10.8
Low-Head Dam

V11.1
Windmills

V11.2
Self-Propelled Lawn Sprinkler

V11.3
Windshield Washer Pump

V11.4
Pelton Wheel Lawn Sprinkler

V11.5
Dental Drill

VA.1
Tornado Simulation

Conversion Factors from BG Units to SI Units[a]

	To convert from	to	Multiply by
Acceleration	ft/s^2	m/s^2	3.048 E − 1
Area	ft^2	m^2	9.290 E − 2
Density	slugs/ft^3	kg/m^3	5.154 E + 2
Energy	Btu	J	1.055 E + 3
	ft·lb	J	1.356
Force	lb	N	4.448
Length	ft	m	3.048 E − 1
	in.	m	2.540 E − 2
	mile	m	1.609 E + 3
Mass	slug	kg	1.459 E + 1
Power	ft·lb/s	W	1.356
	hp	W	7.457 E + 2
Pressure	in. Hg (60 °F)	N/m^2	3.377 E + 3
	lb/ft^2 (psf)	N/m^2	4.788 E + 1
	lb/in.^{2} (psi)	N/m^2	6.895 E + 3
Specific weight	lb/ft^3	N/m^3	1.571 E + 2
Temperature	°F	°C	$T_C = (5/9)(T_F - 32°)$
	°R	K	5.556 E − 1
Velocity	ft/s	m/s	3.048 E − 1
	mi/hr (mph)	m/s	4.470 E − 1
Viscosity (dynamic)	lb·s/ft^2	N · s/m^2	4.788 E + 1
Viscosity (kinematic)	ft^2/s	m^2/s	9.290 E − 2
Volume flowrate	ft^3/s	m^3/s	2.832 E − 2
	gal/min (gpm)	m^3/s	6.309 E − 5

[a]If more than four-place accuracy is desired, refer to Appendix E.